Handbook of Fibrous Materials

Handbook of Fibrous Materials

Edited by
Jinlian Hu
Bipin Kumar
Jing Lu

Volume 1

WILEY-VCH

Editors

Jinlian Hu
The Hong Kong Polytechnic University
Institute of Textiles and Clothing
Room QT715, Q Core, 7/F
Hung Hom
Kowloon 999077
Hong Kong S.A.R.
P.R. China

Bipin Kumar
Indian Institute of Technology Delhi
Department of Textile and Fibre Engineering
TX135, Hauz Khas
New Delhi 110016
India

Jing Lu
The Hong Kong Polytechnic University
Institute of Textiles and Clothing
Room QT807
Hung Hom
Kowloon 999077
Hong Kong S.A.R.
P.R. China

All books published by **Wiley-VCH** are carefully produced. Nevertheless, authors, editors, and publisher do not warrant the information contained in these books, including this book, to be free of errors. Readers are advised to keep in mind that statements, data, illustrations, procedural details or other items may inadvertently be inaccurate.

Library of Congress Card No.:
applied for

British Library Cataloguing-in-Publication Data
A catalogue record for this book is available from the British Library.

Bibliographic information published by the Deutsche Nationalbibliothek
The Deutsche Nationalbibliothek lists this publication in the Deutsche Nationalbibliografie; detailed bibliographic data are available on the Internet at <http://dnb.d-nb.de>.

© 2020 Wiley-VCH Verlag GmbH & Co. KGaA, Boschstr. 12, 69469 Weinheim, Germany

All rights reserved (including those of translation into other languages). No part of this book may be reproduced in any form – by photoprinting, microfilm, or any other means – nor transmitted or translated into a machine language without written permission from the publishers. Registered names, trademarks, etc. used in this book, even when not specifically marked as such, are not to be considered unprotected by law.

Print ISBN: 978-3-527-34220-4
ePDF ISBN: 978-3-527-34256-3
ePub ISBN: 978-3-527-34259-4
oBook ISBN: 978-3-527-34258-7

Cover Design   Adam-Design, Weinheim, Germany
Typesetting   SPi Global, Chennai, India
Printing and Binding   CPI books GmbH, Leck

Printed on acid-free paper

10 9 8 7 6 5 4 3 2 1

Contents

Volume 1

Preface *xix*

1 Fundamentals of the Fibrous Materials *1*
Jinlian Hu, Md A. Jahid, Narayana Harish Kumar, and Venkatesan Harun
1.1 Introduction *1*
1.1.1 What Are Fibrous Materials? *1*
1.2 Historical Evolution of Fibers *2*
1.3 Classification of Fibrous Materials *2*
1.4 Fundamental Characteristics of Fibrous Materials *6*
1.5 Morphological and Structural Properties of Fibrous Materials *7*
1.5.1 Plant or Natural Cellulosic Fibers *7*
1.5.2 Animal or Protein Fibers *7*
1.5.3 Regenerated Cellulosic Fibers *10*
1.5.4 Synthetic or Manufactured (Textile and Non-textile) Fibers *10*
1.6 Essential or Fundamental Properties of Fibrous Materials *10*
1.6.1 Physical Properties *10*
1.6.1.1 Mechanical Behavior of Fibrous Assemblies *11*
1.6.2 Chemical Properties *14*
1.6.3 Biological Properties *14*
1.6.4 Thermal Properties *14*
1.6.5 Other Desirable Properties *15*
1.7 Textile Processing *15*
1.7.1 Spinning *15*
1.7.1.1 Ring Spinning *16*
1.7.1.2 Rotor Spinning *17*
1.7.1.3 Friction Spinning *17*
1.7.1.4 Yarn Numbering System (Count) *17*
1.7.2 Fabric Manufacturing *18*
1.7.2.1 Weaving Process *18*
1.7.2.2 Knitting *19*
1.7.2.3 Nonwoven *19*
1.7.3 Wet Processing Technology *20*
1.7.3.1 Pretreatment Process *20*

1.7.3.2　Dyeing 20
1.7.3.3　Dyeing Methods 20
1.7.3.4　Dyes 20
1.7.3.5　Finishing 21
1.7.3.6　Printing 21
1.8　Textile Applications 21
1.8.1　Advanced Applications of Textile Material 21
1.8.2　Functional Textile 22
1.8.2.1　Water Repellent 22
1.8.2.2　Water Vapor Permeability 23
1.8.3　Applications of High-Performance Fibrous Materials 24
1.8.3.1　Fibers for Automobile Composite 24
1.8.3.2　Fibers as Building Materials 25
1.8.3.3　Fibers for Auxetic Applications 25
1.8.3.4　Fibers for Environmental Protection: Water Purification/Filtration 25
1.8.3.5　Fibers for Optical Applications 26
1.8.4　Application of Sensors/Actuators 26
1.8.4.1　Fibers as Electronic Devices/Wearable Electronics/Energy Materials/Sensors and Actuators 26
1.8.4.2　Fibers for Medical Compression 27
1.8.4.3　Fibers for Health/Stress/Comfort Management 27
1.8.4.4　Fibers for Thermal Protection 28
1.8.4.5　Fibers for Radiation Protection 28
1.8.5　Applications of Integrated Products 29
1.8.5.1　Fibers for Tissue Engineering 29
1.8.6　Sensitive and Smart Materials 30
1.8.6.1　Fibers with Conductive Properties as Industrial Materials 30
1.8.6.2　Memory Fibrous Materials 30
1.8.7　Advantages of Fibrous Materials 30
　　　　References 31

2　Animal Fibers: Wool 37
Xiao Xueliang
2.1　Introduction 37
2.2　Classification of Wools 41
2.2.1　Classification of Wool in Fineness 41
2.2.2　Classification of Wool in Terms of Fiber Structure 41
2.2.3　Classification of Wool in Terms of Fiber Type on Sheep Hair Layer 44
2.2.4　Classification of Wool in Terms of Hair Picking Method and Original Hair Shape 44
2.2.5　Classification of Wool in Terms of Wool Cut Season 45
2.3　Processing of Wool Fibers and Yarns 45
2.3.1　Primary Processing of Wool 45
2.3.2　Yarn Spinning Process of Wool 46
2.4　Chemical Compositions and Structural Characteristics of Wool 48
2.4.1　Compositions of Wool 48
2.4.2　Macromolecular Structure of Wool 49

2.4.3	Morphology and Hierarchical Structure of Wool Fiber	50
2.5	Properties of Wool	52
2.5.1	Fineness of Wool	52
2.5.2	Length of Wool Fiber	53
2.5.3	Crimpness of Wool	55
2.5.4	Friction and Felting Properties of Wool	56
2.5.5	Wool Grease and Impurities	57
2.5.6	Other Properties of Wool Fibers	59
2.6	Quality Inspection and Evaluation of Wools	60
2.6.1	Raw Wools from Sheep	60
2.6.2	Fine Wool from China's Raw and Improved Sheep Breed	60
2.6.3	Tops of Domestic Fine Wool and Its Improved Wool	60
2.6.4	Inspection and Evaluation of Australia Wool	61
2.6.5	Inspection and Evaluation of Wool Fabric	61
2.7	Shape Memory Properties of Wool	63
2.8	Future Trends of Application of Wool Keratin	70
2.8.1	Extraction of Keratin from Wool for New Product Development	70
2.8.2	Industrial Trends Relating to Sustainable Polymers	71
2.8.3	Future Trends	71
2.8.4	Sources of Further Information and Advice	72
	References	72
3	**Animal Fibers: Silk**	**75**
	K. Murugesh Babu	
3.1	Introduction to Silk and Silk Industry	75
3.2	Types of Silk and Their Importance	78
3.2.1	Mulberry	78
3.2.1.1	Types of Mulberry Silk	79
3.2.2	Non-mulberry	80
3.2.2.1	Tasar	80
3.2.2.2	Oak Tasar	81
3.2.2.3	Eri	81
3.2.2.4	Muga	82
3.2.2.5	Anaphe Silk	82
3.2.2.6	Fagara Silk	83
3.2.2.7	Coan Silk	83
3.2.2.8	Mussel Silk	83
3.2.2.9	Spider Silk	83
3.2.3	Fine Structure of Silk	83
3.2.3.1	Longitudinal View	84
3.2.3.2	Cross-Sectional View	84
3.2.4	Amino Acid Composition	85
3.2.5	Properties of Silk Fibers	86
3.2.5.1	Tensile Properties	86
3.2.5.2	Optical Properties	87
3.2.5.3	Viscoelastic Behavior	88
3.2.6	Applications	88

3.2.6.1	Textile and Apparels	*89*
3.2.6.2	Biomedical Field	*90*
3.2.6.3	Fiber-Reinforced Composites	*90*
3.3	Future Trends	*91*
3.4	Summary	*92*
	References	*92*

4 Cellulose Fibers *95*
Feng Jiang
4.1 Introduction *95*
4.2 Structure and Biosynthesis of Cellulose *95*
4.3 Nanoscaled Cellulose Fibers *100*
4.4 Submicron-Scaled Cellulose Fibers *107*
4.5 Macroscaled Cellulose Fibers *114*
4.6 Applications of Cellulose Fibers *118*
4.7 Conclusion and Perspectives *119*
References *119*

5 Chitosan Fibers *125*
Seema Sakkara, Mysore Sridhar Santosh, and Narendra Reddy
5.1 Introduction *125*
5.2 Extraction/Modification of Chitosan *127*
5.3 Fibers from Chitosan *131*
5.3.1 Production of Pure Chitosan Fibers *131*
5.3.2 Chitosan Blend Fibers *135*
5.3.3 Generating Hollow Chitosan Fibers *139*
5.3.4 Microfluidic Method of Producing Chitosan Fibers *140*
5.3.5 Application of Chitosan Fibers *140*
5.4 Electrospun Chitosan Fibers *145*
5.5 Regenerated Chitosan Fibers *151*
5.6 Conclusions *151*
Acknowledgments *152*
References *152*

6 Collagen Fibers *157*
Jinlian Hu and Yanting Han
6.1 Introduction *157*
6.2 Source and Structure of CF *157*
6.3 Isolation of Natural CF *160*
6.4 Spinning of CF *160*
6.4.1 Extraction of Collagen *160*
6.4.2 Electrospinning of CF *162*
6.4.3 Wet Spinning of CF *162*
6.4.4 Microfluidic Spinning of CF *164*
6.4.5 Collagen Composite Fiber *165*
6.5 Application of CF *165*
6.6 Perspectives *171*

References *171*

7   Electrospun Fibers for Filtration *175*
Xia Yin, Jianyong Yu, and Bin Ding
7.1   Introduction *175*
7.2   Fabrication Technologies *176*
7.2.1   Electrospinning *176*
7.2.2   Electro-netting *177*
7.3   Principles and Theories *178*
7.3.1   Fundamental Theory of Electrospinning *178*
7.3.2   Formation Mechanism of the Nanofiber/Net (NF/N) Membranes *180*
7.4   Structure and Properties *181*
7.4.1   Types and Structures of the Nanofiber Membranes *181*
7.4.2   Structures and Species of the Nanofiber/Net (NF/N) Membranes *183*
7.5   Application of Nanofibrous Membranes in Air Filtration *185*
7.5.1   Normal Temperature Filter *185*
7.5.1.1 Electrospun Nanofiber Membranes *185*
7.5.1.2 Electrospun Nanofiber/Net (NF/N) Membranes *191*
7.5.2   Medium-High Temperature Filter *196*
7.6   Future Trends *199*
References *199*

8   Aramid Fibers *207*
Manjeet Jassal, Ashwini K. Agrawal, Deepika Gupta, and Kamlesh Panwar
8.1   Introduction *207*
8.2   Preparation of Aromatic Polyamides *207*
8.2.1   Low Temperature Polycondensation *208*
8.2.2   Direct Polymerization in Solution Using Phosphites *210*
8.2.3   Copolyaramids *210*
8.3   Aramid Solutions *211*
8.4   Spinning of Aramid Fibers *214*
8.4.1   Dry Spinning *214*
8.4.2   Wet Spinning *214*
8.4.3   Dry-Jet-Wet Spinning *215*
8.4.4   Aramid Nanofibers *216*
8.4.5   Fiber Heat Treatment *216*
8.5   Influence of Structure on Properties *217*
8.5.1   Fiber Structure *217*
8.5.2   Fiber Properties *219*
8.5.2.1 Chemical Properties *219*
8.5.2.2 Mechanical Properties *220*
8.5.2.3 Thermal Properties *221*
8.6   Applications *222*
8.6.1   Composites with Soft Materials *222*
8.6.2   Advanced Composites *224*
8.6.3   Ultrafiltration and Hemodialysis *224*
8.6.4   Ropes and Cables *224*

8.6.5	Industrial Protective Apparel	225
8.6.6	Ballistics	225
8.6.7	Permselective Use	225
8.6.8	Electrical Application	226
8.6.9	Communication	226
8.7	Future Trends	227
	References	227

9 Conductive Fibers 233
Tung Pham and Thomas Bechtold

9.1	Introduction	233
9.2	Production of Conductive Fibers: Principles and Technologies	233
9.2.1	Conductivity	233
9.2.2	Metal Fibers: Coating and Deposition	234
9.2.3	Intrinsically Conductive Polymers and Coating	239
9.2.4	Carbon-Based Fibers (Carbon Black, Carbon Nanotubes, Graphene)	242
9.2.5	Combination of Different Techniques	243
9.3	Integration of Conductivity into Textile Structures	244
9.3.1	Fiber Material Selection/Yarn Production	245
9.3.2	Fabric Production	248
9.3.3	Embroidery/Sewing	248
9.3.4	Printing	249
9.4	Applications/Examples	250
9.4.1	Material Selection	250
9.4.2	Washable Sensors (Example: Moisture Bed Sensor)	251
9.4.3	Textile Electrodes	252
9.4.4	Flexible Devices	253
9.4.5	Wearable Electronics	253
9.4.6	Summary and Future Trends	253
	References	254

10 Phase Change Fibers 263
Subrata Mondal

10.1	Introduction	263
10.2	Phase Change Materials (PCMs)	264
10.2.1	Principle of Phase Change Materials	264
10.2.2	Various Types of Phase Change Materials	265
10.2.2.1	Organic PCMs	266
10.2.2.2	Inorganic PCMs	266
10.2.2.3	Eutectic PCMs	267
10.3	Phase Change Fibers	267
10.3.1	How PCM Works with Fibrous Structures	267
10.3.2	Microencapsulation of PCMs	268
10.3.3	Techniques to Prepare Phase Change Fibrous Structures	270
10.3.3.1	Coating of Encapsulated PCM on Fibrous Structure	270

10.3.3.2 PCM Incorporated Fibrous Structure by Conventional Spinning   *271*
10.3.3.3 Phase Change Fibers by Electrospinning   *272*
10.4 Phase Change Fibers for Advanced Material Applications   *274*
10.4.1 Phase Change Fibers for Advanced Textile Applications   *274*
10.4.1.1 Sportswear   *274*
10.4.1.2 Hospital Applications   *274*
10.4.1.3 Beddings and Accessories   *275*
10.4.1.4 Other Applications in Advanced Textiles   *275*
10.4.2 Automotive Industries   *275*
10.4.3 Electrical Applications   *275*
10.4.4 Other Applications   *276*
10.5 Summary   *276*
References   *276*

11 Bicomponent Fibers   *281*
Rudolf Hufenus, Yurong Yan, Martin Dauner, Donggang Yao, and Takeshi Kikutani
11.1 Introduction   *281*
11.2 Bicomponent Fiber Spinning Technologies   *282*
11.2.1 Spin Pack Design   *282*
11.2.2 Cross-Sectional Geometries   *284*
11.2.3 Melt Spinning Equipment   *285*
11.2.4 Special Spin Pack Designs   *286*
11.3 Principles and Theories of Bicomponent Spinning   *288*
11.3.1 Structure Formation During Spinning and Drawing   *288*
11.3.2 Mutual Influence of Components on Orientation and Crystallinity   *288*
11.3.3 Interfacial Adhesion   *290*
11.3.4 Coextrusion Instabilities   *291*
11.3.5 Encapsulation   *292*
11.3.6 Volatiles   *293*
11.3.7 Simulation and Modeling   *293*
11.4 Post-treatment of Bicomponent Fibers   *295*
11.5 Applications of Bicomponent Fibers   *296*
11.5.1 Fibers as Bonding Elements in Nonwovens   *296*
11.5.2 Microfibers   *297*
11.5.3 Fibers with Special Cross Sections   *297*
11.5.4 Fibers with High-Performance Core   *298*
11.5.5 Fibers with Functional Surface   *299*
11.5.6 Biodegradable Fibers   *300*
11.5.7 Polymer Optical Fibers   *301*
11.5.8 Electrically Conductive Fibers   *301*
11.5.9 Liquid-Core Fibers   *302*
11.5.10 Fibers for Fully Thermoplastic Fiber-Reinforced Composites   *303*
11.5.11 Shape Memory Fibers   *304*
References   *304*

12 Superabsorbent Fibers 315
Nuray Ucar and Burçak K. Kayaoğlu

12.1 Introduction 315
12.2 Overview of Superabsorbent Fibers 316
12.2.1 History of Superabsorbent Fibers 316
12.2.2 Main Principle of Superabsorbency 316
12.2.3 Polymer Materials 319
12.2.4 Production Methods 319
12.2.4.1 Mixing the Superabsorbent Material with Hydrophobic/Hydrophilic Material 320
12.2.4.2 Directly Using a Superabsorbent Polymer for Superabsorbent Fiber Production 322
12.2.5 Test Methods 325
12.3 Application 327
12.4 Future Scope and Challenges Ahead 329
12.5 Summary 330
References 331

13 Elastic Fibers 335
Lu Jing

13.1 Introduction 335
13.1.1 Definition 335
13.1.2 Classification 335
13.2 Structure, Principles, and Characteristics 338
13.2.1 Structure and Principle in Elasticity 338
13.2.2 Structure and Principle in Other Performances 343
13.3 New Development of Elastic Fibers 344
13.3.1 Polyurethane Elastic Fiber 344
13.3.2 Bicomponent Elastic Fiber 350
13.3.3 Polyolefin Elastic Fiber 351
13.3.4 Polyether-Ester Elastic Fiber 352
13.3.5 Polyester Elastic Fiber 353
13.4 Evaluation and Application 353
13.5 Future Trends 356
References 357

14 Smart Fibers 361
Dong Wang, Weibing Zhong, Wen Wang, Qing Zhu, and Mu Fang Li

14.1 Introduction 361
14.1.1 Definition 361
14.1.2 Research Status 361
14.1.3 Classification 362
14.2 Raw Materials and Preparation 362
14.2.1 Raw Materials 362
14.2.1.1 Metal Related 363
14.2.1.2 Inorganic Nonmetallic and Polymer Related 364

14.2.2   Preparations   *365*
14.2.2.1 Spinning Methods   *365*
14.2.2.2 Intelligentization Method   *366*
14.3   Structure and Properties   *367*
14.3.1   Primary Fiber Structure   *368*
14.3.1.1 Conventional Fiber Structure   *368*
14.3.1.2 Skin-Core Structure   *369*
14.3.2   Multilayered Structure   *369*
14.3.2.1 Coating with Single Layer   *370*
14.3.2.2 Coating with No Less Than Double Layers   *370*
14.3.3   Three-Dimensional Fiber-Based Structures   *372*
14.3.3.1 Fabrics and Fiber-Based Membrane   *372*
14.3.3.2 Aerogel   *373*
14.4   Principles and Theories   *373*
14.4.1   Shape Memory Fiber   *373*
14.4.2   Fiber-Based Actuator   *374*
14.4.3   Luminescence Fiber   *375*
14.4.4   Color-Changed Fiber   *376*
14.4.5   Thermoregulated Fiber   *377*
14.5   Applications   *377*
14.5.1   Smart Fibers Used in Textiles   *377*
14.5.1.1 Smart Clothing   *377*
14.5.1.2 Wearable Electronics   *378*
14.5.2   Smart Fibers Used in Industrial   *378*
14.5.2.1 Medical Supplies   *378*
14.5.2.2 Sensors   *380*
14.5.2.3 Energy Conversion   *381*
14.5.3   Other Uses   *381*
14.6   Future Trends   *381*
             References   *383*

15   Optical Fibers   *391*
             Hiroaki Ishizawa
15.1   Introduction   *391*
15.2   Fundamentals of Fiber Optics   *392*
15.2.1   Fiber Optics Classification   *392*
15.2.2   Manufacturing Process   *392*
15.2.3   General Characteristics of Fiber Optics   *394*
15.3   Optical Fiber Sensor   *394*
15.3.1   Advantages of Optical Fiber Sensors   *394*
15.3.2   Principles and Applications of Optical Fiber Sensors   *395*
15.3.2.1 Temperature Measurement   *395*
15.3.2.2 Oxygen Concentration Measurement   *395*
15.3.2.3 Strain Measurement by Brillouin Optical Time Domain Reflectometry (B-OTDR) and Fiber Bragg Grating Sensors (FBG)   *395*
15.3.2.4 Biomedical Measurement   *397*

15.3.3   Principles and Application of the Vital Sign Measurement by FBG Sensor *398*
15.3.3.1   FBG Sensor Interrogator to Measure Human Vital Signs *398*
15.3.3.2   Pulse Wave Measurement *399*
15.3.3.3   Pulse Rate Measurement *400*
15.3.3.4   Respiratory Rate Measurement *400*
15.3.3.5   Blood Pressure Measurement by Pulse Transit Time Detection *401*
15.3.3.6   Partial Least Squares Regression Calibration for Blood Pressure Measurement *402*
15.4   Healthcare Monitoring by Using FBG Sensor *404*
15.4.1   FBG Sensor Application to Multi-vital Sign Measurement *404*
15.4.2   Fabrication of Smart Textiles *405*
15.5   Future Trends: As the Summary *406*
  References *407*

16   Memory Fibers *411*
Harishkumar Narayana, Jinlian Hu, and Bipin Kumar
16.1   Introduction *411*
16.2   Morphology and Molecular Mechanism of Memory Polymers *412*
16.2.1   Nature of Transitions in MPs *413*
16.3   Evaluation of Shape Memory Properties *413*
16.4   Memory Polymers As Fibers (MPFs) *414*
16.4.1   Which Fibers Do Have Better Performance, Wet or Melt Spun? *415*
16.4.2   Effect of Post-spinning Operations on MP Fiber Properties *417*
16.4.2.1   Effect of Thermal Setting or Heat Treatment *417*
16.4.2.2   Influence of Drawing Process *420*
16.4.3   Other Type of MPU Fibers *421*
16.4.3.1   Smart Hollow Fibers *421*
16.4.3.2   Electro-responsive Fibers *421*
16.4.3.3   Electrospun Fibers *422*
16.5   Novel Stress Memory Behavior in MPs *422*
16.5.1   What Is Stress Memory? *422*
16.5.2   Mechanism of Stress Memory *423*
16.5.3   Components of Stress Memory *424*
16.6   Stress Memory Behavior in Memory Fibers *425*
16.7   Techniques of Characterization for Memory Fibers *428*
16.8   Potential Application of Stress Memory Fiber/Filaments *428*
16.9   Recent Advances in MP Fibers *430*
16.10   Future Trends *430*
  References *430*

17   Textile Mechanics: Fibers and Yarns *435*
Zubair Khaliq and Adeel Zulifqar
17.1   Introduction *435*
17.2   Fiber *436*
17.2.1   Fiber Consumption *436*
17.3   Strength Contributing Fiber Parameters *438*

17.4 Mechanical Properties of Fiber *438*
17.4.1 Stress–Strain Curve *438*
17.4.2 Elastic Recovery, Work of Rupture, and Resilience *440*
17.4.3 Effects of Time, Temperature, and Moisture *440*
17.5 Yarn Classification *440*
17.6 Yarn Construction *441*
17.6.1 Ring-Spun Yarn *442*
17.6.2 Yarn Structure *443*
17.6.2.1 Carded and Combed Yarns *443*
17.6.2.2 Compact Yarn *444*
17.6.3 Rotor-Spun Yarn *445*
17.6.3.1 Yarn Structure *446*
17.6.4 Air Vortex Spun Yarn *446*
17.6.4.1 Yarn Structure *446*
17.6.5 Frictional Spun Yarn *448*
17.6.6 Mechanical Properties of the Yarn *448*
17.6.7 Important Parameters of Yarn Tensile Strength *448*
17.6.8 Strength–Comfort–Twist Relationship *451*
References *454*

18 Textile Mechanics: Woven Fabrics *455*
Adeel Zulifqar, Zubair Khaliq, and Hong Hu
18.1 Introduction *455*
18.1.1 Woven Fabric Structures *456*
18.1.1.1 Regular Woven Structures (Basic Weaves) *456*
18.1.1.2 Irregular Woven Structures *457*
18.1.2 Weave Factor *457*
18.1.3 The Myth of Crimp *458*
18.2 Woven Fabrics Geometrical Models *458*
18.3 Woven Fabric Mechanics, Theories, and Methodologies *460*
18.3.1 Tensile Deformation of Woven Fabrics *461*
18.3.1.1 Woven Fabric Behavior When Extended in Principal Directions *462*
18.3.1.2 Woven Fabric Behavior When Extended in Bias Direction *463*
18.3.1.3 Effect of Yarn Crimp and Yarn Friction Coefficient on Fabric Mechanical Properties *463*
18.3.1.4 Anisotropy of Woven Fabric Tensile Properties *464*
18.3.2 Compression Deformation of Woven Fabrics *465*
18.3.3 Shearing Deformation of Woven Fabrics *468*
18.4 Mathematical Modeling of Woven Fabric Constitutive Laws *469*
18.4.1 Constitutive Laws of Woven Fabric *470*
18.4.2 Computational Woven Fabric Mechanics *471*
18.4.2.1 Continuum Models *471*
18.4.2.2 Discontinuum Models *471*
18.4.3 Energy Methods for Woven Fabric Mechanics *472*
18.5 Conclusion *473*
References *474*

19	**Fabric Making Technologies** *477*	
	Tao Hua	
19.1	Introduction *477*	
19.2	Weaving *479*	
19.2.1	Weaving Machines *479*	
19.2.2	Woven Structures *481*	
19.2.3	Properties *484*	
19.2.4	Applications *484*	
19.3	Knitting *485*	
19.3.1	Knitting Machines *485*	
19.3.2	Knitted Fabric Structures *486*	
19.3.3	Properties *488*	
19.3.4	Applications *489*	
19.4	Nonwovens *490*	
19.4.1	Manufacture of Nonwovens *490*	
19.4.2	Properties *492*	
19.4.3	Applications *493*	
19.5	Braiding *493*	
19.5.1	Braiding Processes and Machines *493*	
19.5.2	Braided Structures and Properties *494*	
19.5.3	Applications *495*	
19.6	Future Trends *496*	
	References *496*	

Volume 2

Preface *xv*

20 **Chemical Characterization of Fibrous Materials** *499*
Chi-wai Kan and Ka-po Maggie Tang

21 **Soft Computing in Fibrous Materials** *529*
Abhijit Majumdar, Piyali Hatua, and Mirela Blaga

22 **Fiber-Shaped Electronic Devices** *557*
Yang Zhou, Jian Fang, Yan Zhao, and Tong Lin

23 **Fibers for Optical Textiles** *593*
Dana Křemenáková, Jiri Militky, and Rajesh Mishra

24 **Fibers as Energy Materials** *649*
Jiadeng Zhu, Esra Serife Pampal, Yeqian Ge, Jennifer D. Leary, and Xiangwu Zhang

25 **Fiber-Based Sensors and Actuators** *681*
Xiaomeng Fang, Kony Chatterjee, Ashish Kapoor, and Tushar Ghosh

26	**Textile-Based Electronics: Polymer-Assisted Metal Deposition (PAMD)** *721* Casey Yan and Zijian Zheng
27	**Fibers for Medical Compression** *749* Bipin Kumar, Harishkumar Narayana, and Jinlian Hu
28	**Electrospun Nanofibers for Environmental Protection: Water Purification** *773* Hongyang Ma, Christian Burger, Benjamin Chu, and Benjamin S. Hsiao
29	**Fibers for Filtration** *807* Govindharajan Thilagavathi and Siddhan Periyasamy
30	**Fibrous Materials for Thermal Protection** *831* Gouwen Song and Yun Su
31	**Comfort Management of Fibrous Materials** *857* Chengjiao Zhang and Faming Wang
32	**Fibers for Radiation Protection** *889* Boris Mahltig
33	**Fibrous Materials for Antimicrobial Applications** *927* Yue Deng, Yang Si, and Gang Sun
34	**Fibers for Auxetic Applications** *953* Hong Hu and Adeel Zulifqar

Index *973*

Preface

The art of using a fiber material is as old as human civilization. Initially, they were primarily used to make textiles such as yarn or fabric, serving the basic requirement of garment, storage, protection, building, ropes, fishing nets, etc. In textile industry, fibers are characterized by having a length at least 100 times its diameter (ASTM D123-15) and offer several unique features due to its unique configuration – high flexibility, high specific surface area, easy transformability into textile structures, and high load-bearing ability along axial direction – which makes them perfect material for apparel use. Further with the development of advanced materials and their technologies, it is now possible to generate novel fibers, which may not possess the same properties like textile fibers (cotton, PET, Nylon, wool) but have unique functions such as shape memory, superhydrophobicity, phase change, optics, and conductivity that could solve new scientific and technological challenges of different advanced fields. This book focuses on the research and development in fibrous materials and innovative application potentials. Each of the chapters is exclusive and selectively chosen, edited by the internationally recognized experts in the field to cover latest developments and future trends in their fields.

Chapter 1 introduces the different types of fibrous materials including natural and synthetic fibers and their fundamental characteristics. Chapters 2–6 include topics from animal fibers (wool, silk, collagen, chitosan) and plant fibers (cellulose). Chapters 7–16 focus on research and application of different synthetic fibers including electrospun fibers, aramid fibers, conductive fibers, phase change fibers, bicomponent fibers, superabsorbent fibers, elastic fibers, smart fibers, optical fibers, and memory fibers. Chapters 17–19 describes the scientific principles and technologies for converting fibrous materials into textile yarns and fabrics. Chapters 20 and 21 list some characterization methods for analyzing fibrous materials and their structures. More emphasis is given on the wide application potential and scope of advanced fibrous materials in electronics (Chapter 22), optics (Chapter 23), energy (Chapter 24), sensors and actuators (Chapter 25), wearable (Chapter 26), medical (Chapter 27), environmental protection (Chapter 28), filtration (Chapter 29), protection (Chapters 30 and 32), health (Chapter 33), and auxetic (Chapter 34).

The content has been designed to cover different types of advanced fibers, their materials, and devices as well as different properties, diversified functions, and applications.

Finally, we would like to acknowledge the time and efforts of our contributors, who are experts in the respective areas described in this book.

Jinlian Hu
The Hong Kong Polytechnic University

1

Fundamentals of the Fibrous Materials

Jinlian Hu, Md A. Jahid, Narayana Harish Kumar, and Venkatesan Harun

The Hong Kong Polytechnic University, Institute of Textiles and Clothing, Room TQ715, Q Core 7/F, Hung Hom, Kowloon 999077, Hong Kong S.A.R., P.R. China

1.1 Introduction

1.1.1 What Are Fibrous Materials?

A fiber is a material that is defined by Textile Institute [1] as units of matter characterized by flexibility and fineness and a high ratio of length to thickness. In different fields, fibers have very broad meaning such as those for food supplements and fibers in plants or in human body. A fiber is more often referred to the basic unit of making textile yarns and fabrics. Textile fibers should have some specific properties though. For example, cotton plant contains fibers that are strong and soft enough to be spun into yarns that can be woven or knitted into a fabric by textile processing, but human hair is not a textile fiber because it cannot fill up the above properties. So, we can say all textiles are made of fibers, but not all fibers can be used to make textiles. The important requirements for fibers to be twisted into a yarn include a length of at least 5 mm, cohesiveness, flexibility, and sufficient strength, and other important properties include elasticity, fineness, uniformity, luster, and durability. It is also important to remember that all textile fibers are not created equally [2]. Each fiber contains different qualities and will result in a different textile. Some retain heat better than others; some hold dye very well; some are more durable, while some are more comfortable [3].

The origin of fibrous materials may be of organic, inorganic, or metal. They are fine structures that are formed by joining component atoms into molecules. Fibrous materials can be grouped into two major categories: natural and chemical or man-made fibers. The growth of natural fibers is slow and controlled under genetics from fast production of manufactured fibers in terms of structures. Natural fibers include plant or vegetable fibers (such as cotton, flax, ramie, jute, and hemp), animal fibers (such as silk, wool, and hair fibers), and mineral fibers (such as asbestos). Synthetic fibers include regenerated fibers (such as viscose and acetate), synthetics (such as polyester, polyamides, polyolefins), and inorganic fibers such as glass and carbon fibers with completely amorphous or microcrystalline structures [4]. Another class is high-performance fibers that are manufactured ones with improved tensile and other mechanical properties.

Handbook of Fibrous Materials, First Edition. Edited by Jinlian Hu, Bipin Kumar, and Jing Lu.
© 2020 Wiley-VCH Verlag GmbH & Co. KGaA. Published 2020 by Wiley-VCH Verlag GmbH & Co. KGaA.

Finally, smart fibers are an emerging class of fibrous structures that are responsive to stimulus or an environment, tailored with advanced functionalities for many applications.

1.2 Historical Evolution of Fibers

Natural fibers are abundantly available in nature that are biodegradable and sustainable and have played a key role in the human race since nearly 7000 BC [5]. Human beings have been using the fibers from many ages for which there are no records, which itself is a distinct proof that fiber is present from several hundreds of years. The first choice for the human beings to use is natural fibers. Production of synthetic fibers was beginning only in 1910 by commercially producing the rayon fiber, and this was the result of technological development that has not stopped yet on the present era. Natural fibers have been used in many cultures traditionally across the globe in making utilitarian products. Fibrous materials can be obtained naturally from many resources such as plants, leaf, seeds, bark, stem, and grass. Flax is believed to be the oldest fiber and it was obtained in dates back to 6000 BC. Egyptians started to wear the cotton since 4000 BC, and later revolutionary changes occurred to discover ginning and production of different variety of cotton products. Wool fiber was also discovered around 3000 BC, and then around 40 breeds were explored to produce wide range of wool fibers. Dates back to 2500 BC, silk fiber is believed that it was first discovered by a Chinese princess that was obtained from the cocoons of silk worms. The silk processing was found to be kept secret for almost 3000 years and then started to spread all over the world in the later stage.

Even though natural fibers were comfortable and biodegradable, they had some drawbacks such as wrinkling of cotton and flax and delicate handling of silk and wool fibers receptive to shrinkage and moth attack. Growing needs of human race and these disadvantages of fibers paved the way to produce the first man-made fiber with revolutionary technological efforts into rayon, which is known as artificial silk to the present era. In the later stages, other man-made fibers such as acetate, nylon, acrylic, polyester, spandex, polypropylene, lyocell, and microfibers were produced with rapid transformation from science to market and end users around the world. The complete developments in the evolution of human civilization (Figure 1.1) and textile fiber (Figure 1.2) are presented below [6].

1.3 Classification of Fibrous Materials

A comprehensive classification of fibrous materials is depicted in Figure 1.3. Nature offers us abundant resources for getting fibrous materials that grow in various geographical locations and altitudes. Fibrous materials can be further classified into textile and non-textile fibers. The internal characteristics that are suitable for processing from fibers [7] to yarn, fabric, and other utilitarian products can be termed as textile fibers [8, 9]. However, the properties of fibers

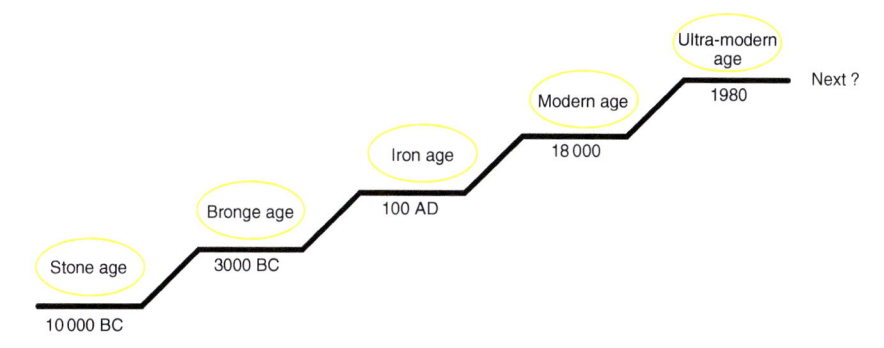

Figure 1.1 Evolution of human civilization.

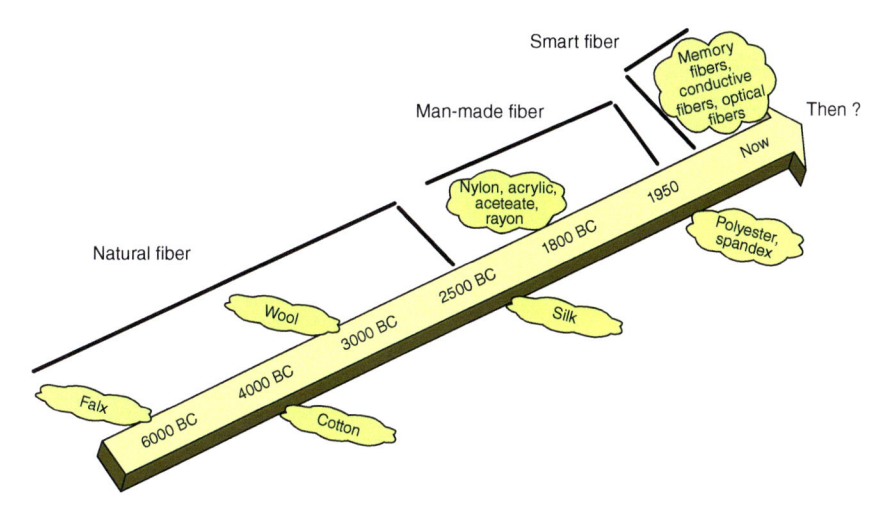

Figure 1.2 Evolution of textile fibers.

that are not suitable for clothing could be considered as non-textile fibers and have been found broad horizon of arenas to fulfill the human desired needs in many ways. Non-textile natural fibers are also used in building materials, animal and human food, ecofriendly cosmetics, agro-fine chemicals, and sources of energy.

Fibers can be classified into two major groups as natural and man-made or manufactured ones. Natural fibrous materials are basically from three origins; they are cellulosic/lignocellulosic (from plants), protein, and mineral, which is decided by the chemical composition and fiber morphology. Cellulosic fibers constitute long-chain molecule joined together with glucose rings by valence bonds and hydroxyl groups linked by hydrogen bonds in its chemical structure. Other cellulosic fibers such as flax, jute, hemp, ramie and sisal differ from cotton in various ways. Normally larger proportion of impurities contains in their structure. They are multicellular fibers composed of microscopic individual cells bonded together and running along the strands in the plant or leaf. Hair fibers are also termed as protein fibers that are morphologically formed by

4 | *1 Fundamentals of the Fibrous Materials*

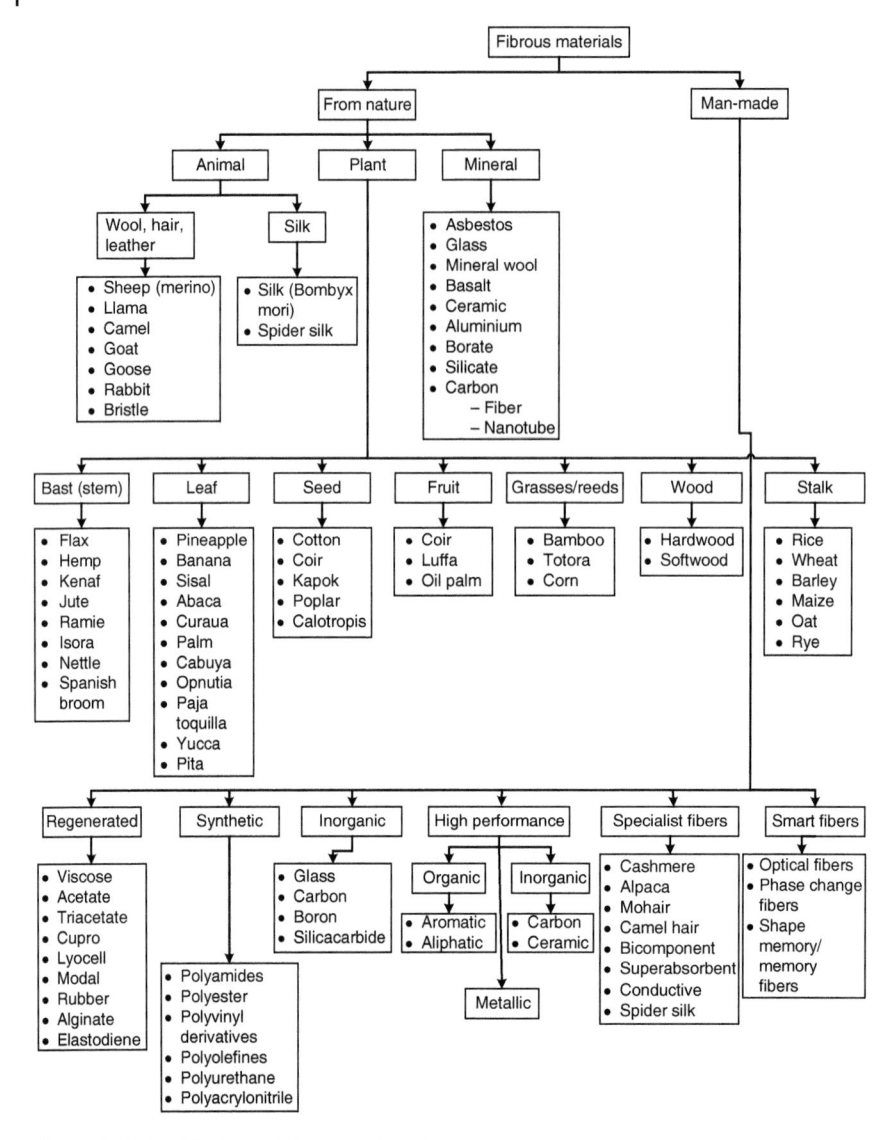

```
                                Fibrous materials
                               /                \
                       From nature            Man-made
                      /      |      \
                 Animal    Plant   Mineral
```

Animal		Plant	Mineral

Wool, hair, leather	Silk		• Asbestos
			• Glass
			• Mineral wool
			• Basalt
			• Ceramic
			• Aluminium
			• Borate
			• Silicate
			• Carbon
			  – Fiber
			  – Nanotube

• Sheep (merino) • Llama • Camel • Goat • Goose • Rabbit • Bristle	• Silk (Bombyx mori) • Spider silk		

Bast (stem)	Leaf	Seed	Fruit	Grasses/reeds	Wood	Stalk
• Flax • Hemp • Kenaf • Jute • Ramie • Isora • Nettle • Spanish broom	• Pineapple • Banana • Sisal • Abaca • Curaua • Palm • Cabuya • Opnutia • Paja toquilla • Yucca • Pita	• Cotton • Coir • Kapok • Poplar • Calotropis	• Coir • Luffa • Oil palm	• Bamboo • Totora • Corn	• Hardwood • Softwood	• Rice • Wheat • Barley • Maize • Oat • Rye

Regenerated	Synthetic	Inorganic	High performance		Specialist fibers	Smart fibers
• Viscose • Acetate • Triacetate • Cupro • Lyocell • Modal • Rubber • Alginate • Elastodiene	• Polyamides • Polyester • Polyvinyl derivatives • Polyolefines • Polyurethane • Polyacrylonitrile	• Glass • Carbon • Boron • Silicacarbide	Organic • Aromatic • Aliphatic Metallic	Inorganic • Carbon • Ceramic	• Cashmere • Alpaca • Mohair • Camel hair • Bicomponent • Superabsorbent • Conductive • Spider silk	• Optical fibers • Phase change fibers • Shape memory/ memory fibers

Figure 1.3 Classification of fibrous materials.

the polymerization of amino acids (NH_2–CHR–COOH) by peptide linkage (–CO–NH–) into long-chain molecules such as silk and wool. Mineral fibers are obtained from the naturally available inorganic substances such as rock, clay, and slag. Another group of fibers that have changed the world with revolutionary scientific discovery of mankind is man-made or manufactured fibers. Different classes of manufactured fibers are grouped in Figure 1.4, which includes fibers from natural polymers (regenerated or artificial or modified celluloses), synthetic polymers, and organic and inorganic substances. Regenerated fibers such as viscose rayon, acetate, and triacetate are produced from the celluloses that are

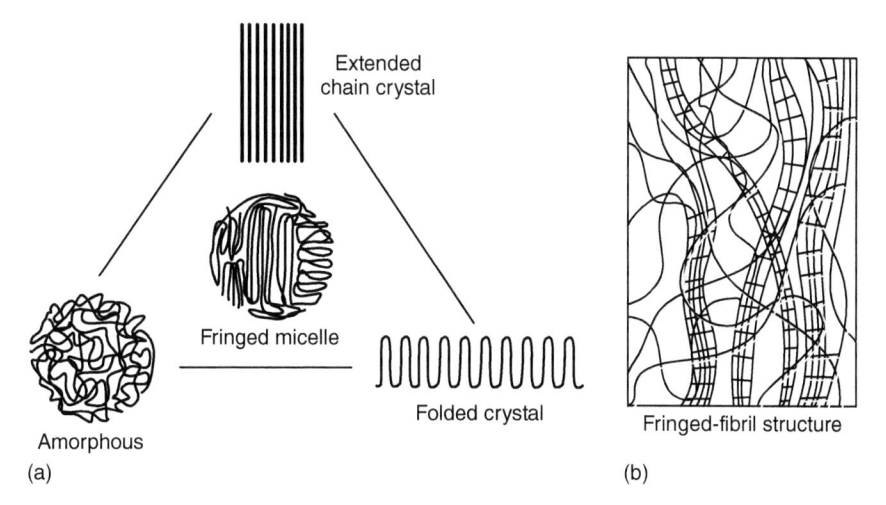

Figure 1.4 Phases of polymeric chains in fibrous materials. Source: Adapted from Morton and Hearle 2008 [2] and Kajiwara 2009 [26].

chemically regenerated from the natural resources such as wood pulp. The major difference in this class of fibers is thinning down or reducing the degree of polymerization by chemical modification of native cellulose. However, alginate fibers are produced from the alginic acid that can be obtained from seaweed. The acid is subjected to degradation and neutralization process to form the metallic salts that form the highly oriented and highly crystalline alginic acid fibers [3]. Synthetic fibers are spun from the polymers that are synthesized using monomers joined together by covalent bonds via process called polymerization. Polymerization could proceed with several variations, and thus they can be classified in various ways as listed in Figure 1.4 under synthetic group. High-performance fibers are of two types, namely, organic and inorganic. Glass is an inorganic with unoriented and amorphous network based fiber, which has a long history as decorative product. Carbon fibers are basis for making composites that are made of acrylic or cellulose precursors via carbonization. The high temperature ceramic fibers are produced by using organosilicon polymers or viscous liquid with aluminum salts. Discovery of liquid crystals led the way to produce fibers with almost perfect crystallinity, almost perfect orientation, and fully extended molecular chains. Aromatic and aliphatic polymers used to produce Kevlar and Nomex that are also called as aramid fibers. Liquid crystals can also be formed from polyethylene to produce ultrahigh molecular weight polyethylene to get highest strength. Specialty fibers are of normally hair fibers such as cashmere, camel hair, alpaca, and mohair. These fibers are generally straighter and smoother than wool [10–12]. Other specialist fibers that are produced by advanced and distinguished technology such as hollow fibers, bicomponent fibers [13, 14], superabsorbent fibers [15, 16], and conductive fibers [17]. Hollow and bicomponent fibers are generally spun using different types of spinnerets. Superabsorbent fibers are produced using acrylic-based products and characterized by improved liquid or fluid transportation that are used in medical, textiles, food packaging, and technical textiles. Other advanced fibers can also be considered as specialist fibers, but they are listed under smart fibers that are stimulus responsive such as optical fibers, phase change fibers, and shape memory fibers. Optical fibers are made of polymer or glass core materials with hollow core along the length for applications such as photonics and sensors, and they are called as bandgap fibers [18]. Phase change fibers are generally composed of thermoplastic polymer and luminescent materials as a composite structure [19]. Another smart fiber is considered as one of the most important class of stimuli responsive materials, i.e. shape memory polymeric (SMP) fibers. SMPs are made of segmented polyurethanes with custom tailorable switching or transition temperatures that can be spun via different spinning methods such as melt spinning, dry spinning, wet spinning, reaction spinning, and electrospinning process [20–23].

Increased demand and population explosion is alarming the world to look up for an alternative material that is sustainable and biodegradable. Researchers and industrial inventors may invent futuristic advanced material-based fibers for upcoming days. This is unstoppable and essential for groundbreaking revolutionary technological advancements for this present competent world to prepare for better future.

1.4 Fundamental Characteristics of Fibrous Materials

The era of synthetic fibers was beginning around ten decades ago, and until that time the usage of natural fibers was most popular owing to their natural properties useful for mankind. These natural properties were being considered as benchmark to consider a fiber to be useful in the production of textile materials [24]. These properties are imperative for textile processing and they can be divided as follows [6]. They are primary properties (such as staple, strength, elasticity, fineness, uniformity, and spinning quality) and secondary properties that are desirable. Fibers that are useful in textile processing should be elastic between 5% and 50% breaking elongation. Other non-textile materials like inorganic fibers such as glass, ceramic, and fully crystalline objects are neither flexible nor extensible. The materials that meet this requirement would be almost partially oriented, partially crystalline, and linear molecular arrangements. Functional applications need high-performance fibers that differ from textile fibers with high strength and low extension. Almost they are linear polymers and added with inorganic networks to integrate the flexibility. Elastomeric fibers are used for high stretch applications. There is also other sort of fibers that are interesting such as those in living organisms, wood fibers used for papers, and various tissues. In addition to general characteristics, there are other important features that characterize the fibers, and they are degree of order, degree of localization of order, length/width ratio of localized units, degree of orientation, size of localized units, and molecular extent in a fiber assembly [3]. Fibrous strand is composed of fibrils that are an assembly of polymer molecules and a primary unit called microfibril that is almost identical in all natural fibers. Man-made fibers also show a similar element in the structure [25]. Hess and Kiessing [27] suggested a schematic fringed micelle model to elucidate the structure of microfibrils of synthetic fibers as shown in Figure 1.4. The microfibril basically constitutes crystalline region (ordered polymer chains along the axis) and amorphous region (randomly oriented or less ordered), and an individual polymeric chain interconnects the two regions (Figure 1.4a). Hearle [27–29] proposed a fringed fibril structure (Figure 1.4b) for low crystalline fibers, which can be seen in stiffer molecular chains such as cellulose. This model combines fibrillar form and the concept in fringed micelle structure having distinct crystalline and non-crystalline portions with molecular chains running through each region continuously. High-performance fibers with highly crystallinity have more closely packed fibrillar structure. Natural fibers have specific fibrillar forms and wool with fibrils separated by various matrices.

1.5 Morphological and Structural Properties of Fibrous Materials

The structure of fibrous materials varies in several ways and depends on how actually they are formed whether naturally or artificially. Natural fibers grow slowly and are controlled under genetics to yield different structures, which vary from fastest passage of synthetic fiber production.

1.5.1 Plant or Natural Cellulosic Fibers

An essential feature of natural fibers is that they are aggregated into finer microfibrillar structures (Figure 1.5a,b). According to biological validation, during the growth glucose rings are joined together to form long-chain molecules of cellulose. The thickness of these natural microfibrils is about 4 nm. The molecular alignment takes place in parallel direction of crystal lattice with no chain folding. The crystallization occurs due to minimization of free energy, and forceful attractions are at the edge of molecules likely to form a ribbonlike structure (Figure 1.5a). Microfibrils are almost flexible and attract each other with strong hydrogen bonding at molecular edges or between the faces by weak van der Waals force. Cotton fibers show irregular, convoluted long ribbonlike structure, covered with cuticle, primary wall, and lumen at the center in a cross-sectional view (bean shape) (Figures 1.5a and 1.6a). Apart from cotton, other cellulosic fibers including jute, ramie, hemp, and flax are also multicellular fibrous structures. Smaller fibrillar cells run along the length from the stem or leaf as shown in Figure 1.5. The fibrils spiral around the fibers and are parallel to axis in all natural cellulosic fibers. However, in bast fibers, the spiral angle is reduced, and structures are highly oriented to fetch high strength with less extensibility.

1.5.2 Animal or Protein Fibers

Silk, wool, or hair fibers are considered as protein fibers originated from the living animals. Groups of amino acids combine into long-chain molecules

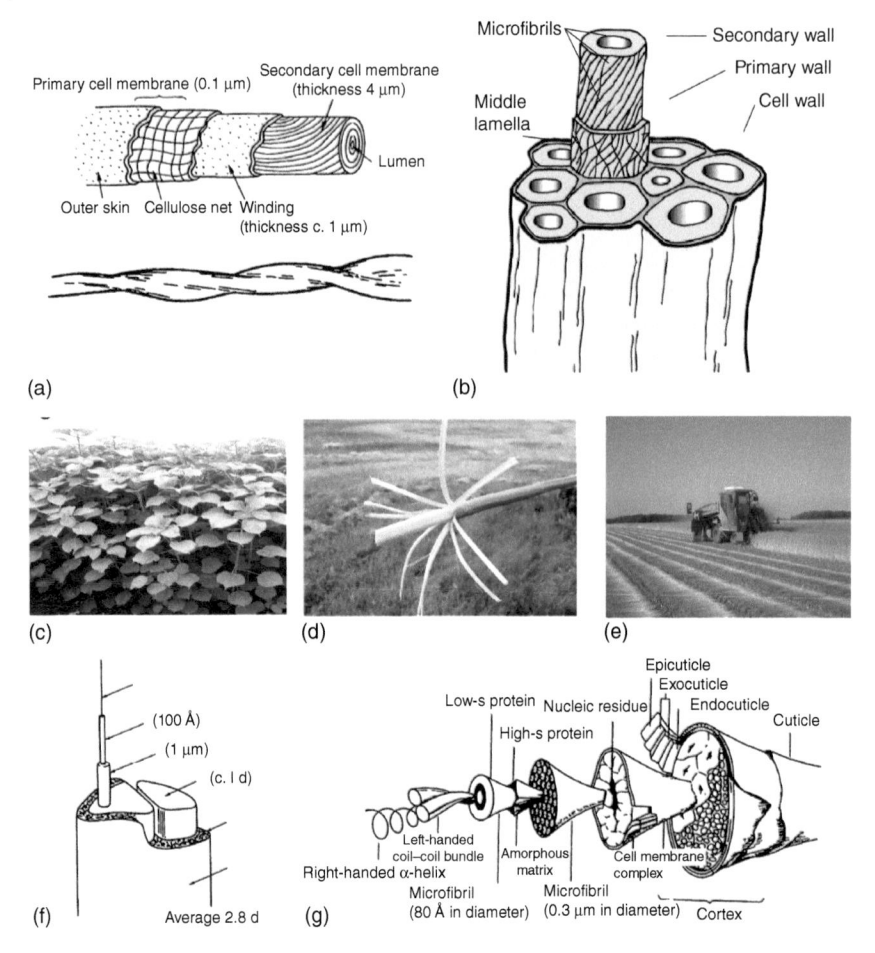

Figure 1.5 Structural and photographical images of natural and animal fibers. (a) Cotton. (b) Jute. (c) Ramie. (d) Hemp. (e) Flax. (f) Silk. (g) Wool. Source: Adapted from Kajiwara 2009 [26], Kozasowski et al. 2012 [30], and Roy and Lutfar 2012 [31, 32].

with polypeptide linkages to form protein fibers. The major protein of silk fiber is fibroin molecule that is 140 nm long with other 17 nm long segments composed of tyrosine and other concentrated side groups, with rest being fully glycine, alanine, and serine acids. Silk is spun by silkworm and constitutes two triangular fibroins covered with a sericin gum (Figures 1.5f and 1.6a). Hair or wool fibers show the most complex structures among any other fibrous materials (Figure 1.5g). The major protein in hair fibers is keratin that consists of several proteins. Wool fibers with α-keratin protein tend to reverse transformation when stretched to form β-keratin. α-Helix consists multiple twisted molecules, two chains makes a dimer, two dimers into a protofilament, two protofilaments a protofibrils, two protofibrils a half filament, and two half filaments gives an intermediate filament having 32 molecules in cross section [3].

1.5 Morphological and Structural Properties of Fibrous Materials | 9

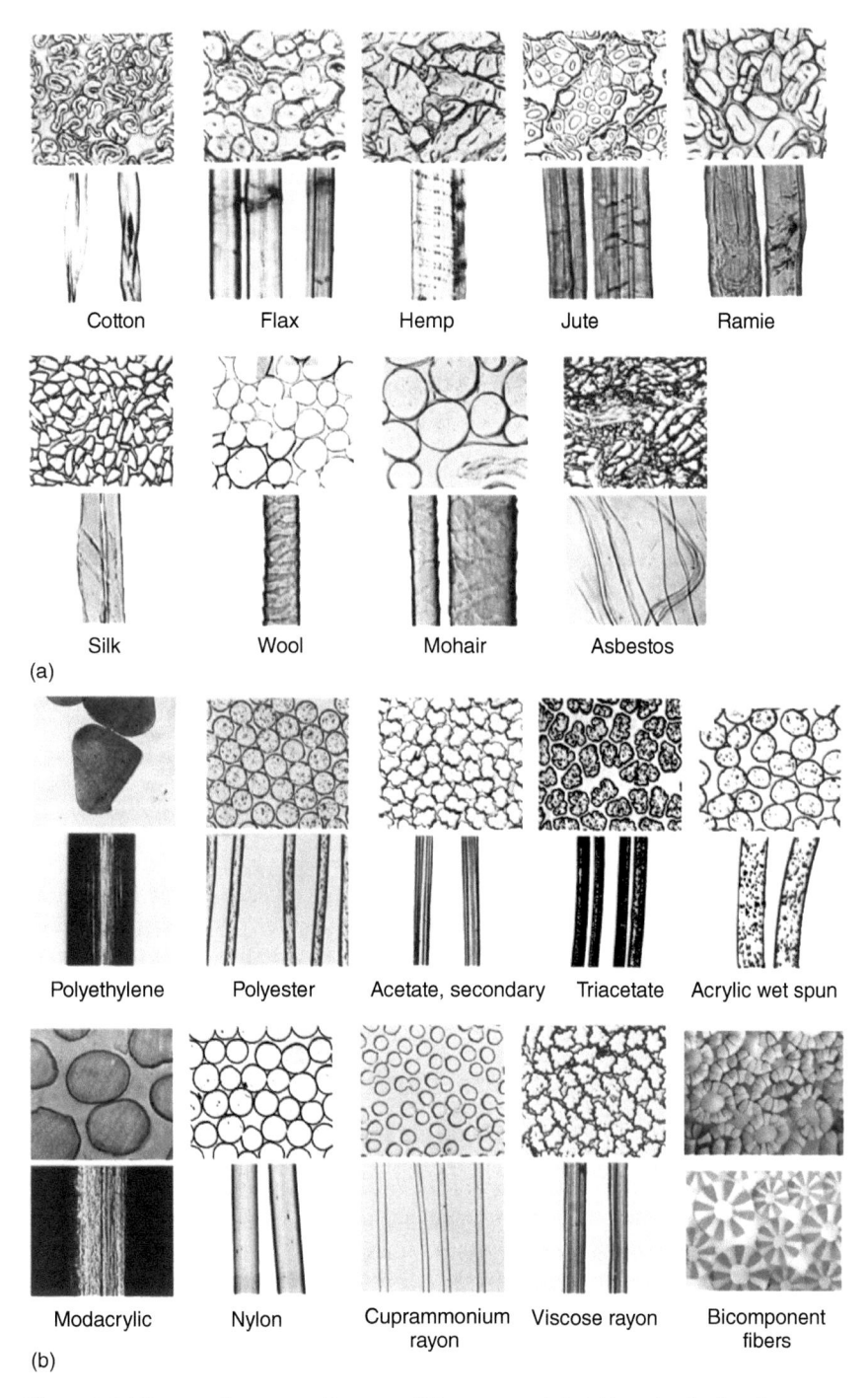

Figure 1.6 Microscopic structural images of fibrous materials. (a) Longitudinal and cross-sectional view of natural fiber. (b) Longitudinal and cross-sectional view of man-made fiber. Source: Adapted from Prahsarn et al. 2016 [14] and AATCC 2013 [34].

1.5.3 Regenerated Cellulosic Fibers

The degree of polymerization is reduced and of the order of 500 compared to native cellulose. The crystalline region is less and of about one third of the total. However, the degree of orientation totally depends on the drawing during spinning. There are diverse forms of rayon with irregular shapes in the cross section (Figure 1.6b). Viscose rayon is formed by intermediate cellulosic xanthate, whereas acetate fibers are treated to replace the hydroxyl groups with acetyl groups. They have good dimensional stability and high softening point and are crease resistant with heat-set ability [3].

1.5.4 Synthetic or Manufactured (Textile and Non-textile) Fibers

Synthetic fibers lack in terms of their structural features. Spinning conditions and the spinneret being used decides the fiber cross section. Melt-spun fibers using circular spinneret shows circular shape and other shapes from shaped spinneret. However, fibers from solution spinning have different shapes such as acrylic fibers (dumbbell shape) due to mass transfer or loss of solvent (Figure 1.6). The longitudinal and cross-sectional views of other synthetic fibers are shown in Figure 1.6b [13].

Other additives such as pigments, delustrants, and antistatic agents can also be included to achieve desired properties. Most of the non-textile fibers that are required for engineering uses need high stiffness and strength and high-performance (high-modulus, high-tenacity [HM-HT]) fibers are the best choice for this. The presence of covalent strong chemical bonding gives high mechanical and chemical resistance to ceramic fibers. Ceramic fibers contain mixture of components such as silica, SiO_2, carbon, Al_2O_3, and B_2O_3. Carbon fibers are composed of graphite crystal lattice, and the carbon atoms are held together by covalent bonds. Aramid (para) fibers such as Kevlar and Twaron are composed of aromatic polyamide, poly(p-phenylene terapthalamide) (PPTA). These fibers are highly oriented with fully extended chains and high crystallinity [3].

1.6 Essential or Fundamental Properties of Fibrous Materials

The fundamental composition and morphological structure vary from fiber to fiber, and there are some important properties to be considered to intelligently utilize them into different applications for human needs. Fundamental properties of fibrous materials are most necessary to distinguish them from each other and find out other desirable properties via standard testing conditions.

1.6.1 Physical Properties

Primary processing of any fibrous material is to be decided by their physical properties such as fiber length, elongation, tenacity, moisture sorption and swelling behavior, color, luster, electrical behavior, specific gravity or density, and solubility. Fiber length of natural fibers varies, and they differ in their staple length except continuous filament such as silk. Physical properties of synthetic fibers can be tailor-made according to end user requirements. Physical properties of fibers materials are shown in Table 1.1.

1.6.1.1 Mechanical Behavior of Fibrous Assemblies

Textile fibrous materials differ from conventional engineering materials in different ways owing to their unique mechanical behaviors. Based on such behaviors, it is possible to predict general fashion and expected behavior to obtain the necessary characteristics. Fibrous materials are highly anisotropic, easy to deform, and inhomogeneous and have larger strains and displacement at lower stress and nonlinear and plastic nature at low stress and in room temperature. Hence they do provide essential characteristics suitable for the movement of human body, perception satisfaction, and other psychological and physiological requirements [1].

It is imperative to comprehend the behavior of fibers in a unit of assembly employed for multidirectional deformations. Figure 1.7 shows the complicated geometrical structure of a fibrous assembly, textile woven fabrics, which is an integration of warp and weft yarns through intersection points. The passage of fibrous strand in the cross-sectional shape is crimped and irregular with protruding the fibers from the strand surfaces. Porus structure of woven fabric is decided by the distance between two adjacent yarns differentiates from the continuum engineering material such as metallic sheets. Pierce has simplified and idealized the cross-sectional shape and physical properties of yarn assemblies through theoretical modeling approach [1].

Nonlinear Stress–Strain Behavior of Textile Fibrous Assembly The mechanical behavior of woven fabric structures is complicated upon deformation, and their typical stress–strain curves are depicted in the Figure 1.8, which shows tensile, bending, and shear deformations. The tensile stretching (Figure 1.8a) shows larger strain even at small force, and this is ascribed to straightening of crimped length of yarns in the woven structure. A typical fabric has a tensile modulus in the order of 10 MPa as compared to steel that is around 2×10^5 MPa. The woven fibrous assemblies are more prone to bending deformations under the transverse loading (Figure 1.8b). The bending stiffness of the fibrous strand to that of a solid rod of the same cross section can be easily determined assuming the slippage between the fibers is unconstrained. That is, $\alpha(\gamma/R)^2$, where α is denoted as porosity ratio and R and γ are the radii of the yarn and its constituent fibers, respectively. For a typical yarn that contains 100 fibers, this ratio is ~1 : 10 000. Thus, it is possible to develop a thick yarn with more flexibility and is advantageous to have less bending stiffness in the woven fibrous structure. As mentioned earlier, the geometry of the woven fabrics is complicated, and when these materials are subjected into deformation, interesting phenomena could be seen, which is distinguished from conventional engineering materials. They are as follows:

(1) *Nonlinear stress–strain trend*: When the woven fibrous material is subjected to small strains, the corresponding stress will be nonlinear, which contrasts

Table 1.1 Properties of different types of textile fiber [6, 26, 30, 31].

Family/group	Fibers	Tensile strength (cN/dtex) Dry	Tensile strength (cN/dtex) Wet	Elongation (%) Dry	Specific gravity (g/cm³)	Young's modulus (cN/dtex)	Moisture regain (%)
Natural fibers	Cotton	2.3–4.3	2.9–5.6	3–7	1.54	60–82	8.5
	Jute	5.4–8.71	—	1–1.8	1.48	—	13.8
	Silk	2.6–3.5	1.9–2.5	15–25	1.33	44–88	11
	Wool	0.9–1.5	0.7–1.4	25–35	1.32	10–22	15
Regenerated fibers	Viscose rayon	1.5–2.0	0.7–1.1	18–24	1.50–1.52	57–75	11
	Triacetate	1.1–1.2	0.6–0.8	25–35	1.30	26–40	3.5
Synthetic fibers	PET	3.8–5.3	3.8–5.3	20–32	1.38	79–141	0.4
	Polyacrylonitrile	2.2–4.4	1.8–4.0	25–50	1.14–1.17	34–75	2
	Polyvinyl chloride	1.11–3.33	1.11–3.33	10–125	1.33–1.40	—	—
	PTFE	1.11–2.22	—	10–30	2.1–2.3	—	0
	Nylon 6	4.2–5.7	3.7–5.2	28–45	1.14	18–40	4.5
	Nylon 6,6	4.4–5.7	4.0–5.3	25–38	1.14	26–46	4.5
	Vinylon (PVA)	3.5–5.7	2.8–4.6	12–26	1.26–1.30	53–79	5
	Polypropylene	4.0–6.6	4.0–6.6	30–60	0.91	35–106	0
	Polyethylene	4.4–7.9	4.4–7.9	8–35	0.94–0.96	—	0
	Polyurethane	0.5–1.1	0.5–1.1	450–800	1.0–1.3	—	1
Aramid fibers	Kevlar	8.8–24.44	7.7–23.33	3–20	1.45	133.3–144.4	3
	Nomex	6.66–13.33	5.55–12.22	20–30	1.38	125.5–133.3	3.5
High-performance fibers	Polycarbonate	4.44–5.55	—	20–45	1.23	—	0.4
	PBI	2.88–3.33	2.33–2.77	25–30	1.43	50–66	15
	Sulfur	3.33–3.88	—	25–35	1.37	33–44	0.6
Inorganic	Carbon	22.2–27.7	—	1–2	1.7–1.8	—	—
	Glass	6.6–11.1	5.5–8.8	—	2.5–2.6	—	0.5

PET, polyester; PTFE, poly(tetrafluoroethylene); PBI, polybenzimidazole.

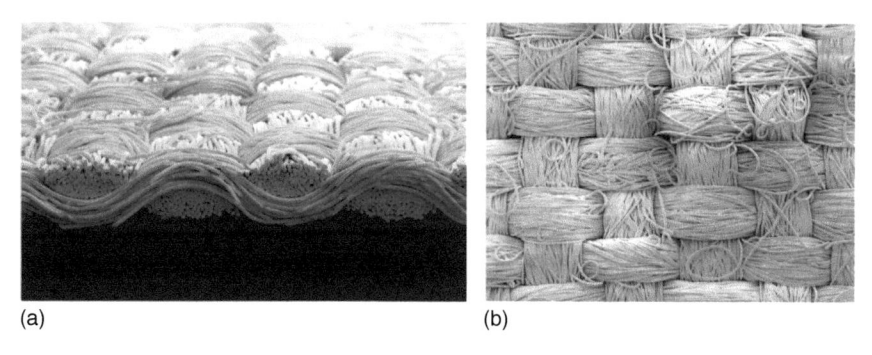

1.6 Essential or Fundamental Properties of Fibrous Materials | 13

(a) (b)

Figure 1.7 Structural photographic images of plain woven fibrous assembly. (a) Cross-sectional image. (b) Surface image. Source: Adapted from Denton and Daniels 2002 [1].

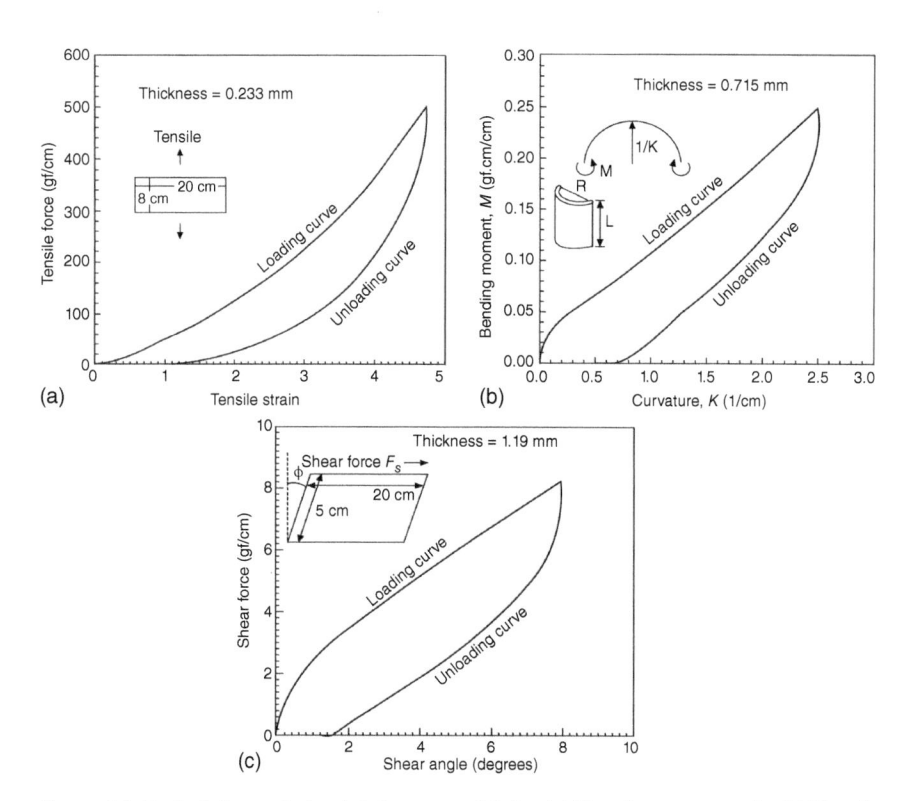

(a) (b) (c)

Figure 1.8 Typical stress–strain plots for woven fabrics. (a) Plot of stress vs. strain. (b) Plot of moment vs. curvature. (c) Plot of shear force vs shear angle. Source: Adapted from Hu 2004 [9].

with that of conventional engineering material with linear trend. The stress–strain curve in fabrics is nonlinear up to small strains, and it becomes linear beyond the critical stress region. This critical region is higher for tensile deformation and very low or nearly zero for bending and shear deformations. Due to porous and crimped nature of fibrous assembly in the woven fabrics, they tend to become straightened upon deformations, thus showing nonlinear trend during the consolidation process. Once the consolidation and reorientation is done, the stress–strain trend shows the linear that is like a solid conventional engineering material.
(2) *Irrecoverable deformation*: Woven textile fibrous assembly shows loops between the loading and unloading curves as shown in the Figure 1.8. It can be seen that, in three of the cases such as tensile, bending, and shear, the deformed length is not able to recover back to their original state. This is denoted as irreversible or irrecoverable deformation occurs at small stresses. This phenomenon differs from the conventional engineering material where the inelastic deformation occurs at high stress level where the failure may happen.

Textile fibrous assemblies do obviously differ from those conventional engineering materials in many ways such as inhomogeneous, anisotropic, and nonlinear behaviors. Complicated mechanical response occurs at low stress and room temperature for fabrics, whereas this happens at high temperature and high stress for engineering materials.

1.6.2 Chemical Properties

Natural and synthetic fibers have different chemical properties such as affinity and inertness to various chemicals. In general, most of the cellulosic fibers are resistant to alkalis and protein fibers to acids. There are some important chemical properties to be considered such as effect of acids, effect of alkalis, effect of oxidizing agents, effect of light or sunlight, and effect of insects.

1.6.3 Biological Properties

Biological properties of fibrous materials play an important role in some very specific applications. These properties include resistance to moths, mildews, and microorganisms, biodegradability, and biocompatibility of fibrous structures. Biodegradable natural or synthetic polymeric fibers such as silk, collagen fibers, and polyurethanes are used for implants and tissue engineering.

1.6.4 Thermal Properties

Fibrous materials are thermoset (natural fibers) or thermoplastic (manufactured) in nature. Thermal properties such as softening point, glass transition temperature, melting temperature, and degradation point or decomposition temperature range are important to be considered for processing or applications.

1.6.5 Other Desirable Properties

Manufactured fibers are modified into broad horizon of materials suitable for vivid smart applications needing desirable properties such as optical performance (photonics), electrical properties (energy storage), phase change behavior, and shape memory properties (memory behavior).

1.7 Textile Processing

Textile manufacturing or processing is a very complex process. The area of textile manufacturing process is too long (Figure 1.9). It starts from fiber to finished products. Fiber is the main material for textile product; properties of textile products mainly depend on fiber type. Textile fiber has few special features that vary between fibers and textile fiber. Textile fiber can be twisted into a yarn or made into a fabric by various methods like weaving, knitting, nonwoven, and braiding. Textile fiber needs to be stronger enough to hold their shape, flexible enough to be shaped into a yarn or fabric, elastic to stretch enough, and durable enough to last [33, 34].

1.7.1 Spinning

Spinning is the first step of textile product processing. The process of making yarns from the textile fiber is called spinning. Spinning is the twisting together of fibers to form yarn. There are different types of spinning process: ring spinning, rotor spinning, friction spinning, etc. Ring spinning is the most commonly and widely used spinning process.

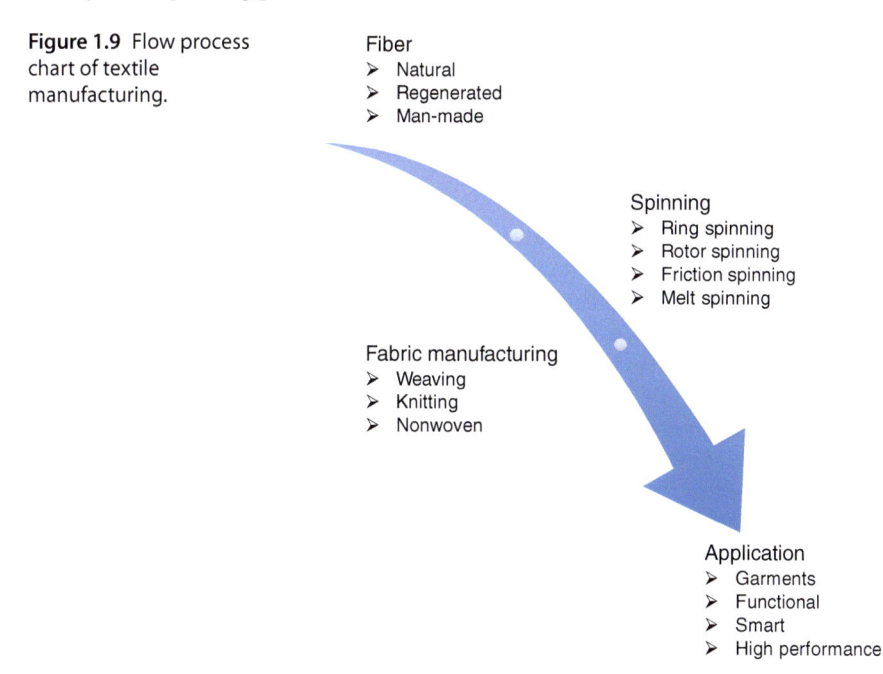

Figure 1.9 Flow process chart of textile manufacturing.

Fiber
➢ Natural
➢ Regenerated
➢ Man-made

Spinning
➢ Ring spinning
➢ Rotor spinning
➢ Friction spinning
➢ Melt spinning

Fabric manufacturing
➢ Weaving
➢ Knitting
➢ Nonwoven

Application
➢ Garments
➢ Functional
➢ Smart
➢ High performance

1.7.1.1 Ring Spinning

Ring spinning is a continuous system of spinning in which twist is inserted into a yarn by using a circulating traveler in ring spinning; the roving is first attenuated by using drawing rollers and then spun and wound around a rotating spindle, which in its turn is contained within an independently rotating ring flyer. Ring spinning is the mostly used spinning process. Ring spinning processes are as follows:

Blow room: The section where the supplied compressed bale is turn into a uniform lap of particular length by opening, cleaning, blending, or mixing is called blow room section. It is the first step of spinning process. It is even feed of material to the card.

Carding: It is the heart of spinning process. This is where the flock from bales will be open into individual fiber. Thus, it will ease to remove the excess impurities on the fiber surface. At this point, short fiber that is not suitable for production in terms of length requirements will be removed. Carding is one of the most important operations in the yarn processing as it directly determines the final features of the yarn. The main objectives of the carding can be summarized as opening the tufts into individual fibers, eliminating all the impurities contained in the fiber that were not removed in the previous cleaning operations, selecting the fibers based on length and eliminating the shortest ones, removing the neps, and parallelizing and stretching the fiber.

Drawing: At this stage, the sliver will be pulled in lengthwise direction over each other. Thus, it will cause it to be stronger and thinner in production, which is very important in evenness of the yarn. Card slivers fed to the drawing machine have some degree of unevenness. The drawing machines are very much important in the yarn manufacturing process. One of the main and most important tasks of the draw frame is improving evenness.

Combing: The cotton sliver produced by the card contains several contaminants that obstruct with the spinning of fine high and good quality yarns. This is a process where the yarn will be straightened again so that they are arranged in parallel manner. While at the same time, the remaining of short fiber will be removed completely from the fiber. The main functions of the comber are to remove a substantial amount of the short fiber as "noil" and eliminate the smallest trash particles and neps.

Roving manufacturing: In this stage, sliver will be further drafted and twist will be inserted to produce roving sliver. Enough twist is given to keep the fibers together but still has no tensile strength. The roving in bobbins is placed in spinning frame where it passes several sets of roller that run at high speed to convert into yarn forms.

Ring spinning: The ring spinning is the most broadly used form of spinning machine due to significant advantages in comparison with the new spinning processes. The ring spinning machine is used in the textile industry to continuously twist staple fibers into yarn and then wind it onto bobbins. The following are the core objectives of ring spinning:
- To draft the roving fed to the ring spinning frame, i.e. to convert roving into very fine strand called yarn.

- To improve strength to the yarn by inserting the necessary amount of twist.
- To collect twisted strand called yarn onto handy and transportable package by winding the twisted thread on a cylindrical bobbin or tube.

Cone winding: In the spinning process, winding is the last step. After winding yarn packages are used for making woven or knitted fabric. Winding process can be defined as the transfer of spinning yarn from one package to another large package (cone). A process of assemble yarn on a package to facilitate the next process is called as winding.

1.7.1.2 Rotor Spinning

In rotor spinning, sliver is fed into the opening end of rotor; then fed sliver is opened into single fibers, and then the singled fibers are reassembled, twisted, and wound on a package. Moreover, sliver from the carding machine goes into the rotor, is twisted into yarn and comes out, wrapped up on a cone shaped bobbin; so there is no roving stage in rotor machine and no need to use the auto-coner to winding the yarn in a cone because output of rotor machine is sliver to yarn into a cone shaped package. Advantages of rotor machine are less labor, less time, and high productivity compared with ring spinning. But the disadvantages are low strength, difficult to produce fine yarn, and difficult to keep spinning condition constant.

1.7.1.3 Friction Spinning

Friction spinning system is an open-end system or a "core-type" spinning system, where the yarn formation takes place by the frictional forces. In friction spinning, fibers are supplied onto the drum surface, which transport it and stack fiber to the fiber bundle circulating between two surfaces passing in opposite directions; then opening or individualization of fibers will happen, followed by reassembling of single fiber and then twisting of assembled fiber by frictional force, winding the final yarn. Advantages of friction spinning are excellent yarn regularity, lower preparation and spinning cost, and very high twist insertion rate, and few disadvantages are low yarn strength, high air consumption leading to high power consumption, and friction spun yarns having higher snarling tendency.

1.7.1.4 Yarn Numbering System (Count)

Count is a numerical value, which expresses the coarseness or fineness (diameter) of the yarn and indicates the relationship between length and weight (the mass per unit length or the length per unit mass) of that yarn. The fineness of the yarn is usually expressed in terms of its linear density or count.

According to Textile Institute, "Count is a number indicating the mass per unit length or the length per unit mass of yarn."

There are a number of systems and units for expressing yarn fineness. But they are classified as follows:

Direct count system: The weight of a fixed length of yarn is determined. The common features of all direct count systems are that the length of yarn is fixed and the weight of yarn varies according to its fineness. In the direct system, the higher the yarn count number, the heavier or thicker the yarn. It is based on the fixed length system. This system is generally used for jute or silk yarn.

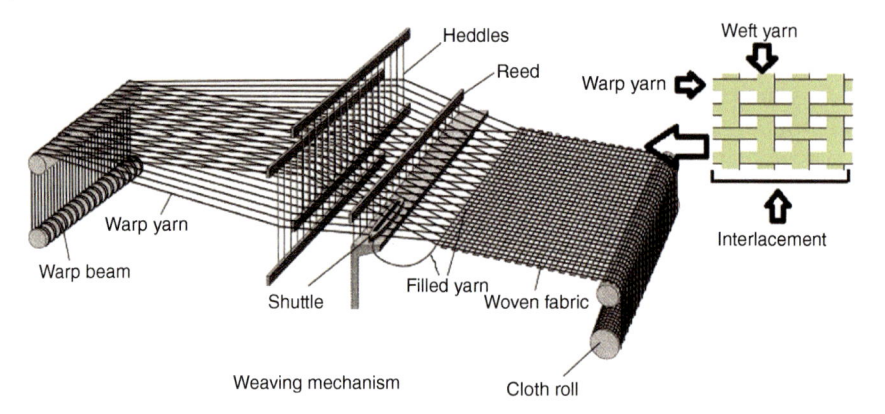

Figure 1.10 Weaving mechanism. Source: Reproduced with permission from https://textilechapter.blogspot.com/2017/03/weaving-loom-principle-mechanism.html.

Indirect count system: In an indirect yarn counting system, the yarn number or count is the number of "units of length" per "unit of weight" of yarn. This means the higher the yarn count number, the finer or thinner the yarn. It is based on the fixed weight system. This system is generally used for cotton, woolen, worsted, and linen yarn [35–37].

1.7.2 Fabric Manufacturing

1.7.2.1 Weaving Process

Weaving is one of the major processes of making fabric. In it, two individual sets of yarns called the warp and weft are interlaced with each other to form a fabric (Figure 1.10). Yarn is a long continuous length of interlocked fibers. The lengthwise yarns that run from the back to the front of the loom are called the warp. The crosswise yarns are the filling or weft. A loom is a device for holding the warp threads in place while the filling threads are woven through them. Yarns made from natural fibers like cotton, silk, and wool and synthetic fibers such as nylon and Orlon are commonly used for weaving textile. Before start weaving we have some work like weaving preparation. In weaving preparation work, we must do the following:

- *Warping*: A process of transferring the warp yarn from the single yarn packages to an even sheet of yarn representing hundreds of ends and then winding onto a warp beam.
- *Sizing/slashing*: Sizing of yarn is carried to improve its strength and produce a smooth surface.

To interlace the warp and weft yarn, there are three operations often called primary motions that are necessary:

Shedding: The process of separating the warp yarn into two layers by raising the harness to form an open area between two sets of warps.

Picking: The process of inserting the filling yarn through the shed by means of the shuttle less while the shed is opening.

Beating: The process of pushing the filling yarn into the already woven fabric at a point known as the fell and done by the reed.

Taking up and letting off: With each shedding, picking, and beating operation, the new fabric must be rolled on the cloth beam that is called "taking up." At the same time, the warp yarns must be open from the warp beam that is called "letting off" [38–40].

1.7.2.2 Knitting

Knitting is a method by which yarn is turned to create a textile or fabric. Knitting is the process through which the yarn is turned into knitted fabric by joining consecutive row of loops (course and wales) using different types of knitting machines. In knitted loop terminology, a course of knit is a mainly horizontal row of needle loops produced by adjacent needles during the same knitting cycle. Wales of knit are a mainly vertical column of interlaced needle loops generally produced by the same needle at successive knitting cycles.

Mainly two different types of knitted fabric are produced. They are as follows.

Warp Knitting The general direction of path of yarn is along the length of the fabric. Warp knit fabrics are constructed with yarn loops formed in a vertical or warp direction. Warp knitting are four types. They are tricot knit, Raschel knit, crochet knit, and milanese knit. Warp knitted fabrics are normally using inner wears (brassieres, panties, sleepwear), sportswear lining, and household (mattress, mosquito nets).

Weft Knitting The general direction of path of yarn is across the length of fabric. Weft knitted fabrics is categorized by a series of horizontal loops formed by horizontally running threads and binds with previously formed series of loops of the same thread. End uses of weft knitting are underwear, T-shirts, sportswear, baby clothes, jumpers, scarves, hats, gloves, etc. https://textilechapter.blogspot.com/2017/03/weaving-loom-principle-mechanism.html [41].

1.7.2.3 Nonwoven

Nonwoven fabric is a fabric-like material made from long fibers, bonded together by chemical, mechanical, heat, or solvent treatment or using proper moisture or heat rather than by spinning, weaving, and braiding. Nonwoven fabric is a molded product with planar structure made by interweaving natural, chemical, and metal fibers according to their mutual characteristics to form a web in the shape of a sheet and bonding them together by mechanical or physical means. Nonwovens may be a limited life, single-use fabric or a very durable fabric. Nonwoven fabrics provide specific functions such as absorbency, liquid repellency, resilience, stretch, softness, strength, flame retardancy, wash ability, cushioning, filtering, bacterial barriers, and sterility [42].

1.7.3 Wet Processing Technology

1.7.3.1 Pretreatment Process

Natural fibers and synthetic fibers both contain primary impurities that are contained naturally and secondary impurities that are added during different process like spinning, knitting, and weaving processes. Textile pretreatment is the series of cleaning operations. All impurities that cause unfavorable effect during dyeing and printing are removed in pretreatment process. Pretreatment processes include desizing, scouring, and bleaching that make dyeing processes easy. Desizing is for the removal of the size coating after weaving. Desizing with enzymes ensures complete removal of starch-based sizes, which means excellent batch-to-batch dyeing reproducibility and evenness. Scouring is a cleaning process used to remove impurities (wax, oil) on fibers, yarns, or cloth. Bleaching is decolorizing the impurities that mask the natural whiteness of fibers to obtain white cloth and increase the ability of the fabric to absorb dyestuffs uniformly.

1.7.3.2 Dyeing

The process of applying color to fiber, yarn, or fabric is called dyeing. Dyeing is the process of giving colors to a textile material through a dye (color). Dyeing can be done at any stage of the manufacturing of textile: fiber, yarn, fabric, or a finished textile product including garments and apparels. The property of color fastness depends upon two factors: selection of proper dye according to the textile material to be dyed and selection of the method for dyeing the fiber, yarn, or fabric.

1.7.3.3 Dyeing Methods

Color is applied to fabric by different methods of dyeing for different types of fiber and at different stages of the textile production process. Mostly used methods are fabric dyeing and yarn dyeing. When a dye is applied directly to the fabric without the aid of an affixing agent, it is called direct dyeing. When dyeing is done after the fiber has been spun into yarn, it is called yarn dyeing. Yarn dyeing is slightly different from woven or knit dyeing. Yarns are dyed in package form by yarn dyeing process.

1.7.3.4 Dyes

Dyes are used for coloring the fabrics. Dyes are molecules that absorb and reflect light at specific wavelengths to give human eyes the sense of color. There are two major types of dyes: natural and synthetic dyes. The natural dyes are extracted from natural substances. Synthetic dyes are made in a laboratory. Chemicals are synthesized for making synthetic dyes. Mostly used dyes are reactive dyes (used for dyeing cellulose fibers and protein fiber), direct dyes (used for dyeing wool, silk, nylon, cotton, rayon, etc.), vat dyes (used for dyeing wool, nylon, polyesters, acrylics, and modacrylics), disperse dyes (used to dyeing nylon, cellulose triacetate, and acrylic fibers), sulfur dyes (used for dyeing cotton and linen, viscose, and jute), basic dyes (used for dyeing cotton, linen, acetate, nylon, polyesters, acrylics, and modacrylics) [43–46].

1.7.3.5 Finishing

In textile manufacturing, finishing refers to the processes that convert the woven or knitted cloth into a usable material and more specifically to any process performed after dyeing the yarn or fabric to improve the look, performance, or hand feel of the finish textile. To impart the required functional properties to the fiber or fabric, it is customary to subject the material to different type of physical and chemical treatments. For example, wash and wear finish for a cotton fabric is necessary to make it crease-free or wrinkle-free. In a similar way, mercerizing, singing, flame retardant, water repellent, waterproof, antistatic finish, peach finish, etc. are some of the important finishes applied to textile fabric [47].

1.7.3.6 Printing

Textile printing is the process of applying color to fabric in definite patterns or designs. In a proper printed fabric, the color is affixed to the fiber, so that it may not be affected by washing and friction. The dyes used for printing mostly include vat, reactive, naphthol, and disperse colors, which have good fastness properties. The pigments, which are not truly dyes, are also used extensively for printing. These colors are fixed to the fiber through resins that are very resistant to laundering or dry-cleaning. Mostly used printing methods are direct printing, discharge printing, resist printing, roller printing, block printing, screen printing, transfer printing, and electrostatic printing [48, 49].

1.8 Textile Applications

Textile structures are now being used for various applications due to their unique advantages. In fact, we can divide their applications into apparel, technical textile, and functional and smart textiles. Apparel manufacturing industries include establishment that process fiber into fabric and fabric into clothing, and other textile products includes shirt, pant, socks, shoes, bra, etc. Technical textiles are indicated to be the quickest improving sector of the textile industry. Functional or technical textiles are the textiles that have been developed to fulfill the high-performance requirement in industries other than conventional clothing [50, 51]. Applications of advanced textiles will be discussed below. The applications of textile materials are presented in Figure 1.11.

1.8.1 Advanced Applications of Textile Material

In this exiting era of advanced materials, we are seeing their widespread contribution in diversified areas, starting from clothing sector to biomedical field, civil engineering, filtration, fiber optics, aerospace, automobile industries, and energy storage and harvesting applications. The convergence of different science and engineering fields is a reason for these astonishing results. Advanced materials are driven by special technical functions that require specific performance properties unique to these materials. Based on their functionality they can be classified into materials for functional, high performance, sensors/actuators,

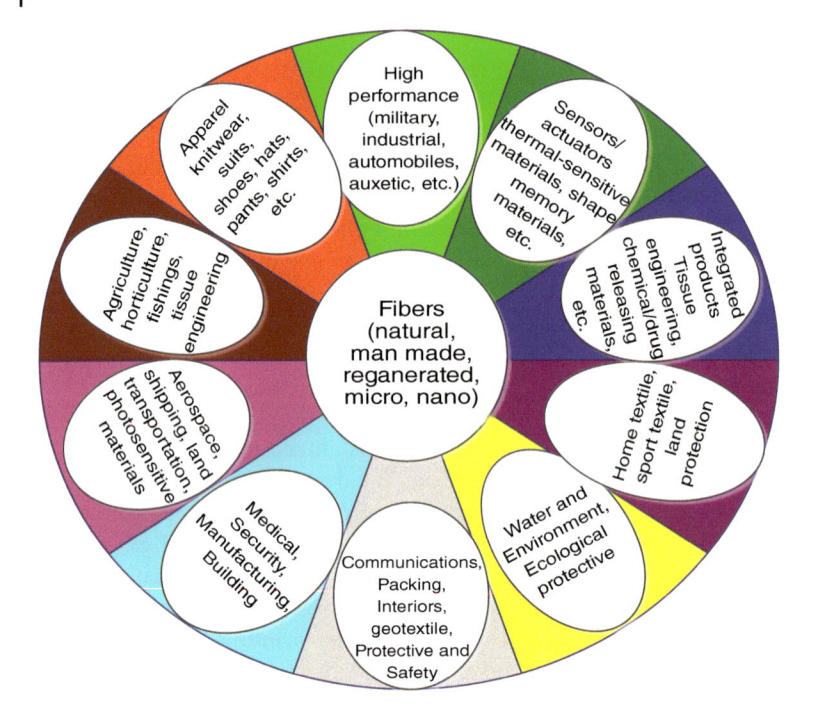

Figure 1.11 Applications of textile materials.

integrated assemblies, and sensitive/smart. Functional materials include water repellent/waterproof, crease resistance, flame retardant, water vapor permeability, etc. High-performance materials usually exhibit high modulus and strength and are used in areas like military, industrial, automobiles, building, auxetics, and land protection applications. The sensors and actuators include fiber optics, nanomaterials, microcapsules, and intelligent membranes/coatings. For integrated products, photonics, tissue engineering, chemical/drug releasing materials are used. Sensitive and smart materials include conductive materials, memory materials, photonic materials, etc. The advanced application of textile materials is shown in Figure 1.12.

1.8.2 Functional Textile

Functional textiles are a class of textiles with integrated property of adjusting textiles according to requirements and functions such as temperature and humidity responsive, water repellent, water vapor permeable, flame retardant, etc. The widely used fibers for functional textiles are viscose, polyester, and polyurethane fibers. The properties of functional textiles are shown in Figure 1.13.

1.8.2.1 Water Repellent
Water repellent is the special type of finishes that repel water, oil, and dry dirt. Water repellent properties are very important for garments, home, and technical textiles. The aim of water-repellent finishes is that the drop of water on the fabric

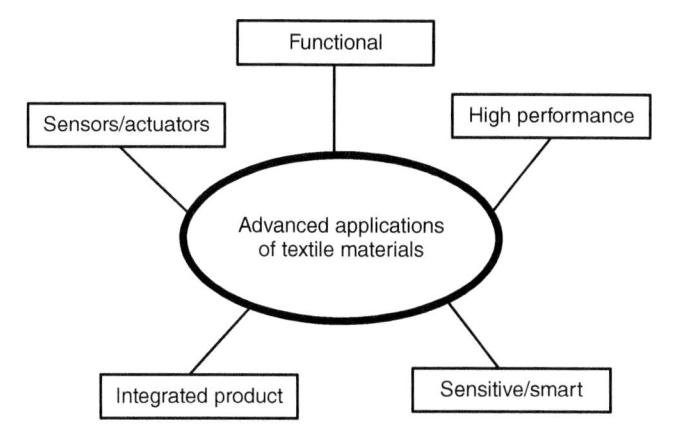

Figure 1.12 Advanced applications of textile materials.

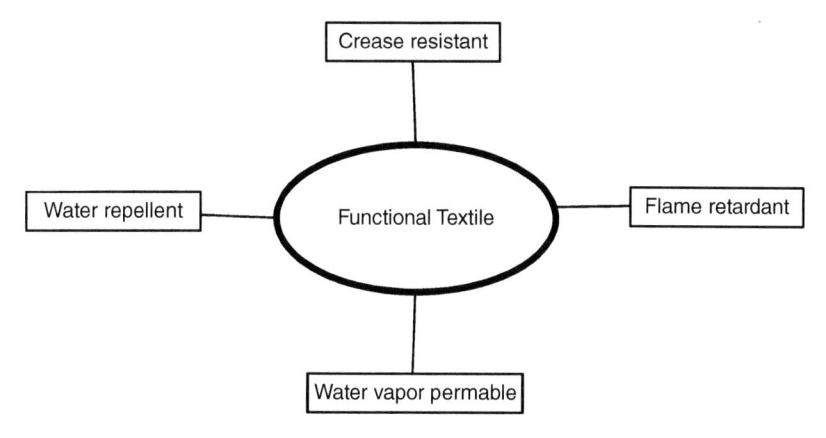

Figure 1.13 Properties of functional textile.

surface should not spread and should not wet the textile; wetting will happen only when the droplet is absorbed by textile. Water repellency is attained by using different products, but oil repellency is achieved only by using fluorocarbon polymers. In water-repellent fabric it is expected that the drop of water will stay on the surface and easily can drips off or it can be brushed off. The air permeability of the finished fabric should not be significantly affected because of water repellency, so water-repellent fabric maintains air permeability. In waterproof breathable fabric, droplet of water should not penetrate into textile, but perspiration should penetrate into textile. In the waterproof breathable fabric coating with micropores, size is small for water droplets to penetrate, and at the same time pores are big enough for perspiration vapor.

1.8.2.2 Water Vapor Permeability

Water vapor permeability is the property of a textile material that allows the passage of water vapor through fabric. It is very important for the textiles used for garment applications to transport the vapour especially when the temperature

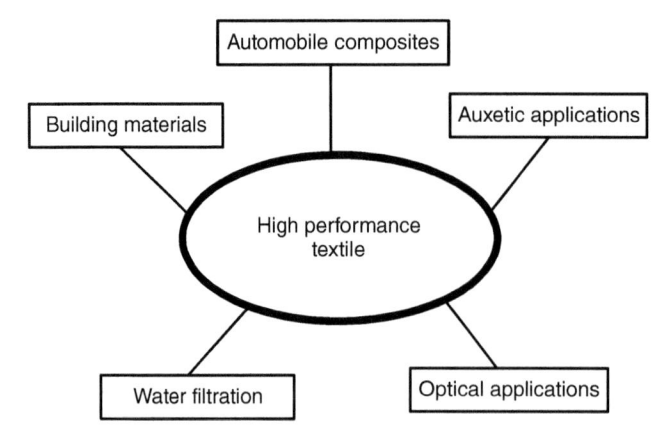

Figure 1.14 Application of high-performance textile.

is high. For getting water vapor (perspiration) permeability properties, we need to coat the fabric by special type of polymer. Water vapor permeability property will help us to avoid physiological problems. Nowadays there are a wide range of woven, laminated, or coated fabrics which are water vapor permeable and waterproof [52–54].

1.8.3 Applications of High-Performance Fibrous Materials

Fiber-reinforced composite materials have been playing an important role for many decades. There are many potential applications of high-performance fibrous materials such as automobile industry, building materials, auxetic application, water filtration, and optical applications. The different applications of high-performance fibrous materials are shown in Figure 1.14.

1.8.3.1 Fibers for Automobile Composite

The polymeric composite materials are widely being used in automotive industry. Fiber mat thermoplastics and compression-molded thermoset polymers are used to manufacture large parts of automotive industries. Composites used for nonstructural functional components are composed of glass fiber-reinforced plastics in the range of 10–50% reinforcement. Such nonstructural composites are used in pedal systems, mirror housing, and so on. To improve the fuel efficiency, hybrid valve lifter used in automotive internal combustion engine was made from carbon fiber/phenolic composite and steel. Semi-structural composites like sheet-molded compounds (SMC) are increasingly used in automotive industry for manufacturing low mass body panels with high strength to weight ratio. The materials that replaced steel with high strength to mass ratio are aluminum, glass SMC, and carbon SMC. Structural polymer composites that are studied have been gradually replacing metal components due to their reduced weight and durability and their resistance to crash. Carbon, glass, aramid, and graphite fiber-reinforced polymers and nanocomposites were studied for impact energy absorption, durability, and crushing behavior.

1.8.3.2 Fibers as Building Materials

Natural and synthetic fibers are used as building materials. They are used in constructing bridges, roads, nonstructural gratings and claddings, structural systems for industrial buildings, roof structures, tanks and thermal insulators, etc. High-performance synthetic fibers are more suitable fibers for building materials. Fibers like nylon, polyester, polypropylene, and aramid fibers are widely used as reinforcements for building materials. Natural fibers like jute, hemp, ramie, and flax are used as reinforcement in building materials. Lower cost of textiles and less energy required to make them most suitable for building materials. Among other natural fibers, jute is widely used as reinforcing fiber in composites for civil engineering applications due to its low specific gravity and high specific modulus when compared with glass.

1.8.3.3 Fibers for Auxetic Applications

The auxetic effect improves the material's mechanical properties, such as enhancement in fracture toughness, shear moduli, and indentation resistance, and allows porosity and permeability variation (under pressure) and dome shape. Due to this wide range of characteristics, auxetic materials can be used in various areas, including medicine, architecture, civil engineering, sport clothing, high-performance equipment, protection against explosives, insulation, and filters. Therefore, the use of high-performance fibers in advanced fibrous architectures like auxetic structures may offer lightweight, excellent mechanical performance and several interesting characteristics that can fulfill the explicit demands imposed by several advanced technical sectors. Auxetic structures for high-performance applications are currently studied using polyester, aramid, carbon, nylon fibers, and shape memory materials [55, 56].

1.8.3.4 Fibers for Environmental Protection: Water Purification/Filtration

There are many examples of using fibers in water treatment process. The advanced fiber technologies led separation membranes made of ultrafine fibers, ion exchange fibers, hollow fiber membranes, and aromatic polyamides in reverse osmosis membrane materials. These membrane technologies are used in many application fields from small-scale home water purification to large-scale filtration of fresh water from seawater sources. Microfine filtration membranes, ultrafine membranes, and reverse osmosis membranes made using the fiber technologies are used to remove small impurities like red rust, bacteria, particles, chemical products and salts from water. Many kinds of natural, inorganic, metallic, and synthetic fibers have been used in filtration. Synthetic fibers have played a significant role in the growth of several segments in the filtration industry. Nonwoven fabrics have been successfully used in the industry as membrane support for microfine, ultrafine, and reverse osmosis filtration. Filters with large surface area and many fine pores are required to filter fine particles. Large surface area can be achieved by using micro- and nanofibers. Hydrophilic or hygroscopic fibers are used to improve or reduce the particle adhesion to the filters. To prevent generation of static charge in filters, conductive fibers like carbon and metallic fibers are used. In hot gas filtration and high temperature filtration and other processes mineral, ceramic or metal fibers are used as filter materials.

1.8.3.5 Fibers for Optical Applications

Optical fibers are those very fine long glass fibers that allow light signals to travel through. The demand for fiber optics has grown enormously. They are used as network carriers; in data transmission, transmitting broadband signals, intelligent transportation systems, and biomedical industry as telemedicine devices for transmitting digital diagnostic images; and in detecting target biomolecules like enzymes, antibodies, and oligonucleotides. It is also used in military and space and automotive sectors. Optical fibers are usually made up of molten solution of silica or silicon dioxide with other materials such as arsenic, quartz, etc. Polymeric optical fibers can be doped with photosensitive material to change the refractive index when exposed to bright light usually in UV range of spectrum. A structured fiber with combination of cores, holes, and electrodes is used for variety of specialty applications. Optical fiber for imaging is made of multiple noninteracting cores, which is used in endoscopes to directly image internal organs. The use of capillary tube-based optical fibers is used for proteomic and genomic studies. Photonic crystal-based fiber optics is used to study light interactions. Fiber-optic biosensors (FOBS) work as a transduction element and depend on optical transduction mechanisms for detecting biomolecules. They are used for applications such as detection of pathogens, medical diagnosis based on protein or cell concentration, and real-time detection of DNA hybridization. Fiber-optic chemical sensors (FOCS) are used in sensing gases and vapors, medical and chemical analysis, marine and environmental analysis, molecular biotechnology, industrial production monitoring and bioprocess control, and automotive industry.

1.8.4 Application of Sensors/Actuators

1.8.4.1 Fibers as Electronic Devices/Wearable Electronics/Energy Materials/Sensors and Actuators

Wearable electronics is one of the widely spoken technologies in smart textile arena. The functionality of the fabric is due to the electronics and interconnections woven between them. Future generation of wearable electronics focuses on the systems to be worn directly on human body. Wearable electronic systems developed using conductive fibers have shown promising results. The technical possibility for building the electronic functions as integral part or on the surface of the fiber is scientifically proven. The development of soft and flexible fiber-based conductive and semiconductive materials is essential. Conductive polymeric fibers are one of the promising materials due to their extremely flexible nature. It also can be blended with variety of composite materials to achieve unique electronic, optical, electrical, and magnetic properties. Π-Conjugated nanoscaled organic molecules was investigated for sensors, transistors, flexible electronic devices, and field emission display in textiles [57]. Polyaniline and polypyrrole-based nanofibers show high conductivity properties at room temperature [58–60]. Poly-(3,4-ethylenedioxythiophene) (PEDOT) is one of the conductive polymers with high conductivity and solution process ability. The application of this polymer as electrodes for wearable capacitors or photodiodes is currently being explored. Carbon-based nanomaterials due to their intrinsic carrier mobility are used as channel materials in transparent electrode and field-effect transistors. Porous carbon materials like graphene, carbon nanotubes (CNTs), carbon fibers, and carbon aerogels are frequently used in wearable electronics due to their large specific surface area and better mechanical properties. Metallic nanoparticles or nanowires possess high conductivity and are very much suitable for wearable electronics. Metallic nanowires are used to develop fiber-based piezoelectric nanogenerator by coating silver on highly stretchable polyamide fabric [57]. Fiber-based electrodes made of CNTs, metals, or alloys like copper and stainless steel are light, foldable, durable, and flexible, thus more suitable for wearable electronics. Integrating sensors with conductive fibers in the fabrics can be used to monitor electrocardiogram (ECG), electromyography (EMG), and electroencephalography (EEG) [61, 62]. Fabric integrated with luminescent elements could be used for biophotonic sensing; shape-sensitive fabrics is used to sense the movement and used with EMG to obtain muscle fitness data. Integrated carbon electrodes in fabrics help to detect environmental features such as oxygen, salinity, moisture, and contaminants. Strain fabric sensing technology uses conductive yarns to record motion or flexing, pressure, and stretching or compression [63]. Merely conductive yarns of specific length are also used to make fabric antennas by stitching them together with nonconductive fabrics [64].

1.8.4.2 Fibers for Medical Compression

Compression garments are used to apply certain pressure on different parts of body. They are specially used for medical treatments, sports, and body shaping purposes. Medical compression garments are used to treat scars, muscular sprains, and low blood pressure and to quicken healing process for deep vein thrombosis patients [65]. Medical compression systems are usually made using fibers like polyester, polyamide elastane. Recent studies about application of shape memory fibers and alloys in the medical compression have gained attention [55, 66–69]. Shape memory materials are actuated using electrical current and body heat to provide compression effect. Segmented polyurethane-based shape memory fibers have shown promising results, and they have gained advantage over shape memory alloys due to their lightweight, flexibility, comfort, and easy actuation without any additional devices [70].

1.8.4.3 Fibers for Health/Stress/Comfort Management

Fibers within the human body serve as an inspiration for development of different kind of advanced materials. Hollow fibers are used in artificial lung and blood vessels. There are several studies which focus on the production of artificial liver, pancreas, skin, muscle, and nervous systems. Lightweight, sweat-absorbent, and easy dry fibers are developed for sportswear application to keep the wearer feel comfortable during workout. Fibers that can retain heat allow the water vapor from the body to pass through and repel water and are developed to protect people from extreme cold weather conditions. Breathable shape memory polyurethanes fibers and films that can respond according to the microclimatic condition between the clothing and body of the wearer are used. Antibacterial fibers are used in clothing, hospital wear, and wound care bandages to prevent contamination and bad odors caused due to bacteria. Fibers used in comfort applications must have the following properties: absorption of sweat, high water absorbency/quick dry bacteria-free/moss-free, microbe controlling, insect/tick repellent, heat retention/storage, moisture retention, moisture absorption, coolness, tranquilizer moisture permeable/water repellent, deodorant, antibacterial, electric control, UV shielding, electromagnetic wave shielding, insulation, lightweight, fitness to skin, high touch, stretch, soil-free/soil release, and shape stability.

1.8.4.4 Fibers for Thermal Protection

Fibers used for flame and thermal protection play a vital role in providing protection to workers engaging in high temperature environments, military clothing, home furnishing materials, and building materials and for firefighters. The protective clothing made using these fibers must have insulation properties and dimensional stability. Flame-resistant viscose fibers are produced by doping the spinning solution before extrusion using phosphorus-based additives (Lenzing FR®), polysilicic acid, and clay nanocomposites (TENCEL®). Flame-retardant polyester and acrylics are produced using FR comonomers during copolymerization, introducing flame-retardant additives during extrusion and flame-retardant finishes. Aromatic polyamides and meta-aramid fibers are used for protective clothing for astronauts, tank crews, and fighter pilots. Nonwoven meta-aramids are used in thermal insulation and hot gas filtration purposes. Para-aramids are specifically used for ballistic and flame protection applications. Polyimide fiber is an aromatic copolyamide sold in the brand name P84®. It has Limiting Oxygen Index (LOI) of 36–38% and is used in the production of protective clothing, hot gas filtration, sealing materials, and aviation materials.

1.8.4.5 Fibers for Radiation Protection

The need for special protective clothing is inevitable to protect people exposed to radioactive rays, UV radiation, and electromagnetic radiations. Woven cotton, polyester/cotton, and polyester/nylon fabrics with twill and sateen weave and nonwoven fabrics are used to protect people from α, β, and γ radiation environments. The fabric acts as a barrier between human skin and nuclear radiation emitted from radioactive source. Closely woven fabrics made up of wool and polyester have high UV absorption and protect skin from UV-A and UV-B radiations. Humans are exposed to different kinds of electromagnetic radiation extending from 1 to 10 000 mHz emitted by various sources, for example, cell phones, microwave ovens, and radar signal communication systems. When exposed at higher levels, these waves will cause abnormal chemical activities in body that produces cancer cells. It also obstructs the capability of cells for regeneration of DNA and RNA. To tackle this problem conductive materials are used to shield electromagnetic and static charges. General textile fibers have sufficient resistivity to shield electromagnetic radiation. The resistivity required for dissipating and shielding electrostatic charges is achieved through conductive coatings on fabric, fibers doped with carbon black, carbon fibers, and metal fibers. The conductive fibers or yarns are incorporated in fabrics to produce conductive garments to resist static charges.

1.8.5 Applications of Integrated Products

1.8.5.1 Fibers for Tissue Engineering

Tissue engineering concept was born when several investigators realized that when cell are placed close enough to each other, they form structures identical with those formed by such cells in a living body. This, apparently, may be achieved thanks to signals that living cells can exchange with the neighboring ones. Various disciplines, such as materials science, cell biology, reactor engineering, and clinical research, are contributors of tissue engineering. Advanced tissue engineering involves the use of polymeric materials implanted at the defective site. The defective sites are usually supported by scaffolds. They provide the framework for the cells to attach, proliferate, and form results in the formation of extracellular matrix. Biodegradable natural and synthetic fibers as well as some nonbiodegradable polymeric fibers are currently used for tissue engineering. Polylactides, polyglycolides, polycaprolactone (PCL), and their copolymers have been often used for the preparation of scaffolds for cartilage tissue engineering. However, their hydrophobicity, the acidity of their decomposition products, and the self-acceleration of their degradation have constituted serious drawbacks. Chitosan and alginate are nontoxic, biodegradable, biocompatible polymer and does not have these disadvantages. The use of natural polysaccharide like alginates that are extracted from algae has shown promising growth in tissue engineering and wound dressing applications. Alginate fibers and dressings as wound care products are resistant to bacterial attack, antiviral, antifungal, nontoxic, high absorbent, hemostatic, non-allergic, breathable, biocompatible and can blended with medicines. Alginate nanofibers were produced by electrospinning technique in the presence of various synthetic polymers and/or surfactants and sometimes in combination with chitosan offer better cell adhesion properties [71, 72]. Bio-artificial scaffolds made of *Chlamydomonas reinhardtii* single-cell green alga mixed with fibrinogen have shown excellent in vitro biocompatibility and photosynthetic activity, and the algae survived for five days in vivo. These kinds of photosynthetic scaffolds can be implanted in full skin defects and can be used in the generation of chimeric tissue composed of mammalian and photosynthetic cells in vivo [73]. Chitosan, a unique biopolymer derived from chitin, exhibits outstanding properties along with excellent biocompatibility and biodegradability. Chitosan and its blends with sodium alginate, tropocollagen, cellulose, sodium hyaluronate, sodium chondroitin sulfate, poly(acrylic acid), and synthetic polymers like PEO,UMHMWPEO, PVA, PLA, and PVP are used to produce nanofiber mats for tissue engineering and wound healing applications [74]. Collagen fibrils and their networks form a highly organized 3D scaffold to surround the cells. It ensures structural and biological integrity of ECM. Collagen can form fibers with high tensile strength and stability. The hollow fiber made using collagen tubing is used for cell culture nerve regeneration [75]. The unique structure of silk, versatility in processing, biocompatibility, availability of different biomaterial morphologies, options for genetic engineering of variations of silks, the ease of sterilization, thermal stability, surface chemistry for facile chemical modifications, and controllable degradation features make silks promising biomaterials for many clinical functions. Silk fibrous materials are used as scaffolds for tissue engineering and sutures.

1.8.6 Sensitive and Smart Materials

1.8.6.1 Fibers with Conductive Properties as Industrial Materials

Conductive fibers are lightweight alternatives to heavy copper wiring in variety of areas where weight is a concern especially in aerospace technologies. Inherently conductive fibers include metallic fibers, carbon fibers, and conjugated polymeric fibers. Treated conductive fibers are conductive-filled fibers and conductive-coated fibers. Metallic fibers are developed from metals or metal alloys. These fibers have very high conductivity, but they possess low flexibility, stiffness, and high weight. Carbon fiber and its composites possess conductivity like that of metals with high strength, stiffness, and lower weight. Conjugated polymers like PEDOT fibers have high conductivity values from 150 to 250 s/cm. Polymeric fibers filled with conductive fillers like metallic powder, carbon black, CNT, graphene, or conjugated polymer powder are used as conductive fibers. The conductive fibers that can be produced by coating insulating materials with highly conductive materials, such as metals, metal alloys, carbon black, carbon nanotubes, and ICPs, are known as conductive-coated fibers. To apply metallic coatings, sputtering, vacuum deposition, electroless plating, carbonizing, and filling or loading fibers are the most extensively used methods. Potential applications of conductive fibers include power lines; aircraft and aerospace wiring systems; harnesses for automotive wiring; wires for missile guidance; electro-textiles for medical, military, and consumer applications; lightweight deployable antennas; thermal blankets and clothing; flexible keyboards; giant-area flexible circuits for energy harvesting; electrostatic charge dissipation; and battlefield monitoring and reporting of vital signs and wound locations on soldiers.

1.8.6.2 Memory Fibrous Materials

Smart textile is the development of textile having sensing, reacting, and adapting capabilities. Materials or structure that sense react to external stimuli or condition such as thermal, chemical, mechanical, or other sources. Shape memory material can move to temporary shape by external stimulus, and it can come back to its original shape with right stimulus. Memory materials include water vapor permeability textile, ventilation textile, medical textile, thermal protective textile, etc.

1.8.7 Advantages of Fibrous Materials

Fibrous materials have been used from many centuries in the application of clothing and other utilitarian products. Natural fibers are abundantly available in different geological locations with various altitudes in the world, and they are easily processable. Synthetic fibers can be produced with customized parameters according to specific end use without much trouble. The revolutionary breakthrough of research in science and technology in the twenty-first century leads to the use of both natural and synthetic fibrous materials in vivid multidisciplinary applications, owing to their several advantages. This is the era of synthetic fibers now, and it offers platform to tailor their properties suitable to put them into proper end use. Synthetic fibers can be produced into micro-, macro-, and nano-sizes with novel functionalities, which is reliable to produce in mass with low material cost. Fibers do offer several advantages over other materials such as lightweight (less density), superior stretch ability, toughness (e.g. regenerated spider silk), high tensile strength (e.g. silk), high specific surface area, vibration damping capability [77], energy storage ability [78], super absorbency [16] and memory behavior (e.g. shape or stress memory) [79, 80], self-healing [81], biodegradability [82], biocompatibility (e.g. tissue engineering and implants) [83, 84], insulation capability [33], chemical inertness, antimicrobial [85], flame retardancy [86], biomimicking [87], etc. These several advantages of fibrous materials enable one to make them into substrates by traditional textile processing technologies such as weaving, knitting, nonwoven webs, and braiding due to their excellent flexibility and strength. These substrates made from fibers have been employed into various multidisciplinary areas including electronics, construction, power harvesting, aerospace, medical, transportation, and industrial high-performance materials. Both natural and synthetic fibers are needed in daily life to the human being, and smart fibers are considered as futuristic material to prepare well for needs of tomorrow's challenging and sustainable days. Hence fibers play a very vital role in fulfilling our requirements on this planet.

References

1 Denton, M.J. and Daniels, P.N. (2002). *Textile Terms and Definitions*, 11e. Manchester: The Textile in Institute.
2 Morton, W.E. and Hearle, J.W.S. (2008). *Physical Properties of Textile Fibers*, Woodhead Publishing in Textiles. England: Woodhead Publishing Limited.
3 Collier, B.J., Bide, M.J., and Tortora, P.G. (2009). *Understanding Textiles*, 7e. Pearson/Prentice Hall.
4 Houck, M.M. (2009). *Identification of Textile Fibers*, Woodhead Publishing in Textiles. Cambridge: Woodhead Publishing Limited.
5 Blackburn, R. (2005). *Biodegradable and Sustainable Fibers*, Woodhead Publishing Series in Textiles. Woodhead Publishing.
6 Mishra, S.P. (2000). *Fiber Science and Technology*. New Delhi: New Age International (P) Limited.
7 Saville, B.P. (1999). 3 - Fiber dimensions. In: *Physical Testing of Textiles*, 44–76. Woodhead Publishing.
8 Hearle, J.W.S. and Backer, P.G.S. (1969). *Structural Mechanics of Fibers, Yarns, and Fabrics*. Wiley-Interscience.
9 Hu, J.L. (2004). *Structure and Mechanics of Woven Fabrics*, Woodhead Publishing in Textiles. CRC Press; Woodhead Publishing.

10 Yang, Q.X. and Li, G.Q. (2014). Spider-silk-like shape memory polymer fiber for vibration damping. *Smart Materials and Structures* 23 (10).

11 Haigh, H.S. (2009). Speciality fibers. *Journal of the Textile Institute Proceedings* 40 (8): 794–813.

12 Das, T. and Ramaswamy, G.N. (2006). Enzyme treatment of wool and specialty hair fibers. *Textile Research Journal* 76 (2): 126–133.

13 Vineis, C., Aluigi, A., and Tonin, C. (2011). Outstanding traits and thermal behaviour for the identification of speciality animal fibers. *Textile Research Journal* 81 (3): 264–272.

14 Prahsarn, C., Klinsukhon, W., Padee, S. et al. (2016). Hollow segmented-pie PLA/PBS and PLA/PP bicomponent fibers: an investigation on fiber properties and splittability. *Journal of Materials Science* 51 (24): 10910–10916.

15 Tallury, S.S., Behnam, P., Melissa, A.P., and Richard, J.S. (2016). Physical microfabrication of shape-memory polymer systems via bicomponent fiber spinning. *Macromolecular Rapid Communications* 37 (22): 1837–1843.

16 Kim, G.H., Youk, J.H., Kim, Y.J., and Im, J.N. (2016). Liquid handling properties of hollow viscose rayon/super absorbent fibers nonwovens for reusable incontinence products. *Fibers and Polymers* 17 (7): 1104–1110.

17 Beskisiz, E., Ucar, N., and Demir, A. (2009). The effects of super absorbent fibers on the washing, dry cleaning and drying behavior of knitted fabrics. *Textile Research Journal* 79 (16): 1459–1466.

18 Lee, T.W., Han, M., Lee, S.E., and Jeong, Y.G. (2016). Electrically conductive and strong cellulose-based composite fibers reinforced with multiwalled carbon nanotube containing multiple hydrogen bonding moiety. *Composites Science and Technology* 123: 57–64.

19 Schmidt, M.A., Argyros, A., and Sorin, F. (2016). Hybrid optical fibers - an innovative platform for in-fiber photonic devices. *Advanced Optical Materials* 4 (1): 13–36.

20 Xi, P., Tianxiang, Z., Lei, X. et al. (2017). Fabrication and characterization of dual-functional ultrafine composite fibers with phase-change energy storage and luminescence properties. *Scientific Reports* 7 (1): 1–9.

21 Meng, Q.H., Hu, J.L., and Yeung, L. (2007). An electro-active shape memory fiber by incorporating multi-walled carbon nanotubes. *Smart Materials and Structures* 16 (3): 830–836.

22 Meng, Q.H. (2009). The influence of heat treatment on the properties of shape memory fibers. II. Tensile properties, dimensional stability, recovery force relaxation, and thermo mechanical cyclic properties. *Journal of Applied Polymer Science* 111 (3): 1156–1164.

23 Meng, Q.H. (2007). Morphology, phase separation, thermal and mechanical property differences of shape memory fibers prepared by different spinning methods. *Smart Materials and Structures* 16 (4): 1192–1197.

24 Meng, Q.H., Hu, J.L., Zhu, Y. et al. (2007). Polycaprolactone-based shape memory segmented polyurethane fiber. *Journal of Applied Polymer Science* 106 (4): 2515–2523.

25 Pan, N. and Gibson, P. (2006). *Thermal and Moisture Transport in Fibrous Materials*. Cambridge: Woodhead Publishing Limited.

26 Kajiwara, K. (2009). Synthetic textile fibers: structure, characteristics and identify cation. In: *Identification of Textile Fibers* (ed. M.M. Houck), 68–87. Cambridge: Woodhead Publishing Limited.
27 Hess, K. and Naturwissenschaft, H.K. (1944). Over-long-term interferences and micellar fiber refinement in fully synthetic fibers (polyamides and polyesters). *Journal of Physical Chemistry* A193 (171): 196.
28 Hearle, J.W.S. (1958). You have full text access to this content fringed fibril theory of structure in crystalline polymers. *Journal of Polymer Science* 28 (117): 432–435.
29 Hearle, J.W.S. (1963). The fine structure of fibers and crystalline polymers. I. Fringed fibril structure. *Journal of Polymer Science* 7 (4): 1175–1192.
30 Kozasowski, R.M., Mackiewicz-Talarczyk, M., and Allam, A.M. (2012). 5 - Bast fibers: flax A2 - Kozłowski, Ryszard M. In: *Handbook of Natural Fibers*, 56–113. Woodhead Publishing.
31 Roy, S. and Lutfar, L.B. (2012). 3 - Bast fibers: jute A2 - Kozłowski, Ryszard M. In: *Handbook of Natural Fibers*, 24–46. Woodhead Publishing.
32 Roy, S. and Lutfar, L.B. (2012). 4 - Bast fibers: ramie A2 - Kozłowski, Ryszard M. In: *Handbook of Natural Fibers*, 47–55. Woodhead Publishing.
33 Koh, E. and Lee, Y.T. (2017). Antimicrobial activity and fouling resistance of a polyvinylidene fluoride (PVDF) hollow-fiber membrane. *Journal of Industrial and Engineering Chemistry* 47: 260–271.
34 AATCC (2013). *Fiber analysis: qualitative*.
35 Edward, L.G. (1966). *Natural and Manmade Textile Fibers: Raw Material to Finished Fabric*, 1e. New York: Duell, Sloan and Pearce.
36 Lawrence, C.A. (2010). *Advances in Yarn Spinning Technology*. Cambridge: Woodhead Publishing Ltd.
37 Mahadevan, M.G. (2009). Textile spinning, weaving and designing. In: *Chandigarh*, 1e. Abhishek Publications.
38 Vasudeo, K.M. (2013). *Fundamentals of Yarn Winding*. New Delhi: Woodhead Publishing.
39 Ormerod, A. and Sondhelm, W.S. (1995). *Weaving: Technology and Operations*. Manchester: The Textile Institute.
40 Abhijit, M. (2017). *Principles of Woven Fabric Manufacturing*. Boca Raton, FL: CRC Press.
41 Belal, S.A. (2009). *Understanding Textiles for a Merchandiser*, 1e. Dhaka: BMN3 Foundation.
42 Chandra, R.S. (2012). *Fundamentals and Advances in Knitting Technology*. New Delhi: Woodhead Publishing India Pvt.
43 Spencer, D.J. (2001). *Knitting Technology*, 3e. Cambridge: Woodhead Publishing.
44 Albrecht, W. and Fuchs, H.W. (2003). *Nonwoven Fabrics*. Weinheim: Wiley-VCH.
45 Gürses, A., Açıkyıldız, M., Güneş, K., and Gürses, M.S. (2016). *Dyes and Pigments*. Switzerland: Springer.
46 Fu, J. (2013). *Dyeing: Processes, Techniques, and Applications*. Hauppauge, NY: Nova Science Publishers, Inc.

47 Clark, M. (2011). *Handbook of Textile and Industrial Dyeing*. Cambridge: Woodhead Publishing.

48 Arthur, B.D. (2001). *Basic Principles of Textile Coloration*. Bradford: Society of Dyers and Colorists.

49 Trotman, E.R. (1975). *Dyeing and Chemical Technology of Textile Fibers*, 5e. London: Griffin.

50 Kate, W. (1997). *Fabric Dyeing & Printing*. Loveland, CO: Interweave Press.

51 Yohanan, P. (1990). *Dyeing and Printing: A Handbook*. London: Intermediate Technology.

52 Thakur, S., Jahid, M.A., and Hu, J.L. (2018). Mechanically strong shape memory polyurethane for water vapour permeable membranes. *Polymer International* 67: 1386–1392.

53 Jahid, M.A., Hu, H.L., and Zhou, H. (2018). *Smart Textile Coatings and Laminates*, Chapter 6, 155–173.

54 Jahid, M.A., Hu, J.L., Wong, K.H. et al. (2018). Fabric coated with shape memory polyurethane and its properties. *Polymers* 10: 681.

55 Nayak, R. and Padhye, R. (2015). *Garment Manufacturing Technology*. Sawston, Cambridge: Woodhead Publishing, an imprint of Elsevier.

56 Koncar, V. (2016). *Smart Textiles and Their Applications*. Duxford: Woodhead Publishing.

57 Horrocks, A.R. and Anand, S.C. (2016). *Handbook of Technical Textiles*, 2e. Cambridge: Woodhead Publishing in association with the Textile Institute, Woodhead Publishing is an imprint of Elsevier.

58 Meng, Q. and Hu, J.L. (2008). A temperature-regulating fiber made of PEG-based smart copolymer. *Solar Energy Materials and Solar Cells* 92 (10): 1245–1252.

59 Zhu, Y., Hu, J.L., and Yeung, K.W. (2009). Effect of soft segment crystallization and hard segment physical crosslink on shape memory function in antibacterial segmented polyurethane ionomers. *Acta Biomaterialia* 5 (9): 3346–3357.

60 Jang, S.Y., Seshadri, V., Khil, M.S. et al. (2005). Welded electrochromic conductive polymer nanofibers by electrostatic spinning. *Advanced Materials* 17 (18): 2177–2180.

61 Huang, K., Wan, M., Long, Y. et al. (2005). Multi-functional polypyrrole nanofibers via a functional dopant-introduced Process. *Synthetic Metals* 155 (3): 495–500.

62 Huang, J. and And Kaner, R.B. (2004). Nanofiber formation in the chemical polymerization of aniline: a mechanistic study. *Angewandte Chemie International Edition* 43 (43): 5817–5821.

63 Zeng, W., Tao, X.M., Chen, S. et al. (2013). Highly durable all-fiber nanogenerator for mechanical energy harvesting. *Energy & Environmental Science* 6 (9): 2631–2638.

64 Coosemans, J., Hermans, B., and Puers, R. (2006). Integrating wireless ECG monitoring in textiles. *Sensors and Actuators A: Physical* 130: 48–53.

65 Custodio, V., Herrera, F.J., Lopez, G., and Moreno, J.I. (2012). A review on architectures and communications technologies for wearable health-monitoring systems. *Sensors* 12 (10): 13907–13946.

66 Pacelli, M., Taccini, N., and Paradiso, R. (2006). Sensing fabrics for monitoring physiological and biomechanical variables: E-textile solutions. *2006 3rd IEEE/EMBS International Summer School on Medical Devices and Biosensors*.

67 Salonen, P. and Hurme, L. (2003). A novel fabric WLAN antenna for wearable applications. *IEEE Antennas and Propagation Society International Symposium. Digest. Held in conjunction with: USNC/CNC/URSI North American Radio Sci. Meeting (Cat. No.03CH37450)*.

68 Wang, L., Felder, M., and Cai, J. (2011). Study of properties of medical compression fabrics. *Journal of Fiber Bioengineering & Informatics* 4 (1): 15–22.

69 Kumar, B., Hu, J.L., and Pan, N. (2016). Smart medical stocking using memory polymer for chronic venous disorders. *Biomaterials* 75: 174–181.

70 Kumar, B., Hu, J.L., and Pan, N. (2016). Memory bandage for functional compression management for venous ulcers. *Fibers* 4 (1): 10.

71 Moein, H. and Menon, C. (2014). An active compression bandage based on shape memory alloys: a preliminary investigation. *Biomedical Engineering Online* 13: 135.

72 Zhu, Y., Hu, J.L., Yeung, K.W. et al. (2007). Effect of cationic group content on shape memory effect in segmented polyurethane cationomer. *Journal of Applied Polymer Science* 103 (1): 545–556.

73 Rinaudo, M. (2014). Biomaterials based on a natural polysaccharide: alginate. *TIP* 17 (1): 92–96.

74 Jeong, S.I., Krebs, M.D., Bonino, C.A. et al. (2010). Electrospun alginate nanofibers with controlled cell adhesion for tissue engineering. *Macromolecular Bioscience* 10 (8): 934–943.

75 Schenck, T.L., Hopfner, U., Chavez, M.N. et al. (2015). Photosynthetic biomaterials: a pathway towards autotrophic tissue engineering. *Acta Biomaterialia* 15: 39–47.

76 Croisier, F. and Jérôme, C. (2013). Chitosan-based biomaterials for tissue engineering. *European Polymer Journal* 49 (4): 780–792.

77 Chattopadhyay, S. and Raines, R.T. (2014). Review collagen-based biomaterials for wound healing. *Biopolymers* 101 (8): 821–833.

78 Deng, J., Ye, Z., Yang, Z. et al. (2015). A shape-memory supercapacitor fiber. *Angewandte Chemie International Edition* 54 (51): 15419–15423.

79 Narayana, H., Hu, J.L., Kumar, B. et al. (2017). Stress-memory polymeric filaments for advanced compression therapy. *Journal of Materials Chemistry B* 5 (10): 1905–1916.

80 Tonazzini, A., Stefano, M., Bryan, S. et al. (2016). Variable stiffness fiber with self-healing capability. *Advanced Materials* 28 (46): 10142–10148.

81 Emmert, M., Patrick, W., Miranda, R.G., and Doris, H. (2017). Nanostructured surfaces of biodegradable silica fibers enhance directed amoeboid cell migration in a microtubule-dependent process. *RSC Advances* 7 (10): 5708–5714.

82 Akbari, M., Tamayol, A., Bagherifard, S. et al. (2016). Textile technologies and tissue engineering: a path toward organ weaving. *Advanced Healthcare Materials* 5 (7): 751–766.

83 O'Connor, R.A. and McGuinness, G.B. (2016). Electro spun nanofibre bundles and yarns for tissue engineering applications: a review. *Proceedings of*

the Institution of Mechanical Engineers, Part H: Journal of Engineering in Medicine* 230 (11): 987–998.
84 Korjenic, A., Zach, J., and Hroudova, J. (2016). The use of insulating materials based on natural fibers in combination with plant facades in building constructions. *Energy and Buildings* 116: 45–58.
85 Shukla, A., Basak, S., Ali, S.W., and Chattopadhyay, R. (2017). Development of fire retardant sisal yarn. *Cellulose* 24 (1): 423–434.
86 Zhang, K., Fan, L., Yan, Z. et al. (2012). Electrospun biomimic nanofibrous scaffolds of silk fibroin/hyaluronic acid for tissue engineering. *Journal of Biomaterials Science Polymer Edition* 23 (9): 1185–1198.
87 Weng, W., Chen, P., He, S. et al. (2016). Smart electronic textiles. *Angewandte Chemie International Edition* 55 (21): 6140–6169.

2

Animal Fibers: Wool

Xiao Xueliang

School of Textiles and Clothing, Jiangnan University, Lihu Road 1800, Binhu, Wuxi 214122, Jiangsu, P.R. China

2.1 Introduction

In animal fibers, natural hairs have a huge number of family members. In terms of sources, there are sheep hair (wool), goat hair (guard hair and cashmere), camel hair (camel guard hair and camel cashmere), alpaca hair (vicuna hair and Peru wool), rabbit hair (e.g. Angora rabbit hair), and other animal hairs (e.g. ox hair, horse hair, yak hair, and deer velvet) rarely mentioned. Among these hairs, sheep hair fiber, namely, "wool," as shown in Figure 2.1, is a kind of the most important natural protein fiber and is also an important textile material that has a number of excellent properties, such as good elasticity, high water absorption, natural wavy crimpness, good warmth retention, difficult stained, gentle luster, etc. These properties endow wools with different unique styles. The high-grade clothing made of wools, like cassimere and medium tweed, can be spring and autumn fabrics with hand feel of slippery waxy, rich bone, good elasticity, surface clean, and natural shine, while all kinds of thick wool coats have a feel of rich texture, plumply, strong warmth retention, etc.

Wool fibers can be used for technical products, such as woolen cloth, felt, blanket, cushion material, etc. Furthermore, many decorative textiles made of wool, such as tapestry and wool carpet, give people a precious gorgeous feel [1]. Wool has unique physical and chemical properties that perform extremely versatile due to the special wool compositions and macromolecule spatial structure especially the molecular conformation. For example, the high temperature of heat setting (>100 °C) on wool products can permanently set the wool shape for the change of molecular conformation that cannot backward the wool original shape ever. Moreover, because of high moisture absorbing properties for the large amount of hydroxyl groups on the hair macromolecular chains, wool can be warm or cool in terms of casual or formal format that makes people feel comfortable to wear actively and passively. Moreover, because wool is a natural protein fiber, the use of such fiber in textile can be considered by consumers to be sustainable, renewable, and environmental friendly. Therefore, like cotton, wool is also one of the most popular natural and

Handbook of Fibrous Materials, First Edition. Edited by Jinlian Hu, Bipin Kumar, and Jing Lu.
© 2020 Wiley-VCH Verlag GmbH & Co. KGaA. Published 2020 by Wiley-VCH Verlag GmbH & Co. KGaA.

Figure 2.1 Wool fibers from sheep, (a) sheep before shearing, (b) fine Merino shearing, (c) champion hogget fleece (Walcha Show), (d) raw wools before processing, and (e) processed wools. Source: From https://ipfs.io/ipfs/QmXoypizjW3WknFiJnKLwHCnL72vedxjQkDDP1mXW-o6uco/wiki/Wool.html.

sustainable fibers by present consumer and retail who demand for all kinds of clothing.

However, due to the decreased number of total sheep around the world, wool's productivity over all the global market share has declined a little during the last 10 years, as shown in the tendency in Figure 2.2. The percentage of wool (now a minority fiber) that takes around only 1.9% of world fiber production [2] decreases slightly in the very large textile fiber market. However, wool is still the number one of protein fibers in weight measure that is used in textile industry. Wool does, however, play a significant role in the market of men's suiting, knitwear, and carpets, as well as the area of unique properties that wool can supply with the excellent appearance, comfort, and durability that cannot be satisfied by other natural and chemical fibers.

Australia produces the highest amount of wool (around 1/3 of world total output) and is also the number one of countries that exports wool to other countries. The second highest production of wool per year in the world is New Zealand, followed by Argentina, Uruguay, South Africa, Russia, and United Kingdom. China's wool production is also the highest one in the world; however, China is also a big consumer country of wool. Every year, China has to import a large amount of wool from Australia [1].

Here, take China's wool as an example to demonstrate wool's category, distribution, and quality because China is a typical country that is so vast to span the climate regions of frigid, temperate, subtropical, and tropical. Wool's type and quality are versatile in comparison with relatively pure types of other countries' wool. It was reported that, since the 1970s, China's number of sheep has been increased to four times higher, indicating that the amount of produced wool has a great development. However, 20% of fine wools (high quality of wool) are still

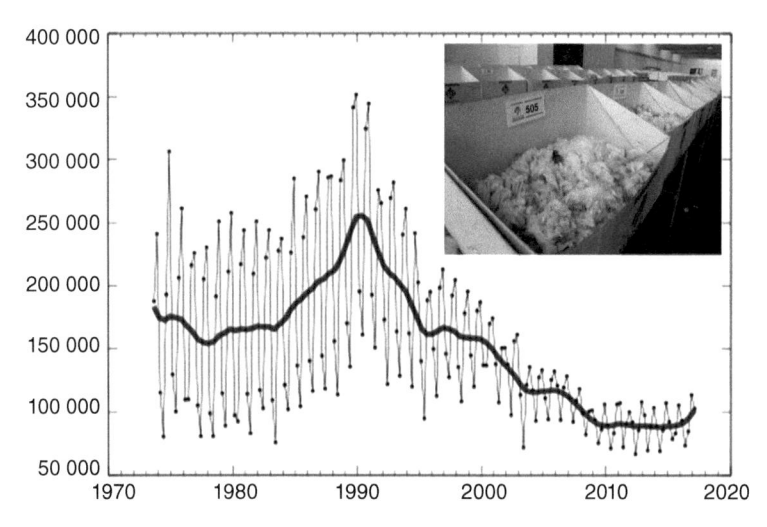

Figure 2.2 Wool received by Australian brokers and dealers (tons/quarter) since 1973 (Toby Hudson, Own work; this image is based on Australia Bureau of Statistics data) and the inset image showing Merino wool samples for sale by auction (Newcastle, New South Wales). Source: From https://ipfs.io/ipfs/QmXoypizjW3WknFiJnKLwHCnL72vedxjQkDDP1mXWo6uco/wiki/Wool.html.

required importation. For domestic wool, most of China's wools are supplied from husbandry areas of Xinjiang, Inner Mongolia zone, and Gansu province. In this case, China's wool has many types according to the wool's fineness (in diameter measure), for example, superfine, semi-fine, and coarse wool. The domestic wool also has home wool and improved wool according to the wool quality, where the latter has improved fine wool and semi-fine wool from hybrid breed of sheep while the former is only from Mongolia, Tibet, and Hakelong types. Table 2.1 compares the quality of China's wool according to the types and wool mechanical properties.

Beyond semi-fine wool, when wool's diameter is larger than 40 μm; this hair is called guard hair. The wools from Inner Mongolia, Tibet, Hazake sheep, etc. have a proportion of guard hairs that have a function of protection of superfine hairs for warmth retention. When producing guard hairs, it is inevitable to include some fine and superfine hairs and even dead wool inside; thus the quality and property of guard hair products are variable. Thus, these hairs are commonly used for wool blanket, carpet, tapestry, etc. In Table 2.1, most fine wools are from improved sheep after several generations of cross breeding. For example, Xinjiang fine wool is the first improved wool from the sheep after hybridization of China's Hazake female sheep and Russian Lanbuliye and Caucasian male sheep. The Xinjiang wool has fineness of below 0.6 tex and length of 6–7 cm and oil/grease content of 7–10%. China's northeast fine wool is from the sheep after

Table 2.1 Quality of main China's wools.

Wool types (production area)		Productivity per year (kg/sheep)		Wool count	Wool fiber length (cm)	Tenacity		Net wool rate (%)
		Male sheep	Female sheep			Strength (cN)	Elongation at break (%)	
Fine wool (fiber diameter <25 μm)	Xinjiang	10.1–11.5	3–4	60–66	6–7	3.9–4.9	30–38	44
	Northeast	13–14	5–6	60	7–8	5.9–9.8	40–47	35
	Inner Mongolia	12	5.5	60–64	6–7	7.8–11.8	25–30	47
	Gansu	9–10	4–4.7	60–64	7–8	5.9	40–45	35
	Merino	14–15	5–6	60–64	7–8	5.9–10.8	45	48
	Qinghai	8–8.5	3–4	60–64	7–8	4.9–7.9	30–35	45
Semi-fine wool (diameter 25–40 μm)	Qinghai	5	3.2	46–58	8–10	9.8–17.6	34–40	55
	Northeast	4.6	3.7	56–58	8–9.5	9.8–11.8	35	53
	Inner Mongolia	—	—	48–60	6–9	9.8–15.7	38	50
	Anhui	7–12	5–6	56–58	9–15	9.8	42	45

several hybridization generations of American Lanbuliye sheep and Inner Mongolia sheep and Lanbuliye and Caucasian male sheep. The related wool has fineness of below 0.55 tex (wool count of 60–64) and length of 7–8 cm, and such wool shows high tenacity and good luster. China's wool from Inner Mongolia is cut from the sheep after several hybridization generations of Xinjiang female sheep, Corriedale male sheep, and Caucasian male merino. The wool has fineness of below 0.55 tex (wool count of 60–64) and length of 6–7 cm. While China's merino seems have better quality with fineness of below 0.6 tex and length of beyond 8 cm, and such wool shows strong tenacity, good luster, crimpness, and appropriate oil content.

In comparison with China's wool in Table 2.1, Table 2.2 lists the wool's type and properties, such as productivity, wool count, wool length, sheep body weight, and fertility, from other countries and regions that produce the most amount of wool in the world. The wool has been classified as fine wool (fiber diameter < 25 μm) and semi-fine wool (fiber diameter is in the range of 25–40 μm) and long wool, which means the fiber length is larger than 10 cm. After comparison, it is found that Australia Merino wool (strong type) shows highest productivity per year, higher wool count, and long wool fiber, indicating the Australia Merino wool is of high importance in wool market.

2.2 Classification of Wools

There are many classification methods of wool in terms of wool fineness, fiber structure, fiber type on hair coat, wool picking and original wool shape, and the wool cut seasons, as shown in the classification branches in Figure 2.3. According to Figure 2.3, the detailed description of the classification has been divided into five sections.

2.2.1 Classification of Wool in Fineness

(1) *Botany wool*: The diameter of botany wool is in the range of 10–25 μm.
(2) *Medium wool*: The diameter of medium wool is in the range of 25–40 μm; the length of such wool has a range below the value of 15 cm.
(3) *Coarse wool*: The diameter of coarse wool is in the range of 40–70 μm; the length of such wool is also less than 15 cm.
(4) *Long wool*: The length of such wool is in the range of 15–30 cm, and the diameter of most of such wool is more than 37 μm.

2.2.2 Classification of Wool in Terms of Fiber Structure

(1) *Down wool hair*: Such wool has layers of only epidermis and cortical without medullary substance, as shown in the cross section of a down wool hair in Figure 2.4a. Because there is no medulla layer, the quality and spinning performance of such wool is excellent due to better elasticity and fineness for high combining force with each other.

Table 2.2 Type and property of wools from the main production country.

Wool types			Productivity per year (kg/sheep)		Wool count	Wool length (cm)	Body weight (kg)		Fertility rate (%)
			Male sheep	Female sheep			Male sheep	Female sheep	
Fine wool (<25 μm)	Australia merino	Fine hair	6.3–9	2.7–4.5	64–90	5.8–10	59–76	36–63	105–135
		Intermediate	8–12	3.6–9	60–75	7.5–11	68–90	45–63	105–135
		Strong type	10–15	4–8	56–70	7.5–13	79–114	54-72	105–135
	Russia merino		10–12	5–7	64–70	7.5–10	90–100	50	135
	Germany merino		7–8	5	60–64	7–10	125–150	100–130	120
	Askanya sheep		10–14	5.5–6	64–70	7.8–8	100–120	60–65	145–160
	South Africa merino		5.6–9.6	4.5–8.2	64–70	7.5–10	70–90	36–55	105–130
	Caucasian merino		10–11	5.8–6.5	64–70	7–8	90–100	55–60	105–110
	Lanbuliye sheep		7–10	4–5	64–80	6–8	70	50–60	130–140
	Tebuli sheep		5.5–11	3.6–7.2	60–70	6.4–9	68–120	57–73	120
Semi-fine wool (25–40 μm)	Corriedale sheep		7.5	4.5	50–70	7.5–10	60–100	30–65	125–140
	Pohl Wass sheep		7.5	4	58–60	10–12	75	55	110–130
	Tsigai sheep		5.5–6	4.3–4.5	46–56	8–9	70–78	52–55	105–120
Long hair wool (length > 10 cm)	Romney Marsh sheep		12–14	6–9	40–46	15–20	130–140	110–120	105–145
	Leicester sheep		4.5		46–56	14–20	100–150	80–90	120–165
	Border area of Leicester sheep		3.5–4.5		44–48	18–23	90–100	60–80	150–200
	Lincoln sheep		5.5		36–44	30	138	110	125–160

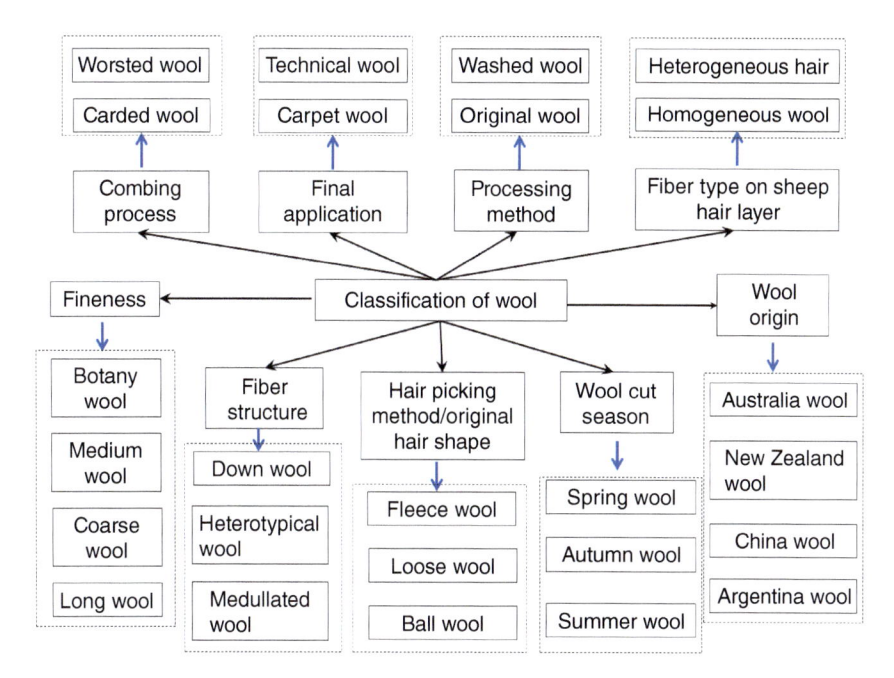

Figure 2.3 Classification of wool in terms of different parameters.

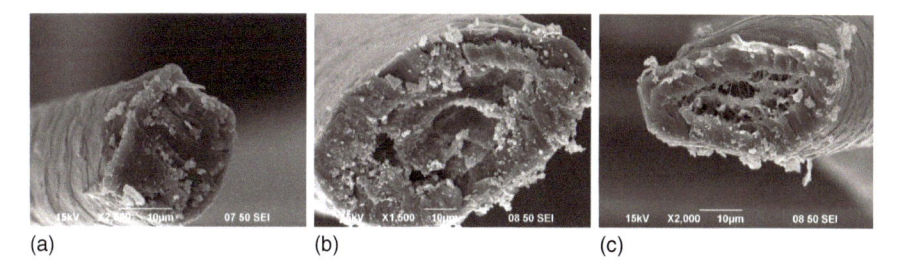

Figure 2.4 SEM images of cross sections of (a) down wool, (b) heterotypical wool, and (c) medullated wool.

(2) *Heterotypical wool fiber*: Such wool fiber has discontinuous medulla, as shown in Figure 2.4b. Meantime, the wool has both the characteristics of down wool hair and coarse wool. In a bundle of such wool fibers, the difference of fiber thickness is large, and the performance of the yarn using heterotypical fiber is worse than the yarn made of down hair because of large deviation of fiber performance.

(3) *Medullated wool*: Such wool fiber has a continuous medulla layer, as shown in Figure 2.4c. With the different content of medullary substance, such wool has many types of performance fibers, such as bristles, wool, hair, dry wool cavity, and dead wool, in which the dry and dead wool have relatively low spinning value for their poor mechanical properties.

2.2.3 Classification of Wool in Terms of Fiber Type on Sheep Hair Layer

(1) *Homogeneous wool*: Such wools from a sheep body are made of the same wool fiber. Thus, the fineness and length of wools are basically the same values. In terms of fineness, homogeneous wool belongs to high-count wool. Generally speaking, the quality of homogeneous wool is rather good compared with other types of wool.

(2) *Heterogeneous hair*: Such wools on a sheep body are made of two or more types of hair fibers with different wool length and fineness. Heterogeneous hair normally has different grades of hairs according to the gross content of coarse cavity. The yarn spinning using such wools requires more factors that should be considered, such as yarn strength, evenness, and hairiness.

2.2.4 Classification of Wool in Terms of Hair Picking Method and Original Hair Shape

(1) *Fleece wool*: Such wool fibers that cut from sheep body connect each other and can form a sheet of hair coat, called "fleece hair," as shown in Figure 2.5. The fleece hair is classified as closed hair and open hair. From outside appearance, the closed fleece hair looks like a complete assembly of wool fibers, while the open fleece hair looks like a bulgy wool braid from the appearance, and the hair bottom connects each other, and the hair top does not link each other within open appearance.

(2) *Loose wool*: The loose wool means that the wool fibers just cut from a sheep body and do not link each other, which does not form a hair sheet.

(3) *Ball wool*: When sheep is in an unhairing season, people would like to comb wool fibers from sheep body by means of comb-like tools. These wool fibers are called "ball wool." Usually, cashmere is a kind of ball wool. Thus, "ball wool" can be used for high-level wool products.

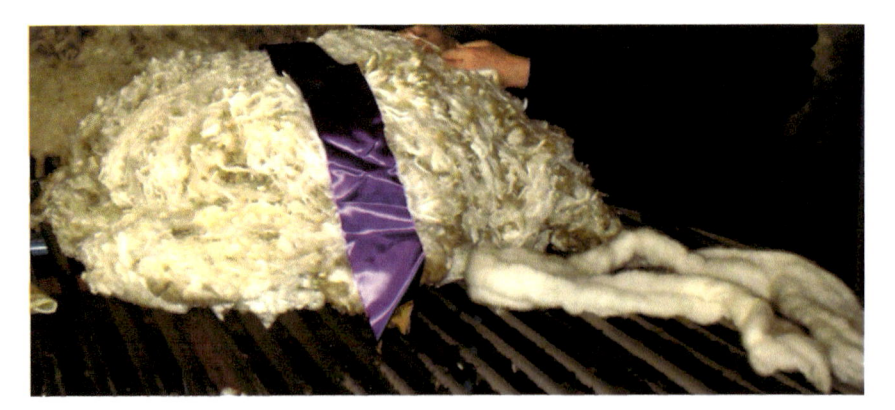

Figure 2.5 Fleece of fine New Zealand Merino wool and combed wool top on a wool table. Source: From https://ipfs.io/ipfs/QmXoypizjW3WknFiJnKLwHCnL72vedxjQkDDP1mXWo6uco/wiki/Wool.html.

2.2.5 Classification of Wool in Terms of Wool Cut Season

(1) *Spring wool*: The wool obtained from sheep in spring season usually is fine and long, the content of fine fluff is high, and the content of fatty oil on fiber surface is also high. The performance of such wool is outstanding for making end-use textiles.
(2) *Autumn wool*: The wool obtained from sheep in autumn season usually is coarse and short, but the luster is good and color is white. The wool quality is followed by spring wool.
(3) *Summer wool*: The wool obtained from sheep in summer season is like autumn wool, fibers are short and coarse, and the content of dead wool is high, and the quality of wool is relatively poor.

In addition, there are many kinds of wool classification methods, for example, wool can be classified as worst wool and carded wool in terms of combing process in yarn spinning. Wool can be classified as carpet wool and technical wool in terms of final application. Wool can be classified as original wool and washed wool in terms of processing method. Wool can be classified as fine wool, semi-fine wool, coarse wool, and long fiber wool in terms of pastoral sheep breed. Wool can be classified as Australia wool, New Zealand wool, China wool, etc. in terms of country of origin.

2.3 Processing of Wool Fibers and Yarns

2.3.1 Primary Processing of Wool

Primary processing of wool fibers include two main steps, namely, the cleaning process and conditioning process, for the post-manufacturing processes along the whole textile engineering, because there are many impurities in hair fibers that just combed from sheep. The primary processing method comprises the following steps: *shearing → sorting and classification → grading and package → wool washing → carbonization (wash hair) → final pack*.

(1) *Wool shearing and wool sorting processes*: The shearing process of sheep is usually carried out in spring. Fine wool and semi-fine wool are usually obtained once a year, while coarse wool can be obtained from sheep twice a year. Normally the shearing process takes place in spring and autumn. The just sheared hairs are connected together to become a whole set of wools. This connection is easily sorted out for post-treatment. The sorting process is for better rationally use of raw materials. In order to use the good selected wool for good end use, the sheared hairs are sorted according to the wool quality, including the division of different grades of wool into different qualities of piled hair separately. In the sorting process of defective wool, two kinds of miscellaneous grass shearing are separated; these works are usually completed manually.

(2) *Washing and weeding of wool*: The purpose of washing wool fibers is to wash the wool fatty film from the wool fiber surface, sheep deposited sweaty spots and sand, and other dirt stuff in wools. The key of washing wools is to remove the fatty materials on fibers (lanolin). The main method of washing process to raw wools is to employ detergent that contains surface active agent. The commonly used detergent is mainly soap alkali and synthetic detergent solutions. After washing process of wool, the cleaned wool although has pressured for removing water, the wools generally still contains about 40% weight of water, therefore, after washing process of wool, the wet cleaned wools must have a drying process for later storage and transportation. At the present, the factory of preliminary wool treatment usually employs the combination of opening and washing processes together for opening wool package, washing fatty films, drying wet washed wools, and then finally obtaining the dry cleaned wools for the post wool manufacturing.

Some tips should be noted in washing process, because some impurities from plants are in raw wools, such as leaves, seeds, and burr, which are referred to as the grass complex, and usually these impurities are entangled with wool fibers. In practice, it is not easy to remove them in opening wool package and washing wool processes. The method of removing plant impurities is usually performed by chemical method that is called "carbonization." The principle of this chemical method is due to the different acid resistance in which wool is a good acid-resistant material, while plant impurities are not such materials. Using this method, raw wool fibers containing grass impurities are soaked into sulfuric acid solution; thus, the impurities are carbonized into inorganic carbons. Then the wool fibers are dried and compressed that the carbonized impurities are fragile to be collapsed and separated from the processed wools. This chemical processing can reduce the impurity content of grass from raw wool significantly. For China's wool, before carbonization, the wool miscellaneous grass carbonization rate is in 1–2% (China short hair 2–4%). For Australasian wool, the value is around 1%; after carbonization, the wool containing grass mixed rate decreased to 0.1% (China short hair below 0.2%).

2.3.2 Yarn Spinning Process of Wool

In woolen spinning, slubbings, which are lightweight continuous strands of staple fibers, instead of slivers are produced directly from carding. These are spun directly into yarns typically using ring or mule spinning systems. The spinning process of wool is relatively complicated compared with cotton spinning, for example, there are more combing steps in wool spinning for high grade of wool products. However, their processing principles are similar, i.e. combing for orientation of fibers along the sliver axis and twisting for improvement of yarn strength. Herein, the worsted spinning process is similar to the combed cotton spinning process in that the yarns are carded, gilled (a process similar to drafting

2.3 Processing of Wool Fibers and Yarns

Figure 2.6 Wool fibers in format of slivers, single yarn, or plied structural yarns. Source: From https://ipfs.io/ipfs/QmXoypizjW3WknFiJnKLwHCnL72vedxjQkDDP1mXWo6uco/wiki/Wool.html.

Wool slivers/tops   Wool yarn (single yarn)   Worsted yarns in plied structure

but using pins to support and comb the fibers as they are drawn), combed, gilled again, formed into a roving, and then spun. Woolen and worsted spinning processes are not constrained to wool fibers but can be used for synthetic fibers of similar length and fineness to wool fibers that would normally be spun using them. Normally, long staple yarns combed and produced using the worsted spinning process are called worsted yarns, and long staple produced on the woolen system without combing are called woolen yarns, as shown in some different wool yarn forms in Figure 2.6.

For example, a worsted spinning process for sliver dyeing product is described as follows (Example (1)). It can be seen that a very complicated process has been shown from wool raw material to final wool yarns (dyed) for weaving or knitting and many gilling and combing processes are arranged during the yarn spinning because of the removal of impurities and short fibers and improvement of orientation of wool fibers for better yarn quality:

(1) Wool selection → opening packing → washing → drying → mixing wools → oiling → carding → first gilling for wool sliver → second gilling of wool tops → third gilling of wool tops → combing → fourth gilling of wool tops → last gilling of wool tops → (wool tops completed) → balling → coiling in cans → dyeing for tops → dehydration → rewashing → combing of mixed tops → first gilling of pre-spinning → third gilling of wool tops → recombing → fourth gilling → mixing tops → last gilling of wool tops (re-combing of wool tops) → first combing of pre-spinning → second → … → fourth combing of pre-spinning → roving → spinning → plying → twisting → streaming → winding (post-spinning) → (automatic spinning → electronic plying → doubling twisting).

(2) Pre-treatment → first carding → first gilling → second combing → second gilling → … → spinning → post-treatment (waxing, bleaching, washing, or coloring).

Example (2) is a simplified process for wool spinning without dyeing before spinning; this is a normal case for most wool mills because of relatively shorter processes. Thus, the wool yarn quality is varied, depending on the number of gilling or effective combing processes.

2.4 Chemical Compositions and Structural Characteristics of Wool

2.4.1 Compositions of Wool

The main composition of wool fiber is a kind of insoluble protein, known as α-keratin. The fundamental chemical composition of α-keratin is elements of carbon (49–52% in weight), oxygen (17.8–23.7%), nitrogen (14.4–21.3%), hydrogen (6.0–8.8%), sulfur (2.2–5.4%), and ash (metal oxide 0.16–1.01%). The α-keratin of wool fiber is made up of many kinds of α-amino acids, and the contents of various α-amino acids are shown in Table 2.3.

Table 2.3 Content (%) of α-amino acids in natural proteins.

Amino acid	Wool fiber	Silk fibroin	Mulberry silk gum	Tussah fibroin	Casein	Soybean protein
Glycine	3.10–6.50	37.5–48.3	1.1–8.8	20.3–24.0	0.5	4.00–7.77
Alanine	3.29–5.70	26.4–35.7	3.5–11.9	34.7–39.4	1.9	4.31–4.85
Leucine	7.43–9.75	0.4–0.8	0.9–1.7	0.4	9.7	7.71–9.60
Isoleucine	3.35–3.75	0.5–0.9	0.6–0.8	0.4	9.7	4.40–5.27
Phenylalanine	3.26–5.86	0.5–3.4	0.3–2.7	0.5	3.9	5.70–6.12
Valine	2.80–6.80	2.1–3.5	1.2–3.1	0.6	8.0	3.93–5.72
Proline	3.40–7.20	0.4–2.5	0.3–3.0	0.3	8.7	5.32–6.78
Lysine	2.80–5.70	0.2–0.9	5.8–9.9	0.2	6.2	4.67–5.56
Histidine	0.62–2.06	0.14–0.98	1.0–2.8	2.2	2.5	1.30–1.63
Arginine	7.90–12.10	0.4–1.9	3.7–6.1	9.2–13.3	3.7	7.46–9.15
Tryptophan	0.64–1.80	0.1–0.8	0.5–1.0	1.8–2.1	0.5	0.12–0.47
Serine	2.90–9.60	9.0–16.2	13.5–33.9	9.8–12.2	5.0	4.32–4.83
Threonine	5.00–7.02	0.6–1.6	7.5–8.9	0.1–1.1	3.5	3.21–4.12
Tyrosine	2.24–6.76	4.3–6.7	3.5–5.5	3.6–4.4	5.4	0.11–0.24
Hydroxyproline	—	1.5	—	—	0.2	0.11–0.24
Aspartic acid	2.12–3.29	0.7–2.9	10.4–17.0	4.2	6.0	10.89–13.87
Asparagine	3.82–5.91	0.7–2.9	10.4–17.0	4.2	6.0	10.89–13.87
Glutamate	7.03–9.14	0.2–3.0	1.0–10.1	0.7	21.6	20.96–24.71
Glutamine	5.72–6.86	0.2–3.0	1.0–10.1	0.7	21.6	20.96–24.71
Cystine	10.84–12.18	0.03–0.9	0.1–1.0	—	0.4	0.00
Cysteine	1.44–1.77	—	—	—	—	0.00
Methionine	0.49–0.71	0.03–0.2	0.1	—	3.3	0.91–1.76

2.4.2 Macromolecular Structure of Wool

Wool is composed of α-keratin proteins. It is found from experimental [1] that wool can be dissolved easily by acid (H^+) or caustic soda (OH^{1-}) solutions, and the final unit after hydrolysis of wool is α-amino acid. It is well known that the molecular formula of α-amino acids is –COOH–CHR–NH–, in which R represents a variety of substituents (pendant group). The groups of hydroxyl (–OH), amino (–NH), carboxyl (–COOH), and disulfide bonds (DBs) (–S–S–) on the backbone of peptide (–CO–NH–) macromolecule chains are the functional element groups in the protein fibrils, which determine the properties of wool fiber, such as resistance of acid and alkali, moisture absorption, and many other physical properties like mechanical properties.

The proportion of α-amino acids in protein fibrils was significantly different in terms of the species of protein type, sheep growth conditions, growth sites, and harvest seasons. In 20 kinds of α-amino acid (main compositions) of wool fibers as listed in Table 2.3, arginine (two amino acids), aspartic acid, proline (two amino acid), glutamic acid, aspartic acid (hydroxy acid), and cystine (sulfur-containing amino acids) are the highest content (>30%) of amino acids in wool. Furthermore, as to the wool keratin backbone, some branches of backbone connect each other with horizontal linkage bonds such as salt bonds, DBs, hydrogen bond, and space. These bonds can enhance the strength of wool fibrils significantly.

The macromolecular compositions and structure of α-keratin proteins in wool fiber are shown in Figure 2.7, where the macromolecule chains may pass across the amorphous area and crystalline area alternatively, in which the crystalline

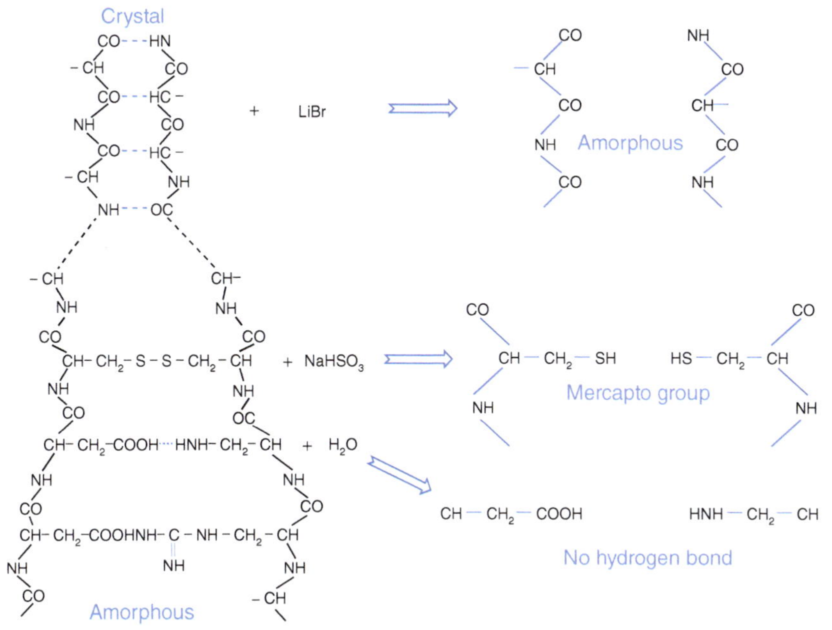

Figure 2.7 Two-dimensional macromolecular structure of wool fiber and interaction with a few chemical solutions.

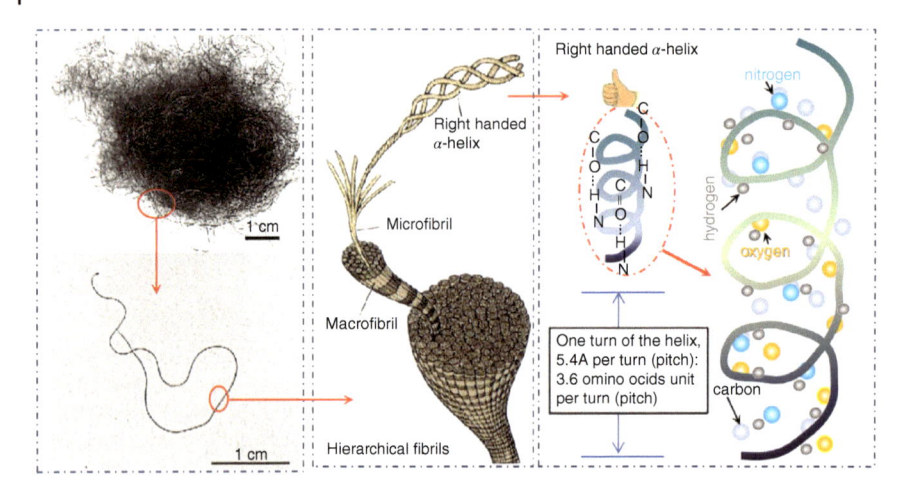

Figure 2.8 The right-hand spatial structure of α-keratin macromolecular chain and chemical compositions along a molecule chain. Source: Adapted from http://www.fzfzjx.com/tech_news/detail-4048.html[4].

phase is made of regular structural molecules and the amorphous phase is made of macromolecule chains and other matrix materials. For the two phases, experimental data [30] showed that LiBr solution can dissolve the wool crystalline phase into amorphous molecule chains and $NaHSO_3$ dilute solution (a reductant) can break down the DBs of two branches and aqueous molecules can break down the hydrogen bonds at intra- and inter-keratin macromolecules, respectively.

The natural protein fiber has two kinds of spatial structure, i.e. α-helical structure and β-sheeted structure. The silk fiber mainly contains β-sheeted structural proteins, which is in linear zigzag shape that has repeatable unit of angles and amino acid unit. While the α-keratin shows the helical conformation, which is one of the most common protein structures that encountered in nature, namely, α-spiral keratin protein as shown in a schematic illustration in Figure 2.8. The helical spatial structure obeys right-hand rule in spatial sequence. In the helical spatial structure, the molecular backbone, salt bond, DB, and hydrogen bond are linked to form a stable spiral structure, in which hydrogen bonds are existed at intra- and intermolecular chains, while DBs only existed between intermolecular chains.

2.4.3 Morphology and Hierarchical Structure of Wool Fiber

The external shape of a wool fiber is like a python that the main part is a long cylinder with scaly surface. The wool root is slightly coarser than the wool head. From any cross section of wool in the middle of hair fiber, the wool has clearly three components that are epidermis (scales), cortex, and medulla layers, where the scaly surface is a very thin layer of tile overlapping structure, as shown in the scale cortex (four layers of cuticles) in Figure 2.9.

Under the scaly surface, the cortex layer takes the main component of wool (95% in weight) that is usually composed of positive cortex and inner

2.4 Chemical Compositions and Structural Characteristics of Wool

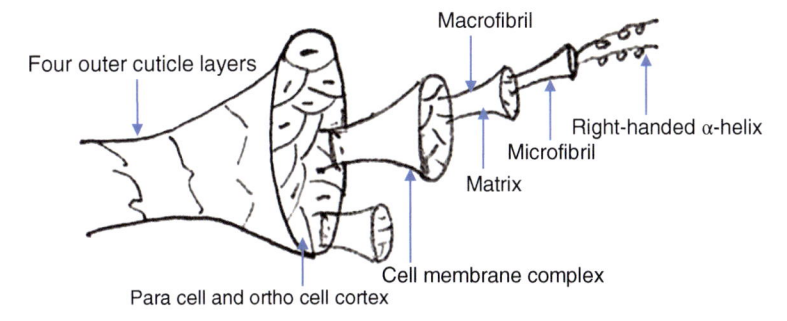

Figure 2.9 A typical hierarchical structure of wool fiber.

cortex. From the view of hierarchical structure, the cortex is in the scale of top macroscale structure that is made of macrofibrils and substance that is in the space between macrofibrils. The macrofibrils are made of fibrils and microfibrils, while the latter are made of photofibrils and, at the final microscale, are coils of macromolecule chains (in helical conformation), as shown in a typical hierarchical structure in Figure 2.9. In the center of each wool fiber, there is usually a center porous medulla. Medulla is a kind of biodegradable cellular materials in the form of porous medium and continuous or discontinuous cavities that benefit the hair warmth retention. The difference of scaly surface, cortex, and medulla displays at the configuration that is resulted from the volume ratio and texture of cortex and medulla. The detailed descriptions of the three components are as follows:

(1) *Cuticle layer*: The cuticle layer, also known as the scaly layer, consists of a piece of keratin cells, coated on the surface of wool fiber. The average thickness of wool cuticle layer is in the range of 0.2–2 µm, width is in the range of 25–30 µm, and height is in the range of 35.5–37.5 µm. The root of cuticle layer is attached to the hair trunk, and the tip extends out of the surface and points to the mouth of the hair when the hair is dry. The main function of the scale layer is to protect the wool from the external environmental conditions. The density and degree of scale attachment have great influence on the gloss and surface properties of wool. The scales of a fine wool are usually arranged closely and covered by a ring, and the protruding end is prominent. The scales of a coarse wool is arranged in a thin and cracked cover layer. In addition, due to the existence of the scale layer, the wool has an evident feature of felting that related wool products are narrowed down easily in size after washing.

(2) *Cortex layer*: The cortical layer is the main component of wool fiber, which determines the physical and chemical properties of wool. The cortical layer is composed of two kinds of different cortical cells, namely, the bilateral structure of orthocortex and partial cortex, and the two cortical cells have conversion position along the fiber axis so that the natural wool fiber crimpness is formed. If there is a large difference in the proportion of the orthocortex and partial cortex, the wool's crimpness is not obvious. The orthocortex (soft

cortex) structure is loose, it is in the outer side of wool coiling arc, and its sulfur content is less. The orthocortex is inactive with reaction of enzyme and chemical reagent, causing wool with good moisture absorption and dyeing ability. However, the partial cortex (cortex) is in a more compact state, which is in the inner side of the crimpness arc, and the sulfur content in partial cortex is high that gives wool an easy affinity of acid dyes and poor response to chemical reagents. The experience tells us [5] that the good grown wool hair would have higher content of cortical layer, leading to the better quality of wool fiber. As a consequence, the strength of crimpness and elasticity of wool are all better than the wool from poorer growth environment. Some wool fibers also have natural pigments in their cortex layer, causing that these wools are difficult to remove the inherent colors before dyeing.

(3) *Medulla layer*: Usually, there is a continuous cavity in wool center that is composed of loosely structured and air-filled keratinocytes. The cells in this layer are poorly linked and dark black. The medulla layer influences the strength, crimpness, and elasticity of wool fiber and affects the consequential spinning process. Generally, the wool fibers in high quality have no medullary layer or only intermittent medullary layer. The coarser the wool fiber, the larger proportion of medulla layer will be. Experience tells us [6, 7] that the more medulla layer of wool is more brittle and easier to break under stretching.

2.5 Properties of Wool

In this section, a few physical properties are discussed in detail to interpret the wool performance in end use of clothing and decorative and technical textiles, including wool's fineness, length, crimpness, friction and felting properties, content of grease and impurities, and other physical properties.

2.5.1 Fineness of Wool

The fineness of wool mainly depends on the factors of variety, age, gender, growth position, and feeding conditions of sheep. The cross section of wool fiber is approximately circular, and the fineness of wool fiber is usually expressed in diameter; the unit is micron (μm), as shown in a well-known diameter range for defining the wool fineness in Figure 2.10b, where the superfine wool is defined of diameter below 20 μm. Most wools have the diameter in the range of 20–50 μm, called fine wool, and fine wool accounts for the most productivity as shown in Figure 2.10a. Wool fineness varies greatly, the minimum diameter of villus is 7 μm, and the maximum diameter can be up to 240 μm. One wool has its diameter great variation usually that can reach up to 5–6 μm; the coarser wool would have more uneven fineness. The normal cross section of fine wool is approximately circular, the aspect ratio of the section is "1 : 1.2" without medullary layer. The cross section of the setae is oval, containing medulla layer, and the aspect ratio is "1.1 : 2.5." The cross section of dead wool is usually flat, and the ratio of the length and width for such section is more than 3.

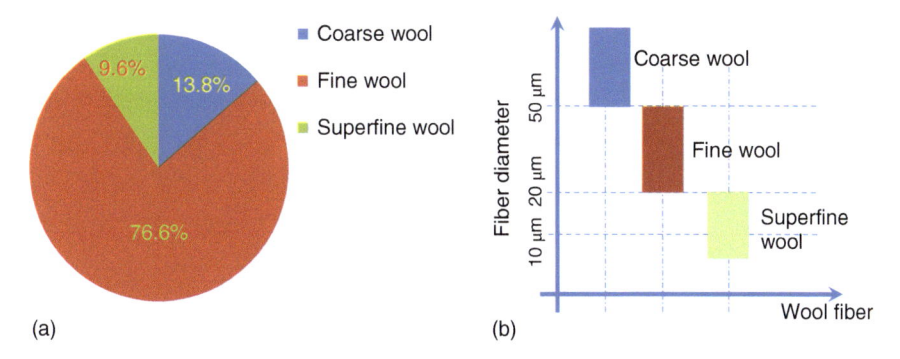

Figure 2.10 Wool productivity and wool fineness distribution. (a) China wool production in 2014. Source: Data from http://big5.qianzhan.com. (b) Diameter range for defining wool's fineness.

Fineness is an important index to determine the quality and application value of wool. Generally, the finer of wool would have smaller dispersion of wool bundles, the higher of relative strength of wool would have greater degree of curvature, the larger of scale density of wool would have the higher lipid and lanoline content, but the length is relatively short. The fineness of wool fiber has a great influence on the quality and style of final wool products. Worsted wool products are usually produced using homogeneous fine wool fibers, while woolen products are normally produced using semi-fine or modified wool fibers. Some wool cords are normally using semi-fine wools in the count range of 46–58, and underwear fabrics are using superfine wools. Experience tells us [5, 8] that the finer wool fibers would lead to more knots in the process of spinning and easier surface pilling on fabric.

Apart from the diameter as an index to measure the fineness of wool, there are still other factors such as linear density, metric count, and quality count to measure the wool fineness. For example, for the wools with the average diameter of 11–70 μm and the coefficient of variation of diameter of 20–30%, the corresponding linear density is in the range of 1.25–42 μm.

The quality count of wool simplified was called "count," which is an important measure index in wool long-term production. At present, the classification of wool in textile industry and international trade and the establishment of the strip processing technology are all based on the wool's count. In the early stage, the quality of wool is mainly assessed by subjective method. According to the situation of that time, fineness of the actual count number of wool spinning was called the quality count, in order to express the real apparent quality of wool. Now the quality count of wool only means the fineness of wool. Different countries have different definitions of quality counts for different wool fibers. In China, the wool count quality and the corresponding average diameter are shown in Table 2.4.

2.5.2 Length of Wool Fiber

Due to the existence of natural crimp, the length of wool fiber is divided into natural length and stretched length. Generally, the length of wool bundle is used to

Table 2.4 Relationship of wool quality count and average diameter of China's wool.

Quality count number	Average diameter (µm)	Yarn count that can be spun
70	18.1–20.0	More than 64
66	20.1–21.5	52–60
64	21.6–23.0	45–52
60	23.1–25.0	45–52
58	25.1–27.0	36–45
56	27.1–29.0	32–34
50	29.1–31.0	28–32
48	31.1–34.0	
46	34.1–37.0	
44	37.1–40.0	
40	40.1–43.0	
36	43.1–55.0	
32	55.1–67.0	

describe the natural length of wool bundle, while the stretched length is employed to evaluate the wool quality. Natural length refers to the straight distance between the two ends of wool under natural crimp. Stretched length refers to the length of wool fiber after the removal of the crimp. Wool fiber length depends on the factors such as sheep breed, age and gender, feeding conditions, shearing numbers, and seasons. The length of fine wool is generally in the range of 6–12 cm, and the length of semi-fine wool is 7–18 cm, and the length of long wool (fleece) is in the range of 15–30 cm. Moreover, from the same sheep, it is found that the wool fiber would be long when it is from the sheep's shoulder, neck, and back, while the hair is short when it is from sheep's head, legs, and abdomen.

The effect of wool length on its yarn quality is not as important as wool fineness. However, when wool fineness is the same value, it is found that the longer the wool would give higher count of spun yarns. Meanwhile, when wool count is constant, the woolen yarn strength is stronger, and yarn is more even with the increase of wool length. In the spinning process, the stretched wool below the length of 3 cm, e.g. short wool, should be controlled strictly; otherwise, the involvement of such short wools in spinning would lead to poor quality of woolen yarns, for example, such wools involvement easily forms yarn slub, thick and thin section, big belly yarn, etc. For the measurement of raw wool fibers, it is normally to take a simple method from the bundle length, for example, to measure the length of 30 wool bundles to calculate the average length, variance, and coefficient of variation of wools, as shown in a typical tool for measuring wool length in Figure 2.11.

On the other hand, for the wool sliver or washed wool hairs after combing process, it is usually to employ a length analyzer to measure the length of wool fibers.

10 mm

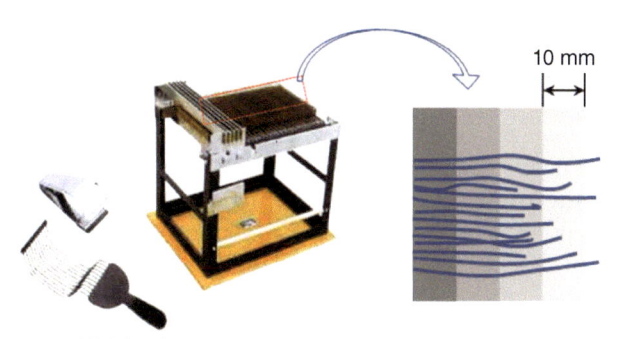

Weighted average method for measuring fiber length

Figure 2.11 A set of typical measure tool for wool length.

For example, for a wool bundle sample of 2 g, from long to short wool bundle in the group of 10 mm, each group of wools is weighed; thus, the values of weighted average length, variance, or coefficient of variation of main hair length and content rate of short hair can be obtained based on such method.

2.5.3 Crimpness of Wool

Wool fiber has natural periodic curly crimpness along the direction of fiber axis. The degree of wool crimpness is related to the factors such as the sheep breeds, wool fineness, and wool fleece location. Thus, the wool crimpness has a great reference to judge the fineness, homogeneity, and uniformity of wool. According to the definition of crimpness, the shape of wool's crimpness is divided into three types: weak crimp, regular crimp, and strong crimp; this can be shown in Figure 2.12. Usually, weak crimpness refers to a crimp less than half a circle, and most part of such wool is in straight state along the fiber length direction; the number of wool coiling is small. The majority of the semi-fine wool fibers belongs to this type of weak crimpness; the ratio of wave width to wave height is usually in the range of "4–5." The regular crimp refers to the curly shape of semicircle. The majority of fine wool belongs to this type of crimpness, which is mostly used in worsted spinning. The ratio of wave width to wave height is in the range of "3–4." Strong crimp of wool has higher amplitude of coils, and the coiling number of wool is high; some abdominal wool hair belongs to this type of crimpness. This kind of wool is usually used for woolen spinning; the ratio of wave width to wave height is less than 3.

Figure 2.12 Illustration of crimped shape of wool fiber along fiber axis.

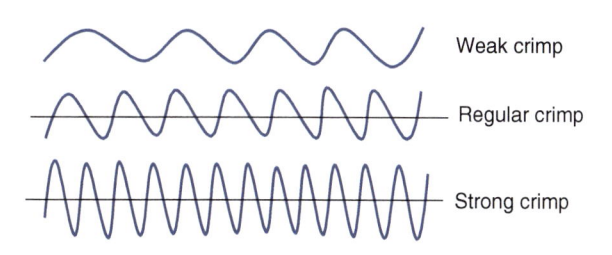

Weak crimp

Regular crimp

Strong crimp

The shape of wool's crimpness refers to wool's distribution of orthocortex and partial cortex. The cortical layer of fine wool is composed of two kinds of cortical cells, which is in bilateral structure and present conversion position along the fiber axis. Thus, this conversion of two cortex cells leads to the crimpness. Scientifically, the number of wool fiber crimps is a measure for the crimpness with unit of number per centimeter; the fine wool usually has such parameter in the range of $6-9\,\mathrm{cm}^{-1}$. There are also other indexes of wool fiber for crimped rate, such as the index of elasticity of wool fiber, the recovery rate, and the elastic recovery rate.

2.5.4 Friction and Felting Properties of Wool

In the normal conditions, the surface of wool covers scales, the root of which attaches to the wool stem. The tip of wool scale is out of the stem and points to the wool fiber head. Due to the characteristic feature of scale, the friction factor is different when sliding takes place in different directions. Common sense tells us that the frictional coefficient is much larger for friction happens reversed to the scale grown direction than the direction takes place along the grown direction of scale. This characteristic is called directional frictional effect and can be denoted by terms of frictional effect and scaly degree.

The frictional property of wool fiber is the basis of wool product felting. Under the conditions of high temperature and moisture or some chemical reagents, the wool products (wool yarns or fabrics) are usually pressed and rubbed repeatedly or periodically by external mechanical forces, for example, wool fabric usually appears frictional behavior after washing process. This is performed of wool products within gradual contraction in size. Some wool fibers are entangled with each other. This property is known as wool "felting." The main reason is, first, directional friction effect of wool fiber. This effect means the aggregation of fibers when subjected to the repeated external forces that the frictional force is much greater when inter-fibers motion is against scale direction than along the scale direction. This inter-fibers motion would always keep a direction to the wool fiber root. Second, the existence of natural crimpness would lead to the mutual movement of wool fibers irregular, and such natural crimpness of wool would cause inter-fibers easily to be entangled. Finally, the wool fiber itself has a good elasticity, which means when an external force is exerted to the wool fiber, the fiber is stretched by force and is contracted when the force is removed. The repeated creep of stretching and contraction of wool fiber results in the fiber curling and winding. Thus, it can be seen that the directional friction effect, coiling, and elasticity are the intrinsic causes of wool fulling. The effects of heat and humidity, chemical reagent, and external force are the main external factors on wool fulling.

After a fulling finishing process of wool fabric, the length of fabric is shrunk; the thickness and the tightness are thereafter increased. The fabric surface is then exposed with a layer of fluff, which causes the fabric with advantages of beautiful appearance, rich and soft hand feeling, and improved thermal insulation performance. To use such wool fulling, we can also put some loose short wool fibers together under certain mechanical forces to make certain shape and density wool felts; this method is called "felting." Some wool products, such as boots,

Figure 2.13 Wool blanket, belt, and fabrics manufactured using wool friction and felting properties.

hat, blankets, and other end-use wool products, are mainly made by using the felting process.

The fulling property leads to wool products with a unique style; on the other hand, fulling is also the main reason of wool fabric in size contraction and deformation after longtime wearing or washing. Here, in washing process, the factors of machine rubbing, warm water, and detergent usually promote the shrinkage of wool fiber and products. Wool knitted fabric in the wearing process is easily produced of felting, fuzzing, and pilling phenomenon at the fabric positions encountering with human sweat and friction easily. Most worsted wool fabrics and knitted fabrics require clear lines and stable shape, which are required to reduce or eliminate the wool fulling property. Therefore, some high-grade wool products require wool shrink-proof processing before they get into market, as shown in some wool products in Figure 2.13.

There are two kinds of methods for wool shrink-proof treatment: "oxidation" and "resin." Oxidation method, also known as the degradation method, is to eliminate the wool surface scales. The corrosion of scale agents usually has oxidants, such as sodium hypochlorite, chlorine, chloramine, potassium hydroxide, potassium permanganate, and other chemical reagents. Among these agents, the most commonly used oxidant is the chlorine oxidation agent, also known as chlorination. Resin method is to coat the wool fiber with a resin thin film that can reduce or eliminate the friction effect between the wool fibers or to make the fibers bonding each other at the crosshead position. This can restrict the mutual movement of wool fibers significantly, making the coated wool fibers loss of fulling. Here, the normally used resins include urea formaldehyde, melamine formaldehyde, silicone, polyacrylic resin, etc. Sometimes, the two mentioned methods are combined to use to enhance the anti-felting effect of treatment.

2.5.5 Wool Grease and Impurities

(1) *Wool grease*: The grease of wool is composed of two parts, namely, lanolin and perspiration, which are secreted by the sebaceous glands and sweat glands in the surface skin of sheep. Grease is a grease coating of wool fiber, which can protect wool from heavy sunshine and rain. Sweat can, to some extent,

Table 2.5 Lipid content (%) of wool fiber.

Wool type	Content of wool grease (%)	Content of wool sweat (%)
China's fine wool	10–20	7–10
China's domestic wool	3–7	8–11
Australian Merino wool	14–25	4–8

prevent wool felting but usually makes wool fibers adhesive to each other, so that prevents the external material into the fleece wool, only left a limited depth of black pollution layer at the tip of wool hair. Lack of wool grease, the wool's hand feel is stiff and rough without normal luster of wool fibers; meanwhile, the wool has poor resistance to wind erosion and is easy to be dyed unevenly. Lanolin is composed of high-level fatty acids and high-level alcohols. The physical and mechanical properties, chemical properties, and amino acid content of wool fibers are related to the content of grease and the color of lanolin. On the other hand, there is a great difference of wool's grease content depending on the variety, age, and growth environment of sheep. And different fineness of wool also has different contents of grease. For example, fine wool with content of fatty grease is more than 20%, while the coarse wool has such content below 10%; moreover, wool from sheep body side has higher content of fatty grease. Table 2.5 shows the contents of grease and sweat of wool fibers from China and Australia.

Furthermore, the color of wool lanolin is different depending on the sheep breeds and contents of grease and sweaty materials. Usually, the best quality of wool grease manifests white and light yellow colors; other colors such as yellow, orange, deep brown, and tea brown show poorer quality. The influence of different grease colors on wool quality is different. According to the color of wool, grease can identify the quality of wool, for example, wool quality with white or light milk color of grease is better than the color of yellow or darker yellow of grease and fatty materials.

The wool grease has strong anti-chemical and antimicrobial performances, which cannot be corrupted, and wool grease can infiltrate into human skin; thus, it is usually used for the matrix of some cosmetics and skin care products. This material can also be used for treatment of human burn injury or be used as anti-rust agent in industry. The wool grease (or lanolin) is a by-product of wool scouring process; it has very high value in potential applications. This valued material is usually recycled from the washed wool detergent. The main components of wool grease and sweat are inorganic salts, for example, potassium carbonate accounts for 78.5–86%, potassium sulfate accounts for 3–5%, potassium chloride accounts for 3–5%, a part of insoluble substances accounts for 3–5%, and other organic matter was 3–5%. The content of wool sweat is generally 4–20%, and its aqueous solution is alkaline.

Figure 2.14 Impurities of raw wool.

(2) *Impurities of wool*: As shown in Figure 2.14, the impurities of wool are those that are attached to the wool, such as muddy sand, dust, manure, and some plant impurities (which is the most harmful to wool, such as alfalfa seeds). The raw wools are the hairs directly cut from sheep body. With impurities in raw wool, the quantity and type of impurities have great differences because of various sheep breeds, rearing conditions, and the local climates.

"Net raw wool rate" refers to the raw wool after processes of cleaning and removal of grease and sweat, vegetable matter, sand, and ash, and the pure wool weight is transferred to certain moisture regain (%), certain concentrate of fatty material and ash, and then the weight percentage of such pure wool to the weight of raw wool. The net raw wool rate is an important index to evaluate the economic value of wool, which is closely related to the cost of wool manufacture factory and the amount of wool used in textile industry. In our country, the raw wool has a high content level of impurity and a low level of net raw wool rate.

2.5.6 Other Properties of Wool Fibers

α-Keratin protein is the main composition of wool fiber; therefore, wool fiber is more acid resistance other than alkali resistant. Experimental data [9] showed that wool in dilute sulfuric acid boiling environment for a few hours has no damage. Under room temperature, the strength of wool is almost not affected in 80 wt% sulfuric acid solution for a short time, and other weak acids such as acetic acid, formic acid, and organic acid are the dyeing accelerant in the dyeing process of wool. The effect of alkali on wool's property is more severe than that of acid. With the increase of alkali concentration, temperature, and prolonging of treatment time, the damage of wool is getting more serious. The alkali normally turns the wool to be yellow, decreased sulfur content, and is partially dissolved. Compared with other natural fibers, the tensile strength of wool fiber is the smallest, and the elongation rate is the largest, indicating that the elastic recovery of wool is the best. Moreover, the hygroscopicity of wool is the highest in natural fibers; its moisture regain reaches up to 15–17% under the general atmospheric conditions. The main reason is a large amount of hydrophilic groups existed in wool macromolecules.

Wool fiber has a relative lower value of thermal conductivity, and natural crimpness of wool increases the storage of still air among wool products. Therefore, the warmth retention of wool textiles is excellent. Wool performs worse than other general fibers in heat resistance, for example, woof would become yellow and stiff under 100–105 °C of dry heat conditions that leads to complete removal of water from hair fiber after drying. When the heating temperature rises to 120–130 °C, the wool fibers begin to collapse. The practice experience [10] shows that the effect of wool fiber can be shape set permanently under certain heat and moisture conditions. In addition, worms like wool fibers and the end-use products very much and damage them after a while of storage, because the wool protein molecules are good food of worms.

2.6 Quality Inspection and Evaluation of Wools

2.6.1 Raw Wools from Sheep

According to national standard (China) "GB1523-93," raw wool cut from sheep should be grading and counting. There are four assessment index to grade the fine raw wools, that is, fineness, length, height of the hair bundle, and content of coarse/dead hair, in which the lowest value of evaluated index is to determine the count of wool. Here, the wool fineness is the main assessment index. The wool from improved sheep breed is usually graded through wool length, coarse cavity hair, and dry-dead hair to be the evaluation criteria, in which the lowest one is defined as the wool grade. The appearance of raw wool is a reference index, which is the hair assessment for middle level of wools.

2.6.2 Fine Wool from China's Raw and Improved Sheep Breed

The fine wool from domestic and modified sheep breeds has grading types, called "count wool." For homogeneous wool, wool is graded by fineness like in 70s, 66s, 64s, and 60s. Another one is called "grade wool," which is based on homogeneous and heterogeneous media, according to the content of coarse medullated hair. The raw wool is divided into grade one, two, three, four A, four B, and five. For washed wools, they are divided into first class and second class. Herein, there are two conditions for the determination of wool class, that is, the content of soil in wool such as the rate of soil contamination and the rate of felting. The lowest value is the grading class for the wool products. There are three items for wool production factory to ensure the grading class, that is, the oil content, moisture regain, and rate of residual alkali.

2.6.3 Tops of Domestic Fine Wool and Its Improved Wool

The tops of China's wool and its improved wool are divided into count wool tops and improved graded wool tops. The count of wool tops is graded according to the wool's fineness; there are 70s (18.1–20.0 µm), 66s (20.1–21.5 µm), 64s (21.6–23.0 µm), and 60s (23.1–25.0 µm). The grade of wool tops is determined according to the content of coarse cavity wool; there are grade one, two, three, four A, four B, and five.

The technical indexes of wool count tops and improved wool tops have physical indexes and appearance defects. The physical indicators include discretion of fineness, content of coarse cavity wool, weighted average length, variation of coefficient of wool length, content of short wool below 30 mm, weight tolerance, and weight unevenness. The appearance defection indexes include wool grain, wool piece, flax, and other grass fiber. According to the test results, the wool tops are divided into grade one, two, and so on.

2.6.4 Inspection and Evaluation of Australia Wool

Australia is the world's largest producer of fine wool. Australia has mature objective inspection and evaluation system, including the feel visual subjective evaluation system before wool packing process and objective evaluation system for wool quality after wool package into commercial circulation (auction). At the present, the subjective evaluation system is being objective gradually, for example, the original subjective measurement of raw wool fiber length and fineness uniformity using DFDA200 method. The objective evaluation system is mainly aimed at wool fineness and cleaning rate (for the first inspection certificate), taking into account the strength of wool, length, and weak section inspection (for the second inspection certificate). Here, the first inspection certificate includes parameters like average fiber diameter, diameter variation coefficient, and the content of crude fiber, as well as wool wash rate (wool matrix), total impurity content, and impurity content of three parties. Currently, the first inspection of certificate has reached a rate of more than 99%. The second inspection certificate includes the index of strength and length of hair bundle, weight of the break point, etc. Currently, the second certificate inspection has reached a test rate of more than 60%.

The abovementioned inspections on Australia wool's quality are carried out in Australian Wool Testing authority (AWTA). All related tests are performed under objective measurements. The test results are given to wool auction center in the format of data table, and a variety of software for data analysis can be used for forecast, including the wool market price and analysis of processing performance.

2.6.5 Inspection and Evaluation of Wool Fabric

Similarly with quality inspection of wool fiber and top, the inspection and evaluation of wool fabric is also based on the wool product test standard or agreement with customer to do the grading evaluation. The grading evaluation of fabric usually has two aspects of evaluations that are apparent faults and inherent quality. The fabric apparent faults are performed under subjective sensory detection method, i.e. a comparison between the fabric sample and standard sample that based on the quality grading standard for marking the fabric faults. The inherent quality of wool fabric has many objective parameters, for example, breaking strength and breaking elongation, warp and weft density, and gram weight per square meter, and these inherent parameters can be measured by specific test device for grading evaluation based on particular quality evaluation standard. For some special performance, functions, and requirements of wool textiles, the evaluation system should be established according to the lowest value during different aspects of inspection and evaluation. Here, take worsted wool fabric as an example to demonstrate the grading system for wool fabrics. Technical requirements for worsted fabrics include safety requirements, product quality, inherent quality, and appearance quality. The safety requirements of worsted fabrics should meet the regulations of GB 18401 of "National Fundamental Safety Technical Standard for Textiles." The product quality of wool worsted fabric includes the fabric surface, hand feel, and luster. The fabric internal quality involves physical indicators and color fastness. The fabric appearance quality includes localized defect and scattered defects.

The quality grading of worsted fabric is divided into high-class product, grade A, grade B, and others below grade B that are classified into offgrades. Fabrics made of combed wool yarns are metering based on the unit of fabric piece. Thus, the fabric quality is assessed and evaluated by three combining factors of fabric weight, inherent quality, and appearance quality and graded finally according to the lowest marking value of them. Here, if two or two more indexes of the three evaluation items are grade B at the same time, the product will be reduced to offgrades.

The length of wool worsted fabric should be longer than 12 m. While the net length of wool worsted fabric beyond 17 m can be composed of two sections, but the shortest part is not less than 6 m. When matching two pieces of fabric, they should be the same grade and the same color and luster.

For sample processing by buyer, the producers should establish seal sample according to the buyer's requirements and confirm the seal by both parties. In inspection, it should be graded carefully by comparing every piece of sealed sample. The samples that apparently worse than grade A should be classified as grade B, and those significantly worse than grade A are classified as grade C. The inherent grades of worsted fabric are assessed comprehensively using physical indicators and dyeing fastness. The lowest value of the two indicators determines the final fabric grade. Physical indicators of worsted fabric are graded in accordance with Table 2.6, and dyeing fastness is in accordance with Table 2.7. The change rate in size of "machine washable" products can be referred to Table 2.8. The appearance of defects, evaluation, and grading of wool worsted products are detailed listed and compared in Table 2.9, where the comparison was performed along warp and weft directions and degree of faults on fabric surface was also classified as four levels, including appearance faults and scatter faults.

In terms of influence on the end use and the appearance of different problems, the appearance defects of wool worsted fabric can be divided into two types of quality issues: local appearance defects and distributed appearance defects. Here, we would not discuss the defects in detail because of its strong professional description. The appearance of defects, evaluation, and grading are listed in Table 2.9.

Table 2.6 Physical requirements of wool worsted fabric.

Item		High-level class	Grade one	Grade two
Deviation of fabric width (cm) ≤		2	2	5
Tolerance of fabric weight per square meter (%)		−4 to +7	−5 to +7	−14 to +10
Static dimension variation rate (%)≥		−2.5	−3.0	−4.0
Fiber content (%)	Content allowance of wool fiber in wool blended products	−3 to +3	−3 to +3	−3 to +3
Pilling (level) ≥	Suede face	3–4	3	3
	Smooth face	4	3–4	3
Breaking strength (N)≥	(7.3tex*2)*(7tex*2) (80S/2*80S/2) and single weft no less than 14.5tex (40S)	147	147	147
	Other	196	196	196
	Normal worsted wool products	15	10	10
	(8.3tex*2)*(8.3tex*2) (70S/2*70S/2) and single weft no less than 16.7tex (35S)	12	10	10
Dimension change rate of fabric under steam (%)		−1 to +0.5	−1 to +0.5	—
Deformation of fabric in water (level)≥		4	3	3
Degree of fabric fracture (mm) ≥		6	6	8

(1) In order to improve the spinning performance and durability of wool products, the "pure" wool products are allowed to blend with 5% of synthetic fiber. For wool products with decorative fiber products (decorative fiber must be visible with adornment effect), the non-wool fiber content should be less than 7%. But the sum content of both high-performance and decorative fibers should not exceed 7%.
(2) Functional fiber and cashmere in the final wool products show the content less than 10%; the decrease of the content should not be higher than that of tagging content of 30%.
(3) The fiber in the linkage of two-layer fabric is not considered in the examination.
(4) Content of embedded threads lower than 5% is not considered in the examination.
(5) The degree of fabric fracture in the leisure clothes is 10 mm.

2.7 Shape Memory Properties of Wool

For thousands of years, wool has been merely considered as a textile fiber with outstanding performances for its excellent elasticity and thermal insulation. These properties are ascribed to the wool hierarchical structure with macro- and microfibrils and helical coils, which are wrapped by cortex and cuticles outside as shown in Figure 2.9. In addition, as shown schematically in Figure 2.7, the structural components of hydrogen and DBs in crystals and amorphous regions play key roles for the advanced properties. A hydrogen bond is an electrostatic attraction between polar molecules. It is not actually a true bond but a particularly strong dipole–dipole attraction. However, DB is a strong covalent cross-linker between hair molecules (can be characterized by Raman

2 Animal Fibers: Wool

Table 2.7 Color fastness requirements of wool worsted fabric.

Item		High-level class	Grade one	Grade two
Color fastness to light ≥	≤1/12 standard depth of color (light)	4	3	3
	≥1/12 standard depth of color (dark)	4	4	3
Color fastness to water ≥	Color change	4	3–4	3
	Staining felt	3–4	3	3
	Other adjacent fabric staining	3–4	3	3
Color fastness to perspiration ≥	Color change (acidic)	4	3–4	3
	Staining felt (acidic)	4	4	3
	Other adjacent fabric staining (acidic)	4	3–4	3
	Color change (alkalinity)	4	3–4	3
	Staining felt (alkalinity)	4	4	3
	Other adjacent fabric staining (alkalinity)	4	3–4	3
Color fastness to ironing ≥	Color change	4	4	3–4
	Staining felt	4	3–4	3
Color fastness to rubbing ≥	Dry friction	4	3–4	3
	Wet friction	3–4	3	2–4
Color fastness to washing ≥	Color change	4	3–4	3–4
	Staining felt	4	4	3
	Other adjacent fabric staining	4	3–4	3
Color fastness to dry washing ≥	Color change	4	4	3–4
	Solvent change	4	4	3–4

(1) Color fastness to washing and color fastness to rubbing are not considered in "dry washing only."
(2) Color fastness to dry washing is not considered for "careful hand washing" and "machine washable."

Table 2.8 Dimensional stability to washing requirements of worsted wool fabric in "machine washable."

		High-level class, grade one, grade two	
Item		Suit, pants, coat, overcoat, one-piece dress, jacket, skirt	Shirt, evening dress
Dimensional change rate to relaxation (%)	Width	−3	−3
	Length	−3	−3
	Washing procedure	1*7A	1*7A
General dimensional change rate (%)	Width	−3	−3
	Length	−3	−3
	Edge	−1	−1
	Washing procedure	3*5A	3*5A

Table 2.9 Appearance of defects, evaluation, and grading of wool worsted products.

Fabric faults		Degree of faults appearing on fabric	Local braiding	Scatter degradation	Remarks
Warp direction	(1) Coarse yarn, fine yarn, doubling yarn, loose yarn, tight yarn, wax yarn, local narrowed area of fabric surface	Obvious 10–100 cm; >100 cm; spread all over per 100 cm; scatter all the pieces	1 1	Grade two Shuffs	
	(2) Oil yarn, dirty yarn, jute, yarn exchange between face and back, selvage press mark, cut mark	Obvious 5–50 cm; >50 cm, per 50 cm; scatter all; spread all over	1 1	Grade two Shuffs	
	(3) No warping yarns, dead crease	Obvious 5–20 cm warp; >20 cm, per 20 cm; scatter all	1 1	Shuffs	
	(4) Barre, crease, barring, print of warping exchange, selvage depth, depth of fabric ends	Obvious 40–100 cm; >100 cm, per 100 cm; spread all over; scatter all the pieces	1 1	Grade two Shuffs	Grade two for (selvage depth 4)
	(5) Streakiness, color tone	Obvious 20–100 cm; >100 cm, per 100 cm; spread all over; scatter all the pieces	1 1	Grade two Shuffs	
Warp direction	(6) Bristle mark	Obviously 20 cm or less; >20 cm per 20 cm; spread all over	1 1	Shuffs	
	(7) Clip mark, broken selvage	Obvious 2–10 cm; >100 cm per 100 cm; spread all over; scatter all the pieces	1 1	Grade two Shuffs	
	(8) Thorn flash, edge friction, hairy edge character incomplete, edge character seriously stained, bleached fabric edge stitch, white edge depth more than 1.5 cm needle, needle embroidery, flounces, dense edge	Obvious 0–100 cm; >100 cm per 100 cm; scatter all	1 1	Grade two	
Weft direction	(9) Coarse yarn, fine yarn, doubling yarn, loose yarn, tight yarn, exchange mark of yarn	Obvious 10 cm to all; spread all over; scatter all the pieces	1	Grade two Shuffs	

Table 2.9 (Continued)

Fabric faults		Degree of faults appearing on fabric	Local braiding	Scatter degradation	Remarks
Weft direction	(10) Mispick, oil yarn, dirty yarn, jute, small braid yarn	Obvious 5 cm to all; spread all over; scatter all the pieces	1	Grade two Shuffs	
	(11) Thick places, weft shallow, hard joining stencil mark, severe press mark, yarn uniformization	Obviously 20 cm and less warp >20 cm; spread all over; scatter all the pieces	1 1	Grade two Shuffs	
	(12) Thin places of fabric, loom bar, wrong construction, mat-up, loose weave, mottled, hole repair mark, hole correction mark, big belly yarn, warp bar	Obviously 10 cm or less warp; >10 cm per 10 cm; spread all over	1 1	Shuffs	Big belly yarn 1 cm for start point 0.5 cm mottled for note 2
	(13) Bore chafed yarn	2 cm (including 2 cm); spread all over	1	Shuffs	
	(14) Combed hair, knot, grass, dead hair, overshot, loose pattern	Spread all over; scatter all the pieces		Grade two Shuffs	
	(15) Bias of fabric face	Plain fabric: 4 cm Plaid fabric: 3 cm, 40–100 cm, >100 cm per 100 cm. Plain fabric: 4–6 cm scatter all; spread over 6 cm. Plaid fabric: 3–5 cm scatter all; spread over 5 cm	1 1	Grade two Shuffs Grade two Shuffs	High-level fabric product plaid fabric from 2 cm; plain fabric from 3 cm

(1) Fabric faults in the area from edge to less than 1.5 cm of fabric (a line refers to the inner cloth within 0.5 cm depth edge of defects) in the identification of fabric goods are not considered as fabric faults; however, the hole, broken edge, edge burr, edge wear, bleaching, and stitch defect edge character on the edge of fabric should be checked; if the defect length extends to the inner part of fabric, this should be considered to inspect together.
(2) For the appearance faults of wool fabric, in the case of exceeding the above provisions of special circumstances, it can be sorted out according to the extent of their impact on the use of similar defects in the evaluation of the relevant provisions of the discretion.
(3) For the scatter appearance faults, especially the serious impact on fabric wearability, it should be set price according to quality.
(4) The top grade of wool products should not have more than 1 cm of broken hole, spider web, rolling shuttle, and serious weft bar.

spectroscopy) [11] that can control the hair elasticity [5, 8]. Keten et al. reported that both hydrogen bonds and DBs in the amorphous regions of animal hair can be reversible under certain environmental conditions such as redox agents, water, and heat [12, 13], which lead to the conversion of temporary and permanent shrinkage macroscopically [14]. In addition, ~40% of the crystallinity in wool is proven to be too hard to cleave and has been studied by textile researchers using X-ray diffraction (XRD) that the crystalline phase of wool is made of very dense hydrogen bonds [7, 15, 16].

Shape memory polymers (SMPs) are a kind of smart polymers characterized by their stimuli-responsive behavior adapting to our human demands. SMPs can be described in terms of netpoint-switch structure as refer to the work [17] where their permanent shape is determined by netpoints and reversible bonds in the amorphous region acting as a switch unit that leads to temporary shape. Even though there are a few latest reports of triple stimuli-responsive SMPs [18–21], SMPs have one stimulus in most existing cases. However, Schattling et al. [22] stressed the rising importance of developing multi-stimuli-responsive polymers that are called "all-in-one talents" in many industries, such as life sciences for comparable natural adaptability and information technology for the parallel writing of information to give a dramatic increase of memory density. They believed that the addition of one more stimulus in a polymer can improve the degree of control, provide more free choice, and increase the level of intricacy, which in turn enhances the adaptability of a material to different environments by more intelligent functions. However, developing such multi-responsive talented polymers requires new technologies that may be beyond our normal imagination. On the other hand, nature is full of wisdom, and we may learn tricks even from a very common subject such as animal hair fiber [3]. For example, an interesting shape memory (SM) phenomenon of a dry wool yarn is usually encountered in textile industry, where the yarn contracts by 20% in water after being pretreated using reducing agent (NaHSO$_3$) and pre-stretched to 30% strain [23]. It is believed that this is the SM of wool responsive to redox and water.

For hydrogen bonds in polymers, they can be very strong acting as dense physical cross-linking, which are difficult to cleave such as in a hard polymer crystal phase. This is well known, and they are widely used as netpoints in many SMPs where hydrogen bonding-based crystal acts as netpoint for the purpose of shape recovery [24, 25]. When they are not that strong, they may be reversibly dissociated and re-formed under conditions such as heat and water [26–28], which can be used as switch in SMPs. For example, it is found that the glass transition temperature (T_g) of wool is decreased with the increase of water content (Figure 2.15a) [6], due to the collapse of a large amount of HBs in the amorphous region for the cause of aqueous molecules. It is also observed that the collapsed HBs in the amorphous region can be re-formed when the wet wool is dried. This indicates the HBs are reversible, with and without water alternatively [29]. Thus, HB is the key behind water-responsive SM and plays a main role in the water-driven shape recovery process [27, 30, 31]. Furthermore, Jung and So observed a water-responsive SM in a poly(ethylene glycol) (PEG)-based polyurethane where the crystalline segment was dissolved by water due to the disappearance of HBs [32]. Using high molecular weight of

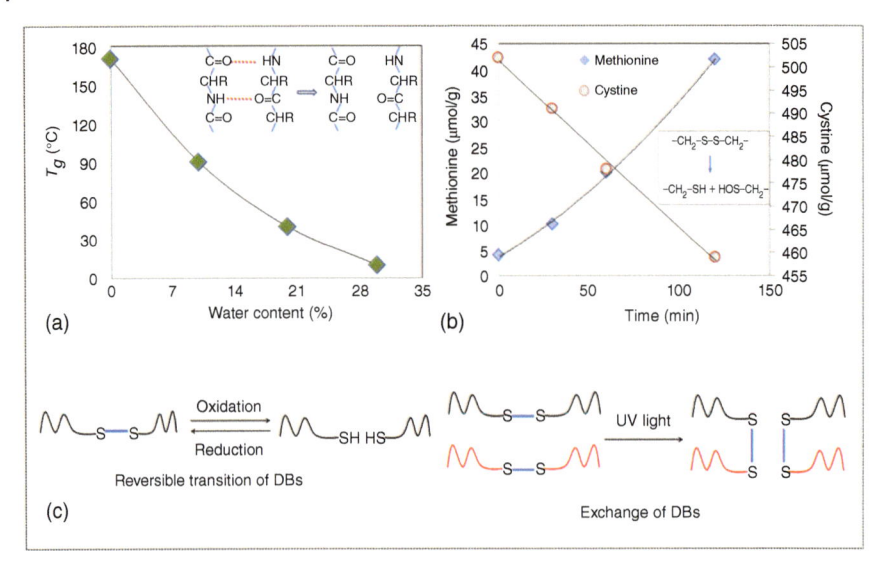

Figure 2.15 Physical properties of wool. (a) Effect of water content on wool glass transition temperature. (b) Effect of reducing reaction on amount of disulfide bonds (DBs). (c) Reversible transition of DBs in redox reaction and exchange of DBs among macromolecules under UV light.

PEG, another shape recovery of 99% copolymer was synthesized through HB switches [33]. A simple and controllable triple-SM supramolecular composite was also developed successfully through HB switches between a polymer and mesogenic units [34] where the HBs enable broad and independent control of both T_g and cross-link density [35].

For DBs, it is found that it can be broken when reducing agent is applied because a decreased amount of cysteine and increased methionine were observed throughout the reducing reaction as shown in Figure 2.15b [1]. This is because DBs can be cleaved to form two thiol groups in reducing solution. In turn, the broken DBs can be reversed by an oxidation reaction [36]. DBs are a type of dynamic covalent bonds responsive to redox and different light triggers, which form the basis as switch in SMPs. As shown in Figure 2.15c, reversible change happens between DBs and thiols where DBs are cleaved to thiol groups by a reduction reaction and re-formed by an oxidation reaction [37]. This reversible change can be found in responsive polymer capsules and micelles and gels for applications such as drug delivery [38–40]. DBs have another type of reversibility: exchange reaction that can lead to the healing ability of a polymer driven by UV and visible light as shown in Figure 2.15c [41–44]. As far as SMPs are concerned, there are only two reports in literature: a thermally responsive SMP was reported with semicrystalline and covalently cross-linked network where DBs were used for self-healing function only under UV light [45]. Thus, only one SMP reported in the literature has utilized DBs as a switch responsive to redox treatment where cellulose derivatives with cross-linkable mercapto groups were used [46]. Strictly speaking, the use of DBs in SMPs as netpoints has not been reported yet, but they are strong chemical bonds that are stable under some conditions and can be used as netpoints in SMPs as evidenced in wool, where they control its elasticity under normal environmental conditions [5]. Moreover, it is also clear that DBs could act as reversible switches, having self-healing capability [47] when exchange reactions take place among macromolecules under light, particularly in UV light conditions.

In a report, wool was investigated on its multi-stimuli-responsive SM behaviors and for corresponding mechanisms in molecular and structural networks. The innate shape of a wool hair can be recovered in varying degrees through different stimuli effect on the fixed deformation, which is a typical multi-stimuli SM material. High temporary shape fixation (>0.8) and shape recovery (>0.5) of wool exposure on water and redox agent demonstrated that wools are smart α-keratin fibers stimulated by water and redox. However, low temporary shape fixation under heat (0.55) and low shape recovery (0.23) under UV illuminating indicates that wools have less SM abilities under stimuli of heat and UV light. The single tensile of wool interpreted well for the effect of each stimulus on the switch (assuming netpoint intact after stimuli) open and close. XRD and DSC characterizations of wool under four stimuli showed that the crystalline phase can be viewed as an invariable component during SM program. However, Raman spectra showed the DBs in breakage when wool faces UV light and reductant ions and in variation under the stimuli of water and heat, indicating these bonds act as switch unit for stimuli of UV light and redox agent and as netpoint for water and heat. FTIR characterization for HBs of wool through the variations of characteristic peaks in wavenumber shifting and intensity ratio showed that these bonds could act as switch for the stimuli of heat, water, and redox agent in each SM behavior. Based on this, a twin-netpoint-switch structure model for wool was proposed, as shown in Figure 2.16, for interpreting the different SM abilities of the hair when exposed to different external stimuli, where a twin-netpoint/single-switch structure is for the stimulus of water, heat,

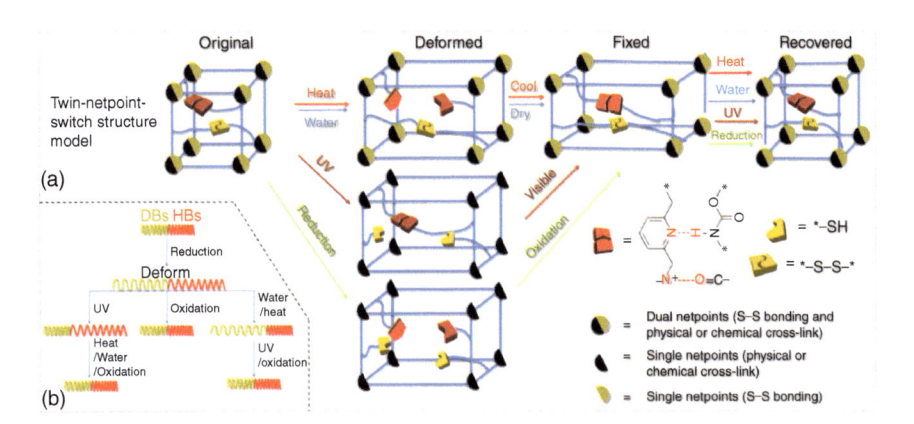

Figure 2.16 (a) Twin-net-switch structural model of wool for SM abilities. (b) Variation of three key components under four types of stimuli where reductant can open both switches (DBs and HBs), HBs can be closed using oxidant, water, and heat, and DBs can be re-formed using UV and oxidant.

and UV light and a single-netpoint/twin-switch structure is for the stimulus of redox agent.

2.8 Future Trends of Application of Wool Keratin

When considering the application of wool beyond textiles, it is of importance to first consider the materials from which wool is made up, such as the inner components for SM applications. Once the macroscopic wool as a fiber is not fundamental in defining the application, then it comes down to the base materials from which the fiber is formed to define the relevant applications.

From the view of macroscopic and microscopic scales, both characteristics of wool have been extensively reviewed [48, 49]. By far the most important material in this consideration is keratin protein, defined by the US National Library of Medicine as "A class of fibrous or sclera proteins important both as structural proteins and as keys to the study of protein conformation." The family represents the principal constituent of epidermis, hair, nails, horny tissues, and the organic matrix of tooth enamel. It is clear from the definition that keratin proteins, in many forms, are widely present in biological systems, in the hair, skin, nails, and other tissues.

This widespread occurrence in nature, as a tough, physical material from which to build structures, is of significance when considering the advanced applications of wool. The tenacity of wool, the resilience of our nails, and the barrier characteristics of our skin tell us much about the features that we can look for in advanced applications of wool-derived keratins. While keratin may account for up to 95% of the dry matter of a wool, the other components are also of note for their interesting characteristics. Considered as an ordered collection of elongated cells, rich in multiple types of keratin proteins, the wool structure is brought together by the cell membrane complex, the material that defines the cell boundaries and a key element in keeping the cells together in a strong and ordered fibrous alignment.

2.8.1 Extraction of Keratin from Wool for New Product Development

Beyond the fiber, advanced applications of wool utilize the keratin protein from which the fiber is constructed. The first step, then, in development of these materials is isolation of that protein in a versatile form that can be further manipulated into the desired applications. Wool, by its nature, is a tough insoluble material evolved by nature to be robust and resistant to degradation. Its conversion to a state where it can be widely manipulated is not straightforward, and this has been a key reason that keratin protein has not been exploited commercially as a material in the way that other biopolymers have been, such as collagens, alginates, celluloses, and starches. The simplest way to isolate a soluble from of keratin from the source fiber is to apply conventional methods of hydrolysis, using acid, alkali, or enzymatic digestion of the peptide bonds to create a mixture of peptides and amino acids that have retained none of the secondary or tertiary structure of the core protein.

In most cases, these processes also destroy many of the core amino acids. Although this harsh processing destroys many of the unique characteristics of protein, it has provided a route for several developments, such as the historic industrial uses explored below. In order to progress beyond these limited applications of keratin-derived peptides, hydrolysates, and amino acids, a series of processes for the isolation of keratin proteins from wool that maintain core features of the protein and its structure have been developed [50, 51]. Most of the processes do not hydrolyze the peptide linkages, and this allows the keratin proteins that are produced to retain a form and function similar to the native keratins. By maintaining the structural and chemical complexity of the native proteins, by having substantial fibrous proteins that naturally form robust and ordered materials, and with chemically active amino acid components such as cystine present in abundance, then advanced applications in a range of areas may be developed. The development of technology to achieve these forms of keratin is a key to the creation of advanced applications of wool.

2.8.2 Industrial Trends Relating to Sustainable Polymers

Advanced applications of wool keratins can be considered in a similar manner to developments undertaken elsewhere into biopolymer- or biologic-derived materials. The factors influencing the developments come primarily from the commercial arena, which is in turn driven by opportunities created by innovation and importantly environmental and legislative factors. Global interest in sustainably produced materials and full life cycle analysis of products has driven interest and investment in biological materials. Biopolymers, as alternatives to synthetic plastics, have been the promise of white biotechnology for several decades, with Biopol™ from Monsanto as an early example of a promising technology that has not yet had the appropriate alignment of technological, environmental, and legislative factors to drive enough commercial interest to lead to success. NatureWorks™ from Cargill Dow, used from packaging to fiber applications, is an example of a biopolymer material for which there is rapidly increasing consumer interest. The development was made possible by advanced biotechnology, but its continued development is driven by a commercial need to find biological alternatives to oil-derived synthetic polymers because of legislative and economic pressure on oil. While price signals are important, there is increasing pressure on landfill that is being progressively occupied by synthetics derived from oil. These same drivers create strong opportunities for other natural biopolymers with an established biological function as a material. Advanced applications from wool keratin are emerging to capture this opportunity.

2.8.3 Future Trends

The future advanced uses of wool keratins may be in two distinct areas. The first is medical technology. Developments will be driven by science and improved healing outcomes. Recent academic research into the previously unknown biological role of keratins in body was identified through the work of Coulombe, Omary, and others [52, 53]. This has determined that keratin in body plays more than a physical role but is essential in the transportation of signaling peptides crucial for the growth and differentiation of cells. Coupled with the developing technology for the creation of medical devices from keratin, substantial advanced uses of wool will develop in the medical area. This is already commencing in the areas of wound care and orthopedics, and the natural evolution into other areas of medical care will follow.

The second area of advanced use is in industrial technology and will be driven by legislation and commercial necessity. With the price of oil on a sustained sharp increase over recent years, and a flow-on effect to all oil-derived materials, coupled with harsher environmental legislation being established globally relating to both carbon neutrality and source and waste management, the focus on biopolymers will naturally increase sharply. Keratin is one of nature's most successful materials; it was evolved to be versatile, tough, and protective, and as a natural protein, there are wide opportunities in a range of industrial areas. Robust materials, such as plastics, to the less obvious industrial speciality chemicals that are also derived from a synthetic backbone are all a likely development for advanced uses of wool. None of these future applications requires the fiber to maintain its fibrous form, but all utilize the complex fundamental structure that the wool fiber is based on.

2.8.4 Sources of Further Information and Advice

Further information on the structural components of wool can be gained from Marshall et al. [49] and Gillespie [48]. The biological role of keratin in the skin and other tissues is the subject a variety of publications including Kim et al. [53] and Omary and Ku [52]. The use of keratin in applications beyond the conventional fiber is the subject of numerous patent applications (see, for example, [51], and [50]). Biopolymer trends and the importance of both technology and legislation is well documented by the Greentech organization (www.greentech.org).

References

1 Yao, M. (2009). *Textile Materials*. Beijing: China Textile Press.
2 IWTO (2016). World's fibers production and distribution, Forward – The Economist, 2018. https://www.qianzhan.com/analyst/detail/220/180704-ede538d7.html.
3 Xiao, X. and Hu, J. (2016). *Scientific Reports* 6: 26393.
4 *The Macromolecular Structure of Wool Fiber*. http://www.fzfzjx.com/tech_news/detail-4048.html (accessed 22 June 2010).
5 Harris, M., Mizell, L.R., and Fourt, L. (1942). *Industrial and Engineering Chemistry* 34 (7): 833–838.
6 Wortmann, F.J., Rigby, B.J., and Phillips, D.G. (1984). *Textile Research Journal* 54: 6–8.
7 Lotay, S.S. and Speakman, P.T. (1977). *Nature* 265: 274–276.
8 Lindley, H. (1957). *Textile Research Journal* 27: 690–695.

9 Hearle, J.W. (2000). *International Journal of Biological Macromolecules* 27 (2): 123–138.
10 Okajima, S. and Kikuchi, K. (1956). *Fiber* 12 (12): 899–902.
11 Akhtar, W. and Edwards, H.G.M. (1997). *Spectrochimica Acta Part A: Molecular and Biomolecular Spectroscopy* 53 (7): 1021–1031.
12 Chapman, B.M. (1970). *Journal of the Textile Institute* 61 (9): 448–457.
13 Keten, S., Chou, C.C., Van Duin, A.C.T., and Buehler, M.J. (2012). *Journal of the Mechanical Behavior of Biomedical Materials* 5: 32–40.
14 Astbury, W.T. and Woods, H.J. (1934). *Philosophical Transactions of the Royal Society of London, Series A: Mathematical, Physical and Engineering Sciences* 232: 333–394.
15 Wu, B., Yi, Y., Xu, T., and Lei, H. (2012). *2nd International Conference on Electronic & Mechanical Engineering and Information Technology (EMEIT-2012)*, 1405–1408. Paris, France: Altantis Press.
16 Samways, C. and Hastings, G.W. (1970). *Nature* 225: 634–635.
17 Hu, J.L. and Chen, S.J. (2010). *Journal of Materials Chemistry* 20 (17): 3346–3355.
18 Chen, J., Zhang, S., Sun, F. et al. (2016). *Polymer Chemistry* 7: 2947–2954.
19 Kumpfer, J.R. and Rowan, S.J. (2011). *Journal of the American Chemical Society* 133: 12866–12874.
20 Wang, L., Yang, X., Chen, H. et al. (2013). *Polymer Chemistry* 4: 4461–4468.
21 Tao, Z., Peng, K., Fan, Y. et al. (2016). *Polymer Chemistry* 7: 1405–1412.
22 Schattling, P., Jochum, F.D., and Theato, P. (2014). *Polymer Chemistry* 5: 25–36.
23 Hu, J.L., Zhu, Y., Huang, H.H., and Lu, J. (2012). *Progress in Polymer Science* 37 (12): 1720–1763.
24 D'hollander, S., Van Assche, G., Van Mele, B., and Du Prez, F. (2009). *Polymer* 50: 4447–4454.
25 Zhang, T., Wen, Z., Hui, Y. et al. (2015). *Polymer Chemistry* 6: 4177–4184.
26 Ratna, D. and Karger-Kocsis, J. (2008). *Journal of Materials Science* 43: 254–269.
27 Zhang, S., Yu, Z.J., Govender, T. et al. (2008). *Polymer* 49 (15): 3205–3210.
28 Huang, W.M., Yang, B., An, L. et al. (2005). *Applied Physics Letters* 86: 114101–114103.
29 Eaves, J.D., Loparo, J.J., Fecko, C.J. et al. (2005). *Proceedings of the National Academy of Sciences of the United States of America* 102 (37): 13019–13022.
30 Fan, K., Huang, W.M., Wang, C.C. et al. (2011). *Express Polymer Letters* 5 (5): 409–416.
31 Xiao, X. and Hu, J. (2016). *International Journal of Chemical Engineering* 1 (12): UNSP 4803254
32 Jung, Y.C. and So, H.H. (2006). *Journal of Macromolecular Science, Part B: Physics* 45 (6): 1189–1190.
33 Liu, G., Guan, C., Xia, H. et al. (2006). *Macromolecular Rapid Communications* 27: 1100–1104.
34 Chen, H., Liu, Y., Gong, T. et al. (2013). *RSC Advances* 3: 7048–7056.
35 Ware, T., Hearon, K., Lonnecker, A. et al. (2012). *Macromolecules* 45: 1062–1069.

36 Wedemeyer, W.J., Welker, E., Narayan, M., and Scheraga, H.A. (2000). *Biochemistry* 39 (15): 4208–4216.
37 Zheng, M., Aslund, F., and Storz, G. (1998). *Science* 279: 1718–1721.
38 Dailing, E.A., Nair, D.P., Setterberg, W.K. et al. (2016). *Polymer Chemistry* 7: 816–825.
39 Kakizawa, Y., Harada, A., and Kataoka, K. (1999). *Journal of the American Chemical Society* 121: 11247–11248.
40 Yan, Y., Wang, Y., Heath, J.K. et al. (2011). *Advanced Materials* 23: 3916–3921.
41 Fairbanks, B.D., Singh, S.P., Bowman, C.N., and Anseth, K.S. (2011). *Macromolecules* 44 (8): 2444–2450.
42 Otsuka, H., Nagano, S., Kobashi, Y. et al. (2010). *Chemical Communications* 46: 1150–1152.
43 Canadell, J., Goossens, H., and Klumperman, B. (2011). *Macromolecules* 44 (8): 2536–2541.
44 Amamoto, Y., Otsuka, H., Takahara, A., and Matyjaszewski, K. (2012). *Advanced Materials* 24: 3975–3980.
45 Michael, B.T., Jaye, C.A., Spencer, E.J., and Rowan, S.J. (2013). *ACS Macro Letters* 2: 694–699.
46 Aoki, D., Teramoto, Y., and Nishio, Y. (2007). *Biomacromolecules* 8: 3749–3757.
47 Chang, R., Huang, Y., Shan, G. et al. (2015). *Polymer Chemistry* 6: 5899–5910.
48 Gillespie, J.M. (1983). *Biochemistry and Physiology of the Skin*, 1e (ed. L.A. Goldsmith), 475. Oxford University Press.
49 Marshall, R.C., Orwin, D.F.G., and Gillespie, J.M. (1991). *Electron Microscopy Reviews* 4: 47–49.
50 Smith, R.A., Blanchard, C.R., and Timmons, S.F. (1997). US Patent 5, 932, 552
51 Kelly, R.J., Worth, G.H., Roddick-Lanzilotta, A.D. et al. (2002). US Patent 7, 148, 327.
52 Omary, M.B. and Ku, N.O. (2006). *Nature* 441 (7091): 296–298.
53 Kim, S., Wong, P., and Coulombe, P.A. (2006). *Nature* 441 (7091): 362–365.

3

Animal Fibers: Silk

K. Murugesh Babu

Bapuji Institute of Engineering and Technology (Affiliated to Visvesvaraya Technological University, Belgaum), Department of Textile Technology and Research Centre, Davangere, 577004, Karnataka, India

3.1 Introduction to Silk and Silk Industry

Silk is one of the oldest fibers known to man. Silk is an animal fiber produced by certain insects to build their cocoons and webs. Although many insects produce silk, only the filament produced by the mulberry silk moth *Bombyx mori* and a few others in the same genus is used by the commercial silk industry [1]. The silk produced by other insects, mainly spiders, is used in a small number of other commercial capacities, for example, weapon and telescope cross hairs and other optical instruments [2].

Silks belong to a group of high molecular weight organic polymers characterized by repetitive hydrophobic and hydrophilic peptide sequences [3]. Silks are fibrous protein polymers that are spun into fibers by some arthropods such as silkworms, spiders, scorpions, mites, and fleas [3, 4]. There are thousands of silk-spinning insects and spiders, yet only a few have been investigated in detail. Silks differ in composition, structure, and properties depending on their specific source and function [3, 5, 6]. They are naturally produced by spiders or insects, such as *Nephila clavipes* and *B. mori*, respectively [7, 8]. Silkworm fibers are classified as domestic and wild silk. Wild silks are produced by caterpillars other than the mulberry silkworm, and they differ from the domesticated varieties in color, size, and texture. The cocoons gathered in the wild usually have been damaged by the emerging moth before the cocoons are gathered, so the silk thread that makes up the cocoon has been torn into shorter lengths. Domestic silkworms like *B. mori* are commercially reared, and the pupae are killed by dipping them in boiling water before the adult moths emerge, allowing the whole cocoon to be unraveled as one continuous thread. There are other commercially exploited silkworms other than *B. mori* that can be used to rear the silk known as non-mulberry silkworms like *Antheraea mylitta* (tasar),

Philosamia ricini (eri), and *Antheraea assama* (muga). Silks are also produced by spiders and insects that secret glycine-rich silks characterized by their unique synthesis and processing features, as well as strength and extensibility.

The silk filament contains 72–81% fibroin, 19–28% sericin, 0.8–1.0% fat and wax, and 1.0–1.4% coloring matter and ash of the total weight. The silkworm extrudes the liquid fiber from the two excretory canals of sericteries that unite in the spinneret in its head, each of them termed as brin. The two brins cemented together in the spinneret by sericin become a single continuous fiber called the bave or filament. The seric bave is thus made by the union of two brins held together by sericin [9]. Fibroin is a valuable protein along with sericin, which acts as a glue to fix fibroin fiber together in a cocoon. Sericin and fibroin protein is useful because of its properties and has been found to possess various biological functions.

The molecular weight of sericin ranges from 10 to 310 kDa, and that of fibroin ranges from 300 to 450 kDa [10, 11], but silk produced by spiders is mechanically superior to any insect silk. However, spider silk differs in many aspects from those produced by mulberry silkworms. The fibroin is a protein dominated in composition by the amino acids glycine, alanine, and serine, which form antiparallel β sheets in the spun fibers [12–14]. But only the silk of moth caterpillars has been used widely for textile manufacture and other applications. The main difference between spider and mulberry silkworm silks is that dragline spider silks are mechanically superior, but better water resistance and availability make silkworm silks more prominent. The choice of silk for a particular application must therefore include careful consideration of the desired properties.

Genetically, silks are characterized by a combination of highly repetitive primary sequences that lead to significance in secondary structure, which provides unique mechanical properties. Because of these impressive properties, combined with their biocompatibility and relative environmental stability, silks have been used as an important set of material options in the fields of controlled release and for biomaterials, in addition to their uses as textile materials [15].

The structure of silk is shown in Figure 3.1. Silk of *B. mori* is composed of the proteins fibroin and sericin and matters such as fats, wax, sand pigments, and minerals. Fibroin in the *B. mori* comprises a high content of the amino acids glycine and alanine, 42.8 and 32.4 g, respectively, as shown in Table 3.1. The key amino acids in sericin are serine (30.1 g), threonine (8.5 g), aspartic acid (16.8 g), and glutamic acid (10.1 g).

Table 3.1 shows that there are different proportions of amino acid residues in fibroin. Fibroin has high proportion of alanine, glycine, and serine. A small amount of cystine residues give a very small amount of sulfur in the fiber. Fibroin contains only a small amount of amino acid side chains. The amounts of acid and alkali that can be absorbed by silk are relatively lower than those absorbed by wool (c. 0.2 equiv/kg of silk). The isoelectric point of silk fiber is around pH 5. There is a low proportion of amino acid residue with large side chain. In silk hydrogen bonding is important in fibroin. An X-ray structure analysis of the crystalline domain of fibroin shows that the peptide chains pack in fully extended forms [16] (Figure 3.2).

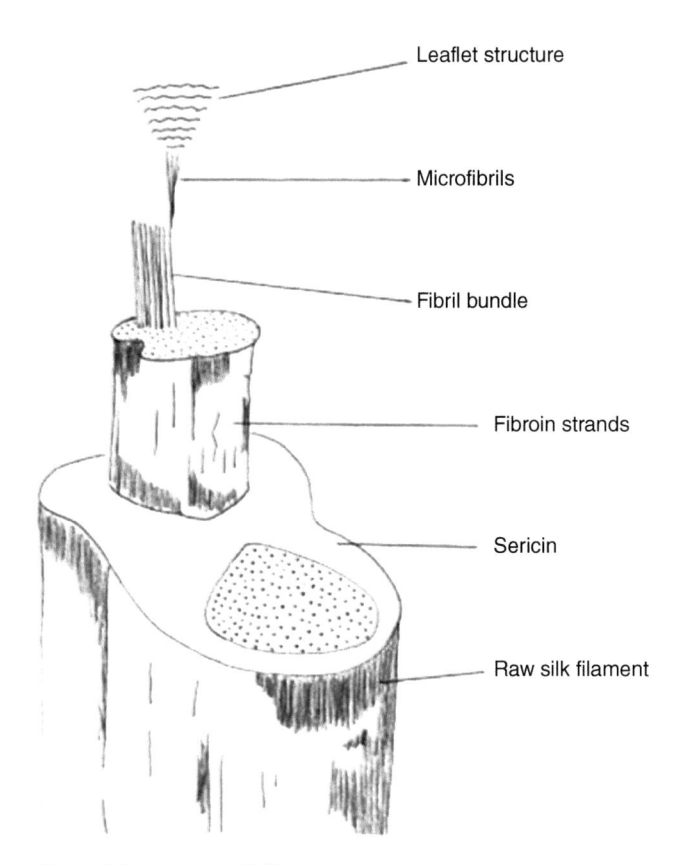

Figure 3.1 Structure of silk.

Table 3.1 Amino acid composition of fibroin and sericin.

Amino acids	Fibroin	Sericin	Amino acids	Fibroin	Sericin
Glycine	42.8	8.8	Glutamic acid	1.7	10.1
Alanine	32.4	4.0	Serine	14.7	30.1
Leucine	0.7	0.9	Threonine	1.2	8.5
Isoleucine	0.9	0.6	Phenylalanine	1.2	0.6
Valine	3.0	3.1	Tyrosine	11.8	4.9
Arginine	0.9	4.2	Proline	0.6	0.5
Histidine	0.3	1.4	Methionine	0.2	0.1
Lysine	0.5	5.5	Tryptophan	0.5	0.5
Aspartic acid	1.9	16.8	Cystine	0.1	0.3

Values are given as gram of amino acid per 100 g of protein.
Source: Tanaka et al. 1999 [10].

Figure 3.2 Crystal structure of the polypeptide bond in the silk fibroin. Source: Guljarni 1992 [31]. Reproduced with permission of John Wiley & Sons.

3.2 Types of Silk and Their Importance

Silk is mainly divided into two types:

- Mulberry silk (*B. mori*), also called cultivated silk.
- Non-mulberry silk.

3.2.1 Mulberry

Mulberry silk is also known as cultivated silk and bombyx silk, but mulberry silk is the most commonly used term. It is also sometimes referred to it by its type, which is thrown or reeled silk [17]. Figure 3.3 shows the mulberry worm and the cocoons.

The bulk of the commercial silk produced in the world comes from mulberry silk. Mulberry silk comes from the silkworm *B. mori*, which solely feeds on the leaves of mulberry plant. These silkworms are completely domesticated and reared indoors. In India, the major mulberry silk producing states are Karnataka, Andhra Pradesh, West Bengal, Tamil Nadu, and Jammu Kashmir, which together

Figure 3.3 Mulberry silkworms and cocoons.

Table 3.2 World raw silk production (tonnes).

Country	2002	2003	2004	2005	2006	2007	2008	% Share
China	68 600	94 600	102 560	105 360	130 000	108 420	98 620	81.24
India	16 319	15 742	16 500	17 305	18 475	18 320	18 370	15.13
Japan	394	287	263	150	150	105	95	0.08
Brazil	1607	1563	1512	1285	1387	1220	1177	0.97
Korea Republic	154	150	150	150	150	150	135	0.11
Uzbekistan	1260	950	950	950	950	950	865	0.71
Thailand	1510	1500	1420	1420	1080	760	1100	0.91
Vietnam	2200	750	750	750	750	750	680	0.56
Others	3814	1500	1500	1500	1000	500	350	0.29
Total	95 858	117 042	125 605	128 870	153 942	131 175	121 392	100.00

Figures of India is for financial year April to March.
Source: ISC website www.inserco.org SS:08-09-2009.

accounts for 92% of country's total mulberry raw silk production [17]. World raw silk production is shown in Table 3.2.

3.2.1.1 Types of Mulberry Silk

B. mori, the domesticated silkworm, has been reared for over 2000 years. During this long history, many mutations have occurred. The mutants have been further classified with each other resulting in a combination of various genes producing a large number of silkworm races. The silkworm races are classified on the basis of:

(1) Place of origin
(2) Voltinism
(3) Molting

Place of Origin

Indian Races These races are aboriginal in India and Southeast Asia. The larval stage is longer and they are robust against high temperature and humidity. The size of the cocoon and larvae is small. In many cases, the cocoon is spindle shaped, and the cocoon color is green, yellow, or white. The cocoon shell is thin with less shell percentage. They are mainly multivoltines.

Japanese Races These races are aboriginal in Japan. The larvae are robust. The cocoon is in peanut shape. The sizes of the larvae do not correspond with the long larval duration. The cocoon color is usually white, but few are also green or yellow in color. The ratio of double cocoons is more. The quality of the silk filament is inferior and it is thick and short. They are univoltine or bivoltine.

Chinese Races These races are aboriginal in China. The larvae are robust against high temperature but weak against high humidity. The larvae are plain and active. They voraciously eat mulberry leaves and grow quickly. The cocoon shape is, in many cases, elliptical spherical and spindle shaped in few cases. The cocoon color is white, golden yellow, green, red, or beige. The cocoon filament is fine and reelability is good. They are univoltine, bivoltine, and multivoltine.

European Races These races are aboriginal in Europe and Central Asia. Larval duration is long and they actively eat mulberry leaves. Larvae are weak against high temperature and high humidity. The cocoon size is big with a little constriction. The cocoon reelability is good.

Voltinism According to voltinism, silkworms are classified into (a) univoltines, (b) bivoltines, and (c) multi- or polyvoltines.

Univoltines These races have only one generation in a year. The body size of larvae is large. The cocoon weight, shell weight, shell ratio, and cocoon filament weight are high. The cocoon filament quality is good.

Bivoltines These races have two generations in a year; the silkworms are more uniform and strong. The cocoon weight, cocoon shell weight, shell ratio, and cocoon filament weight are less compared with univoltines. Larvae are robust compared with univoltines.

Multivoltines The life cycle is short. They have many generations in a year. Larvae are robust and can withstand high temperature. The size of cocoons is small. The cocoon weight, shell ratio, and cocoon filament weight are less than bivoltines. Cocoon filament is fine.

Molting Based on molting, silkworms are classified into (a) trimolters, (b) tetramolters, (c) pentamolters, rarely bimolters, and hexamolters. Tetramolters are mainly reared for commercial purposes.

Thai silk is only one of the mulberry silkworm (*B. mori*) silks, but it differs somewhat in appearance and is yellower in color. The filament is coarser and has more silk gum (e.g. up to 37%) than normal mulberry silk (e.g. 20–25%) [11, 18]. These characteristics cause Thai silk to have its own style after weaving. Thai silk products are mainly produced by domestic industries in the northern and north east part of Thailand.

3.2.2 Non-mulberry

A large number of species (400–500) are used in the production of non-mulberry silks, but only about eighty have been commercially exploited in Asia and Africa, chiefly in tribal communities. The major varieties of non-mulberry silk are described below [1].

3.2.2.1 Tasar

Tasar silk, popularly known as "Dasali Pattu," is a wild silk reeled from tasar cocoons. Tasar (tussah) is a copperish, coarse silk mainly used for furnishings and interiors. It is less lustrous than mulberry silk, but has its own feel and appeal.

Figure 3.4 Tasar silkworm and cocoon.

Tasar silk is generated by the silkworm *A. mylitta*, which mainly thrive on food plants Asan (*Terminalia tomentosa*) and Arjun (*Terminalia arjuna*). The rearing is conducted in nature on the trees in the open areas. In India, tasar silk is mainly produced in the states of Bihar, Jharkhand, Chhattisgarh, and Orissa, besides Maharashtra, West Bengal, and Andhra Pradesh. Tasar culture is the mainstay for many a tribal community in India [1]. Figure 3.4 shows the tasar worm and the cocoons.

3.2.2.2 Oak Tasar
Oak tasar is a finer variety of tasar generated by the silkworm *Antheraea proylei* J in India, which feeds on natural food plants of oak, found in abundance in the sub-Himalayan belt of India, covering the states of Manipur, Himachal Pradesh, Uttar Pradesh, Assam, Meghalaya, and Jammu and Kashmir. China is the major producer of oak tasar in the world, and this comes from another silkworm known as *Antheraea pernyi*.

3.2.2.3 Eri
Eri silk is known as "Ahimsa silk." Also known as endi or errandi, eri is a multivoltine silk spun from open-ended cocoons, unlike other varieties of silk. Eri silk is the product of the domesticated silkworm *P. ricini*, which mainly thrive on food plants castor and tapioca, kessera, papaya, etc. Eri cocoon is open mouthed; the filament is discontinuous and thus can be used only for spinning. Like tasar, the cocoon varies in color, size, and softness. The soft cocoons are better for mechanical spinning and slightly hard and bigger cocoons for hand spinning [19]. A picture of the worm and the cocoons is presented in Figure 3.5.

Eri silk has the unique dual characteristics of the softness of silk and the warmth of the wool and holds a lot of promise as a commercial venture. Ericulture is a household activity practiced mainly for protein-rich pupae, a delicacy for the tribal. Resultantly, the eri cocoons are open mouthed and are spun. The silk is used indigenously for preparation of chaddars (wraps) for own use by these tribals. In India, this culture is practiced mainly in the northeastern states and Assam. It is also found in Bihar, West Bengal, and Orissa [18].

The eri silk production in India is 1483 Metric Tonnes per annum mainly with 90% from northeast states. It provides gainful employment to the people with

Figure 3.5 Eri silkworm and cocoons.

Figure 3.6 Muga silkworm and cocoons.

emphasis on women folk in rural areas where poverty and unemployment are concentrated.

3.2.2.4 Muga
This golden yellow-colored silk is prerogative of India and the pride of Assam state. It is obtained from semidomesticated multivoltine silkworm *Antheraea assamensis*. These silkworms feed on the aromatic leaves of Som and Soalu plants and are reared on trees similar to that of tasar. Muga culture is specific to the state of Assam and an integral tradition and culture of that state. The muga silk and high value product is used in products like sarees, mekhalas, chaddars, etc. Figure 3.6 shows the muga silkworm and the cocoons.

3.2.2.5 Anaphe Silk
This silk of southern and central Africa is produced by silkworms of the genus *Anaphe*: *A. moloneyi* Druce, *A. panda* Boisduval, *A. reticulata* Walker, *A. carteri* Walsingham, *A. venata* Butler, and *A. infracta* Walsingham. They spin cocoons in communes, all enclosed by a thin layer of silk. The tribal people collect them from the forest and spin the fluff into a raw silk that is soft and fairly lustrous. The silk obtained from *A. infracta* is known locally as "book," and that from *A. moloneyi* as "tissnian-tsamia" and "koko." The fabric is elastic and stronger than that of mulberry silk. Anaphe silk is used in velvet and plush.

3.2.2.6 Fagara Silk

Fagara is obtained from the giant silk moth *Attacus atlas* L. and a few other related species or races inhabiting the Indo-Australian biogeographic region, China, and the Sudan. They spin light-brown cocoons nearly 6 cm long with peduncles of varying lengths (2–10 cm).

3.2.2.7 Coan Silk

The larvae of *Pachypasa otus* D., from the Mediterranean biogeographic region (southern Italy, Greece, Romania, Turkey, etc.), feed primarily on trees such as pine, ash, cypress, juniper, and oak. They spin white cocoons measuring about 8.9 × 7.6 cm. In ancient times this silk was used to make the crimson-dyed apparel worn by the dignitaries of Rome; however, commercial production came to an end long ago because of the limited output and the emergence of superior varieties of silk.

3.2.2.8 Mussel Silk

Where the non-mulberry silks previously described are of insect origin, mussel silk is obtained from a bivalve, *Pinna squamosa*, found in the shallow waters along the Italian and Dalmatian shores of the Adriatic. The strong brown filament or byssus is combed and then spun into a silk popularly known as "fish wool." Its production is largely confined to Taranto, Italy.

3.2.2.9 Spider Silk

Spider silk is another non-insect variety. It is not only soft and fine but also strong and elastic. The commercial production of this silk comes from certain Madagascan species, including *Nephila madagascariensis*, *Miranda aurantia*, and *Eperia*. The spinning tubes (spinerules) are in the fourth and fifth abdominal part to a frame, from which the accumulated fiber is reeled out four or five times a month.

Because of the high cost of production, spider silk is not used in the textile industry; however, durability, resistance to extremes of temperature, and humidity make it indispensable for cross hairs in optical instruments [1].

3.2.3 Fine Structure of Silk

Silk fibers (*B. mori*) spun out from silkworm cocoons consist of fibroin in the inner layer and sericin in the outer layer. Each raw silk thread has a lengthwise striation, consisting of two fibroin filaments of 10–14 m each embedded in sericin. The chemical compositions are, in general, silk fibroin of 75–83%, sericin of 17–25%, waxes of about 1.5%, and others of about 1.0% by weight. Silk fibers are biodegradable and highly crystalline with well-aligned structure. It has been known that they also have higher tensile strength than glass fiber or synthetic organic fibers, good elasticity, and excellent resilience. Silk fiber is normally stable up to 140 °C, and the thermal decomposition temperature is greater than 1500 °C. The densities of silk fibers are in the range of 1320–1400 kg/m^3 with sericin and 1300–1380 kg/m^3 without sericin. Silk fibers are also commercially available in a continuous fiber type.

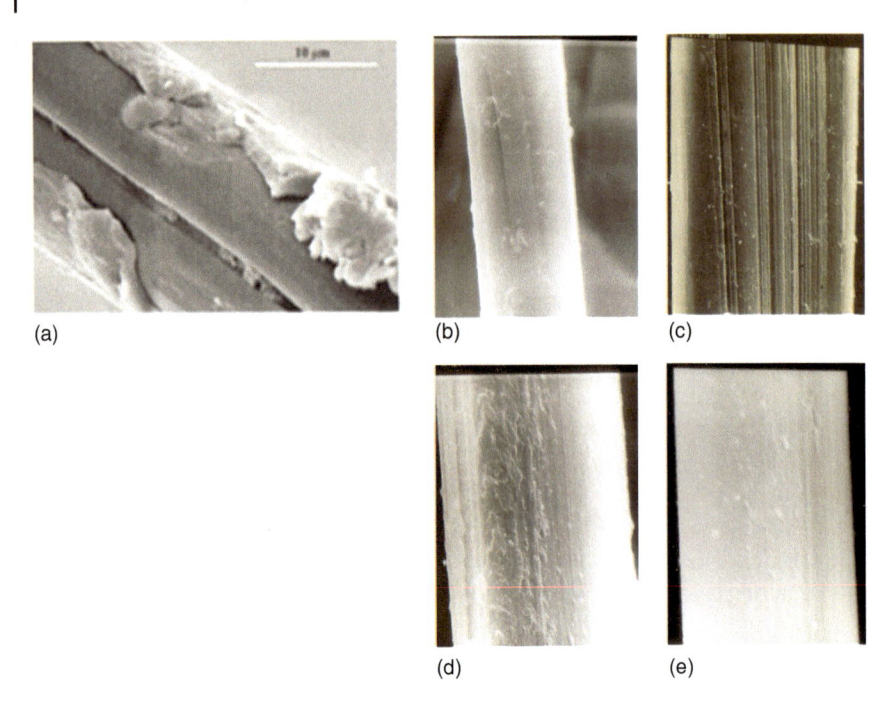

Figure 3.7 (a) Longitudinal view of silk fibers (undegummed). Longitudinal view of silk fibers (degummed). (b) Mulberry. (c) Tasar. (d) Muga. (e) Eri.

3.2.3.1 Longitudinal View

Scanning electron micrographs of longitudinal views of undegummed and degummed silk fibers are presented in Figure 3.7a,b, respectively.

It may be observed that mulberry silk shows a more or less smooth surface (Figure 3.7b), whereas non-mulberry silks such as tasar, muga, and eri (Figure 3.7c–e) all have striations on their surface compared with mulberry.

3.2.3.2 Cross-Sectional View

Scanning electron micrographs of cross section of silk fibers are presented in Figure 3.8. The cross section of silk fiber, which is made up of two types of protein, namely, sericin and fibroin, is shown in figure. It is found that two strands of fibroin filaments are enveloped by nonfibrous sericin. When a strand of fibroin filament is enlarged for its inner structure, it appears like a bundle of fibrils in which a large number of fibrils are accumulated [20].

There are variations depending upon the variety of silkworms and also among the individual cocoons. It may be observed that, in this respect, the mulberry and non-mulberry silks exhibit an altogether different cross-sectional morphology. The mulberry silks show a more or less triangular cross section and a smooth surface (Figure 3.8a). Among the non-mulberry varieties, tasar and muga exhibit an elongated rectangular- or a wedge-shaped cross section and a large cross-sectional area (Figure 3.8b,c). The eri silk has a more or less triangular shape (Figure 3.8d).

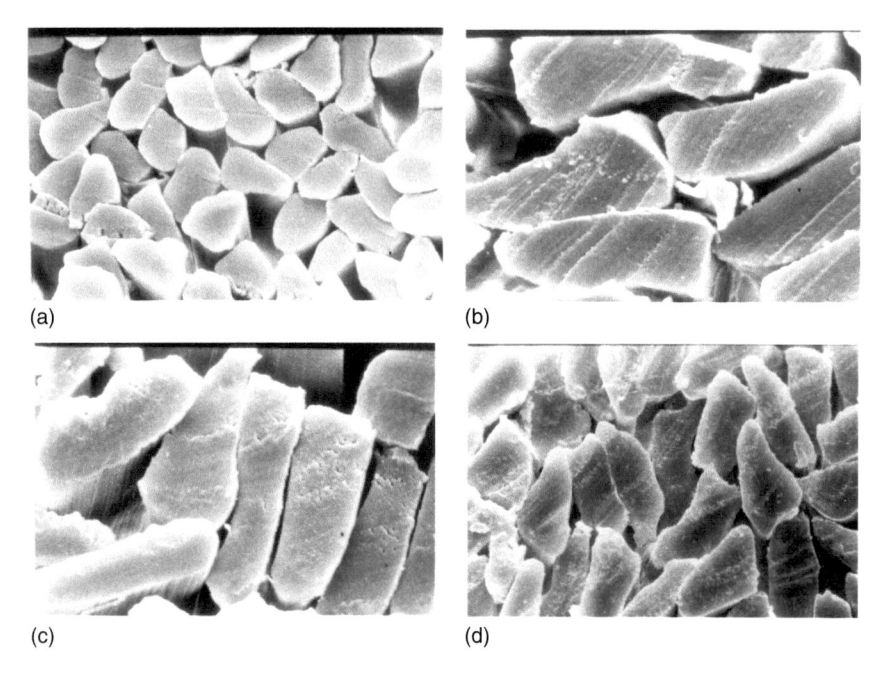

Figure 3.8 Cross-sectional view of silk fibers (degummed). (a) Mulberry. (b) Tasar. (c) Muga. (d) Eri.

Usually in the case of the cocoon fibers of domestic silkworms like mulberry and eri, the cross section is irregular ranging from triangular shape to circular shape. Moreover, even in the same fibroin filament, there are variations in the cross-sectional area depending upon the level of the cocoon layer.

3.2.4 Amino Acid Composition

The amino acid composition varies in different varieties of silk. Three major amino acids such as serine, glycine, and alanine may be found in mulberry and non-mulberry varieties. Among the other major amino acids present are tyrosine and valine. In general, in mulberry silks, glycine, alanine, and serine together constitute about 82%, of which about 10% is serine. Tyrosine and valine may be considered next to these at about 5.5% and 2.5%, respectively. The overall composition of acidic amino groups (i.e. aspartic and glutamic acids) in the mulberry variety is greater than that of the basic amino acids. The other important aspect is the composition of amino acids with bulkier side groups. The presence of bulky side groups can hamper close packing of molecules and hinder crystallization process. In general, a large portion of the mulberry fibroin is made up of simple amino acids such as glycine and alanine, suggesting a favorable condition for crystallization [21].

Compared with the mulberry silks, the total amount of glycine, alanine, and serine constitute about 73% in the non-mulberry variety, less by about 10%. All the non-mulberry silks exhibit a high proportion of alanine compared with that in the mulberry variety. The proportion of alanine is about 34% in tasar, 36% in eri,

Table 3.3 Amino acid composition of silk fibers.

Amino acid	Amino acid composition (mol%)			
	Bombyx mori (mulberry)	Antheraea mylitta (tasar)	Antheraea assama (muga)	Philosamia ricini (eri)
Aspartic acid	1.64	6.12	4.97	3.89
Glutamic acid	1.77	1.27	1.36	1.31
Serine	10.38	9.87	6.11	8.89
Glycine	43.45	27.65	28.41	29.35
Histidine	0.13	0.78	0.72	0.75
Arginine	1.13	4.99	4.72	4.12
Threonine	0.92	0.26	0.21	0.18
Alanine	27.56	34.12	34.72	36.33
Proline	0.79	2.21	2.18	2.07
Tyrosine	5.58	6.82	5.12	5.84
Valine	2.37	1.72	1.5	1.32
Methionine	0.19	0.28	0.32	0.34
Cystine	0.13	0.15	0.12	0.11
Isoleucine	0.75	0.61	0.51	0.45
Leucine	0.73	0.78	0.71	0.69
Phenylalanine	0.14	0.34	0.28	0.23
Tryptophan	0.73	1.26	2.18	1.68
Lysine	0.23	0.17	0.24	0.23

and 35% in muga. This value is consistent but is lower than that particularly for muga (~44%). On the other hand, the glycine content in these varieties is about 27–29%, which is lower than that found in the mulberry varieties (~43%).

In addition, the non-mulberry varieties have a substantial proportion of amino acids with bulky side groups, especially aspartic acid (4–6%) and arginine (4–5%), which means that not only the acidic but also basic amino acid levels are greater. It is interesting to note the presence of sulfur-containing amino acids (i.e. cystine and methionine) in all the varieties of silk. Methionine content in non-mulberry silks is slightly higher (0.28–0.34%) compared with that found in mulberry varieties (0.11–0.19%), whereas the cystine content is comparable [21]. Amino acid composition of different varieties of silk is presented in Table 3.3.

3.2.5 Properties of Silk Fibers

3.2.5.1 Tensile Properties
The tensile properties of different varieties of silks in terms of tenacity, elongation at break, and initial modulus have been determined by a number of workers [22–26].

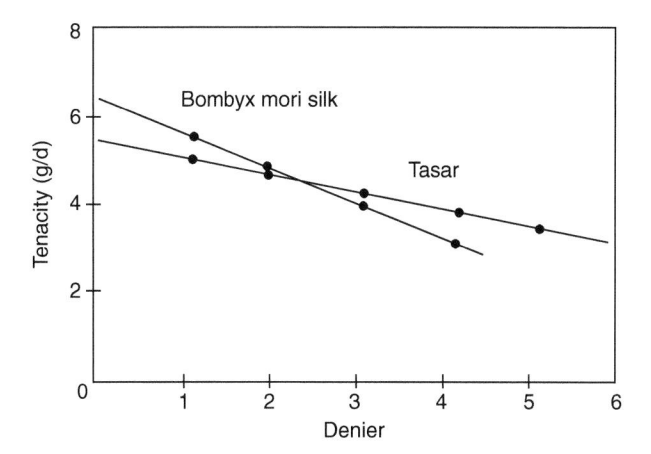

Figure 3.9 Tenacity vs. denier thread relationship.

Studies conducted on some mulberry and non-mulberry varieties by E. Iizuka et al. reveal that the tenacity, elongation at break, and modulus are all dependent on the linear density of the filament and the linear density or the mean size in turn depends on the silkworm race. The tenacity is found to be linearly related to the linear density of the filament (Figure 3.9). The correlation is negative, i.e. as the linear density increases, tenacity decreases. Similar trend has been observed for modulus too. Elongation on the other hand increases with an increase in linear density.

The tenacity ranges from 2.5 to 4.82 g/d for Japanese and Chinese mulberry varieties, 2.4–4.32 g/d for Indian mulberry varieties, and 3.74–4.6 g/d for Indian tasar varieties [25]. In a study on chemical structure and physical properties of *A. assama* (muga) silk, it has been reported that the tenacity of muga varies from 3.2 to 4.95 g/d [22, 24]. Another important non-mulberry variety, eri, showed lowest tenacity value, ranging from 2.3 to 4.0 g/d [26].

Elongation at break, on the other hand, showed a higher value for all the non-mulberry silks compared with mulberry varieties. The values range from 31% to 35% for tasar, 34% to 35% for muga, and 29% to 34% for eri silks, respectively. The elongation values for mulberry varieties ranged from 19% to 24%. Some of the mechanical properties of different varieties of silk are summarized in Table 3.4.

3.2.5.2 Optical Properties

Silk fibroin extracted from silkworm cocoons is a unique biopolymer that combines biocompatibility and implantability, along with excellent optical properties. Silk may be used as an optical material for applications in biomedical engineering, photonics, and nanophotonics. Silk can be nanopatterned with features smaller than 20 nm. This allows manufacturing of structures such as (among others) holographic gratings, phase masks, beam diffusers, and photonic crystals out of a pure protein film. The properties of silk allow these devices to be "biologically activated" offering new opportunities for sensing and biophotonic components.

Table 3.4 Mechanical properties of different varieties of silk.

Variety	Sex	Dynamic modulus (10^{10} dyn/cm^2)	tan δ	Tenacity (g/d)	Elongation (%)
Shunreix shougetsu (mulberry)	M	1.847	—	5.265	20.36
	F	1.808	—	5.207	21.48
A. mylitta (tropical tasar)	M	1.132	0.030	3.412	31.36
	F	1.087	0.035	3.256	31.12
A. proylei (temp. tasar)	M	1.305	0.023	4.123	31.45
	F	1.087	0.025	4.128	31.48
A. assama (muga)	M	1.205	0.020	3.170	34.83
	F	1.230	0.023	3.823	34.10

Many interesting bio-optical devices can be fabricated by doping silk films with fluorescent materials (such as quantum dots shown above). Ways can be explored to enhance the light emission by patterning the silk film surfaces, as well as making tunable wavelength devices and printing specific patterns on silk film surfaces.

Luster associated with silk is due partly to the influence on the pattern of light – reflection of its triangular shape. In an attempt to understand the optical properties of silk, many researchers have determined the refractive index and birefringence of fibers. The refractive index of silk generally varies through its cross section. The birefringence (n) value varies from 0.051 to 0.0539, mulberry silk and 0.030 to 0.034 for non-mulberry silks [27].

3.2.5.3 Viscoelastic Behavior

Silk fiber exhibits viscoelastic behavior. Time-dependent mechanical properties of silk fibers such as stress relaxation, creep, creep recovery, etc. have also been the subject of interest. Creep is a phenomenon associated with time-dependent extension under an applied load. The complementary effect is stress relaxation under a constant extension.

Creep and stress relaxation behavior of silk has been reported [28]. The instantaneous extension and secondary creep are both higher for tasar silk compared with those for mulberry silk. The stress relaxation was also found to be more in non-mulberry silks than in mulberry silk.

Silk has also been shown to exhibit inverse stress relaxation phenomenon [28]. The inverse relaxation could be observed for both mulberry and tasar silks when the level of strain was maintained below a certain value. Inverse relaxation becomes higher with the increase in peak tension. Cyclic loading has been found to reduce the extent of inverse relaxation.

3.2.6 Applications

Silk is one of the most beautiful fabrics available, with a long and colorful history and changing applications in the world today. Be it for gowns, medical use, home decor, and more, the use of silk is a wide and varied topic.

3.2.6.1 Textile and Apparels

Silk's good absorbency makes it comfortable to wear in warm weather and while active. Its low conductivity keeps warm air close to the skin during cold weather. It is often used for clothing such as shirts, blouses, formal dresses, high fashion clothes, negligees, pajamas, robes, skirt suits, sundresses, and kimonos. Silk's elegant, soft luster and beautiful drape makes it perfect for many furnishing applications. It is used for upholstery, wall coverings, window treatments (if blended with another fiber), rugs, bedding, and wall hangings.

Silk has several uses in daily life. It is a common cloth used for high-end garments, including wedding gowns and blouses. It appears frequently in accessories like handbags and headbands and scarves. Because of its luster and texture, it also is commonly used for home decoration, especially as sheets, bedding, curtains, and cushions. It is also used for warmth and skiing garments because of its remarkable ability to retain heat; paradoxically, it also is a popular summer cloth because it keeps the wearer cool in warmer weather.

Silk continues to be used as a material to produce fine dresses. Chinese women continue to have their wedding cheongsam dresses made out of silk. They choose this material because it is one of the finest materials known in ancient Chinese culture. Delicately woven dragons, flowers, and butterflies are sewn into the silk dresses. The material is thick so it allows a woman to look leaner and have cleaner lines. Additionally, the shininess of the material is very flattering and alluring. Women's evening gowns are also made of silk. It also falls well and is a material that is slightly warmer allowing the women's body to feel a bit of warmth, even as she wears a sleeveless gown in the winter.

Silk shirts remain popular among women. Especially for women's blouses, silk is a great material. It is seen as more classic and flattering since it drapes a woman's body. Additionally, the thinness of the material allows women to easily tuck the shirt into their skirts or wear a jacket on top without having the jacket look bulky. For men, silk shirts are becoming more and more popular. Men have started wearing silk short sleeve shirts as well as long sleeve shirts. Men used to only wear silk shirts for the evening, but the lightness of the material has made it a popular alternative to cotton.

Silk is also used very frequently for pajamas. Silk pajamas feel soft and smooth and very comfortable while sleeping. Both men's and women's pajamas are now made of this fine thread. For women, silk is often used to make lingerie as well. Using this fine thread has become a great option since it is smooth, tends not to catch on your hair, and is very breathable for the body.

Silk scarves have also been around for a long time. The number of gorgeous patterns and designs on the scarves has made these items very expensive. Women tend to tie the scarf around their necks for warmth and for style. Men tend to carry one in their pocket as an alternative to the cotton handkerchief. Be prepared to pay more for a silk scarf with a fancy design.

More and more, manufacturers of bedding have started to make silk sheets and silk pillowcases. The health benefits of silk are starting to be more widely known. Sleeping on them helps prevent coughing and sneezing, especially for those allergic to dust mites who do not like silk. Additionally, sleeping on silk sheets is sometimes beneficial to women's hair, helping it to have less tangles and less split ends. Beyond the solid color sheets, many makers of luxurious bedding are now offering their own version of sheets with lots of different patterns and designs in stores.

Silk has come a long way since the Silk Road. It is still a highly sought-after material. Having pieces of fine clothing or a scarf or even a silk sheet is often seen as the ultimate extravagance.

3.2.6.2 Biomedical Field

Silk, and especially *B. mori* silk, has a very long history in biomedical applications. In recent years, the reported exceptional nature of silk leads to increased interest in silk for biomedical applications [29]. Silk fibroin has been increasingly studied for new biomedical applications due to the biocompatibility, slow degradability, and remarkable mechanical properties of the material. In addition, the ability to now control molecular structure and morphology through versatile processability and surface modification options has expanded the utility for this protein in a range of biomaterial and tissue engineering applications. Silk fibroin in various formats (films, fibers, nets, meshes, membranes, yarns, and sponges) has been shown to support stem cell adhesion, proliferation, and differentiation in vitro and promote tissue repair in vivo. In particular, stem cell-based tissue engineering using 3D silk fibroin scaffolds has expanded the use of silk-based biomaterials as promising scaffolds for engineering a range of skeletal tissues like bone, ligament, and cartilage, as well as connective tissues like skin [30]. More recent studies with well-defined silkworm silk fibers and films suggest that the core silk fibroin fibers exhibit comparable biocompatibility in vitro and in vivo with other commonly used biomaterials such as polylactic acid and collagen [3].

Silk from the silkworm *B. mori* has been used as biomedical suture material for centuries. The unique mechanical properties of these fibers provide important clinical repair options. Silk has been used in native fiber form as sutures for wound ligation and became the most common natural suture, surpassing collagen over the past 100 years [3]. Silk sutures are used in ocular, neural, and cardiovascular surgeries, as well as a variety of other tissues in the body. Silk's knot strength, handling characteristics, and the ability to lie low to the tissue surface make it a popular suture in cardiovascular applications. The above features of silk fibroin have led to the recent emergence of silk-based biomaterials for a wide range of cell and tissue studies such as scaffolds, films, sponges, hydrogels, bone materials, cardio vascular devices, etc.

3.2.6.3 Fiber-Reinforced Composites

Silk is a fiber with remarkable mechanical properties. This unique characteristic of silk has led its use in fiber-reinforced composites for various applications. Silk yarn is easily available as the waste product of textile industry, so the composite is cost effective and perfect utilization of waste product. Though the silk is extensively used as a valuable material for textile purposes, in recent years it is being widely considered as a reinforcing material for composites made from epoxy and other biodegradable biopolymeric resins. The organization of silk fibers can contribute significantly to the impact resistance by ensuring either or both a sufficient strength of the composite and a good deformability of the composite.

3.3 Future Trends

With improved analytical techniques, together with the tools of biotechnology, a new generation of silk-related materials is envisioned that depart from traditional textiles and medical sutures, which are currently the focus for these materials. These proteins are already finding broadened applications in medical fields as biomaterials. The unique and remarkable mechanical properties of silks have already prompted studies for their application in high-performance materials, as well as modes to mimic these features through more traditional synthetic organic polymer chemistry routes. Since the specifics of the silks can be modulated through genetic manipulation, through processing and through choice of starting protein, a great deal of control of structure, morphology, and functional attributes in materials derived from these proteins can be obtained. This control is already exploited in the biomaterials arena and suggests that this family of novel proteins can serve as an important blueprint for understanding structure–function relationships as well as for new and novel materials. A number of consumer products are already promulgated based on silks, including cosmetics, hair replacements, and shampoos, among others. Sutures, biomaterials for tissue repairs, wound coatings, artificial tendons, bone repair, and related needs may be possible applications since silks are biocompatible and slowly degrade in the human body. Genetically engineered silks have also been commercialized as cell culture plate coatings to improve cell adhesion. Genetic variants of silks are also actively pursued as controlled-release systems to deliver pharmaceuticals in a variety of systems.

New trends in production and applications of silk have to be deployed for maintaining a sustainable market for silk. As a result researchers in Japan have developed fluorescent silks. The world's first silks exhibiting fluorescence and other pioneering properties have been successfully developed as a result of transgenic silkworm research conducted by Japanese researchers. Fluorescent silk thread is made by pulling threads from cocoons made by silkmoth larvae (silkworms). Researchers have now developed three lines of transgenic silkworms. The first line produces silk threads that emit green, red, or orange fluorescent light. These threads were created by introducing into silkworm egg genes that promote the generation of fluorescent proteins. It has been possible to achieve green fluorescence using genes extracted from jellyfish, a technique developed by Nobel Prize-winning chemist Osamu Shimomura, and red and orange fluorescence with genes extracted from coral, a technique that has already been used in commercial applications. The fluorescent silk threads have great potential for use in the fashion industry, and there is expected to be considerable demand for them from producers of high-end apparel.

Though the traditional practices make silk to construct only textiles, the new approach extends its application toward nutritional, cosmetic, pharmaceutical, biomaterial, biomedical and bioengineering, automobile, house building, and art craft applications. The inclination rightly suits to silk due to faster production rate and increasing global demand for its variable eco-friendly composites and viable contributing impact on value, employment, and environmental safety. This move, however, requires more of awareness among stakeholders, trainings, and idea exchange between entrepreneurs, besides service accessibility to consumers. The celebration of International Year of Natural Fibres 2009 will create awareness on insect-based natural fiber, silk for its continued use, and innovative research during the year and beyond.

3.4 Summary

Although substitution of natural materials has been underway for several decades, natural silks remain commercially important because of their unique properties, as well as their relative environmental stability and biocompatibility and consumer preferences. The industries should look at the utilization of silk fiber in total for innovative marketable products of modern society's application and appreciation, which facilitate its all-round development. Though, traditionally, silk is used for making textiles, the new approaches must be attempted to extend the application of silk in nutritional, cosmetic, pharmaceutical, biomedical and bioengineering, automobile, house building, and art craft products. During the past decade a great deal of progress has been made in understanding silk genetic and protein structures. On the other hand, materials scientists have long been fascinated by structure–function relationships in silk proteins. Since the exploration of biomaterial applications for silks, aside from sutures, is only a relatively recent advance, the future for this family of structural proteins to impact clinical needs appears promising.

References

1 Jolly, M.S., Sen, S.K., Sonwalker, T.N., and Prasad, G.K. (1979). Non-mulberry silks. In: *Manual on Sericulture* (eds. G. Rangaswami, M.N. Narasimhanna, K. Kashivishwanathan, et al.), 1–178. Rome: Food and Agriculture Organization of the United Nations.
2 Spring, C. and Hudson, J. (2002). *Silk in Africa*. Seattle, WA: University of Washington Press.
3 Altman, G.H., Diaz, F., Jakuba, C. et al. (2003). Silk-based biomaterials. *Biomaterials* 24: 401–416.
4 Craig, C.L. (1997). Evolution of arthropod silks. *Annual Review of Entomology* 42: 231–267.
5 Craig, C.L., Hsu, M., Kaplan, D., and Pierce, N.E. (1999). A comparison of the composition of silk proteins produced by spiders and insects. *International Journal of Biological Macromolecules* 24: 109–181.
6 Sheu, H.S., Phyu, K.W., Jean, Y.C. et al. (2004). Lattice deformation and thermal stability of crystals in spider silk. *International Journal of Biological Macromolecules* 34: 325–331.
7 Becker, N., Oroudjev, E., Mutz, S. et al. (2003). Molecular nanosprings in spider capture-silk threads. *Natural Materials* 2: 278–283.
8 Bell, F.I., McEwen, I.J., and Viney, C. (2002). Fibre science: supercontraction stress in wet spider dragline. *Nature* 416: 37.

9 Carboni, P. (1952). *Silk, Biology, Chemistry and Technology*. London: Chapman & Hall Ltd.
10 Tanaka, K., Kajiyama, N., Ishikura, K. et al. (1999). Determination of the site of disulfide linkage between heavy and light chains of silk fibroin produced by Bombyx mori. *Biochimica et Biophysica Acta (BBA) - Protein Structure and Molecular Enzymology* 1432: 92–103.
11 Zhou, C.Z., Confalonieri, F., Medina, N. et al. (2000). Fine organization of Bombyx mori fibroin heavy chain gene. *Nucleic Acids Research* 28: 2413–2419.
12 Asakura, T. and Kaplan, D.L. (1994). Silk production and processing. In: *Encyclopedia of Agricultural Science*, vol. 4 (eds. C.J. Arntzen and E.M. Ritter), 1–11. New York: Academic Press.
13 Asakura, T., Yao, J., Yamane, T. et al. (2002). Heterogeneous structure of silk fibers from *Bombyx mori* resolved by ^{13}C solid-state NMR spectroscopy. *Journal of the American Chemical Society* 124: 8794–8795.
14 He, S.J., Valluzzi, R., and Gido, S.P. (1999). Silk I structure in *Bombyx mori* silk foams. *International Journal of Biological Macromolecules* 24: 187–195.
15 Yao, J. and Asakura, T. (2004). Silks. In: *Encyclopedia of Biomaterials and Biomedical Engineering*, 1363–1370. Marcel Dekker Inc.
16 Sonthisombat, A. and Speakman, P.T. (2004). *Silk: Queen of Fibres - The Concise Story by Speakman*. Department of Textile Engineering, Faculty of Engineering, Rajamangala University of Technology Thanyaburi (RMUTT).
17 Mahadevappa, D., Halliyal, V.G., Shankar, D.G., and Bhandiwad, R. (2001). *Mulberry Silk Reeling Technology*. India: Oxford and IBH Publication Co.
18 Dhavalikar, R.S. (1962). Amino acid composition of Indian silk fibroins and sericins. *Journal of Science and Industrial Research* 21(C): 261–263.
19 Sonwalker, T.N. (1969). Investigations on degumming loss and spinning performance of pierced and cut cocoons in silk worm, *Bombyx Mori-L. Indian Journal of Sericulture* 7 (1): 61.
20 Minagawa, M. (2000). *Structure of Silk Yarn*, vol. I (ed. N. Hojo), 185–208. New Delhi: Oxford & IBH Publishing Co. Pvt. Ltd.
21 Sen, K. and Murugesh Babu, K. (2004). Studies on Indian silk. I. Macrocharacterization and analysis of amino acid composition. *Journal of Applied Polymer Science* 92: 1080–1097.
22 Freddi, G., Gotoh, Y., Mori, T. et al. (1994). Chemical structure and physical properties of *antheraea assama* silk. *Journal of Applied Polymer Science* 52: 775–781.
23 Iizuka, E. (1994). Physical properties of antheraea silks. *International Journal of Wild Silkmoth and Silk* 1 (2): 143–146.
24 Iizuka, E., Okachi, Y., Shimizu, M. et al. (1993). Chemo and bioassay of high yielding triploids and diploids of mulberry (*Morus* Spp.). *Indian Journal of Sericulture* 32 (2): 175–183.
25 Iizuka, E. (1995). Size dependency of the physical properties of *Bombyx* silk. *Journal of Sericultural Science of Japan* 65 (2): 102–108.
26 Iizuka, E. and Itoh, H. (1997). Physical properties of eri **silk**. *International Journal of Wild Silkmoth and Silk* 3: 37–42.

27 Tsukada, M., Freddi, G., Minoura, N., and Allara, G. (1994). Preparation and application of porous silk fibroin materials. *Journal of Applied Polymer Science* 54: 507–514.
28 Das, S. (1996). Studies on tasar silk. PhD thesis. IIT, Delhi.
29 Hakimi, O., Knight, D.P., Vollrath, F., and Vadgama, P. (2007). Spider and mulberry silkworm silks as compatible biomaterials. *Composites Part B: Engineering* 38: 324–337.
30 Wang, Y., Kim, H.-J., Vunjak-Novakovic, G., and Kaplan, D.L. (2006). Stem cell based tissue engineering with silk biomaterials. *Biomaterials* 27: 6064–6082.
31 Guljarni, M.L. (1992). Degumming of Silk. *Review of Progress in Colouration.* 22: 79–89.

4

Cellulose Fibers

Feng Jiang

University of British Columbia, Department of Wood Science, Sustainable Functional Biomaterials Lab, 2424 Main Mall, Vancouver, BC, Canada

4.1 Introduction

Cellulose, as the most abundant polymeric material on Earth with approximately 75 billion tons annual production, originates from varied sources including plants, marine animals (tunicates), bacteria, fungi, algae (*Cladophora* green algae), and protozoans, with plants being the dominant source [1, 2]. The term "cellulose" was first introduced in 1839 by scientists in French Academy of Sciences to describe the solid fibrous substance in plant tissues that is insoluble in acids, ammonia, alcohol, and ether, which was first discovered by the French chemist Anselme Payen in the prior year [3]. Among all fibrous substance, cellulose fiber is one of the earliest materials being utilized in daily life, either as building and clothing materials thousands of year ago or as chemical raw materials for the past 150 years. In fact, cellulose nitrate through esterification reaction with nitric acid has been widely recognized as the first thermoplastic polymer that, when plasticized by camphor, was marketed by Hyatt Manufacturing Company as "celluloid" in 1870 [3, 4].

4.2 Structure and Biosynthesis of Cellulose

Although cellulose has been discovered and utilized as chemical raw material in the nineteenth century, its structure was not completely revealed until 1920s when Hermann Staudinger discovered that small molecules could link with each other via primary valence bonds to form macromolecules [5, 6]. Cellulose, with the molecular formula of $C_6H_{10}O_5$, has been well accepted as a type of linear homopolymer consisting of β-D-glucopyranose repetitively linked together through C4 and anomeric C1 hydroxyl groups via β-1-4 glycosidic linkage, irrespective of its origins [3, 7–9]. In cellulose chain, the glucopyranose unit adopts a $^4C_1$ chair conformation with adjacent unit rotating 180° with each other, forming a cellobiose repeating unit with 10.3 Å distance (Figure 4.1). The linear chain structure is stabilized by two types of intramolecular hydrogen

Handbook of Fibrous Materials, First Edition. Edited by Jinlian Hu, Bipin Kumar, and Jing Lu.
© 2020 Wiley-VCH Verlag GmbH & Co. KGaA. Published 2020 by Wiley-VCH Verlag GmbH & Co. KGaA.

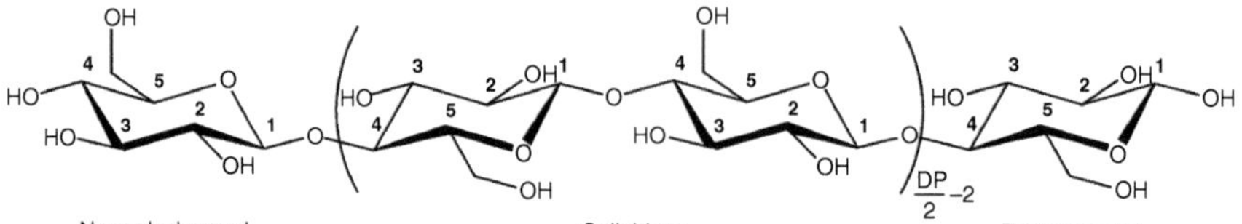

Figure 4.1 Molecular structure of cellulose showing repeating cellobiose unit, reducing and nonreducing ends. Source: Habibi et al. 2010 [2]. Reprinted with permission of American Chemical Society.

bonds, originating from O(3′)-H hydroxyl to O(5) ring oxygen and O(2)-H hydroxyl to O(6′) hydroxyl. The degree of polymerization (DP) of cellulose is highly dependent on the source, varying from 500 to 2000 for wood pulp, 10 000–15 000 for cotton, and as high as 44 000 for *Valonia* [5, 10]. Besides, the DP is also subject to change from chemical treatment, rendering lowered value of 250–500 for regenerated cellulose and 150–300 for acid-hydrolyzed microcrystalline cellulose [3, 11]. Condensation polymerization between C_4 and anomeric C_1 hydroxyls linked glucose unit in a unidirectional fashion that the cellulose chain is chemically asymmetric with one end adopting a reducing hemiacetal group (C_1) and another bearing a nonreducing hydroxyl group (C_4), regardless of the chain length.

Cellulose biosynthesis has been proposed to through multiple subunits cellulose synthases according to Preston's "ordered granule hypothesis" in 1964 [12], which was not confirmed until Brown and Montezinos observed linearly organized terminal complexes (TCs) at the end of cellulose microfibrils in the freeze fracture of plasma membrane from green algae *Oocystis apiculata* in 1976 [13]. These linear TCs are mostly found in varied algae species, such as *Valonia*, *Boergesenia*, *Glaucocystis*, *Erythrocladia*, and *Vaucheria* [14]. From the electron microscopy images of freeze-fracture surface of *Zea mays* plasma membrane, Mueller et al. successfully observed globular TCs composed of subunits associated at the end of cellulose microfibrils [15]. Rosette-shaped TCs were further confirmed, presenting as sixfold symmetrical subunits aggregates in the corn mesocotyl plasma membrane [16]. The rosette TC is predominantly present in all land plants including mosses, ferns, higher plants, and advanced algae [14]. Each rosette TC is composed of six subunits, arranged in hexagonal shape with average diameter of 25–30 nm in the plasma membrane, with each subunit having a diameter of approximately 8 nm (Figure 4.2a) [17, 19, 20]. Cellulose biosynthesis in rosette TC is through cellulose synthase enzyme (CeSA), which polymerizes glucan chains at the nonreducing end (Figure 4.2b). A hexamer of CeSAs is present within one rosette subunit, and with a sixfold subunit rosette, 36 glucan chains could be extruded and bound with each other to form cellulose microfibrils [18, 20]. The width of the microfibrils with 36 glucan chains is around 3.5 nm for higher plant [20], whereas 1200–1400 perfectly packed glucan chains assembled into 28 nm wide single crystal has also been reported in *Valonia macrophysa* [21].

Glucan chains, once being extruded from CeSA via polymerization of UDP-α-glucose, do not present as individual molecules or fold into more thermodynamically stable cellulose II crystalline structure. Instead, the glucan

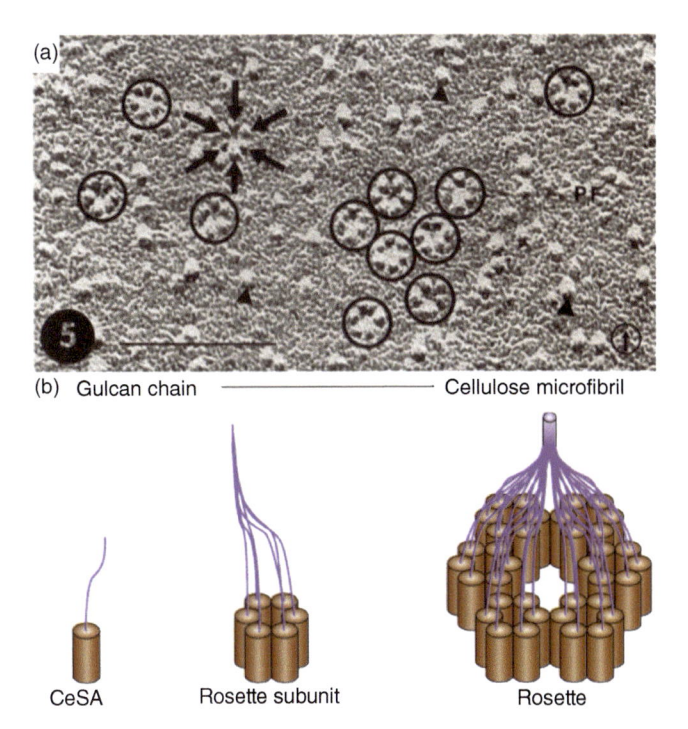

Figure 4.2 (a) Plasmatic fracture face of plasma membrane of a developing xylem vessel elements in root of cress, *Lepidium sativum* L., showing rosette TCs with six subunit (arrows) and irregular clustering (encircled rosettes) (magnification: 430 000×). Source: Herth 1985 [17]. Reprinted with permission of Springer-Verlag. (b) Schematic of cellulose synthesis from rosette TCs. Each rosette TC is comprised of six subunits, each of which contains hexamer of CeSA. A single β-(1,4)-glucan chain is synthesized from each CeSA, which is bound into six-chain fibrils. Within the rosette TC, 36 glucan chains could assemble into cellulose microfibrils. Source: Cosgrove 2005 [18]. Reprinted with permission of JSPP.

chains synthesized within the same TCs tend to crystallize with each other to form cellulose I crystalline structure. The preferential adoption of less stable cellulose I structure over the thermodynamically more stable cellulose II in native cellulose suggests that the crystalline structure could be confined by the synthetic TCs, limiting chain mobility by immediate crystallization following chain polymerization [7]. Depending on the source of cellulose, the native cellulose I crystalline structure includes two types of polymorphs of Iα and Iβ [22, 23]. In general, Iα-rich structure could be found in cellulose secreted by algae [23, 24] and bacteria [23, 25], accounting for as high as 64% of the crystalline polymorphs [25, 26]. However Iβ is predominantly present in higher plant [27] or the exclusive polymorph in tunicate [28]. Cellulose Iα is less thermodynamically stable than cellulose Iβ and could be transformed to Iβ by annealing at high temperature (260 °C) steam [29], NaOH solution [24], or varied organic solvents [30].

Both cellulose Iα and Iβ coexist in most of native cellulose, except in tunicate [31] or hydrothermally annealed *Valonia* cellulose [32] where pure Iβ crystal structure could be observed. Electron diffraction of tunicate or annealed *Valonia* cellulose showed symmetric diffraction spots patterns in the four quadrants, corresponding to a two-chain P2$_1$ monoclinic unit cell for cellulose Iβ [31]. The unit cell of Iβ has dimensions of $a = 0.801$ nm, $b = 0.817$ nm, c (chain axis) $= 1.036$ nm, $\alpha = \beta = 90°$, $\gamma = 97.3°$, cell volume $= 0.6725$ nm^3, and density of 1.599 g/cm^3 (Figure 4.3a,b) [34]. However, electron diffraction of *Valonia* and *Microdictyon tenuous* cellulose showed extra diffraction spots other than the monoclinic unit cells observed in tunicate, asymmetrically distributed in the four quadrants in every layers except for the equator [31].

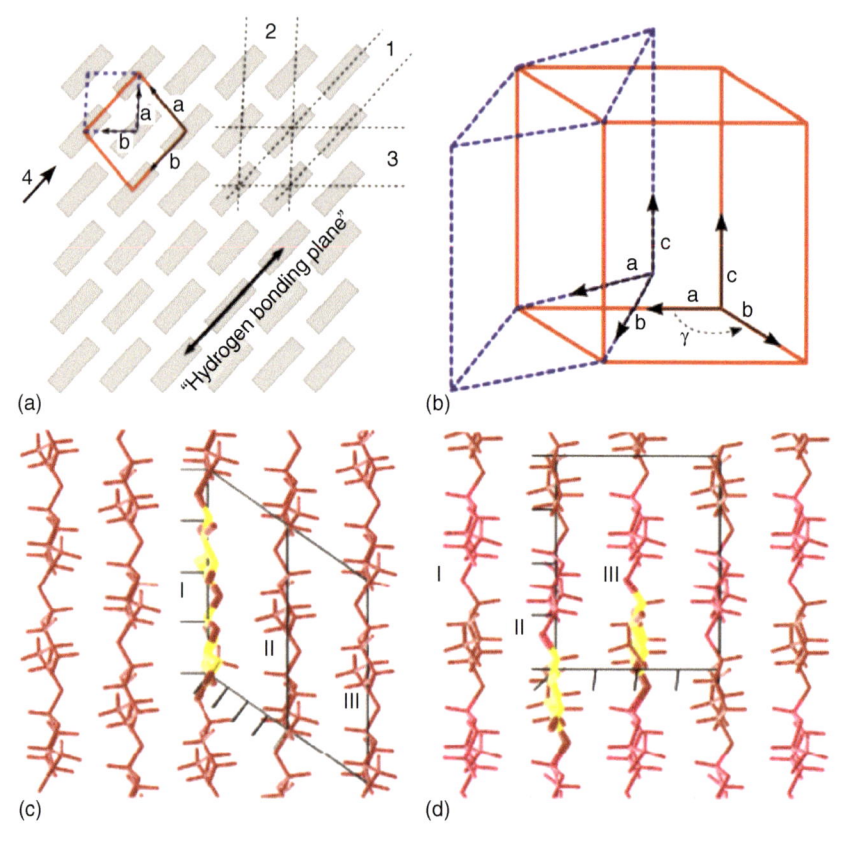

Figure 4.3 Schematic of unit cells for cellulose Iα (triclinic, dashed line) and Iβ (monoclinic, solid line). (a) Cross-sectional projection along the c-axis of a 36-chain cellulose I crystals showing parallelogram shape of both unit cells. The glucan chains in cellulose I crystals are arranged in the same manner when looking down the c-axis, sharing the same lattice planes of 1, 2, and 3 with corresponding d-spacings of 0.39, 0.53, and 0.61. The lattice planes for 1, 2, and 3 are designated as (110)$_t$, (010)$_t$, and (100)$_t$ for Iα and (200)$_m$, (110)$_m$, and ($\bar{1}$10)$_m$ for Iβ. (b) Relative configuration of one-chain triclinic unit cell (Iα, blue dashed line) and two-chain monoclinic unit cell (Iβ, red solid line). Source: Moon et al. 2011 [1]. Reprinted with permission of Royal Society of Chemistry. (c) Crystal structure of cellulose Iα. (d) Iβ perpendicular to the chain axis and in the plane of hydrogen-bonded sheet (110 for Iα and 200 for Iβ). The cellulose chains are represented by red skeletal models. Source: Nishiyama et al. 2003 [33]. Reprinted with permission of American Chemical Society.

These asymmetrical diffraction spots is from a one-chain P1 triclinic unit cell for cellulose Iα, with corresponding dimensions of $a = 0.674$ nm, $b = 0.593$ nm, c (chain axis) $= 1.036$ nm, $\alpha = 117°$, $\beta = 113°$, $\gamma = 81°$, cell volume $= 0.3395$ nm^3, and density of 1.582 g/cm^3 (Figure 4.3a,b) [34]. All glucan chains within cellulose I crystals are assembled in the so-called "parallel-up" fashion, indicating that the chains are oriented with the $1 \rightarrow 4$ glycosidic linkage pointing to the same positive c-axis direction of the unit cell. Although glucan chains packing within cellulose Iα and Iβ are nearly identical when looking down the c-axis, they have different displacement along the c-axis that helps to differentiate their structures, which could be visualized by looking perpendicular to the c-axis and along the hydrogen-bonded planes (Figure 4.3c,d). The second hydrogen-bonded sheet (II) in both cellulose Iα and Iβ shifted one quarter of c-axis "up" relative to the first sheet (I), whereas the third sheet (III) continuously shifted "up" and "down" relative to sheet II in cellulose Iα and Iβ, respectively [33]. Therefore, the cellobiose repeat units demonstrated a diagonally shifted arrangement in Iα and staggered arrangement in Iβ.

Besides the native cellulose I structure, there also exist another five different polymorphs (II, III$_I$, III$_{II}$, IV$_I$, and IV$_{II}$) that could be interconverted [7]. As mentioned in Section 4.2 on cellulose biosynthesis, cellulose I is not a thermodynamically stable state that the glucan chains prefer to form if not being regulated by the synthetic TCs. Instead, cellulose II is the most stable polymorph that could be irreversibly converted from cellulose I by either dissolution and regeneration or mercerization by sodium hydroxide solution [2, 3]. Cellulose II adopts a two-chain P2$_1$ monoclinic unit cell structure with dimensions of $a = 0.810$ nm, $b = 0.903$ nm, c (chain axis) $= 1.031$ nm, $\alpha = \beta = 90°$, and $\gamma = 117.10°$ [35]. In contrast to the parallel-up chain arrangement, cellulose II adopts antiparallel chain arrangement that could favor more stronger hydrogen-bonding patterns, leading to stable structure [7, 36]. Cellulose III$_I$ and III$_{II}$ could be obtained by treating the respective cellulose I and II with ammonia or amines, which could be further converted to cellulose IV$_I$ and IV$_{II}$ by heating up to 260 °C in glycerol [2].

The crystalline structure of native cellulose I is held together by both intra- and interchain hydrogen bonding. The intrachain O3–H···O5 hydrogen bonds are prevalent in both cellulose Iα and Iβ structure, but with different types of alteration in H···O bond distance and O–H···O bond angles (Figure 4.4) [33]. Within the two-chain monoclinic unit cell in cellulose Iβ, the alteration could be observed between origin and center chains, with respective bond distances of 1.966 and 1.752 Å and bond angles of 137.08° and 162.23°. However, in the one-chain triclinic unit cell structure of cellulose Iα, the alteration is down the same chains with the values of 2.072 and 1.954 Å for bond distances, as well as 137.59° and 163.94° for bond angles. Intrachain O2–H···O6 hydrogen bonds are also present in both cellulose Iα and Iβ, but with a shorter bond distance in the former case. These intrachain hydrogen bonds help to lock the glucosyl units in the opposite position to form a flat-ribbon chain configuration [37]. Interchain hydrogen bonds involving O6–H···O3 and O2–H···O6 could help to link the glucan chains edge to edge to form a flat sheet. However, the occurrence of these interchains hydrogen bonds is found to be at different chain position in cellulose Iβ (i.e. both are present in the center chains, whereas only O6–H···O3 is present

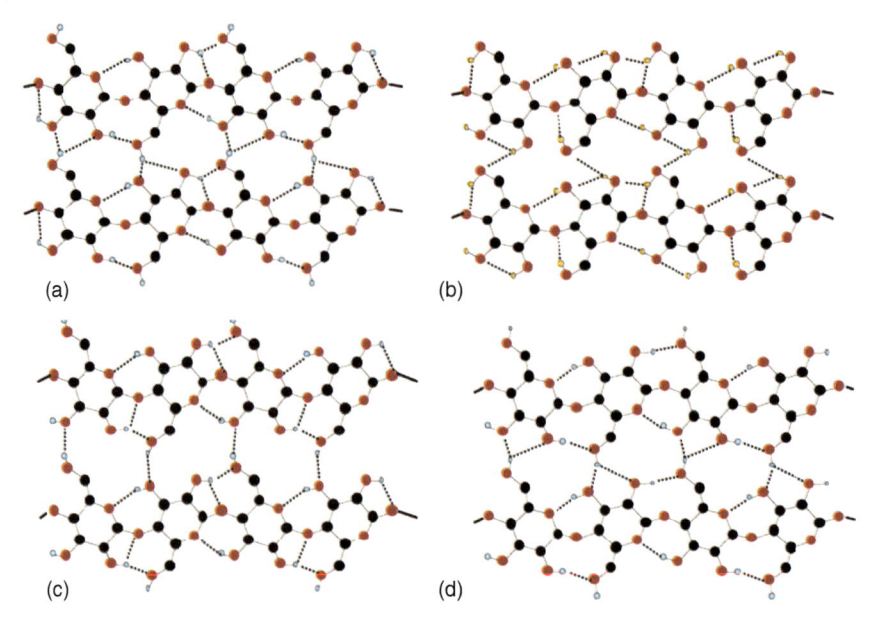

Figure 4.4 (a, b) The two alternative hydrogen-bond networks in cellulose Iα [33]. (c, d) The dominant hydrogen-bond network in cellulose Iβ: (c) chains at the origin of the unit cell; (d) chains at the center of the unit cell. Source: Sturcova et al. 2004 [37]. Reprinted with permission of American Chemical Society.

in the origin chains) or at dynamical balance in the same chains in cellulose Iα [38]. As all hydroxyls are located in the equatorial position of the glucopyranose rings, only intrachain and intrasheet hydrogen bonding could be formed in cellulose I, without signs of intersheet O–H···O hydrogen bonding [39]. Therefore, the hydrogen-bonded sheets are assembled together by hydrophobic interactions and weak C–H···O bonds [38, 40].

4.3 Nanoscaled Cellulose Fibers

Cellulose fiber represents a hierarchical structure consisting of fibers with progressively increased diameters, including the finest elementary fibrils of 1.5–3.5 nm wide, microfibrils of 10–30 nm wide, microfibrils bundles of 100 nm, and the macroscopic cellulose fibers on the order of microns (Figure 4.5) [3, 8]. With the aid of negative contrast electron microscopic technique, Muhlethaler has successfully observed finest fibrils with averaged 3.5 nm width along the axis [44], which was regarded as elementary fibrils following Frey-Wyssling's terminology [45]. Elementary fibrils have been considered as a universal structural unit in native cellulose that has been observed in varied sources including cotton, ramie, jute, and wood [43]. Microfibrils are assembled from the coalescence of these elementary fibrils to reduce the surface free energy, with general diameter of less than 35 nm [46]. The cellulose microfibrils are characterized as paracrystalline structures consisting of numerous cellulose I

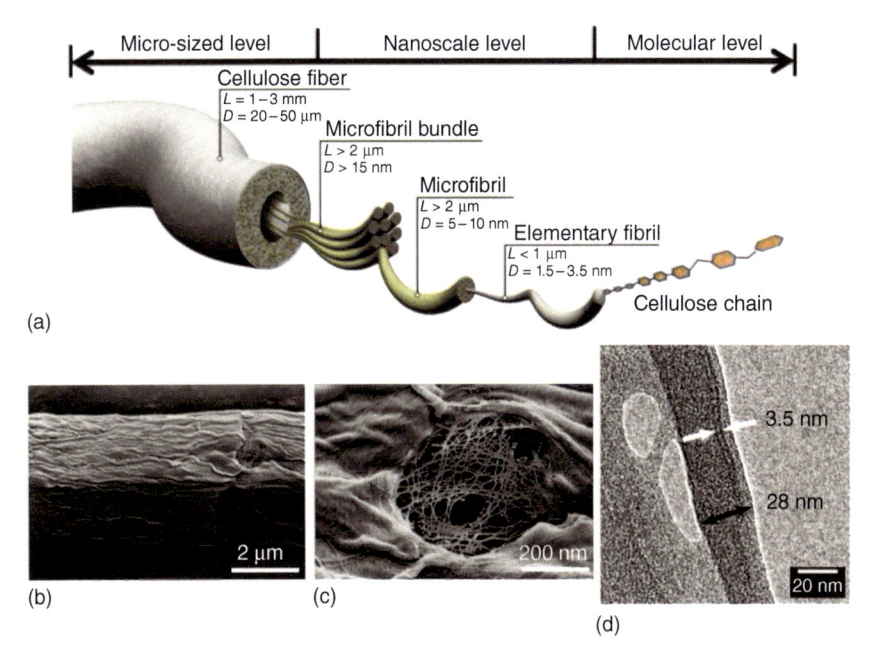

Figure 4.5 (a) Schematics of hierarchical structure of wood cellulose fibers. Source: Zhu et al. 2014 [41]. Reprinted with permission of Royal Society of Chemistry. SEM images of *Eucalyptus* fibers showing the (b) gross fiber surface and (c) microfibrils on the surface. Source: Chinga-Carrasco et al. 2011 [42]. Reprinted with permission of Microscopy Society of America. (d) TEM image of an individual microfibril of *Pinus radiata*. The black arrow indicates the boundaries of a microfibrils of 28 nm wide, and the white arrows indicate a single elementary fibrils of 3.5 nm wide. Source: Chinga-Carrasco 2011 [43]. Reprinted with permission of Springer.

crystallites stringed together by the amorphous cellulose chains. Both the degree of crystallinity and crystallite dimensions vary from different cellulose sources, with the algae cellulose generally having higher crystallinity and larger crystallite dimensions (Table 4.1).

Owing to the intermittent crystalline and amorphous regions, the cellulose microfibrils demonstrated superb flexibility owing to the amorphous chains, as well as high stiffness due to the highly crystalline crystallites. Young's modulus of crystalline cellulose Iβ has been reported in the range of 120–220 GPa, depending not only on the sources of origin but also the characterization techniques [1, 47, 48]. Young's modulus of long cellulose microfibrils or elementary fibril are also source and measurement technique dependent, with reported value of ~150 GPa for the tunicate microfibrils (8 nm thick and 20 nm wide) from three-point bending measurement by atomic force microscopy (AFM) [49] and 78 and 114 GPa for bacterial cellulose (BC) (30–90 nm wide) using the same AFM bending techniques [50] and Raman spectroscopic techniques [51], respectively. Although elastic modulus has been well characterized by either theoretical or experimental methods since the 1930s [52], mechanical strength of the cellulose microfibrils has been scarcely reported. The strength of cellulose microfibrils has only been recently estimated based on the model

Table 4.1 Degrees of crystallinity (x_c), crystallite sizes ($D_{(hkl)}$), and lateral dimensions (d) of microfibrils of native cellulose.

Cellulose source	x_c (%)	Crystallite sizes (nm)			d (nm)
		$D_{(1\bar{1}0)}$	$D_{(110)}$	$D_{(020)}$	
Algal cellulose	>80%	10.1	9.7	8.9	10–35
Bacterial cellulose	65–79	5.3	6.5	5.7	4–7
Cotton linters	56–65	4.7	5.4	6.0	7–19
Ramie	44–47	4.6		5.0	3–12
Flax	44 (56)[a]	4–5	4–5	4–5	3–18
Hemp	44 (59)[a]	3–5	3–5	3–5	3–18
Dissolving pulp	43–56			4.1–4.7	10–30

a) Degree of crystallinity relative to cellulose.
Source: Klemm et al. 2005 [3]. Reproduced with permission from John Wiley & Sons.

for sonication-induced fragmentation method as 1.6–3 GPa for wood cellulose and 3–6 GPa for tunicate cellulose [53]. The mechanical properties of cellulose microfibrils are comparable to those of multiwalled carbon nanotube and synthetic Kevlar fibers [1, 53]. However, the biological origin of these sustainable nanomaterials has driven researchers to explore varied techniques to obtain these nanoscaled cellulose fibers.

As cellulose microfibrils are intimately imbedded in lignocellulosic matrix with tight association with each other as well as the lignocellulosic matrices, isolation of individual cellulose microfibrils needs to be assisted with various means to disrupt these interfibrils association, i.e. via a "top-down" route to individualize the microfibrils from the large cellulose fibers. Currently, several techniques have been proven effective in isolating these cellulose microfibrils, including (I) mechanical shearing to physically disrupt the interfibril bondings (microfluidization, homogenization, ultrasonication, high-speed blending, aqueous counter collisions [ACC], twin-screw extruding, and grinding), (II) chemical treatments to reduce the hydrogen bonding capacities between the microfibrils (oxidation, quaternization, acid hydrolysis), (III) biological treatment to breakdown cellulose chains (enzymatic hydrolysis by cellulase), and (IV) a combination of these techniques for more efficient and synergistic results [54].

Although the terminology of cellulose microfibrils is most commonly used in describing the bundled elementary fibrils of 3–10 nm wide formed during biosynthesis in cell wall, "microfibrillated cellulose (MFC)" was firstly used to describe the cellulose fibrils disintegrated from plant cell walls by homogenization treatment, which demonstrated gel-like characteristics in water with pseudoplastic and thixotropic properties [55, 56]. Therefore, "microfibrillated cellulose" should not be confused with the "cellulose microfibrils," as the former usually has diameters of around 20–40 nm or even greater, consisting of multiple "microfibrils" packed together [57]. MFCs are generally defibrillated from cellulose fibers by the high shearing force created by repetitive mechanical treatment to break the hydrogen bonds and disintegrate into microfibril bundles.

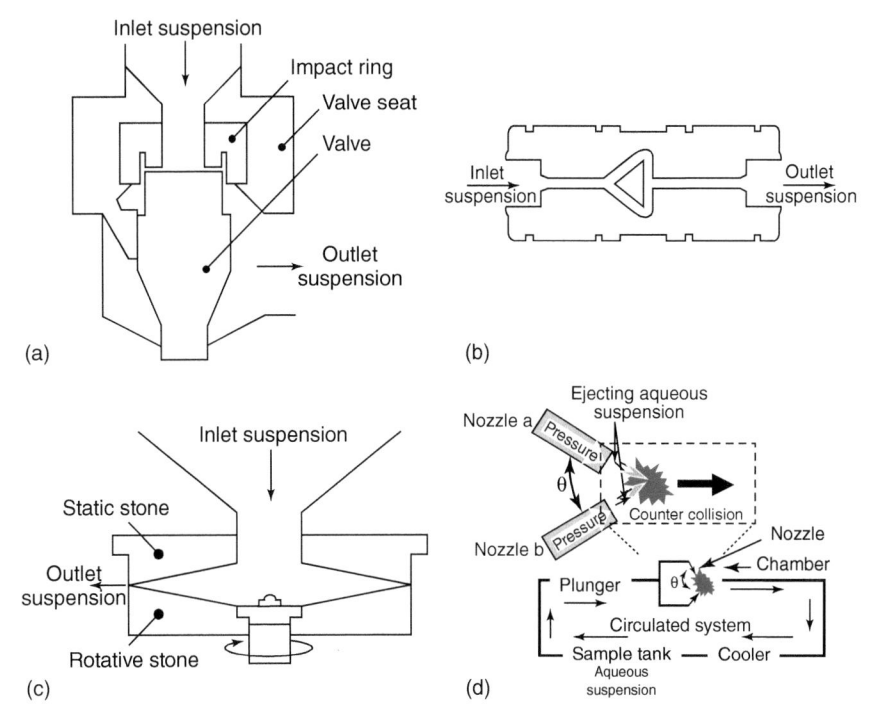

Figure 4.6 Schematics of the working principle of (a) homogenizer, (b) microfluidizer, (c) grinder; Source: Nechyporchuk et al. 2016 [59]. Reprinted with permission of Elsevier, and (d) aqueous counter collisions. Source: Kondo et al. 2014 [60]. Reprinted with permission of Elsevier.

High-pressure homogenization is firstly used to produce MFC back in the 1980s and remains being actively used nowadays [55, 58]. During homogenization, high shear and impact forces are exerted on cellulose fiber slurries when passing through the small gaps between the impact ring and spring-loaded valve (Figure 4.6a), achieving microfibril defibrillation [59]. Microfluidizer represents another technique for processing MFC by passing the cellulose suspension through a Z- or Y-shaped chamber under high pressure (150–210 MPa), exerting strong shear and impact forces against the channel walls (Figure 4.6b) [59, 61, 62]. As another commonly used technique, grinding processes MFC by passing the cellulose slurry through the static and rotative stones with adjustable inter-disk spacing, which generates shearing forces to disintegrate the microfibrils (Figure 4.6c) [59, 63, 64]. Recently, Kondo and coworkers developed another setup called "aqueous counter collision" to produce MFC, which creates high-speed head-on collision of two jets of aqueous cellulose suspension to pulverize the large cellulose fibers into smaller MFCs (Figure 4.6d) [60, 65, 66]. The high energy consumption on the order of 100 kWh/kg has been considered as the major barrier for mass production of MFC, especially with the high-pressure homogenization process [67]. However, it is found that microfluidization, ACC, and micro-grinding processes are more energy efficient, consuming significantly less energy than homogenization [60, 64, 66, 68]. Besides those most widely used

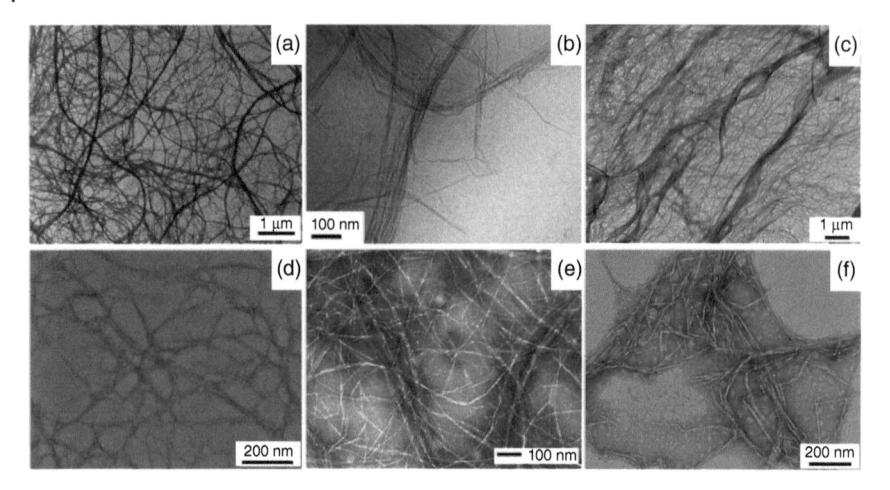

Figure 4.7 TEM images of mechanically derived microfibrillated cellulose: (a) high-speed homogenizer; Source: Dufresne et al. 1997 [73]. Reprinted with permission of John Wiley & Sons, Inc., (b) microfluidizer; Source: Taipale et al. 2010 [62]. Reprinted with permission of Springer, (c) grinding; Source: Jonoobi et al. 2012 [64]. Reprinted with permission of Elsevier, (d) ACC; Source: Jiang et al. 2016 [66]. Reprinted with permission of American Chemical Society, (e) ultrasonication; Source: Chen et al. 2014 [74]. Reprinted with permission of John Wiley & Sons, Inc., and (f) high-speed blending. Source: Jiang and Hsieh 2013 [71]. Reprinted with permission of Elsevier.

techniques, MFC production has also been reported using other technique such as ultrasonication [69, 70] and high-speed blending [71, 72].

The morphologies of MFC defibrillated using the common mechanical means are present in Figure 4.7, showing ultralong cellulose fibrils with heterogeneous diameter distribution ranging from 10 to 50 nm. The exact fibril diameters depend on the source of cellulose, treatment parameters such as pressure and treatment cycles/times, and separation procedures (centrifugation and filtration). Nevertheless, MFCs are considered as cellulose microfibril bundles containing multiple elementary fibrils bonded together via interfibril hydrogen bonding. As no chemical is used during mechanical defibrillation, MFCs show several advantages such as no surface modification, high aspect ratios, and exceptional thermal stability comparable to the original cellulose fibers [66, 71]. For more complete review on the preparation and properties of MFC, the readers could refer to recent reviews by Lavoine et al. [75] and Nechyporchuk et al. [59].

Although microfibril bundles of several tens of nanometers have been successfully isolated with the aid of mechanical means in the 1980s, due to the presence of numerous strong interfibril hydrogen bonds, isolation of elementary fibrils has remained a grand challenge until the beginning of this century. The prerequisite for isolating the finest elementary fibrils is to break the interfibril hydrogen bonds by varied chemical treatments, among which selective surface oxidation by 2,2,6,6-tetramethyl-1-piperidinyloxy (TEMPO) treatment represents the most effective one [67, 76]. TEMPO-mediated oxidation of native cellulose is commonly carried out at elevated pH of 10–11 using NaClO as the primary oxidant and TEMPO/NaBr as catalyst. This reaction has been proven to be highly

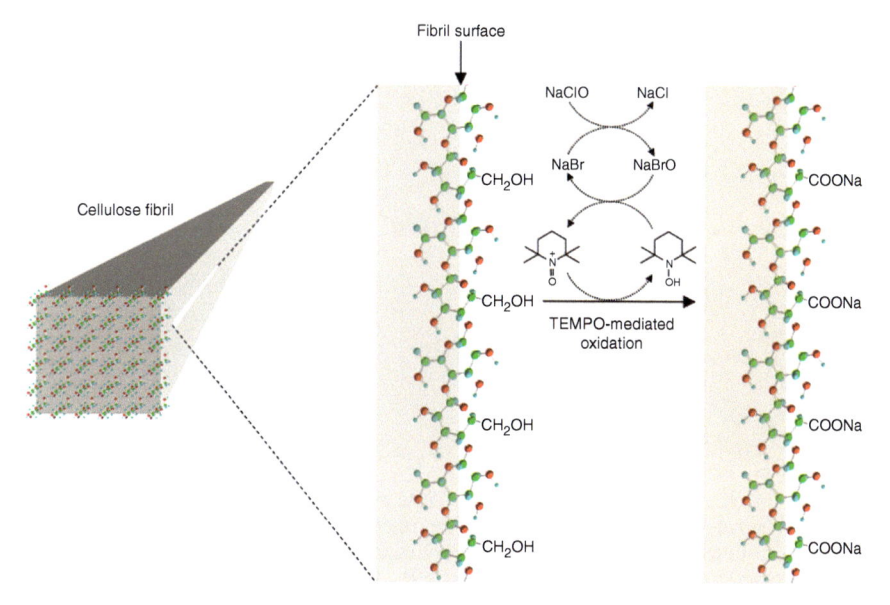

Figure 4.8 Schematic model of oxidation of C6 primary hydroxyl on cellulose microfibril surfaces by TEMPO oxidation. Source: Okita et al. 2010 [77]. Reprinted with permission of American Chemical Society.

regioselective that could only oxidize the C6 hydroxymethyl groups of cellulose (Figure 4.8) [9]. However, unlike regenerated cellulose where clear solution of cellouronic acids is obtained after oxidation [78], TEMPO oxidation could hardly convert the native cellulose I into soluble products [71, 79]. Therefore, further mechanical treatment is necessary to obtain individualized cellulose nanofibrils (CNFs), including magnetic stirring [80], high-speed blending [79, 81], and ultrasonication [82], but at significantly reduced energy consumption as compared with mechanical defibrillation alone [67].

The CNFs obtained from TEMPO oxidation appear as homogeneous long fibrils with lateral dimensions depending on the cellulose sources and extent of oxidation (Figure 4.9a–c). CNFs from wood showed uniform width of a few nanometers, whereas tunicin and BC demonstrated significantly wider fibril dimensions. Nevertheless, the lateral dimension of the CNFs derived from TEMPO oxidation is almost similar to the lateral crystallite dimension as determined from the wide- and small-angle X-ray diffraction measurements [79, 81]. Successful isolation of the elementary fibrils requires breaking the interfibril hydrogen bonds, which could also be achieved by other chemical or biological treatments, other than the mostly investigated TEMPO oxidation. Carboxymethylation of softwood dissolving pulp using monochloroacetic acid could liberate CNFs of 5–15 nm wide and over 1 μm long (Figure 4.9d), due to the electrostatic repulsion caused by the negatively charged surface carboxymethyl groups [83]. Besides these anionization reactions, cationization reaction can introduce positive charges on the surface of elementary fibrils, breaking the hydrogen bonds and facilitating its

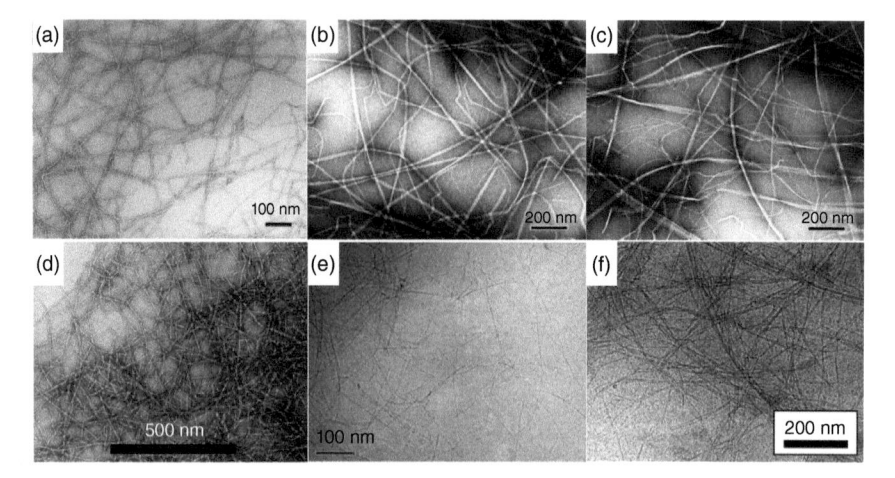

Figure 4.9 Transmission electron microscopy images of cellulose nanofibrils defibrillated via combined chemical–mechanical treatment. TEMPO oxidation of (a) wood, (b) tunicin, and (c) bacterial cellulose; Source: Saito et al. 2006 [81]. Reprinted with permission of American Chemical Society, (d) carboxymethylation; Source: Wagberg et al. 2008 [83]. Reprinted with permission of American Chemical Society, (e) quaternization; Source: Olszewska et al. 2011 [84]. Reprinted with permission of Springer, and (f) enzymatic hydrolysis. Source: Paakko et al. 2007 [85]. Reprinted with permission of American Chemical Society.

liberation by mechanical treatment. Quaternization of dissolving pulp with N-(2-3-epoxypropyl)trimethylammonium chloride has shown to introduce 354 µEq/g of positively charged quaternary amine to the elementary fibrils, which have been liberated as individual fibrils of 2.5–3 nm, close to the elementary fibrils from Norway spruce 2.5 ± 0.2 nm obtained from TEM and small- and wide-angle X-ray scattering [84, 86]. Besides introducing charged species on the microfibril surface, enzymatic hydrolysis by endoglucanase has also found to be able to introduce interfibril defects that could facilitate mechanical defibrillation, yielding individual nanofibrils of c. 5 nm and thicker bundles of 10–20 nm wide [85].

All of these aforementioned defibrillation methods include a top-down process that involves liberating the preassembled plant cellulose fibers into the individual elementary fibrils. Due to the strong interfibrils and interspecies association, chemical and energy-intensive mechanical treatments are required to remove the lignin/hemicellulose as well as to individualize the cellulose microfibrils. On the other hand, chemically pure cellulose free of lignin, pectin, and hemicellulose could be obtained by biosynthesis from the gram-negative, aerobic microorganism (*A. xylinum*) using D-glucose as the carbon sources or a "bottom-up" process (Figure 4.10) [3, 87]. The first discovery of BC synthesized by *A. xylinum* was made in 1886 by Brown, who found a gelatinous translucent membrane formed during vinegar fermentation on the surface of medium broth, which has been identified with the same chemical structure of plant cellulose [88]. The synthesis of BC starts with glucose, which could be transformed into UDP (guanosine triphosphate)-glucose via glucose-6-phosphate and glucose-1-phosphate, followed by polymerization reaction mediated by cellulose synthase to add the

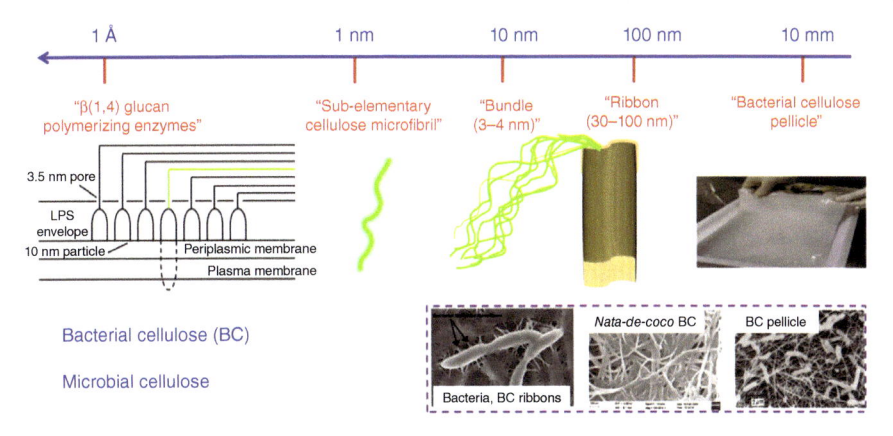

Figure 4.10 Hierarchical structure of bacterial cellulose. Source: Lin and Dufresne 2014 [52]. Reprinted with permission of Elsevier.

UDP-glucose to the end of glucan chains by cellulose synthase (TCs) [87]. Similar to plant cellulose, BC is also synthesized through TCs. However, in contrast to the rosette-shaped TCs found in the plasma membrane of higher plant, the BC TCs are linearly arranged on the cell surface of bacteria, with a basket-shaped structure. Each basket-shaped structure could synthesize 16 glucan chains through the catalytic sites, assembling into sub-elementary fibrils containing four mini-sheets, approximately 1.5 nm wide. One linear TC contains three basket-shaped subunits, linearly positioned parallel to the longitudinal axis of the bacteria. Therefore, the secreted sub-elementary fibrils could assemble with each other to form a ribbon like elementary microfibrils, which further assemble into larger ribbon-like fibers of approximately 40–50 nm wide [87, 89]. The flat-ribbon-like BC nanofibers intertwine with each other to form BC pellicles to float on the oxygen-rich surface of the media, with the pellicle protecting the cells from lethal effects of ultraviolet radiation and other harsh environments [87, 90]. BC contains both Iα and Iβ, with over 80% being dominant by cellulose Iα and crystallinity as high as 84–89% [91]. BC has extremely long fibers of approximately 1–9 µm, with DP as high as 14 000–16 000 [92], depending on the cultivation conditions and bacterial strains. Due to its high crystallinity and extraordinary hydrogen bonding systems, BC is extremely strong with Young's modulus of 78 GPa measured on a single BC fiber using three-point bending AFM technique [50]. BC nanofibers could also be further broken down into smaller nanofibrils by ACC [65] or TEMPO oxidation treatments [93], showing reduced diameters to 30 and 15–40 nm, respectively. Further reduction in the diameters of bacterial CNFs into elementary fibrils has not been reported, possibly due to the strong interfibril hydrogen bonding.

4.4 Submicron-Scaled Cellulose Fibers

Precise dimensional control of the linear cellulose fibers is intriguing considering that numerous properties of cellulose fibers are dependent on the fiber dimensions, such as specific surface area, surface functionality, porosity, permeability, and mechanical properties. However, in nature, except for BC where the strands of cellulose nanofilaments secreted by different bacteria are well separated with each other within the pellicles, all plant cellulose fibers lose such delicacy of well-separated nanocellulose. Instead, due to the high specific surface areas of cellulose elementary microfibrils and the large availability of surface polar hydroxyl groups, they readily assemble with each other to form large macroscopic fibers of hundreds of microns wide. Therefore, it has drawn great attention in designing macroscopic cellulosic fibrous materials with imbedded submicron or even nanoscaled cellulose fibers to harness both the intrinsic nanoscale properties and the easy-to-handle macroscopic properties. Several approaches have been developed to design this type of materials, mainly involving controlled assembly of the cellulose chains into continuous fibrous structure. In this section, two of the most adopted approaches will be reviewed, including electrospinning and self-assembly.

Since its invention, electrospinning has evolved as an efficient technique in fabricating continuous polymeric micro/nanofibers, with the fiber diameters ranging from less than 100 nm to up to a few microns [94]. During electrospinning, the polymer solution was subject to electric filed, which charges the solution to induce charge repulsion. This charge repulsion, together with the contraction of the surface charges to the counter electrode, exerts a force on the surface of polymer solutions to overcome the surface tension. At sufficient large electric field, the jet of polymer solution could be ejected and fly toward the metal collector, forming polymer nanofiber membranes once the solvent evaporates [95]. The prerequisite of electrospinning polymeric nanofibers is to form a homogeneous solution, which represents the greatest challenge for cellulose, considering the strong intramolecular and intermolecular hydrogen bonding that prevent it dissolving into common solvents or melting. Several solvent systems, which have been successfully used in cellulose fiber spinning, were found to be suitable for electrospinning of native cellulose, including N-methylmorpholine-N-oxide/water (NMMO/H_2O) [96, 97], lithium chloride/dimethylacetamide (LiCl/DMAc) [96, 98], ionic liquid [99–101], and ionic liquid cosolvent system [100, 102, 103]. The common problem with these solvent systems is their low volatility that cannot evaporate completely when traveling from the spinneret to the collector [104]. Besides, some solvent systems also contain nonvolatile salts such as LiCl and ionic liquids that could not be evaporated from the electrospun fibers. Therefore, specific collector design is required to effectively remove the solvents, including using a heated collector (~100 °C) for LiCl/DMAc to enhance evaporation of DMAc, followed by coagulation with water to remove the LiCl salt, resulting a dry and stable mat of cellulose fibers with diameters as small as 150 nm (Figure 4.11a) [96, 98]. As a contrast, electrospinning of cellulose nanofibers from NMMO/H_2O system requires a lower temperature collector (9–10 °C) to promote fast solidification of fibers, followed by the same coagulation with water bath to remove the solvent, obtaining uniform cellulose fibers of less than 1 μm wide (Figure 4.11b) [96]. Nontoxic room temperature ionic liquid 1-ethyl-3-methylimidazolium acetate ([C_2min][CH_3CO_2]) was also used to dissolve cellulose to electrospun

Figure 4.11 Electrospun pure cellulose nanofibers from different solvent systems: (a) 3% (w/w) cellulose/8% LiCl/DMAc; Source: Kim et al. 2005 [98]. Reprinted with permission of John Wiley & Sons, Inc., (b) 9 wt% cellulose/NMMO/H_2O, T_{nozzle} = 50 °C; Source: Kim et al. 2006 [96]. Reprinted with permission of Elsevier, (c) 8 wt% cellulose in [C_2min][CH_3CO_2] ionic liquid; Source: Freire et al. 2011 [99]. Reprinted with permission of Royal Society of Chemistry, (d) 5 wt% cellulose in AMMIMCl/DMSO (1 : 2 ratio); Source: Xu et al. 2008 [103]. Reprinted with permission of Elsevier, (e) 6.3 wt% cellulose in [C_2min][OAC]/DMF (1 : 1 ratio); Source: Ahn et al. 2012 [102]. Reprinted with permission of Elsevier, and (f) 6.3 wt% cellulose in [C_2min][OAC]/DMAc (1 : 1 ratio). Source: Ahn et al. 2012 [102]. Reprinted with permission of Elsevier.

fibers using grounded water bath as collector to remove ionic liquid in order to prevent fiber fusion (Figure 4.11c) [99]. The average diameter of as-spun cellulose nanofibers was 470 nm, which could be further reduced to 120 nm with addition of a surface active ionic liquid 1-decyl-3-methylimidazolium chloride as cosolvent. Due to its high viscosity and nonvolatility, electrospinning cellulose in pure ionic liquid is a great challenge. Therefore, the common practice is to add a volatile cosolvent to lower the surface tension and viscosity of ionic liquid. Adding DMSO as a cosolvent to 1-allyl-3-methylimidazolium chloride (AMIMCl) has found to increase the conductivity and decrease the surface tension and viscosity, which therefore could produce stable fibers of 500–800 nm (Figure 4.11d) [103]. Besides DMSO, DMF and DMAc were also mixed with ionic liquid (1-ethyl-3-methylimidazolium acetate) to assist cellulose electrospinning [102]. It was found that cellulose dissolved in pure ionic liquid could not form a stable Taylor cone, a prerequisite in obtaining nanofibers. Adding either DMF or DMAc could significantly improve the spinnability, showing uniform cellulose nanofibers with average diameters less than 600 nm at 1 : 1 solvent ratio (Figure 4.11e,f). Recently, a new solvent system NaOH/urea was developed to dissolve cellulose and explored on the suitability in electrospinning thereafter. However, it was found that electrospinning cellulose dissolved in NaOH/urea could not produce fibers, and nanofibers could only be formed when other fiber-forming facilitating polyol binders such as polyethylene glycol (PEG) and polyvinyl alcohol (PVA) [105]. As direct electrospinning cellulose nanofibers from dissolved cellulose solution remains a challenge due to the limited solvent system and their low volatility or nonvolatility, it is much easier to obtain regenerated cellulose nanofibers by post-hydrolysis of cellulose acetate nanofibers. Cellulose acetate nanofibers has been successfully electrospun from mixed acetone/DMAc or acetic acid/DMAc solvent, which could be further converted to regenerated cellulose by deacetylation to remove the acetyl groups [106].

These submicron-wide fibers derived from electrospinning requires complicated processes including dissolution, electrospinning, and post-coagulation, requiring tremendous amount of solvents and energy. Besides, these dissolution and regeneration treatment of cellulose disrupt the native hydrogen bonding between cellulose chains, rendering change of crystal structure from native cellulose I to regenerated cellulose or cellulose II. Therefore, it has drawn a great interest in searching for a more facile and greener method in deriving continuous submicron cellulose fibers with well-preserved cellulose I crystal structures. This search has been fruitful during the past decade, with the booming of researches in nanocellulose. Essentially, the native cellulose I nanocellulose (including CNFs and cellulose nanocrystals [CNCs]) has found to be able to assemble with each other during freezing and freeze-drying, forming assembled submicron-wide fibers or nanofibers. The pioneering work on nanocellulose assembly was reported by Paakko et al. in 2008, showing that the enzymatic hydrolyzed cellulose I nanofibers (5–10 nm wide) could aggregate into hierarchical structures containing interconnected fibrillar networks of approximately 30 nm wide (Figure 4.12) [107].

The intriguing phenomenon of nanocellulose assembly has been extensively investigated, and the formation of ice crystals during freezing and its subsequent sublimation has been identified as the driving force for the assembly. Starting from homogeneously dispersed aqueous nanocellulose suspension, the assembly of nanocellulose takes advantages of ice nucleation and crystal growth, which could segregate the nanocellulose into isolated regions. The locally concentrated high specific surface nanocellulose could thereafter assemble with each other via hydrogen bonding and van der Waals force into fibrous or

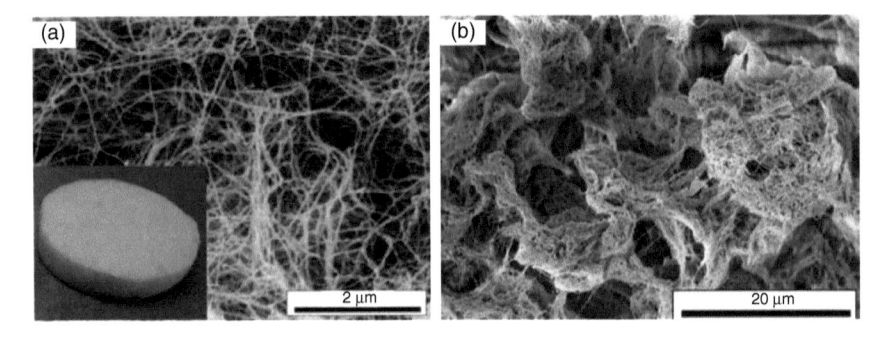

Figure 4.12 SEM micrographs of the aerogels as prepared by quickly freezing the aqueous nanofibrillar gel in liquid propane and extracting the water from the frozen state. (a) Higher magnification showing the fibrillar aerogel skeleton. Inset: 3 mm thick aerogel specimen. (b) Lower magnification showing the hierarchical porosity. Source: Paakko et al. 2008 [107]. Reprinted with permission of Royal Society of Chemistry.

dense film-like structures, depending on a series of factors, such as freezing temperature, nanocellulose concentration, aspect ratio, surface chemistry, crystal structure, and dispersing medium [108–111]. Among them, assembled fibrous forms are commonly obtained by extremely low freezing temperature such as liquid nitrogen (−196 °C) or liquid propane (−180 °C), where smaller ice crystals could be formed to segregate nanocellulose into thinner bundles between ice crystals to form submicron fibers. In contrast, relative higher freezing temperature such as the freezing in the freeze-dryer cooling chamber (−50 °C) or freezer (−20 °C) could lead to nanocellulose assembly into larger two dimensional sheets, due to the growth of large ice crystals at higher freezing temperature [107, 111, 112]. The formation of submicron fibers is also highly dependent on the starting concentration of nanocellulose suspensions, which shows that submicron fibers could only be obtained when the nanocellulose concentration is lower than certain critical concentrations [113]. Essentially, at lower concentration, the nanocellulose has lower local concentration and is well separated by the ice crystals, confining the lateral assembly of nanocellulose into submicron fibers. When the concentration is high, higher local concentration between ice crystals is also expected, which could reduce the nanocellulose spacing and promote more lateral association between nanocellulose to form 2D sheetlike structures (Figure 4.13). The critical concentration in forming submicron fibers shows significant variation for different types of nanocellulose, primarily due to the surface chemistry of the nanocellulose. It was found that CNCs tend to form submicron fibers at concentrations below 0.5 wt%, whereas the TEMPO-oxidized CNFs could only form submicron fibers at concentrations below 0.05 wt%, at least 1 magnitude lower than that for CNCs [111]. The reason that CNFs tend to assemble into film-like structures at much lower concentration is due to the stronger hydrogen bonding capacity of the surface carboxylate groups as compared with the hydroxyl groups for pure cellulose. The same concentration-dependent assembling properties were also observed

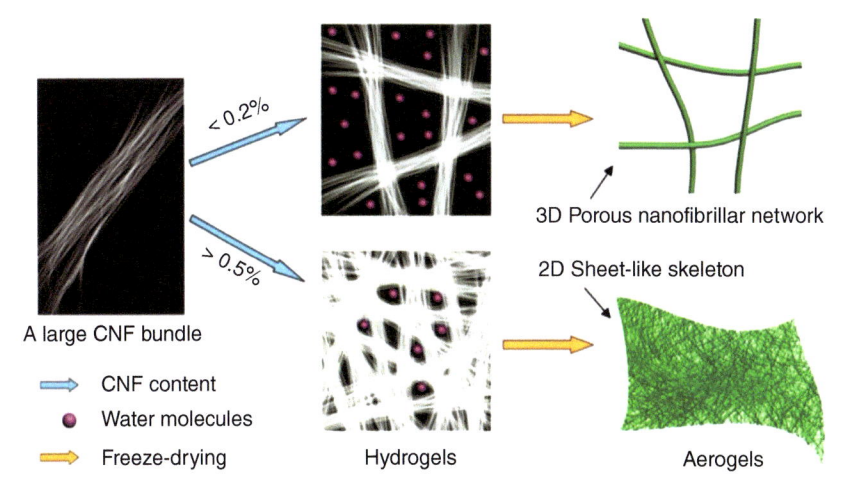

Figure 4.13 Schematics of the effect of CNF content on the assembled morphologies. Source: Chen et al. 2011 [108]. Reprinted with permission of Royal Society of Chemistry.

Figure 4.14 SEM image of self-assembled submicron fibers from varied nanocellulose. (a) CNCs at 0.1 wt%; Source: Jiang and Hsieh 2013 [71]. Reprinted with permission of Elsevier, CNFs at 0.1 wt% with (b) 0.59 and (c) 1.29 mol/g surface carboxyl groups. Source: Jiang et al. 2013 [79]. Reprinted with permission of Royal Society of Chemistry. (d) CNFs from high-speed blending at 0.05 wt%. Source: Jiang and Hsieh 2013 [71]. Reprinted with permission of Elsevier. (e) CNFs from aqueous counter collision treatment at 0.1 wt%. Source: Jiang et al. 2016 [66]. Reprinted with permission of American Chemical Society. (f) CNFs from ultrasonication. Source: Chen et al. 2014 [74]. Reprinted with permission of John Wiley & Sons, Inc.

for ultrasonication-treated CNFs that resemble to pure cellulose without surface chemistry modification, showing that submicron fibers could be obtained at concentration below 0.2 wt% [108].

Regardless of the type of nanocellulose and their respective surface chemistries, submicron fibers could be exclusively obtained when freezing at extremely low temperature (−196 °C using liquid nitrogen) and nanocellulose concentration (preferably <0.1%). Freeze-drying of various nanocelluloses leads to similar white and fluffy fibrous mass, which differ in microscopic morphologies with regard to the nanocellulose characteristics. Due to the higher crystalline and rigid rodlike structure, CNCs usually assemble into straight fibers of 300–400 nm wide (Figure 4.14a) [71]. The surface of the assembled CNC fibers appeared to be undulated with aligned CNC bundles and showed exceptional stability in water by handshaking or gentle mechanical stirring, which could be ascribed to the tight association of the CNCs through abundant inter-crystal hydrogen bonds [114]. Unlike highly crystalline rigid rodlike CNCs, the CNFs isolated either from TEMPO oxidation or mechanical shearing contain both crystalline and amorphous regions, showing significantly enhanced flexibility with higher aspect ratios. Therefore, curved and flexible submicron fibers could be obtained by assembling these longer CNFs (Figure 4.14b–f). The self-assembled morphologies of TEMPO-oxidized CNFs depend highly on the amount of surface charges as well as the degree of surface carboxyl protonation, i.e. the ratios of carboxylate to carboxylic acid [79, 115]. With increased amounts of surface carboxylate groups or degree of protonation at fixed surface carboxylate groups, the diameters of the assembled submicron fibers tend to increase or even evolve into film-like structure, due to the stronger hydrogen bonding capacity of the protonated carboxylic acid groups. Ultrafine submicron fibers with average diameters of 120–140 nm have been successfully assembled from TEMPO-oxidized CNFs with low surface charge density of 0.59 mmol/g [79] or with less than 10% of protonated carboxylic groups at high surface charge density of 1.29 mmol/g [115]. Mechanically defibrillated CNFs preserve the surface hydroxyls without any chemical alteration, therefore most resembling the elementary cellulose microfibrils in plants. These uncharged CNFs were found to be able to assemble into most uniform submicron fibers, with lateral dimensions ranging from 120 to 150 nm (Figure 4.14d–f) [66, 69, 74].

Due to the prevalent surface polar groups on the nanocellulose, such as hydroxyls and carboxyls, nanocelluloses extensively assemble with each other into continuous submicron fibers with diameters ranging from 100 to 500 nm, signifying a facile process in obtaining fibrous materials. However, the extensive assembly inevitably reduces the specific surface area of these nanomaterials, which is considered the most interesting properties of the nanocellulose. In order to preserve these nanoscale dimensions, *tert*-butanol is commonly used as a cosolvent to inhibit the interfibril hydrogen bonding, therefore leading to much finer assembled nanofibers [111, 116]. With addition of respective 93 and 50 vol% *tert*-butanol, both CNCs and CNFs could assemble into relatively homogeneous nanofibers of approximately 40 nm wide (Figure 4.15) [111]. Due to the less interfibril association, these assembled nanofibers are found to be much more

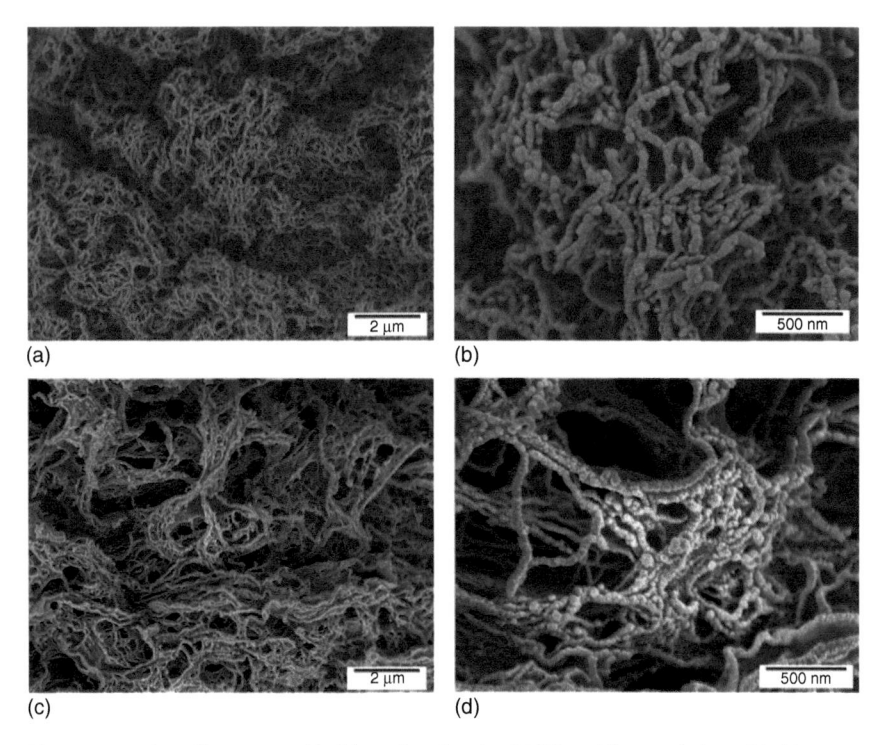

Figure 4.15 Nanofibers assembled from freezing (−196 °C) and freeze-drying of (a, b) 0.1% CNC in 93/7 *tert*-butanol/water and (c, d) 0.1% CNF in 50/50 *tert*-butanol/water. Source: Jiang and Hsieh 2014 [111]. Reprinted with permission of American Chemical Society.

easily redispersed into various solvent systems. The clearly reduced assembled fiber diameters could be ascribed to the different hydrogen bonding capacity between *tert*-butanol and water with nanocellulose and the steric hindrance of *tert*-butanol-bound nanocellulose surfaces. Each *tert*-butanol molecule could hydrogen bond to only one nanocellulose surface hydroxyl or carboxyl, preventing it from further hydrogen bonding or polar interaction, while the three bulky methyl groups add steric hindrance to limit inter-nanocellulose association. As a result of this inhibition on the self-assembly, the specific surface area of the fibrous mass could be significantly improved, showing as high as 350 m^2/g [110, 117].

In essence, submicron-wide cellulose fibers could be facilely obtained through either electrospinning of dissolved cellulose in carefully selected solvent systems or by direct self-assembly of aqueous dispersion of nanocellulose suspension. Compared with the electrospinning process, freeze-drying-assisted self-assembly demonstrates several advantages including less solvent consumption, easier procession, nontoxic solvent system, better control on the fibers' morphologies (ranging from several tens of nanometers to a few hundred nanometers), and preservation of the native cellulose I crystalline structure.

4.5 Macroscaled Cellulose Fibers

Macroscaled continuous cellulose fibers play a critical role in modern society, finding practical applications in textile, membrane, filtration, and reinforced composites. As a naturally abundant biopolymer, cellulose fibers have been used in clothing, construction, papermaking, and fishing for thousands of years. Until nowadays, cotton, a type of highly crystalline plant cellulose fibers, is still one of the major raw materials for textile industry. As the majority of cellulose on Earth is short fibers from trees, it is necessary to further process these short fibers into continuous filaments for more practical applications. Traditional cellulosic fiber process produces regenerated cellulose relying on viscose and cuprammonium rayon and Lyocell processes, all requiring heavy hazardous chemical dosage to derivatize or dissolve cellulose (xanthogenation for viscose, dissolution in cuprammonium and *N*-methylmorpholine-*N*-oxide for rayon and Lyocell, respectively), as well as regenerate into cellulose fibers [118, 119]. Therefore, these processes pose great environmental effects, and much greener and more environmental benign processes are being sought to process macroscaled cellulose fibers. Recently, a novel NaOH/urea aqueous solvent systems was developed to successfully dissolve cellulose, which is much greener compared with the previous organic solvent system [118, 120]. Wet spinning of cellulose dissolved in NaOH/urea generated continuous cellulose fibers with circular cross section and smooth surface, significantly different from the lobulated shape of the viscose rayon, but similar to cuprammonium rayon and Lyocell fibers [118]. Besides, the cellulose fibers spun from NaOH/urea solution showed homogeneous smooth surface without signs of fibrillation, a common problem for Lyocell fibers. Drawing has shown to improve the cellulose molecular orientation and tensile strength, with the mechanical properties being approximate to the viscose rayon. However, this dissolution–regeneration process inevitably converted the native cellulose I to cellulose II structure, which could lead to deteriorated mechanical properties. Besides, stringent processing conditions are generally required for this process due to the narrower temperature window for dissolution, making scaling up a great challenge.

In contrast to dissolved cellulose where the crystalline structures are destroyed through interactions with solvent molecules, both CNCs and CNFs well preserve the native cellulose I crystalline structure, which lead to extraordinary mechanical properties including high modulus of 120–220 GPa [47–49] and strength of 3–6 GPa [53]. Therefore, great efforts have been devoted in fabricating macroscaled fibers out of these mechanically strong nanocellulose, aiming at transferring the exceptional mechanical properties of cellulose I nanocellulose to their assembled macrofibers. Wet extrusion of CNF aqueous suspensions into coagulation bath containing water miscible organic solvent (such as acetone, ethanol, tetrahydrofuran [THF], and dioxane) has been applied to fabricate macrofibers (Figure 4.16a,b). During this process, the aqueous CNF suspensions solidify when contacting the non-dispersing exchange solvent, forming a skin at the surface of the extruded fibers. Shearing force induced by extrusion could lead to preferential orientation of CNFs along the macrofibers, showing modest birefringence when being observed under crossed polarizers (Figure 4.16c). However, pure shearing force induced by extrusion is insufficient for complete nanofibril alignment, therefore resulting in moderate mechanical properties

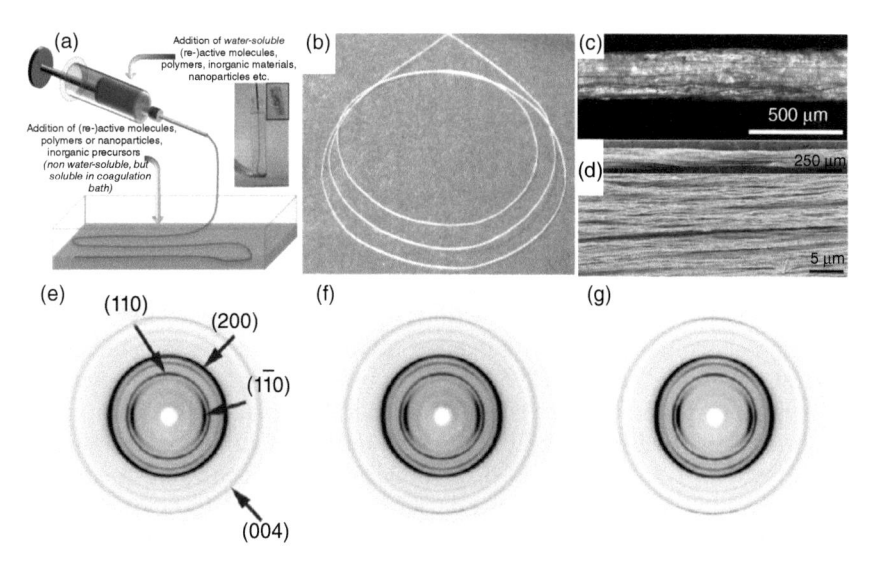

Figure 4.16 (a) Preparation of NFC macrofibers based on extrusion of NFC hydrogels into a coagulation bath and drying. (b) Photograph of a 1 mm long macrofiber of c. 0.2 mm diameter. (c) Optical microscopy image of the macrofibers with cross polarizers. (d) SEM images of macrofibers. Source: Walther et al. 2011 [121]. Reprinted with permission of John Wiley & Sons, Inc. X-ray fiber diagrams of the tunicate fibers spun at rates of (e) 0.1, (f) 1, and (g) 10 m/min. Source: Iwamoto et al. 2011 [122]. Reprinted with permission of American Chemical Society.

of 22.5 GPa for Young's modulus and 275 MPa for tensile strength, yet still 1 order of magnitude higher than typical rayon fibers. Although post-drawing was proposed in this study to improve the alignment of nanofibrils, it was not experimentally carried out to examine this hypothesis. Besides the less preferential alignment along the fiber axis, the macrofibers showed estimated porosity of 10%, with numerous nanopores in the size up to 25 nm being observed on the surface of SEM imaging (Figure 4.16d), which could also lead to the inferior mechanical properties. In another contemporary work, similar wet-spinning strategy was used to spin macrofibers out of TEMPO-oxidized CNFs in an acetone coagulation bath, focusing more on the shear force-induced alignment by varying the spinning speed from 0.1 to 100 m/min [122]. The X-ray diffractograms of the spun fibers showed equatorial arcs corresponding to (110), (1$\bar{1}$0), and (200) diffractions and a meridian arc corresponding to (004) (Figure 4.16e–g). The orientation index of the wood fibers increased with increasing spinning speed, indicating better alignment of the nanofibers along the fiber axis. Correspondingly, the tensile strength and Young's modulus also increased with increasing orientation index, reaching 321 MPa and 23.6 GPa, respectively. This positive correlation between orientation index and mechanical properties is due to the anisotropy in mechanical properties for cellulose nanofibers, with estimated 150 GPa [49] and 18–50 GPa [48] Young's modulus at the respective longitudinal and transverse directions.

Wet-stretching-induced CNF orientation was investigated on wet-spun TEMPO-oxidized cellulose macrofibers coagulated in THF [123]. The wet stretching was conducted using a home-built computer-controlled device (Figure 4.17a), leading to a stretch ratio up to 0.3. Better CNF alignment could be clearly confirmed from the SEM image of the cross sections (Figure 4.17b), as well as increased orientation index from the 2D wide-angle X-ray diffraction. Quadruple increment in Young's modulus from 8.2 to 33.7 GPa and more than doubled increase of tensile strength from 118 to 289 MPa could be obtained by wet stretching, indicating simultaneous stiffening and strengthening with increasing stretching ratio (Figure 4.17c). Although wet stretching could lead to significant increase in mechanical properties owing to improved fibril alignment, the 289 MPa tensile strength is only moderately higher than the previously reported 275 MPa [121], possibly due to the variation of the starting materials. However, wet stretching indeed enhances the stiffness of the macrofibers as evidenced from the over 10 GPa Young's modulus increase. Hydrodynamic alignment was applied to force the CNF aligned in the flow direction, which was immediately fixed by electrolyte diffusion to aligned structure, as shown in the schematic in Figure 4.17d [124]. To obtain homogeneous and smooth gel fibers, the timing for ion diffusion-induced gel fixation has to be fit between the hydrodynamic alignment and the relaxation caused by Brownian diffusion. Compared to the previously reported porous fiber structures, this hydrodynamically aligned and fixed fibers showed void-free cross sections, and individual nanofibers could not be observed, suggesting a much denser and compact structure (Figure 4.17e,f). Due to the dual effects of alignment and fixation, the fibers are mechanically strong and stiff with tensile strength as high as 490 MPa and Young's modulus of 18 GPa (Figure 4.17g), with tensile strength surpassing

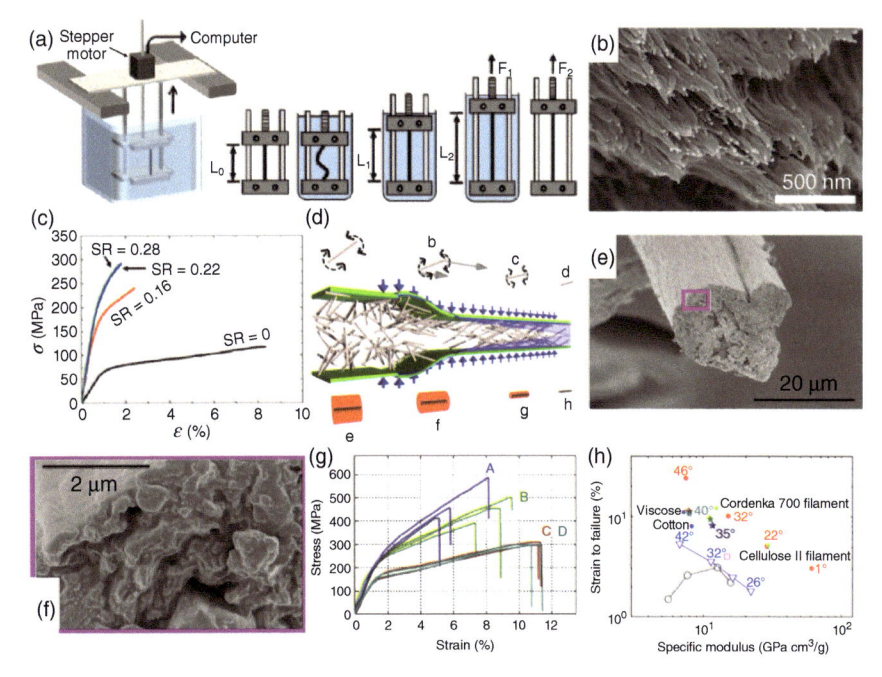

Figure 4.17 (a) Schematic drawing for the computer-controlled wet-stretching device into which a fiber can be clamped and immersed into a liquid and stretched using controlled strain rates; the right-hand side shows the various states during wet stretching. (b) SEM image of cross sections of stretched (SR = 0.28) NFC macrofibers. (c) Stress–strain curves of NFC macrofibers at different SR. Source: Torres-Rendon et al. [123]. Reprinted with permission of American Chemical Society. (d) Illustration of the hydrodynamic assembly process. Source: Hakansson et al. 2014 [124]. Reprinted with permission of Macmillan Publishers Limited. (e, f) SEM images of the filament. (g) Stress–strain plot of filaments. (h) Strain-to-failure versus specific Young's modulus for cellulose pulp fibers, filaments, and films made from CNF and the hydrodynamically assembled filaments of this study. The angle represents the angle of the CNF toward the filament or fiber axis in the case of filaments and toward the tensile testing axis in the case of films. Source: Reprinted with permission of Hakansson et al. 2014 [124]. https://www.nature.com/articles/ncomms5018. https://creativecommons.org/licenses/by/3.0/. Licensed under CCBY 3.0.

all the previously assembled cellulose nanofibers. The cellulose fiber is stronger than the strongest commercially available cellulose filament, owing to the much stiffer cellulose I structure as compared with the cellulose II structure in the regenerated cellulose filaments (Figure 4.17h).

Although wet-spinning CNF is much greener than the predominant spinning regenerated cellulose, it generally involves solvent in the coagulation step. In recent years, dry spinning has been reported to directly spin concentrated CNF aqueous suspension followed by drying in air [125]. The mechanical properties of these dry-spun CNF filaments are determined to be around 222 MPa for tensile strength and 12.6 GPa for Young's modulus, comparably lower than the wet-spun cellulose filaments. However, these dry-spun fibers are more uniform and clearly advantageous considering no chemical coagulation is needed.

4.6 Applications of Cellulose Fibers

With the booming of nanocellulose isolation and advanced characterization techniques, exploring and expanding the applications of these nanoscaled cellulose fibers have been exponentially grown during the past decade, which has expanded essentially in every aspects of human life, including energy, electronics, environmental remediation, and biomedical applications. In this section, these emerging applications for cellulose fibers will be summarized, and due to the limited pages of this chapter, the readers will be directed to respective review papers for more comprehensive analytical summaries on these aspects.

With the rapid development of portable and wearable electronics, energy storage devices have become an essential part of human life. The application of cellulose fibers in energy storage devices has offered several advantages that include biodegradability, flexibility, and lightweight. Therefore, this has sparked tremendous research interests by using cellulose fibers in varied energy storage devices such as lithium ion battery [126] and supercapacitors [127]. The applications of cellulose fibers in supercapacitors has recently been reviewed by Aleman and coworkers [128], showing the promising performance of cellulose fibers as supporting materials, template, separator, and electrolytes.

With increasing environmental deterioration due to the anthropogenic activities, it is of critical importance to remediate the deteriorated natural resources, such as water, air, and soil, by using naturally occurred cellulosic materials. Due to the large surface area and abundantly available surface functional groups, nanocellulose fibers have been widely used in water contaminant removal for water regeneration, including oil contaminant [113, 129], heavy metals [130], and dyes [74, 131], as well as antibacterials [132]. Beyond water treatment, air pollution has also been remediated by nanocellulose-based fibrous materials, such as for use in CO_2 absorption in response to global warming issues [133], and air filter for particle removal [134]. For more comprehensive review on using cellulose nanomaterials on water treatment, the readers could refer to the recent reviews by Wiesner and coworkers [135] and Liu et al. [136].

Due to the excellent biocompatibility, hemocompatibility, biodegradability, and low toxicity, nanocellulose fibers have generally been considered as a safe nanomaterial to be applied in biomedical applications. The applications of nanocellulose fibers in biomedical areas have spanned from the molecular level to macroscopic level, with most works having been extensively reviewed by Lin and Dufresne [52]. From the molecular or nanoscale level, the nanocellulose has been used for fluorescent labeling, cellular culture, drug delivery, enzyme/protein immobilization, and DNA carrier by utilizing the readily available surface functional groups. Besides, the nanocellulose has also been used as tissue bioscaffolds, tissue substitutes, tissue repair materials, and antibacterial wound dressing. The application of macroscaled nanofibrillated cellulose threads decorated with human stem cell has been demonstrated as a suitable surgical bionanomaterials for treating postoperative inflammation and chronic wound healing issues [137].

4.7 Conclusion and Perspectives

Cellulose fibers are the oldest fibrous materials being used by human beings and have been the subject of technical revolution throughout the human history. Currently, cellulose fibers are still widely used in various areas including textile, paper, printed electronics, constructions, and fiber-reinforced composite. Traditionally, continuous cellulosic fibrous materials are predominantly fabricated out of regenerated cellulose or cellulose derivatives, which requires copious amount of solvents to dissolve or derivatize cellulose, creating serious environmental side effects. Besides, the strong and stiff native cellulose I crystalline structures were destroyed during these solvent processing, thus limiting the mechanical properties of the final materials. With the continuous advancement of nanotechnologies in bio-based materials, nanocellulose with well-preserved native cellulose I crystalline structures has drawn great attentions in fabricating fibrous mass. The nanocelluloses are generally a few nanometers wide, approaching the size of single elementary fibrils, and up to micrometers long. The mechanical properties of these nanocelluloses are exceptionally high, with respective Young's modulus of 150 GPa and strength of 3–6 GPa, approaching to or even higher than carbon nanotube or Kevlar fibers. Therefore, it draws great interests in transforming these nanoscale building blocks to the macroscopic fibrous materials, including 2D macrofibers and 3D fibrous mass or aerogels. Although success has been well documented in fabrication of macroscopic fibrous materials from nanocellulose, numerous favorable traits of the nanocellulose have been lost during the transformation, such as the high specific surface area and excellent mechanical properties, just to name a few. The best reported tensile strength of the cellulose filaments is less than 500 MPa, much lower than the 3–6 GPa of the nanocellulose building block. Therefore, in the perspective of fabricating advanced fibrous materials from nanocellulose, several approaches warrant further investigations, including isolating nanocellulose with high mechanical properties and fewer defects, modulating the nanocellulose assembly to transfer the nanoscale characteristics to the macroscopic counterparts and to develop hybrid materials out of nanocellulose and other functional materials. Besides, further expanding the potential applications of the cellulose fibers at different scales is needed to justify the designing of these naturally occurred cellulosic fibrous materials for more functionalities.

References

1. Moon, R.J., Martini, A., Nairn, J. et al. (2011). *Chemical Society Reviews* 40: 3941.
2. Habibi, Y., Lucia, L.A., and Rojas, O.J. (2010). *Chemical Reviews* 110: 3479.
3. Klemm, D., Heublein, B., Fink, H.P., and Bohn, A. (2005). *Angewandte Chemie International Edition* 44: 3358.
4. Roy, D., Semsarilar, M., Guthrie, J.T., and Perrier, S. (2009). *Chemical Society Reviews* 38: 2046.

5 Zugenmaier, P. (2008). *Crystalline Cellulose and Cellulose Derivatives: Characterization and Structures*, 7. Berlin Heidelberg: Springer-Verlag.
6 Staudinger, H. (1920). *Berichte der Deutschen Chemischen Gesellschaft* 53: 1073.
7 Osullivan, A.C. (1997). *Cellulose* 4: 173.
8 Zhu, H.L., Luo, W., Ciesielski, P.N. et al. (2016). *Chemical Reviews* 116: 9305.
9 Habibi, Y. (2014). *Chemical Society Reviews* 43: 1519.
10 Wertz, J.-L., Bedue, O., and Mercier, J.P. (2010). *Cellulose Science and Technology*, 21. Lausanne, Switzerland: EPFL Press.
11 Battista, O.A., Coppick, S., Howsmon, J.A. et al. (1956). *Industrial and Engineering Chemistry* 48: 333.
12 Preston, R.D. (1964). *Formation of Wood in Forest Trees* (ed. M.H. Zimmerman), 169. Academic Press.
13 Brown, R.M. Jr., and Montezinos, D. (1976). *Proceedings of the National Academy of Sciences of the United States of America* 73: 143.
14 Brown, R.M. Jr., (1996). *Journal of Macromolecular Science, Part A Pure and Applied Chemistry* 33: 1345.
15 Mueller, S.C., Brown, R.M., and Scott, T.K. (1976). *Science* 194: 949.
16 Mueller, S.C. and Brown, R.M. (1980). *Journal of Cell Biology* 84: 315.
17 Herth, W. (1985). *Planta* 164: 12.
18 Cosgrove, D.J. (2005). *Nature Reviews Molecular Cell Biology* 6: 850.
19 Saxena, I.M. and Brown, R.M. (2005). *Annals of Botany* 96: 9.
20 Herth, W. (1983). *Planta* 159: 347.
21 Sugiyama, J., Harada, H., Fujiyoshi, Y., and Uyeda, N. (1985). *Planta* 166: 161.
22 Atalla, R.H. and VanderHart, D.L. (1999). *Solid State Nuclear Magnetic Resonance* 15: 1.
23 Atalla, R.H. and Vanderhart, D.L. (1984). *Science* 223: 283.
24 Yamamoto, H. and Horii, F. (1993). *Macromolecules* 26: 1313.
25 Yamamoto, H. and Horii, F. (1994). *Cellulose* 1: 57.
26 Vanderhart, D.L. and Atalla, R.H. (1984). *Macromolecules* 17: 1465.
27 Kataoka, Y. and Kondo, T. (1999). *International Journal of Biological Macromolecules* 24: 37.
28 Helbert, W., Nishiyama, Y., Okano, T., and Sugiyama, J. (1998). *Journal of Structural Biology* 124: 42.
29 Horii, F., Yamamoto, H., Kitamaru, R. et al. (1987). *Macromolecules* 20: 2946.
30 Debzi, E.M., Chanzy, H., Sugiyama, J. et al. (1991). *Macromolecules* 24: 6816.
31 Sugiyama, J., Persson, J., and Chanzy, H. (1991). *Macromolecules* 24: 2461.
32 Sugiyama, J., Okano, T., Yamamoto, H., and Horii, F. (1990). *Macromolecules* 23: 3196.
33 Nishiyama, Y., Sugiyama, J., Chanzy, H., and Langan, P. (2003). *Journal of the American Chemical Society* 125: 14300.
34 Sugiyama, J., Vuong, R., and Chanzy, H. (1991). *Macromolecules* 24: 4168.
35 Langan, P., Nishiyama, Y., and Chanzy, H. (2001). *Biomacromolecules* 2: 410.
36 Sarko, A. and Muggli, R. (1974). *Macromolecules* 7: 486.
37 Sturcova, A., His, I., Apperley, D.C. et al. (2004). *Biomacromolecules* 5: 1333.

38 Nishiyama, Y., Langan, P., and Chanzy, H. (2002). *Journal of the American Chemical Society* 124: 9074.
39 Qian, X.H., Ding, S.Y., Nimlos, M.R. et al. (2005). *Macromolecules* 38: 10580.
40 Jarvis, M. (2003). *Nature* 426: 611.
41 Zhu, H.L., Fang, Z.Q., Preston, C. et al. (2014). *Energy & Environmental Science* 7: 269.
42 Chinga-Carrasco, G., Yu, Y.D., and Diserud, O. (2011). *Microscopy and Microanalysis* 17: 563.
43 Chinga-Carrasco, G. (2011). *Nanoscale Research Letters* 6: 417.
44 Muhlethaler, K. (1967). *Annual Review of Plant Physiology* 18: 1.
45 Frey-Wyssling, A. (1954). *Science* 119: 80.
46 Peterlin, A. and Ingram, P. (1970). *Textile Research Journal* 40: 345.
47 (a) Diddens, I., Murphy, B., Krisch, M., and Muller, M. (2008). *Macromolecules* 41: 9755. (b) Sturcova, A., Davies, G.R., and Eichhorn, S.J. (2005). *Biomacromolecules* 6: 1055.
48 Lahiji, R.R., Xu, X., Reifenberger, R. et al. (2010). *Langmuir* 26: 4480.
49 Iwamoto, S., Kai, W.H., Isogai, A., and Iwata, T. (2009). *Biomacromolecules* 10: 2571.
50 Guhados, G., Wan, W.K., and Hutter, J.L. (2005). *Langmuir* 21: 6642.
51 Hsieh, Y.C., Yano, H., Nogi, M., and Eichhorn, S.J. (2008). *Cellulose* 15: 507.
52 Lin, N. and Dufresne, A. (2014). *European Polymer Journal* 59: 302.
53 Saito, T., Kuramae, R., Wohlert, J. et al. (2013). *Biomacromolecules* 14: 248.
54 Abitbol, T., Rivkin, A., Cao, Y.F. et al. (2016). *Current Opinion in Biotechnology* 39: 76.
55 (a) Turbak, A.F., Snyder, F.W., and Sandberg, K.R. (1983). *Journal of Applied Polymer Science, Applied Polymer Symposium* 37: 815. (b) Herrick, F.W., Casebier, R.L., Hamilton, J.K., and Sandberg, K.R. (1983). *Journal of Applied Polymer Science, Applied Polymer Symposium* 37: 797.
56 Klemm, D., Kramer, F., Moritz, S. et al. (2011). *Angewandte Chemie International Edition* 50: 5438.
57 Svagan, A.J., Samir, M., and Berglund, L.A. (2008). *Advanced Materials* 20: 1263.
58 (a) Nakagaito, A.N. and Yano, H. (2004). *Applied Physics A: Materials Science & Processing* 78: 547. (b) Nakagaito, A.N. and Yano, H. (2005). *Applied Physics A: Materials Science & Processing* 80: 155. (c) Iwamoto, S., Abe, K., and Yano, H. (2008). *Biomacromolecules* 9: 1022.
59 Nechyporchuk, O., Belgacem, M.N., and Bras, J. (2016). *Industrial Crops and Products* 93: 2.
60 Kondo, T., Kose, R., Naito, H., and Kasai, W. (2014). *Carbohydrate Polymers* 112: 284.
61 (a) Zimmermann, T., Pohler, E., and Geiger, T. (2004). *Advanced Engineering Materials* 6: 754. (b) Zimmermann, T., Bordeanu, N., and Strub, E. (2010). *Carbohydrate Polymers* 79: 1086.
62 Taipale, T., Osterberg, M., Nykanen, A. et al. (2010). *Cellulose* 17: 1005.
63 (a) Iwamoto, S., Nakagaito, A.N., and Yano, H. (2007). *Applied Physics A: Materials Science & Processing* 89: 461. (b) Abe, K. and Yano, H. (2009).

Cellulose 16: 1017. (c) Abe, K., Iwamoto, S., and Yano, H. (2007). *Biomacromolecules* 8: 3276. (d) Wang, Q.Q., Zhu, J.Y., Gleisner, R. et al. (2012). *Cellulose* 19: 1631.

64 Jonoobi, M., Mathew, A.P., and Oksman, K. (2012). *Industrial Crops and Products* 40: 232.

65 Kose, R., Mitani, I., Kasai, W., and Kondo, T. (2011). *Biomacromolecules* 12: 716.

66 Jiang, F., Kondo, T., and Hsieh, Y.L. (2016). *ACS Sustainable Chemistry & Engineering* 4: 1697.

67 Isogai, A., Saito, T., and Fukuzumi, H. (2011). *Nanoscale* 3: 71.

68 Spence, K.L., Venditti, R.A., Rojas, O.J. et al. (2011). *Cellulose* 18: 1097.

69 Chen, W.S., Yu, H.P., and Liu, Y.X. (2011). *Carbohydrate Polymers* 86: 453.

70 Chen, W.S., Yu, H.P., Liu, Y.X. et al. (2011). *Cellulose* 18: 433.

71 Jiang, F. and Hsieh, Y.-L. (2013). *Carbohydrate Polymers* 95: 32.

72 Uetani, K. and Yano, H. (2011). *Biomacromolecules* 12: 348.

73 Dufresne, A., Cavaille, J.Y., and Vignon, M.R. (1997). *Journal of Applied Polymer Science* 64: 1185.

74 Chen, W., Li, Q., Wang, Y. et al. (2014). *ChemSusChem* 7: 154.

75 Lavoine, N., Desloges, I., Dufresne, A., and Bras, J. (2012). *Carbohydrate Polymers* 90: 735.

76 Isogai, A. (2013). *Journal of Wood Science* 59: 449.

77 Okita, Y., Saito, T., and Isogai, A. (2010). *Biomacromolecules* 11: 1696.

78 Isogai, T., Yanagisawa, M., and Isogai, A. (2009). *Cellulose* 16: 117.

79 Jiang, F., Han, S., and Hsieh, Y.-L. (2013). *RSC Advances* 3: 12366.

80 Saito, T., Kimura, S., Nishiyama, Y., and Isogai, A. (2007). *Biomacromolecules* 8: 2485.

81 Saito, T., Nishiyama, Y., Putaux, J.L. et al. (2006). *Biomacromolecules* 7: 1687.

82 (a) Li, Q. and Renneckar, S. (2009). *Cellulose* 16: 1025. (b) Li, Q.Q. and Renneckar, S. (2011). *Biomacromolecules* 12: 650.

83 Wagberg, L., Decher, G., Norgren, M. et al. (2008). *Langmuir* 24: 784.

84 Olszewska, A., Eronen, P., Johansson, L.S. et al. (2011). *Cellulose* 18: 1213.

85 Paakko, M., Ankerfors, M., Kosonen, H. et al. (2007). *Biomacromolecules* 8: 1934.

86 Jakob, H.F., Fengel, D., Tschegg, S.E., and Fratzl, P. (1995). *Macromolecules* 28: 8782.

87 Klemm, D., Schumann, D., Udhardt, U., and Marsch, S. (2001). *Progress in Polymer Science* 26: 1561.

88 (a) Brown, A. (1886). *Journal of the Chemical Society* 49: 172. (b) Brown, A. (1886). *Journal of the Chemical Society* 49: 432.

89 Brown, R.M. (1996). *Journal of Macromolecular Science Pure and Applied Chemistry* A33: 1345.

90 Huang, Y., Zhu, C.L., Yang, J.Z. et al. (2014). *Cellulose* 21: 1.

91 Czaja, W., Krystynowicz, A., Bielecki, S., and Brown, R.M. (2006). *Biomaterials* 27: 145.

92 (a) Wu, Z.Y., Liang, H.W., Chen, L.F. et al. (2016). *Accounts of Chemical Research* 49: 96. (b) Tahara, N., Tabuchi, M., Watanabe, K. et al. (1997). *Bioscience Biotechnology and Biochemistry* 61: 1862.

93 Yao, J., Chen, S., Chen, Y. et al. (2017). *ACS Applied Materials & Interfaces* 9: 20330.
94 Doshi, J. and Reneker, D.H. (1995). *Journal of Electrostatics* 35: 151.
95 Huang, Z.M., Zhang, Y.Z., Kotaki, M., and Ramakrishna, S. (2003). *Composites Science and Technology* 63: 2223.
96 Kim, C.W., Kim, D.S., Kang, S.Y. et al. (2006). *Polymer* 47: 5097.
97 Kulpinski, P. (2005). *Journal of Applied Polymer Science* 98: 1855.
98 Kim, C.W., Frey, M.W., Marquez, M., and Joo, Y.L. (2005). *Journal of Polymer Science Part B: Polymer Physics* 43: 1673.
99 Freire, M.G., Teles, A.R.R., Ferreira, R.A.S. et al. (2011). *Green Chemistry* 13: 3173.
100 Quan, S.L., Kang, S.G., and Chin, I.J. (2010). *Cellulose* 17: 223.
101 Viswanathan, G., Murugesan, S., Pushparaj, V. et al. (2006). *Biomacromolecules* 7: 415.
102 Ahn, Y., Hu, D.H., Hong, J.H. et al. (2012). *Carbohydrate Polymers* 89: 340.
103 Xu, S.S., Zhang, J., He, A.H. et al. (2008). *Polymer* 49: 2911.
104 Frey, M.W. (2008). *Polymer Reviews* 48: 378.
105 Qi, H.S., Sui, X.F., Yuan, J.Y. et al. (2010). *Macromolecular Materials and Engineering* 295: 695.
106 Liu, H.Q. and Hsieh, Y.L. (2002). *Journal of Polymer Science Part B: Polymer Physics* 40: 2119.
107 Paakko, M., Vapaavuori, J., Silvennoinen, R. et al. (2008). *Soft Matter* 4: 2492.
108 Chen, W.S., Yu, H.P., Li, Q. et al. (2011). *Soft Matter* 7: 10360.
109 Han, J.Q., Zhou, C.J., Wu, Y.Q. et al. (2013). *Biomacromolecules* 14: 1529.
110 Sehaqui, H., Zhou, Q., and Berglund, L.A. (2011). *Composites Science and Technology* 71: 1593.
111 Jiang, F. and Hsieh, Y.-L. (2014). *ACS Applied Materials & Interfaces* 6: 20075.
112 Jiang, F. and Hsieh, Y.-L. (2014). *Journal of Materials Chemistry A* 2: 350.
113 Jiang, F. and Hsieh, Y.-L. (2014). *Journal of Materials Chemistry A* 2: 6337.
114 Lu, P. and Hsieh, Y.L. (2012). *Carbohydrate Polymers* 87: 564.
115 Jiang, F. and Hsieh, Y.L. (2016). *ACS Sustainable Chemistry & Engineering* 4: 1041.
116 Jiang, F. and Hsieh, Y.-L. (2015). *Carbohydrate Polymers* 122: 60.
117 Saito, T., Uematsu, T., Kimura, S. et al. (2011). *Soft Matter* 7: 8804.
118 Cai, J., Zhang, L.N., Zhou, J.P. et al. (2004). *Macromolecular Rapid Communications* 25: 1558.
119 Lundahl, M.J., Klar, V., Wang, L. et al. (2017). *Industrial and Engineering Chemistry Research* 56: 8.
120 Cai, J., Zhang, L.N., Zhou, J.P. et al. (2007). *Advanced Materials* 19: 821.
121 Walther, A., Timonen, J.V.I., Diez, I. et al. (2011). *Advanced Materials* 23: 2924.
122 Iwamoto, S., Isogai, A., and Iwata, T. (2011). *Biomacromolecules* 12: 831.
123 Torres-Rendon, J.G., Schacher, F.H., Ifuku, S., and Walther, A. (2014). *Biomacromolecules* 15: 2709.
124 Hakansson, K.M.O., Fall, A.B., Lundell, F. et al. (2014). *Nature Communications* 5: 4108.

125 (a) Hooshmand, S., Aitomaki, Y., Norberg, N. et al. (2015). *Acs Applied Materials & Interfaces* 7: 13022. (b) Shen, Y.F., Orelma, H., Sneck, A. et al. (2016). *Cellulose* 23: 3393.

126 (a) Huang, F.L., Xu, Y.F., Peng, B. et al. (2015). *ACS Sustainable Chemistry & Engineering* 3: 932. (b) Chiappone, A., Nair, J.R., Gerbaldi, C. et al. (2011). *Journal of Power Sources* 196: 10280.

127 (a) Yang, X., Shi, K.Y., Zhitomirsky, I., and Cranston, E.D. (2015). *Advanced Materials* 27: 6104. (b) Shi, K.Y., Yang, X., Cranston, E.D., and Zhitomirsky, I. (2016). *Advanced Functional Materials* 26: 6437. (c) Nystrom, G., Marais, A., Karabulut, E. et al. (2015). *Nature Communications* 6.

128 Perez-Madrigal, M.M., Edo, M.G., and Aleman, C. (2016). *Green Chemistry* 18: 5930.

129 Jiang, F. and Hsieh, Y.-L. (2017). *ACS Applied Materials & Interfaces* 9: 2825.

130 (a) Zhu, H., Yang, X., Cranston, E.D., and Zhu, S.P. (2016). *Advanced Materials* 28: 7652. (b) Ma, H.Y., Hsiao, B.S., and Chu, B. (2012). *ACS Macro Letters* 1: 213.

131 Pei, A., Butchosa, N., Berglund, L.A., and Zhou, Q. (2013). *Soft Matter* 9: 2047.

132 Wang, M.S., Jiang, F., Hsieh, Y.-L., and Nitin, N. (2014). *Journal of Materials Chemistry B* 2: 6226.

133 (a) Gebald, C., Wurzbacher, J.A., Tingaut, P. et al. (2011). *Environmental Science and Technology* 45: 9101. (b) Sehaqui, H., Galvez, M.E., Becatinni, V. et al. (2015). *Environmental Science and Technology* 49: 3167.

134 Nemoto, J., Saito, T., and Isogai, A. (2015). *ACS Applied Materials & Interfaces* 7: 19809.

135 Carpenter, A.W., de Lannoy, C.F., and Wiesner, M.R. (2015). *Environmental Science and Technology* 49: 5277.

136 Liu, H.Z., Geng, B.Y., Chen, Y.F., and Wang, H.Y. (2017). *ACS Sustainable Chemistry & Engineering* 5: 49.

137 Mertaniemi, H., Escobedo-Lucea, C., Sanz-Garcia, A. et al. (2016). *Biomaterials* 82: 208.

5

Chitosan Fibers

Seema Sakkara, Mysore Sridhar Santosh, and Narendra Reddy

Centre for Incubation, Innovation, Research and Consultancy, Jyothy Institute of Technology, Thathaguni Post, Bengaluru 560082, India

5.1 Introduction

Chitosan (derived from the raw material chitin) is one of the most abundantly available biopolymers, next only to cellulose. Worldwide, about 22 000 tons of chitosan are produced and the quantity is increasing each year. The market value of chitosan is expected to reach US$4.4 billion by 2020. Typically, industrial grade chitosan having acetylation degrees ranging from 75% to 85% sells at a price of about US$3.50–4.00 kg^{-1}. However, high purity medical grade chitosan is reported to sell at prices up to US$25,000 kg^{-1}. Although most chitosan is derived from chitin found in shrimp, crab, and shells of other marine animals, some fungi such as *Aspergillus niger*, *Mucor rouxii*, and *Penicillium notatum*; microorganisms such as algae, yeast, and others; and insects such as scorpions and spiders (Table 5.1) [1] have also been used to produce chitosan. In terms of structure (Figure 5.1), chitosan has typical features of polysaccharides [2]. The basic unit in chitosan consists of a β-(1-4)-linked D-glucose units. However, unlike cellulose, the hydroxyl at position C-2 of cellulose is replaced by acetamide group, and chitosan contains about 5–8% of nitrogen due to the acetamide group [3]. An FTIR spectrum of chitosan is given in Figure 5.2. An absorption band at 3439 cm^{-1} is seen due to the stretching vibration of the –OH and –NH$_2$ bonds, and the peaks at 1659, 1379, and 1321 cm^{-1} are due to the presence of tertiary amine. The bands in the region of 1460–1200 cm^{-1} and 1160–1000 cm^{-1} are from the COH and COC groups, respectively [4]. Stretching vibrations of amide I and II and bending vibrations from amide III bonds are responsible for the bands at 1659, 1599, and 1321 cm^{-1}.

One of the most notable characteristics of chitosan is its solubility in acidic solutions [1]. Some of the common solvents used to dissolve chitosan are listed in Table 5.2. Recently, water-soluble chitosan has also been developed. Primary applications of chitosan are in agrochemicals, water treatment, medicine,

5 Chitosan Fibers

Table 5.1 Sources used for production of chitosan.

Sea animals	Insects	Microorganisms
Annelida	Scorpions	Green algae
Mollusca	Spiders	Yeast
Coelenterata	Brachiopods	Fungi
Crustaceans	Ants	Mycelia penicillium
• Lobster	Cockroaches	Brown algae
• Crab	Beetles	Spores
• Shrimp		*Chytridiaceae*
• Prawn		Ascomydes
• Krill		*Blastocladiaceae*

Figure 5.1 Structure of chitosan.

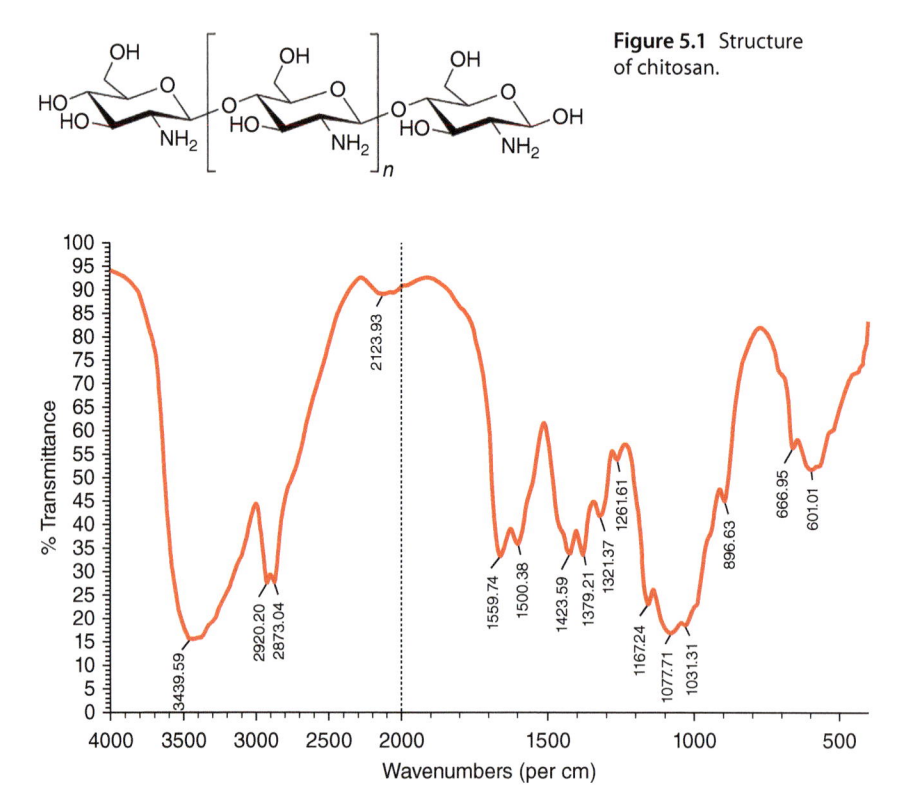

Figure 5.2 FTIR spectrum of chitosan shows the distinct peaks. Source: Chen et al. 2011 [4]. Reproduced with permission from Elsevier.

cosmetics, and others (Table 5.3) [5]. Since chitosan is widely available, is easily soluble, and has unique properties, it has been made into various products and considered for several applications. In addition to films, hydrogels, sponges, etc., considerable efforts have also been made to produce solution spun, electrospun, and regenerated chitosan fibers. Since fibrous materials are preferred for medical applications and chitosan has inherent antibacterial properties, most

Table 5.2 Common solvent/solvent mixtures used to dissolve chitosan.

Solvents
Aqueous acetic acid
Trifluoroacetic acid
Aqueous acetic acid/dimethylformamide
Aqueous acetic acid/dimethyl sulfoxide
1,1,1,3,3,3,-Hexafluoroisopropanol (HFIP)
HFIP/formic acid
Tetrahydrofuran/dimethylformamide
$CHCl_3$
Lithium hydroxide (LiOH)
Glycine chloride

Table 5.3 Various forms and applications of chitosan.

Form of chitosan	Application
Beads, microspheres, nanoparticles	Drug delivery, enzyme immobilization, gene delivery
Coatings	Surface modifications, textile finishes
Fibers, nanofibers, nonwovens	Medical textiles, suture, bone regeneration, nerve tissue engineering, wound healing
Films	Wound care, dialysis membrane, wound dressing and antimicrobial membranes
Powder	Absorbent for pharmaceutical and medical, surgical glove powder, enzyme immobilization
Sponge	Mucosomal hemostatic dressing, drug delivery, enzyme entrapment and artificial skin
Solution	Cosmetics, bacteriostatic, hemostatic agents, anticoagulants and antitumor agents
Gels	Implants, coating, tissue engineering
Tablets and capsules	Disintegrating agent, delivery vehicles

of the studies on chitosan fibers have been intended for medical applications. Also, in chitosan, there are one primary amino and two free hydroxyl groups for each building (C6) unit, making chitosan positively charged. Hence, negative charge-carrying metal ions such as cobalt are easily attracted and make chitosan an ideal material for separation of metals from wastewater [4].

5.2 Extraction/Modification of Chitosan

Chitosan is extracted from chitin using various physical, chemical, and other approaches. Typically, chitin is deacetylated to various degrees using alkaline conditions. In the conventional approach, chitin is treated with 3 M hydrochloric

Table 5.4 Comparison of the apparent viscosities (cP) of extracted and commercial chitosan at different temperatures.

Temperature (°C)	Extracted chitosan	Commercial chitosan
20	56.5 ± 0.14	38.2 ± 0.03
25	46.1 ± 0.07	31.9 ± 0.00
30	38.2 ± 0.02	27.1 ± 0.01
35	31.7 ± 0.06	23.3 ± 0.0
40	26.7 ± 0.01	20.4 ± 0.0

Source: Kucukgulmez et al. 2011 [8]. Reproduced with permission from Elsevier.

acid to remove calcium carbonate. Later, the materials are treated with 10% sodium hydroxide at 80 °C for two hours to remove proteins. Deacetylation of the chitosan is achieved by treating with 50% sodium hydroxide at 100 °C for 2.5 hours. Under similar conditions, chitosan with antibacterial and antifungal properties was extracted from two Tunisian crustacean species [6]. Shrimp shells were used as a source of chitosan and treated with 1 M NaOH for 24 hours and later demineralized using 1 M HCl and deprotonized using 1 M NaOH for 24 hours. Later, chitosan was discolored using $KMnO_4$ and oxalic acid to get the chitin powder [7]. Up to 89% deacetylation could be achieved. The extracted chitosan was grafted onto rayon to improve the antibacterial and antimicrobial properties [6]. Using a similar approach, chitosan was extracted from *Metapenaeus stebbingi* shells. About 17.5% of the weight of the shells could be obtained as chitosan with properties shown in Tables 5.4 and 5.5. It was suggested that the shells could be an inexpensive source to produce chitosan at high efficiency [8]. Although alkalis and acids have been extensively used to extract keratin, the type of acid and extraction conditions determine the properties of the chitosan obtained [9]. This was substantiated when chitosan was extracted from 15-day biomass of *Agaricus* sp., *Pleurotus* sp., and *Ganoderma* sp. [9]. Similarly, fungal-based chitosan was extracted, purified, and characterized from *Aspergillus brasiliensis*. A degree of deacetylation of 81.3%, molecular weight of 310 000 Da, and 96% solubility in 1% acetic acid was obtained for the chitosan [10].

Table 5.5 Characteristics of the chitosan extracted from *Metapenaeus stebbingi* shells compared with commercial chitosan.

	Extracted chitosan	Commercial chitosan
Yield (%)	17.5 ± 0.6	—
Moisture (%)	1.3 ± 0.08	1.07 ± 0.09
Ash (%)	0.6 ± 0.03	0.59 ± 0.07
Deacetylation degree (%)	92.2 ± 2.7	86.92 ± 1.62
Molecular weight (kDa)	2.2 ± 0.03	3.52 ± 0.02
Water binding capacity (%)	713 ± 12.0	493 ± 10.10
Fat binding capacity (%)	531 ± 12.3	383 ± 10.02

Source: Kucukgulmez et al. 2011 [8]. Reproduced with permission from Elsevier.

Many attempts have been made to develop alternative technologies for extraction of chitosan since the conventional process of extracting chitosan consumes considerable time and energy. Microwave irradiation was considered as a solution to extract chitosan rapidly at low cost. Using similar chemicals, chitin was heated for eight minutes in a microwave to demineralize, deprotenize, and deacetylate the chitin. Compared with the conventional approach that takes six hours and 30 minutes, the entire conversion from chitin to chitosan with same level of deacetylation took only 24 minutes [11]. A degree of deacetylation of 82.7% was achieved similar to that of conventional heating. However, chitosan obtained using microwave heating had higher molecular weight but similar crystallinity. It was suggested that microwave heating was an environmentally friendly and inexpensive approach to extract chitosan from shrimp shells (Tayel et al. 2016). Microwave extraction was also used to obtain chitosan from marine sources in the Arabian Gulf [12]. Figure 5.3 shows the schematic of the process used for the microwave extraction. Similar to previous studies on using microwave extraction, the chitosan obtained from the microwave method had higher molecular weight and crystallinity.

Micelia waste from *A. niger* was used as a source of chitosan and extracted using enzymes. Instead of using acid or alkali, lysozymes, snailase, neutral protease, and chitin deacetylase were used for the extraction. Enzymatic method produced chitosan with considerably higher molecular weight but similar level of deacetylation (Table 5.6), leading to an environmentally friendly approach [13].

Chitosan has been extensively modified to make it soluble in water and other common solvents, improve properties, and make it suitable for various medical applications [14]. Carboxymethylation is one of the most common approaches of modifying chitosan. Some of the chitosan derivatives and brief method to obtain carboxymethylated chitosans are shown in Figure 5.4, and the properties of the modified chitosans are in Table 5.7. In addition to carboxymethylation, chitosan has been acylated, alkylated, and grafted with various chemicals and

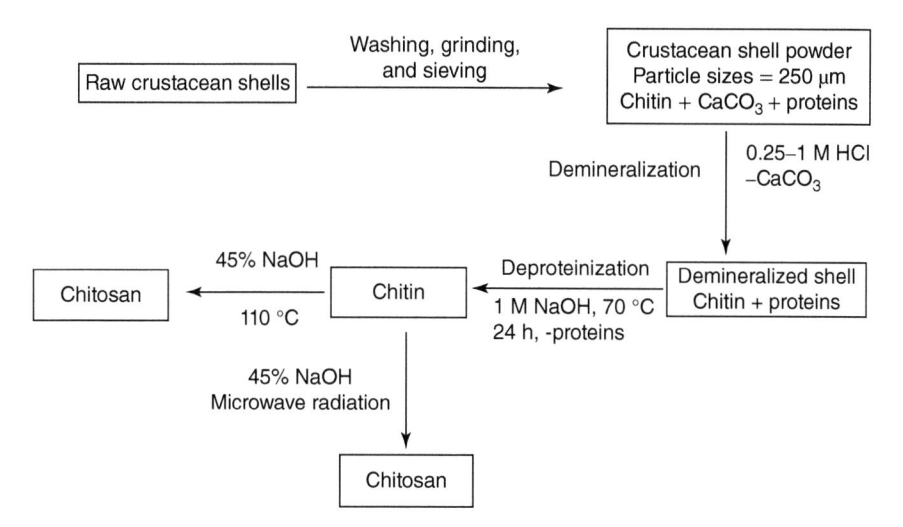

Figure 5.3 Schematic of the process of extraction of chitosan from chitin using microwave heating. Source: Sagheer et al. 2009 [12]. Reproduced with permission from Elsevier.

5 Chitosan Fibers

Table 5.6 Comparison of the properties of chitosan obtained using the conventional chemical approach and the enzymatic method.

Parameter	Chemical method	Enzymatic method
Degree of deacetylation (%)	76.8	73.6
Molecular weight (kDa)	84.04	268
Content of glucosamine (%)	82.7	94.4
Ash (%)	0.370	0.283

Source: Cai et al. 2006 [13]. Reproduced with permission from Elsevier.

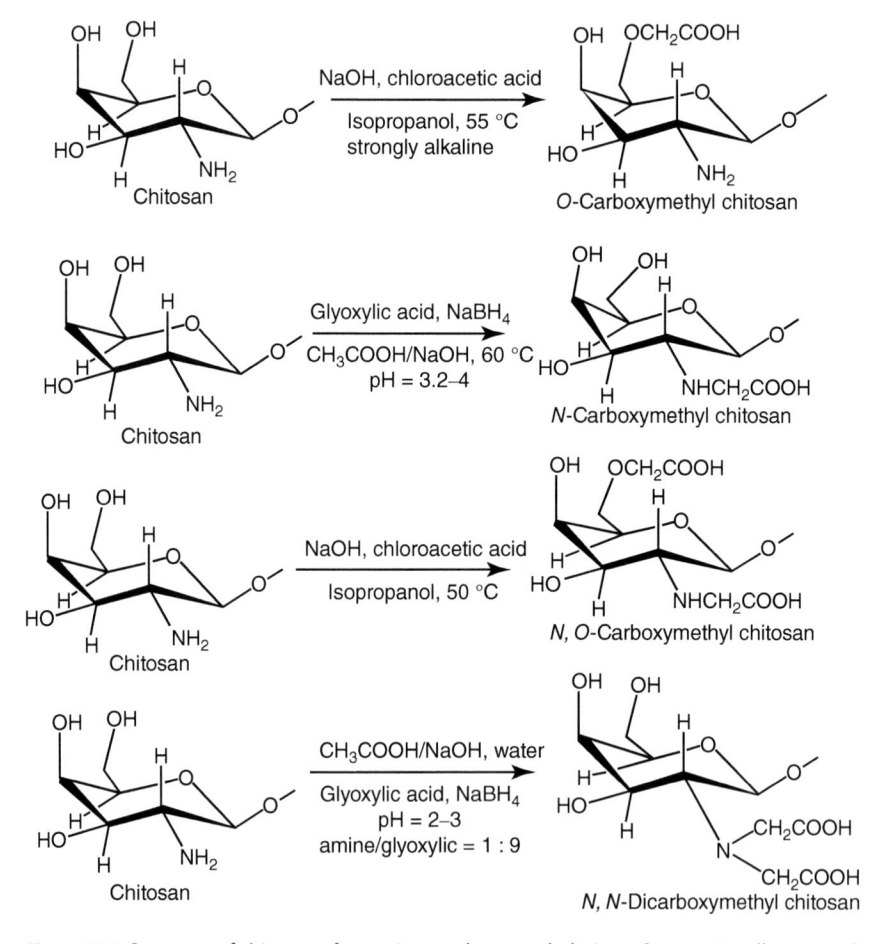

Figure 5.4 Structure of chitosan after various carboxymethylations. Source: Upadhyaya et al. 2013 [14]. Reproduced with permission from Springer Nature.

Table 5.7 Properties of various forms of chemically modified chitosans.

Derivatives	Subtypes	Important properties
Quaternized chitosan	N,N,N-Trimethyl chitosan (TMC)	Good absorption efficiency, mucoadhesion
Carboxyalkyl chitosan	N,O-Carboxymethyl chitosan (N,O-CMC)	Better gelling property, with high water retentivity and less immunogenicity
Hydroxyalkyl chitosan	Hydroxyethyl chitosan	Hydrophilic nature, increase in moisture absorption rate
	Hydroxypropyl chitosan	High inhibition rate of bacteria
Thiolated chitosan	Chitosan-4-thio-butylamidine	Increase in porous structure and solubility
	Chitosan-thioglycolic acid	Excellent gelling property
Sulfated chitosan	2-N, 6-O-sulfated chitosan (26SCS)	Inhibits blood coagulation
Semisynthetic resins of chitosan	Copolymer of chitosan with methyl methacrylate	Good mechanical and nanoporous structure used in drug delivery
Sugar derivatives of chitosan	Glucosamine chitosan	Increase in water solubility and antimicrobial property
Glutaraldehyde chitosan		Form a biocompatible and biostable complex
Chitosan-ascorbate conjugate		Good antioxidant property with scavenging and chelating capacity

Source: Upadhyaya et al. 2013 [14]. Reproduced with permission from Elsevier.

polymers [14]. Other than these conventional approaches, low temperature plasma and electron beam including γ-irradiation have been done to modify the properties of chitosan. Treating with plasma enhanced the medical applications of chitosan, whereas γ-irradiation reduced the molecular weight and also affected the ability to promote cell attachment and growth [15].

5.3 Fibers from Chitosan

5.3.1 Production of Pure Chitosan Fibers

Chitosan has been made into fibers using various techniques. In one of the earliest studies on producing fibers from chitosan, about 5% chitosan was dissolved in 2% acetic acid solution and extruded into a coagulation bath consisting of NaOH. Coagulated fibers were drawn to various ratios to improve their mechanical properties. Mechanical properties and fineness of the fibers was dependent on the draw ratio. Fineness decreased from 14.7 to 6.4 tex, and tenacity varied between 0.17 and 0.19 N/tex, and elongation increased substantially from 4.9% to 8.7%. Chitosan fibers were reconverted into chitin fibers by treating with acetic anhydride [16]. To further improve the antimicrobial properties of chitosan, silver-containing compound was mixed and extruded with the fibers. The silver particles were found to be uniformly distributed along the fiber and increased the antimicrobial activity substantially.

Common approach of developing chitosan fibers is to dissolve the polymer in acetic acid and extrude into NaOH coagulation to form the fibers. In an alternative technique, chitosan having a viscosity average molecular weight of 3.0×10^6 Da and 65% of deacetylation was dissolved using ionic liquids (7.2% LiOH) and 6.0 wt% urea. Dissolved solution was extruded into a coagulation bath consisting of 6–8% sulfuric acid/aqueous ethanol at 40–50 °C and later into a bath containing water. Fibers obtained using the ionic dissolution method had higher thermal stability with a decomposition temperature of 304 °C compared with 295 °C for the normal fibers. Morphologically, the fibers had a rough surface and dense structure. Tensile strength of the fibers was 134 MPa, and elongation was 11.8%, about 20% higher strength compared with the fibers developed using the conventional approach [17]. In another study, glycine chloride was used as an ionic liquid to dissolve chitosan and produce fibers with improved properties [17]. Fibers were extruded into a sodium sulfate/ethanol bath and freeze-dried immediately. Chitosan I crystal form was retained, and fibers produced had tenacity of 3.8 cN/dtex compared with 0.9 cN/dtex obtained for chitosan fibers dissolved using acetic acid. Considerable variations were observed in the tensile properties depending on the conditions used to produce the fibers (Table 5.8). In addition, the ionic liquid could be recovered completely, leading to a green process of producing chitosan fibers.

Chitosan fibers were produced by dissolving in acetic acid–methanol mixture and extruding into various coagulations baths. The chitosan was also modified using vanillin and other aldehydes. Some of the properties of the unmodified and modified chitosan fibers are listed in Table 5.8. Tensile strength and elongation varied considerably when the molecular weight or the coagulation conditions or

Table 5.8 Variations in the properties of chitosan filaments with change in spinning conditions and extent of stretching.

Solvent	Coagulation bath	Stretching	Filament fineness (dtex)	Tenacity (g/tex)	Elongation (%)
2% AcOH + MeOH	10% NaOH + 30% AcONa	No	10.0	1.27	13.0
2% AcOH	10% NaOH + 30% AcONa	No	14.9	0.68	12.6
2% AcOH + MeOH	10% NaOH + 30% AcONa	Yes	4.2	1.43	17.3
2% AcOH	10% NaOH + 30% AcONa	Yes	8.4	0.87	23.0
2% AcOH + MeOH	10% NaOH + 30% Na_2SO_4	No	9.0	0.73	13.6
2% AcOH	10% NaOH + 30% Na_2SO_4	No	8.8	0.98	21.4
2% AcOH + MeOH	10% NaOH + 30% Na_2SO_4	Yes	7.1	1.38	16.5
2% AcOH	10% NaOH + 30% Na_2SO_4	Yes	10.0	1.17	28.1
2% Oxalic acid	10% NaOH + 30% AcONa	No	6.2	0.85	15.1

Source: Hirano et al. 1999 [18]. Reproduced with permission from Elsevier.

Table 5.9 Tensile properties of chitosan films in the dry and wet condition.

Concentration of sodium acetate (%)	Condition of measurement	Tenacity		Elongation		Initial modulus at 2% extension	
		Mean (g/d)	CV (%)	Mean (%)	CV (%)	Mean (g/d)	CV (%)
0	Dry Wet	1.30 1.21	9.3 5.6	15.4 14.6	10.6 9.9	31.9 23.9	9.7 4.3
5	Dry Wet	1.35 1.25	8.9 9.5	15.4 14.3	13.2 12.8	32.5 24.2	4.3 9.5
10	Dry Wet	1.43 1.33	6.7 5.6	13.2 14.6	10.9 12.8	35.8 24	5.4 7.0
15	Dry Wet	1.42 1.31	7.9 4.2	13.2 14.2	11.2 11.8	36.1 25.4	8.5 9.9
20	Dry Wet	1.44 1.35	6.6 10.3	13.6 14.7	10.7 14.4	36.3 28.6	7.8 9.4

Source: Lee et al. 2007 [19]. Reproduced with permission from Elsevier.

the type of aldehyde modifications were changed [18]. Also, stretching caused large variations in the fiber diameters. For instance, increasing molecular weight of chitosan from 14 to 24×10^4 kDa increased tenacity from 0.8 to 1.1 g/dtex, but the elongation did not show any consistent trend. Post-treatment of chitosan with aldehydes was suggested to be a better approach for producing chitosan fibers with improved properties [18]. Chitosan dissolved in acetic acid and having a viscosity of 52 700 cP at 25 °C was extruded into a coagulation bath of 1 M KOH and later into methanol. Dried fibers were immersed into cross-linking solutions (glyoxal or glutaraldehyde) of various concentrations. Considerable changes in the mechanical properties and decrease in swelling were observed after cross-linking (Table 5.9). It was suggested that a Schiff's base reaction occurred between chitosan and the aldehydes and specifically between the chitosan free amine and a hemiacetal [20].

In another study, glyoxal was used as a cross-linking agent for chitosan fibers obtained using 3.5% chitosan solution to improve properties of the fiber [21]. Considerable increase in tensile properties was observed depending on the processing conditions used. A tenacity of about 2.1 cN/dtex (1.0 cN/dtex before cross-linking) was obtained using a glyoxal concentration of 4%. Acetalization and Schiff base reaction were considered to be the reasons for the improved mechanical properties. A one-step fiber production and cross-linking was achieved by extruding chitosan solution into sodium acetate solution containing epichlorohydrin as the cross-linking agent [19]. Increasing the concentration of sodium acetate continually increased the dry and wet strength of the chitosan fibers (Table 5.10). Heat of decomposition also showed substantial increase, and fibers had a smooth cross section after cross-linking. In an earlier study on using epichlorohydrin to cross-link chitosan, strength of the fibers increased from 1.3 to 1.5 g/den with substantial decrease in elongation from 10.4% to 8.7%. However, there was considerable increase in wet strength of the fibers from 0.7 g/den before cross-linking to as much as 1.2 g/den after cross-linking. In another study, chitosan with deacetylation of 95% was made into fibers using acetic acid as the solvent. Fibers obtained were modified by treating with solutions containing phthalate and phosphate ions. Properties of the fibers were

Table 5.10 The variation of dry fiber mechanical properties with concentration of dialdehyde (1 hour, 25.8 °C).

Concentration of dialdehyde[a] (mol/dl), pH	Modulus (g/den)[b],[c]	Tenacity (g/den)[d]	Elongation at break (%)
Glutaraldehyde control (water)	45.19 ± 0.77	1.44 ± 0.02	27.61 ± 2.53
7.64 E-5, 5.5	49.94 ± 1.94	1.71 ± 0.02	11.94 ± 1.05
1.46 E-4, 5.0	51.73 ± 0.68	1.94 ± 0.03	10.67 ± 0.77
6.40 E-4, 4.5	59.78 ± 1.38	2.09 ± 0.06	6.75 ± 0.76
1.27 E-2, 4.0	77.99 ± 1.35	2.19 ± 0.06	5.94 ± 0.28
2.44 E-2, 3.9	78.58 ± 4.46	2.23 ± 0.02	3.71 ± 0.25
5.22 E-2, 3.6	78.27 ± 4.37	2.21 ± 0.00	3.32 ± 0.13
9.88 E-2, 3.3	75.18 ± 0.23	1.64 ± 0.04	2.40 ± 0.09
2.20 E-1, 3.3	64.62 ± 2.70	1.22 ± 0.02	2.32 ± 0.13
Glyoxal control (water)	49.22 ± 0.77	1.44 ± 0.01	28.69 ± 2.25
1.71 E-4, 5.5	48.78 ± 2.05	1.68 ± 0.03	11.40 ± 0.30
3.32 E-4, 5.1	52.01 ± 2.19	1.93 ± 0.04	9.00 ± 1.47
6.29 E-4, 4.7	56.56 ± 4.38	1.96 ± 0.14	7.33 ± 1.30
1.38 E-3, 4.4	71.42 ± 2.73	2.20 ± 0.04	5.88 ± 0.30
2.10 E-3, 4.1	72.76 ± 0.38	2.20 ± 0.21	5.02 ± 1.46
2.75 E-3, 3.9	71.93 ± 5.16	2.35 ± 0.14	5.45 ± 0.87
5.25 E-3, 3.8	79.03 ± 6.22	2.42 ± 0.09	4.92 ± 1.05
2.58 E-2, 3.4	86.66 ± 5.14	2.46 ± 0.01	4.81 ± 0.09
5.08 E-2, 3.1	90.18 ± 5.86	2.53 ± 0.1	4.35 ± 0.32
1.00 E-1, 2.8	92.18 ± 1.16	2.46 ± 0.02	4.28 ± 0.99
2.17 E-1, 2.6	89.13 ± 1.90	2.35 ± 0.00	3.21 ± 0.00

a) Testing conditions were at 65% RH and 20 °C.
b) Fiber denier 5 20.7 dtex or 18.6 denier.
c) Gauge length was set at 2.3 cm.
d) Breaking strength, in grams force, is calculated by multiplying the tenacity by the denier.
Source: Knaul et al. 1999 [20]. Reproduced with permission from John Wiley & Sons.

highly dependent on the pH, immersion time and temperature [22]. Further, drawing of the fibers decreased the denier and elongation but considerably increased the strength due to the better alignment/orientation of the crystals in chitosan.

Common methods of wet spinning produce chitosan fibers that are several magnitudes higher than 10 μm in diameter. To produce fibers of less than 10 μm, a novel method of extruding chitosan solution into a bath consisting of sodium chitosan and STPP leading to formation of stable fibers that had compressive strength of 2400 kPa in dry and 280 kPa in the wet condition considered to be suitable for medical applications was developed [23]. In a similar approach called electro-wet spinning, submicron fibers were produced by dissolving 2% chitosan

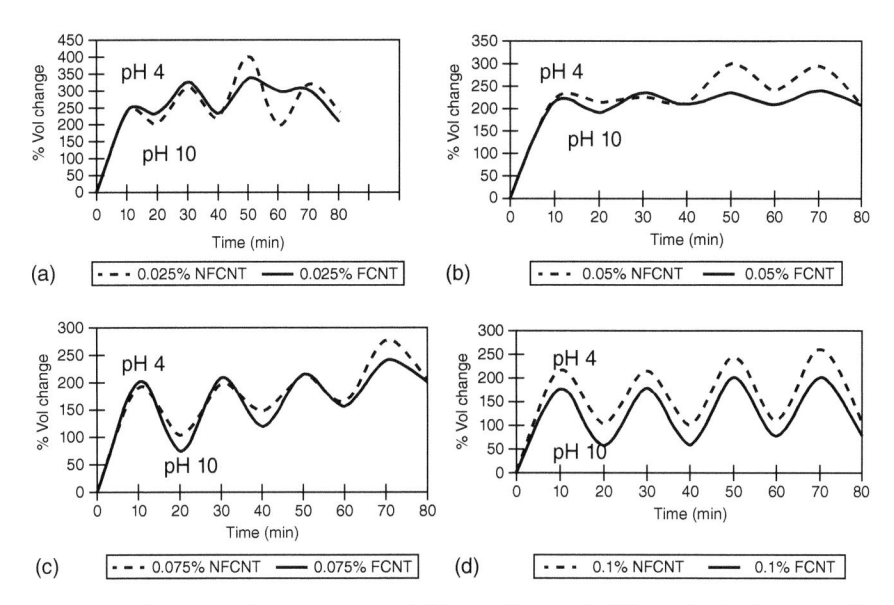

Figure 5.5 Changes in the pH response of chitosan fibers with different levels of single-walled carbon nanotubes. Source: Ozarkar et al. 2008 [25]. Reproduced with permission from Institute of Physics. (a) 0.025% NFCNT and 0.025%FCNT, (b) 0.05% NFCNT and 0.05% FCNT, (c) 0.075% NFCNT and 0.075% FCNT, and (d) 0.1% NFCNT and 0.1% FCNT.

in 2% acetic acid and extruding the solution into a coagulation bath consisting of ethyl alcohol and sodium hydroxide under an applied electric field [24]. Chitosan fibers developed were electroactive and had response time of less than 10 seconds and also showed reversible electroactuated contraction at low voltages (1.5 V) [24]. Better biocompatibility and stimuli-responsive properties of these materials were considered to be suitable for use as artificial muscles, biosensors, or artificial organs. Chitosan fibers reinforced with single-walled carbon nanotubes were also found to have electroresponsive properties. Reinforcing chitosan fibers with the carbon nanotubes increased the strength of the fibers from 96 to 226 MPa even at a nanotube concentration of 0.2 wt% [25]. Electric response and swelling of the fibers was dependent on the pH and other parameters (Figure 5.5).

5.3.2 Chitosan Blend Fibers

Since pure chitosan has limited applications and properties mainly due to its inferior mechanical properties, several researchers have attempted to produce blends of chitosan with various polymers. Many methods such as cross-linking and addition of inorganic or organic fillers have also been used to increase the mechanical properties of chitosan fibers. In a different approach, chitin nanocrystals made through acid treatment of chitin and having high surface area and modulus and biocompatibility were incorporated into chitosan solution and wet-spun into fibers. Nanocrystals were found to be uniformly distributed in the fibers and substantially increased the strength and modulus but decreased the elongation up to 5 wt% in the fibers [26]. Highest strength of 2.2 cN/dtex and modulus of 146 cN/dtex was reported.

Bicomponent electrospun fibers were developed using a combination of chitosan (82.5% deacetylated) and PVA. Chitosan dissolved in acetic acid was mixed with PVA solution in various ratios and electrospun into fibers. However, the presence of chitosan hindered fiber formation, and it was not possible to obtain uniform fibers above a chitosan content of 25%. To overcome this limitation, chitosan was hydrolyzed using sodium hydroxide to substantially decrease the molecular weight and increase the chitosan content in the fibers. By decreasing the molecular weight of chitosan from 1500 dkDa to 800 kDa through alkaline hydrolysis, it was possible to obtain uniform bicomponent fibers (20 nm in diameter), having chitosan content up to 50%. Further, the PVA component in the fibers was removed by treating with NaOH, resulting in porous fibers.

A composite fibrous scaffold suitable for bone tissue engineering was fabricated by combing chitosan, polylactic acid, and hydroxyapatite. Both chitosan and HA were uniformly distributed in the PLA matrix, leading to improved the degradation of PLA. Bone mesenchymal stem cells had good affinity and grew along the direction of the chitosan fibers suggesting their suitability for bone tissue engineering [27]. Chitosan fibers blended with hyaluronan were produced using wet spinning method for potential use as scaffolds for ligament tissue engineering [28]. Chitosan used for fiber production had a molecular weight of 600 000 and deacetylation degree of 81%. Solution was extruded into methanol to form the fibers. Inclusion of HA in chitosan increased mechanical properties and also biological activity, making the fibers suitable for ligament tissue engineering.

Composite fibers composed of chitosan and O-hydroxyethyl chitosan fibers were prepared by blending O-hydroxyethyl chitosan xanthate and cellulose xanthate using the viscose production process [29]. The chitosan ester was prepared using sodium hydroxide and carbon disulfide. Chitosan ester prepared was blended with various ratios of cellulose xanthate and wet-spun into fibers. Properties of the fibers and their activity against *Escherichia coli* were measured. Properties of the blend fibers (Table 5.11) were similar to that of viscose rayon except wet elongation, which increased significantly as the chitosan ratio was increased. The presence of chitosan imparted excellent antibacterial activity against *E. coli* and also increased the antistatic properties [29].

Table 5.11 Comparison of the properties of cellulose/O-hydroxyethyl chitosan fibers with viscose rayon [29].

Fiber	Tensile strength (cN/dtex)		Elongation at break (%)		Moisture absorption (%)	Mass specific resistance (Ωg/cm, $\times 10^8$)
	Dry	Wet	Dry	Wet		
Viscose rayon	3.06	2.70	19.0	17.9	13.01	4.7
CHCFs-3	3.04	2.66	21.2	19.4	13.23	4.3
CHCFs-4	3.01	2.61	22.5	21.2	13.40	4.2
CHCFs-6	2.97	2.55	22.9	24.5	13.55	4.0

Source: Xu et al. 2010 [29]. Reproduced with permission from Elsevier.

Table 5.12 Properties of papain purified using the chitosan/nylon hybrid fibers.

Protein content (mg)	Specific activity (U/mg)	Purification (fold)
15.1	6.67	—
1.3	6.17	0.93
1.1	7.08	1.06
0.7	9.76	2.46
0.4	16.41	4.79
0.3	31.95	2.87
0.2	13.27	1.99

Source: Zhang et al. 2010 [30]. Reproduced with permission from Springer Nature.

Chitosan/nylon 6 hybrid fibers with diameters in the range of 80–310 nm were developed through electrospinning by dissolving the polymers using a 90/10 (v/v) mixture of HFIP/formic acid. Membranes obtained were immobilized with cibacron Blue F3G through nucleophilic reaction. Ability of the immobilized membranes to purify papain was investigated [30]. An equilibrium adsorption capacity of 93.5 mg/g of papain was obtained, and most of the adsorbed papain could be eluted depending on the pH and other conditions (Table 5.12). It was suggested that the membranes could be used for industrial purification of papain due to their high affinity and regeneration potential.

To exploit the advantages of natural and synthetic polymers for medical applications, chitosan was blended with poly(caprolactone) and made into nanofibers. The two polymers were blended by dissolving in mixture of 3/7 acetic acid/formic acid. It was observed that addition of PCL substantially increased the electrospinnability of chitosan [31] due to favorable change in viscosity. Blend fibers had diameters of about 154 nm and an ultrafine nanofiber web could be obtained. In a similar approach, chitosan was blended with poly(lactic acid) in a core–shell form using coaxial electrospinning [32]. Nonwoven fibrous mats were obtained with PLA as the core and chitosan as the sheet. Blend fibers had an average diameter of 236 nm and exhibited excellent antibacterial activity up to 100% against E. coli.

To improve the mechanical properties and applicability of chitosan, chitosan fibers modified with hydroxyapatite were combined with nanoparticles of hydroxyapatite (HAp) and tricalcium phosphate (β-TCP). Fibers were produced by extruding the mixture into aqueous NaOH solution and later washed using ethanol–water solution [33]. Some of the properties of the chitosan fibers containing various levels of the nanoparticles are listed in Tables 5.13 and 5.14. Inclusion of the individual nanoparticles increased the linear density and decreased the tenacity of the fibers, whereas the nanoparticle complex did not decrease the tenacity. However, it was reported that the presence of the nanoparticles facilitated smoother production of the fibers during extrusion [33]. In a similar approach, various (2, 4, or 6%) graphene oxides were added into 10% chitosan dissolved in acetic acid and made into fibers using the wet spinning

Table 5.13 Some properties of chitosan solutions containing hydroxyapatite (HAp) and tricalcium phosphate (ß-TCP) particles.

Solution code	Percentage of solution used (%)			Concentration of				Dynamic viscosity/ temperature (Pa/°C)
	A	B	C	ß-TCP (wt%)	HAp (wt%)	Acetic acid (wt%)	Chitosan (wt%)	
Chit 58	100	—	—	—	—	3.00	5.16	19 000/52
MCT 6	83.3	16.7	—	0.333	—	2.75	4.50	7500/49
MCT 7	83.3	—	16.7	—	0.330	2.75	4.63	9250/49
MCT 8	71.4	14.3	14.3	0.286	0.283	2.36	4.46	4500/52
MCT 11	62.5	25.0	12.5	0.707	0.252	2.10	4.01	1750/51

Source: Wawro et al. 2011 [33]. https://www.mdpi.com/1422-0067/12/11/7286 and http://creativecommons.org/licenses/by/3.0/. Licensed under CCBY 3.0.

Table 5.14 Influence of HAp, β-TCP, and HAp/β-TCP on the mechanical properties of chitosan fibers.

Parameter	Chitosan	MCT6	MCT7	MCT8	MCT11
Linear density (dtex)	4.39	4.48	5.14	5.41	4.16
Breaking force (cN)	3.61	3.51	2.46	2.91	3.35
Tenacity (cN/tex)	8.22	7.83	4.79	5.38	8.05
Elongation at break (%)	17.0	22.0	9.9	11.0	12.0
Tenacity, wet (cN/tex)	6.38	4.93	3.46	4.58	6.86
Elongation, wet (%)	7.8	7.3	7.8	6.1	8.0

Source: Wawro et al. 2011 [33]. https://www.mdpi.com/1422-0067/12/11/7286 and http://creativecommons.org/licenses/by/3.0/. Licensed under CCBY 3.0.

approach [34]. Considerable changes in the morphology and tensile properties were observed due to the presence of graphene oxide. Tensile strength increased from 96 to 168 MPa when the graphene oxide content was increased from 0% to 6%. However, elongation of the fibers decreased with increasing graphene. The blend fibers spontaneously absorbed fuchsin acid dye through an exothermic process. It was suggested that the graphene oxide-containing chitosan fibers would be suitable for removal of dyes from aqueous solutions [34].

Composite chitosan-poly(lactide-co-glycolide) (PLGA) fibers were developed by dissolving the two polymers in a blend of methylene chloride and HFIP. Dissolved polymers were extruded into a methanol bath. Similar approach was also used to produce PLGA-dispersed chitosan blend fibers that were formed into a mat. The fibers obtained had diameters of 30 μm, and the micropores formed in the fibers had diameters in the range of 150–200 μm [35]. Addition of methylene chloride was necessary to completely dissolve chitosan in HFIP. Combing PLGA and chitosan resulted in desired degradability and mechanical properties. Chondrocytes loaded on the blend scaffolds showed active growth and potential for tissue engineering applications [35].

In another study, chitosan and poly(vinyl alcohol) blend fibers were prepared by dissolving chitosan in acetic acid and PVA in water. The two polymers were combined in various ratios and then extruded into a coagulation bath consisting of 6% sodium hydroxide [36]. Blends containing 60% chitosan had highest strength, moisture absorption, and antibacterial properties against *E. coli* and *Staphylococcus aureus*. Instead of a binary blend, a tertiary blend of chitosan/PLA and collagen was prepared for potential use for cartilage tissue engineering. A highly porous scaffold was obtained with good potential for cell attachment and viability [37].

5.3.3 Generating Hollow Chitosan Fibers

Hollow chitosan fibers were prepared as potential carriers for various biomolecules [38]. To obtain the hollow structure, outside layer of the fibers was coagulated first, and later, the core of the fiber was removed by passing air and then treating with 1 M NaOH solution. Internal diameter of the hollow fibers was $400 \pm 20\,\mu m$ (Figure 5.6) with a wall thickness of about $60 \pm 5\,\mu m$, and standard length was 0.5 m. Fibers were immersed in a solution containing various biomolecules, and the diffusion properties were studied. pH was found to have major effect on the diffusion compared with permeate concentration or flow rate. Diffusion of biomolecules such as tryptophan, vitamin B12, and chloramphenicol was dependent on the molecular weight. Since the fiber was hollow, the structure collapsed upon immersion in the media and had limited permeability. The hollow fibers were considered to be suitable for delivery of enzymes, catalysis, and nerve regeneration [38]. A dry–wet spinning approach was used to prepare hollow chitosan fibers, and the fibers obtained were cross-linked using glutaraldehyde [39]. In this method, a 5.5% chitosan solution was extruded through a spinneret having an outer and inner diameter of 2 and 1 mm, respectively, and passed into a coagulation bath containing 10% NaOH. Two different flow rates of bore fluid were used to create the hollow structure. Fibers obtained were cross-linked with glutaraldehyde, which helped to improve the tensile properties (strength up to 50 MPa) up to a glutaraldehyde

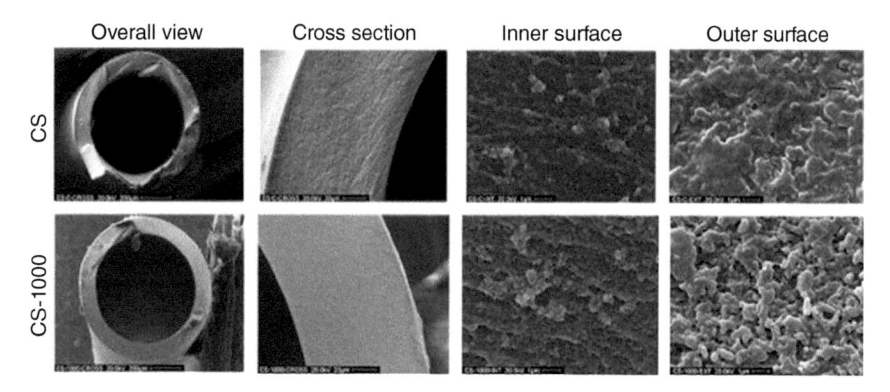

Figure 5.6 Scanning electron images of the morphology of pure and cross-linked chitosan fibers. Source: Tasselli et al. 2013 [39]. Reproduced with permission from Elsevier.

concentration of 200 g/l. Figure 5.11 shows the morphology of the chitosan fibers before and after cross-linking [39]. Cross-linking made the fibers rougher but increased the resistance to swelling. Sorption capacity of the fibers for dyes decreased with increasing level of cross-linking since the available sites would be consumed for cross-linking. However, the fibers could be regenerated using NaOH and reused for dye sorption [39].

5.3.4 Microfluidic Method of Producing Chitosan Fibers

Conventional approach of producing chitosan fibers used harsh chemicals that will kill cells. To incorporate cells into fibers, a microfluidic approach was used, and chitosan–alginate fibers intended for tissue engineering applications were developed [40]. Figure 5.7 shows a schematic process of producing chitosan–alginate fibers loaded with HepG2 cells. Diameter of the fibers could be controlled within 100 μm. Cells could be easily encapsulated within the fibers, and the presence of alginate promoted the attachment, growth, and proliferation of cells. Microfluidic approach of producing chitosan fibers also allowed spatial control of the spreading of the cells that would be beneficial for tissue engineering [40].

5.3.5 Application of Chitosan Fibers

Chitosan fibers have been extensively used for medical, textile, environmental, and cosmetic applications mainly due to their antimicrobial property. Novel

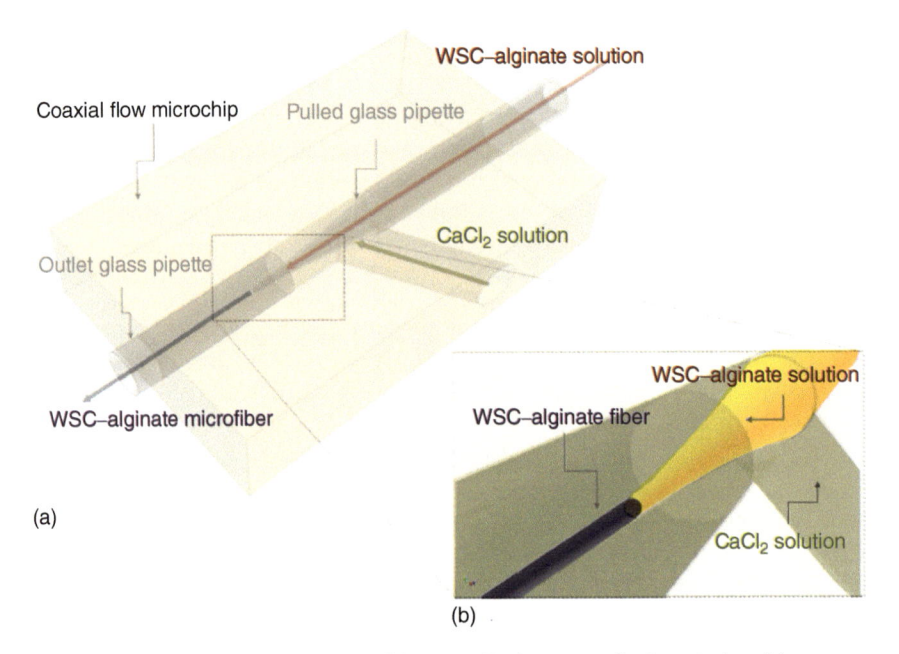

Figure 5.7 Schematic representation of the microfluidic approach of producing chitosan fibers. (a) Schematic for the microfluidic device and (b) detailed view of fiber forming section. Source: Lee et al. 2011 [40]. Reproduced with permission from American Institute of Physics.

chitosan fibers with ability to release various fragrances were prepared by treating with various aldehydes [41]. Strength and elongation of the fibers decreased due to the treatment with aldehydes, but controlled release of the fragrance could be felt in the atmosphere [41]. These fibers were suggested to be suitable for cosmetics, air filters, and textiles. To improve the antimicrobial properties, chitosan fibers were treated with zinc and silver, and the activity against *E. coli*, *S. aureus*, and *Bacillus subtilis* were tested. Up to 100% reduction in bacterial activity was achieved with silver-containing fibers providing the highest resistance [42]. Silver and zinc ions in the fibers could be released into solution, and the extent of release could be controlled depending on the loading and unloading conditions.

An attempt was made to produce chitosan fibers suitable for medical applications without the use of cytotoxic cross-linking agents. Production parameters of chitosan fibers were varied to obtain optimum conditions and desired mechanical properties. Instead of the conventional approach, fibers were immersed in 10% or 25% ammonia solution [43]. Further, fibers were cross-linked using heparin and annealed to improve mechanical properties and stability. Extrusion into ammonia solution increased the crystallinity and hence tensile strength by more than 200%. Changes in the tensile properties of the fibers at various conditions are shown in Figure 5.6. In addition to providing desired mechanical properties, the fibers were also found to support the attachment and growth of valvular interstitial cells (VIC) after 10 days of culture [43]. It was suggested that fibrous chitosan scaffolds with tunable mechanical properties could be produced using this approach [43].

Chitosan fibers were carboxymethylated by treating with chloroacetic acid to impart chelating properties and improve sorption of Cu (II) ions [44]. Degree of carboxymethylation varied from 12% to 40% depending on the conditions used during modification. Sorption capacity of Cu(II) ions was also dependent on the time, pH, temperature, amount of sorbent, etc. (Table 5.15) [44]. Up to 148 mg of Cu(II) was sorbed per mg of fiber when the carboxymethylation was 40%. To improve moisture absorption and retentivity, N-acyl chitosans were prepared using N-acetylated, N-maleyl, and N-succinyl chitosan in glycine chloride [34]. Changes in the moisture sorption of chitosan after different chemical modifications are depicted in Figure 5.8. Moisture sorption capacity of the fibers was dependent on the degree of substitution with high DS providing better sorption. Ability to control the mechanical properties and moisture sorption was suggested to be suitable for use in hernioplasty, saturation. or wrinkle filling [34].

Chitosan fibers having a linear density of 1.23 mg/cm, strength of 1.46 g/d, and extensibility of 12.1% were made into a scaffold (Figure 5.9) for attachment, spreading, and differentiation of osteoblasts [45]. Fibers in the scaffold had average diameter of 20 μm with a flaky surface. The scaffolds were coated with collagen to improve cell attachment and growth. Excellent attachment and proliferation of murine osteoblasts was observed, and it was suggested that the scaffolds were suitable for culture of human normal and stem cells [45].

Wound dressings are one of the most common applications of chitosan fibers and fabrics due to the inherent antibacterial, biocompatible, and excellent

Table 5.15 Effect of various conditions on the removal efficiency of Cu(II) by the chitosan fibers.

	Cu(II) concentration after treatment (mg/l)	Cu (II) removal (%)	Absorption capacity (mg Cu(II)/g fiber)
Temperature (°C)			
25	7.7	95.2	76.1
37	64.2	59.9	47.9
55	127.6	20.2	16.2
70	83.2	48.0	38.4
95	21.2	86.8	69.4
Amount of fiber (g)			
0.050	85.9	46.3	148.1
0.100	34.2	78.6	125.7
0.150	8.7	94.5	100.8
0.200	14.3	91.1	72.9
0.250	8.1	95.0	60.8
pH			
2	49.4	69.1	55.3
4	22.2	86.1	68.9
7	22.2	86.1	68.9
9	33.6	79.0	63.2
11	24.9	84.4	67.5
Time (h)			
0.25	62.2	61.1	48.9
0.5	75.3	53.0	42.4
1	50.4	68.5	54.8
2	44.2	72.3	57.9
5	7.7	95.2	76.1
8	9.8	93.9	75.1
15	17.0	89.4	71.5
24	14.2	91.1	72.9

Source: Qin et al. 2006 [44]. Reproduced with permission from John Wiley & Sons.

biosorption properties. To further improve the antibacterial properties of chitosan, colloidal silver nanoparticles were incorporated into fibers [46]. Amount of silver in the fibers was controlled at 130 and 167 mg/kg. Properties of the fibers including tensile strength and elongation were affected due to the inclusion of silver. Increasing amount of silver in the fibers increased the bactericidal activity against *S. aureus* and *E. coli*. The fibers were also considered as nonconductive and suitable for preparation of bandage and other medical materials [46]. Chitosan-based heart valve scaffolds were developed using chitosan fibers

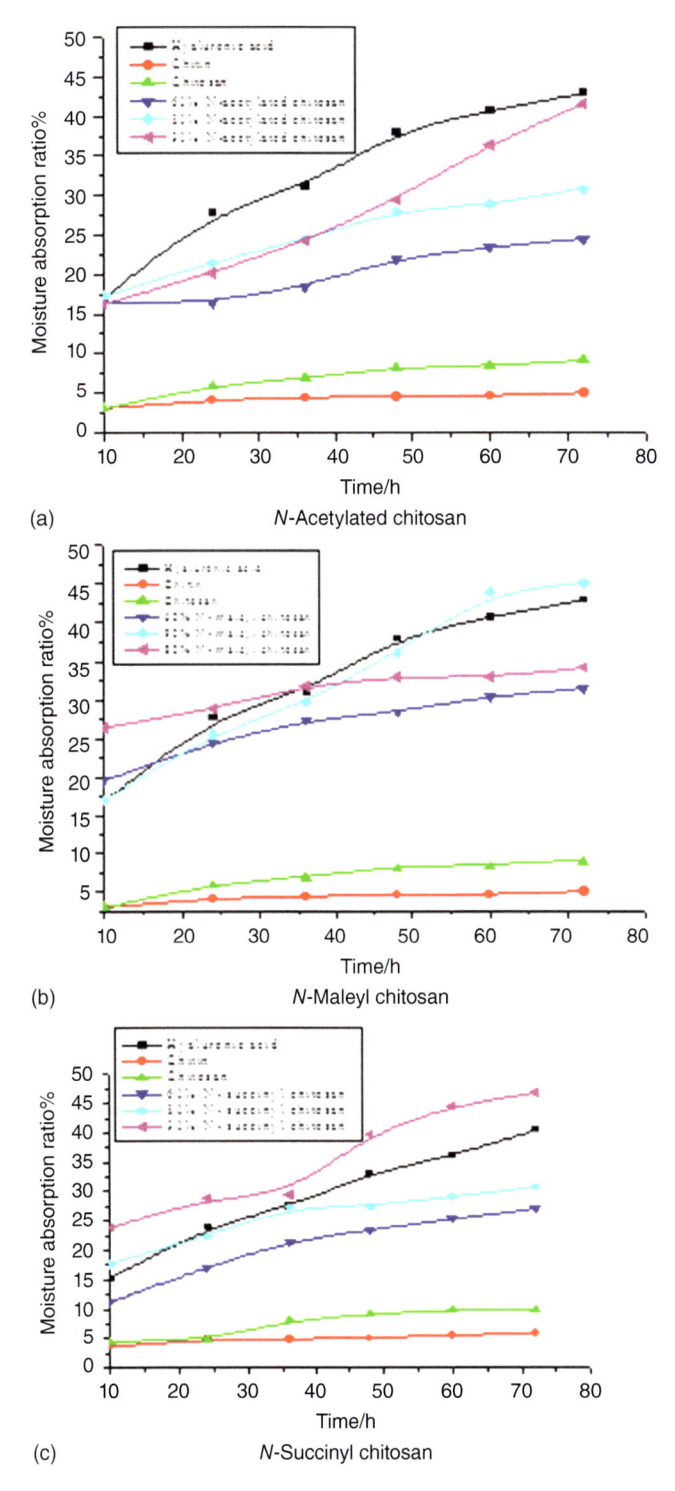

Figure 5.8 Changes in the moisture absorption capacity of chitosan fibers modified using various approaches. (a) N-acetylated chitosan, (b) Nmaleyl chitosan, and (c) N-succinyl chitosan. Source: Li et al. 2013 [34]. Reproduced with permission from John Wiley & Sons, Inc.

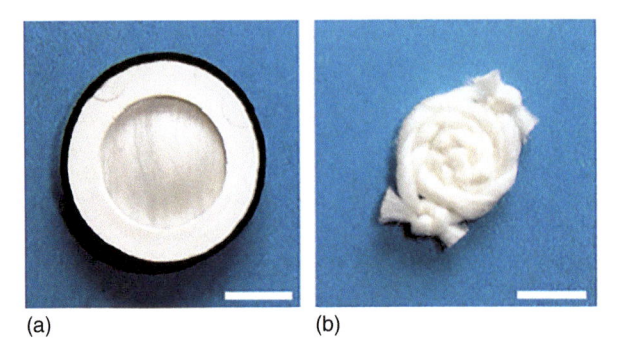

Figure 5.9 Digital image of a scaffold made using chitosan fibers and suitable for tissue engineering. Left image shows chitosan fibers fixed inside a holder and right image is stand-alone chitosan fibers. (a) Chitosan fibers fixed inside a holder and (b) stand-alone chitosan fibers. Source: Heinemann et al. 2008 [45]. Reproduced with permission from American Chemical Society.

cross-linked with heparin [43]. Addition of fibers into the scaffold increased the strength (220 kPa) and stiffness equivalent to radial values of human pulmonary valve leaflets. Unreinforced scaffolds had strength of 58 kPa compared with 750 kPa for the reinforced scaffolds. Similarly, the strain of the fibers was 20% compared with 90% for the scaffold. It was suggested that the mechanical properties of the scaffold could be controlled as per requirement. Chitosan-based composite nanofibrous membranes were prepared by blending chitosan, PLA, and silk nanoparticles and electrospinning. N-carboxymethylation of chitosan was done to make chitosan soluble in water and compatible with PVA. Strong interactions were observed between the components in the fibers, and inclusion of silk nanoparticles promoted attachment and growth of mouse fibroblasts. Hence, the scaffolds were considered suitable as wound dressings [47].

Fiber meshes intended for tissue engineering were prepared from 85% deacetylated chitosan. Dissolved chitosan was extruded into a coagulation bath containing sodium sulfate and sodium hydroxide and later drawn to increase the orientation. The fibers were subject to bioactivity, biocompatibility, and cytotoxicity tests [48]. Morphological analysis showed that the fibers were oriented in one direction and had smooth but striated surface. Fibers had a tensile strength of 205 MPa and elongation of 8.5%. When immersed into body fluids, the fibers were able to initiate the formation of calcium phosphate suggesting bioactivity. Osteoblasts seeded on the fiber meshes had excellent attachment, growth, and proliferation (Figure 5.10).

Various flavonoids were grafted onto chitosan fibers using tyrosinase as the cross-linking agent and studied for their potential to improve antimicrobial and antioxidant properties [49]. Compared to ungrafted chitosan, higher antioxidant activity ranging from 0.5% to 88% was determined depending on the type of flavonoid and assay method used [49] (Table 5.16). The use of an enzyme and natural antioxidant to impart antibacterial and antioxidant properties was considered highly suitable for medical applications.

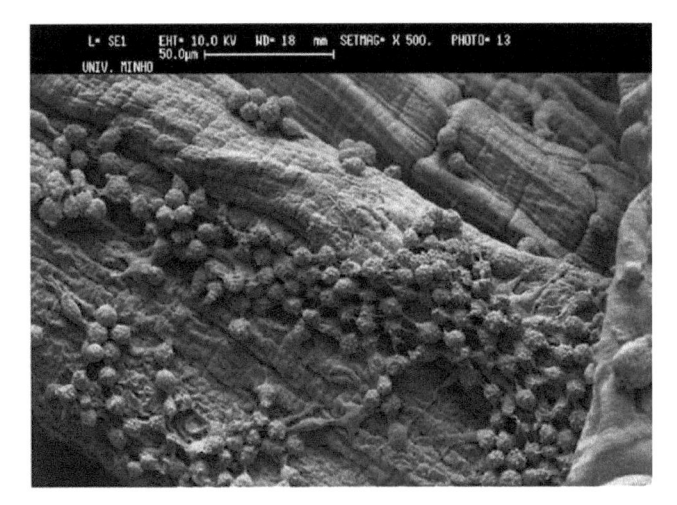

Figure 5.10 SEM images shows substantial proliferation of osteoblast cells on the surface of the chitosan fiber meshes after seven days of culture. Source: Tuzlakoglu et al. 2004 [48]. Reproduced with permission from John Wiley & Sons.

5.4 Electrospun Chitosan Fibers

Nanofibers have high surface area and can hence sorb higher levels of chemicals and pollutants. Also, nanofibers mimic the nature of the extracellular matrix in the human body and hence are preferred for medical applications. Chitosan has been widely studied for developing nanofibers for various applications (Table 5.17). Nanofibers with average diameter as low as 66 nm were developed by dissolving 75% deacetylated chitosan in a mixture of trifluoroacetic acid and dichloromethane [51]. Electrospinning conditions and the ratio of the two solvents significantly affected the properties of the electrospun fibers. A chitosan concentration of 8% and acetic acid to dichloromethane ratio of 80 : 20 produced fiber mats with the lowest thickness when a voltage of 30 kV and needle diameter of 0.5 μm was used [51]. In a process termed "electrohydrodynamic processing," chitosan grafted with L-lactide was made into fibrous substrates for potential use as tissue engineering scaffolds [52]. Grafted chitosan was dissolved using ethyl acetate and 2-butanone and electrospun into fibers. Hydrophilicity and rate of biodegradation of the scaffolds could be controlled by varying the L-lactide to chitosan ratio. Viability and proliferation of cells was higher on the grafted scaffolds. However, higher levels of lactic acid reduced the pH of the media, resulting in cytotoxicity [52]. Chitosan nanofibers have also been used as sorbent for various metals. To produce the fibers, chitosan was dissolved in HFIP and extruded using a voltage of 8 kV and tip to collector distance of 7 cm. Fiber mats were immersed in metal solutions, and various sorption conditions were studied [53]. Sorption of metal ions by the chitosan fibers was dependent on the sorption conditions, and up to 100% sorption could be achieved. The fibers also showed selectivity toward various metal ions with the following order Cu > Fe > Ag > Cd and were found to be suitable for use in solid-phase extraction cartridges [53].

Table 5.16 Antioxidant activity of chitosan fibers grafted with various flavonoids.

Antioxidant activity	Catechin	Epicatechin	Epigallocatechin	Epigallocatechin gallate	Fisetin	Quercetin	Rutin	Hesperidin	Daidrein
Total antioxidant activity	13.9±0.02	10.6±0.03	69.9±0.5	15.1±0.05	—	—	3.1±0.05	4.2±0.05	20.7±0.03
DPPH/free radical scavenging	87.4±0.03	88.5±0.04	—	53.9±0.03	41.1±0.1	24.2±0.08	6.6±0.1	7.9±0.02	2.5±0.01
ABTS radical cation assay	8.5±0.04	21.0±0.03	20.0±0.05	26.8±0.02	3.4±0.0	25.5±0.03	17.9±0.01	21.3±0.05	26.9±0.03
Superoxide anion scavenging activity	36.9±0.05	1.1±0.05	73.9±0.08	35.1±0.03	32.9±0.3	38.9±0.07	—	48.9±0.02	19.7±0.04
Ferrous metal ion chelating activity	0.46±0.03	4.18±0.04	10.9±0.05	—	—	8.8±0.04	3.6±0.04	12.2±0.05	—

Source: Sousa et al. 2009 [49]. Reproduced with permission from Elsevier.

Table 5.17 Examples of methods used to process chitosan into nanofibers and their potential applications [50].

Polymer	Solvents	Average fiber diameter (nm)	Application
Chitosan	aq AA	130	—
Chitosan	TFA/DCM	130 ± 10	—
Chitosan	TFA/DCM	126 ± 20	Tissue engineering
Chitosan (neutralized with K_2CO_3 solution)	TFA	235	Filtration
Chitosan/PVA (removing PVA with 0.5 M NaOH)	aq AA	80–150	Enzyme immobilization
Chitosan/PVA	aq AA	99 ± 21	Wound dressings
Chitosan/PEO	aq AA	80 ± 35	Filtration
Chitosan/UHMWPEO (10%)	aq AA/DMSO	138 ± 15	—
Chitosan/UHMWPEO (20%)	aq AA/DMSO	102 ± 14	—
Chitosan/PET	TFA	500–800	Wound dressings
Chitosan/PCL	HFIP	450 ± 110	Bone tissue engineering
Chitosan/nylon-6	HFIP/FA	80–310	Filtration
Chitosan/PVA-PLGA (by multi-jet electrospinning method)	aq AA THF/DMF	275 ± 175	Tissue engineering
Chitosan/collagen/PEO (cross-linked by GA vapor)	aq AA	398 ± 76	Wound dressings
H-CS	$CHCl_3$	640–3930	Skin tissue engineering
Q-CS/PVP (cross-linked by UV irradiation)	Water	2400 ± 640	Wound dressings
CS-HOBt/PVA	Water	190–282	Drug delivery
Chitosan/HAp/PVA	aq AA/DMSO	100–700	Bone tissue engineering
Chitosan/gelatin/AgNPs	aq AA	220–400	Wound dressings
CS-g-AA/PaNPs	aq AA	70–200	Catalyst

THF, tetrahydrofuran; aq AA, aqueous acetic acid solution; DMSO, dimethyl sulfoxide; DMF, dimethylformamide; HFIP, 1,1,1,3,3,3-hexafluoro-2-propanol.

Low (70 000 Da), medium (190 000–310 000 Da), and high molecular weight (500 000–700 000 Da) chitosan was dissolved in trifluoroacetic acid and electrospun into fibers. High molecular weight chitosan produced fibers with diameter of 108 nm compared with about 77 nm for the other two molecular weight range. Although the fibers had acceptable mechanical properties, they instantly dissolved in water or acetic acid. To improve the stability, the fibers were cross-linked using glutaraldehyde vapors [54]. Cross-linking resulted in substantial decrease in strength (from 4 to 1.2 MPa) and elongation of the fibers.

However, the cross-linked fibers were found to be stable in water, acetic acid, and NaOH solutions [54]. Glutaraldehyde vapors were considered suitable for cross-linking chitosan nanofibers, but the effect of cross-linking on cells and hence biocompatibility was not assessed. Schiffman et al. also demonstrated that a one-step cross-linking and electrospinning of chitosan could be done that was 25% faster and also produced finer fibers. Glutaraldehyde solution was combined with the chitosan solution and electrospun. Cross-linking occurred during electrospinning, thereby avoiding the need for an additional cross-linking step [55]. Cross-linked fibers were stable to acetic acid, water, and NaOH for up to 72 hours. However the biocompatibility/cytotoxicity of these fibers was not evaluated.

Highly aligned nanofibers desirable for use in textile, energy, environment, and bioengineering were developed from chitosan using a novel centrifugal electrospinning system (Figure 5.11) [56]. Increasing speed produced fibers with better alignment that were also shown to have better piezoelectric properties for PVDF fibers. No results were presented for chitosan fibers.

To prepare chemically stable chitosan membranes, carbon black having an average size of 67 nm was mixed with chitosan (190–310 kDa) dissolved using trifluoroacetic acid. Solutions containing up to 62% carbon black were electrospun to form membranes, which were cross-linked using glutaraldehyde. Inclusion of the carbon black changed the color of the membranes (Figure 5.12) but did not change the morphology of the fibers [57]. Although uncross-linked membranes dissolved immediately, the cross-linked membranes had excellent stability in acidic, basic, or aqueous solutions for up to 20 days. Conductivity of

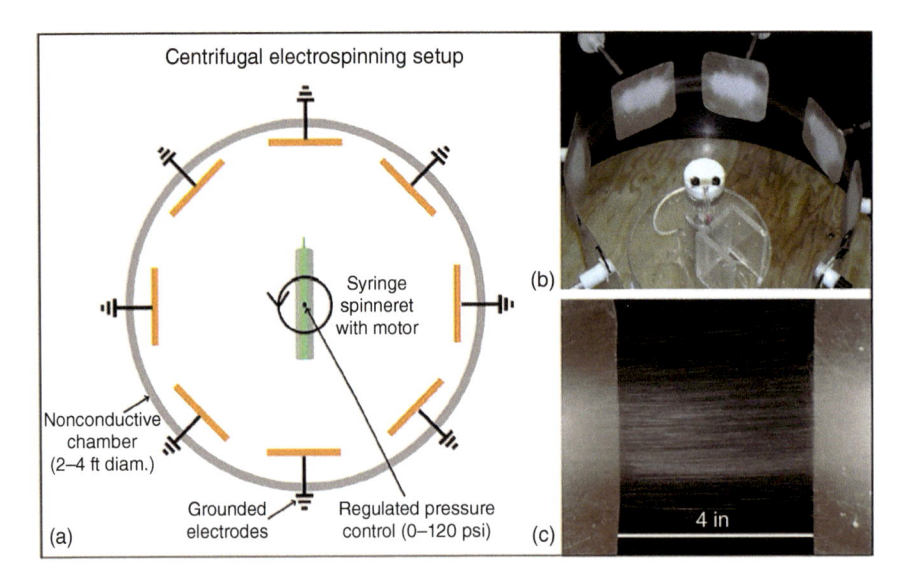

Figure 5.11 New method of centrifugal electrospinning using different collectors and rotating needle. (a) Schematic representation of the system configuration, (b) photograph of the CE system with deposited PVDF nanofibers, and (c) electrospun PVDF fibers deposited across a four-inch gap between two grounded electrodes. Source: Edmondson et al. 2012 [56]. Reproduced with permission from Royal Society of Chemistry.

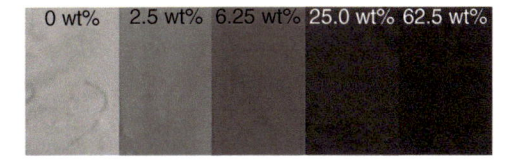

Figure 5.12 Images of chitosan membranes containing various levels of carbon black shows considerable change in color due to the addition of the carbon black. Source: Schiffman et al. 2011 [57]. Reproduced with permission from Elsevier.

Figure 5.13 Digital image of the morphology of the fibers and picture of the fibrous chitosan mat produced by electrospinning. Source: Frohbergh et al. 2012 [58]. (a) Electrospun chitosan microfibers and (b) fibrous mat from the chitosan fibers. Reproduced with permission from Elsevier.

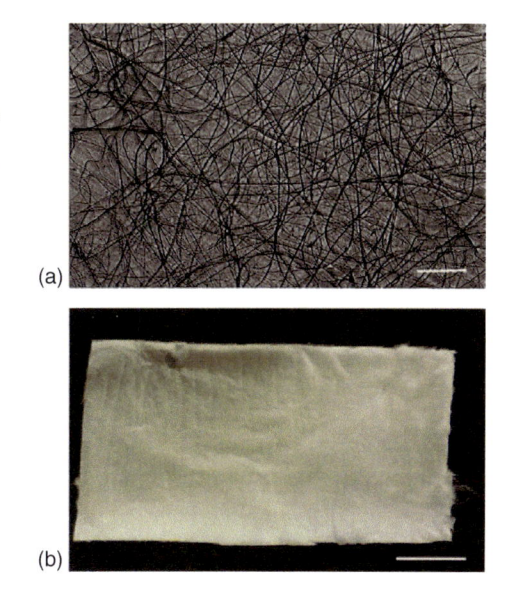

the membranes increased nearly three times when 2.5% or 6.25% carbon black was present, but the tensile strength and elongation decreased substantially. Good chemical stability and ability to control the conductivity were considered to be desirable for electrochemical applications [57].

Composite electrospun fibers intended for bone tissue engineering were made by adding hydroxyapatite nanocrystals into chitosan solution [58]. Electrospun mats obtained (Figure 5.13) were cross-linked with genipin to increase water stability. The scaffolds were soaked in the genipin solution in PBS for 24 hours for the cross-linking to occur. The presence of hydroxyapatite significantly promoted the formation of bone tissue, and cross-linking increased the modulus up to fourfolds (142 ± 13 MPa). Modulus of the fibrous scaffold was considered to be similar to that of natural periosteum. Attachment, growth, proliferation, and mRNA expression of osteoblast cells suggested that the scaffolds could be useful for repair and regeneration of maxillofacial defects and injuries [58]. Since electrospun fibers are more susceptible to moisture and disintegrate in aqueous systems, it is necessary to cross-link or modify them with chemical and physical approaches as discussed earlier. Instead of the conventional covalent cross-linkers, Kiechel et al. used glycerol phosphate (GP), tripolyphosphate (TPP), and tannic acid (TA) as new set of noncovalent cross-linkers [59]. Type of cross-linker used influenced the stability, fiber diameters, and hence application. Cross-linking could be done in a one- or two-step process depending on the cross-linker. The cross-linked nanofiber membranes were suggested to be suitable for tissue engineering and drug delivery applications [59].

Three-dimensional electrospun chitosan mats were prepared using trifluoroacetic acid as the solvent. The fiber mats obtained were treated with fibronectin to promote the attachment and growth of primary ventricular cardiomyocytes and also co-cultured with microvascular endothelial cells or fibroblasts [60]. It was found that the 3D chitosan scaffolds supported the attachment and growth of cells and could form cardiac tissue constructs similar to native heart tissue. It was suggested that co-culture of cells could maintain the properties of the individual cells in long-term use [60].

Core–shell polylactic acid/chitosan fibers were produced by electrospinning using sodium dodecyl sulfate [17]. Chitosan was modified using SDS and later dissolved along with PLA using dichloromethane and dimethyl sulfoxide, and the solution was electrospun into fibers. Cross-linking of the fibers was done using glutaraldehyde vapors, and the scaffolds obtained were used as substrates for mouse fibroblast cells (L929). SDS on the fibers was removed to create porous substrates that promoted the attachment, growth, and proliferation of the cells. Shape and proliferation of the cells suggested that the fibers would be suitable for tissue engineering and wound healing [17]. Instead of using cross-linking agents, Pangon et al. demonstrated that addition of chitin whiskers into chitosan/PVA blends substantially improved the mechanical properties. In addition, biomineralization with hydroxyapatite leads to the promotion of the attachment and growth of osteoblast cells. It was suggested that the chitosan/chitin/PVA hydroxyapatite blends would be ideal for bone tissue engineering [61]. In a similar approach, biomimetic scaffolds mimicking the structure of the extracellular matrix were prepared by combining chitosan with collagen. Various ratios of collagen and chitosan were dissolved in a 90/10 mixture of 1,1,1,3,3,3-hexafluoroisopropanol/trifluoroacetic acid and electrospun into fibers and later cross-linked with glutaraldehyde (Chen et al. 2010). Fibers in the blend had average diameter of 434–691 nm, and cross-linking stabilized the morphology of the scaffolds. Both endothelial and smooth muscle cells had excellent proliferation on the nanofibers suggesting their suitability for tissue engineering and regeneration of blood vessel and nerve tissues (Chen et al. 2010).

Conducting nanofibers were obtained by blending plasma-modified chitosan with poly(3,4-ethylenedioxythiophene) using PVA as a supporting polymer. Plasma-modified chitosan and PEDOT were dissolved using 0.5 M acetic acid and later combined with PVA dissolved in water to form the blend solution for electrospinning [62]. Plasma modification provided better antibacterial activity and formation of finer fibers. Electrochemical studies showed that the blend fibers had electroactive properties and could be useful for biomedical applications [62] (Figure 5.14).

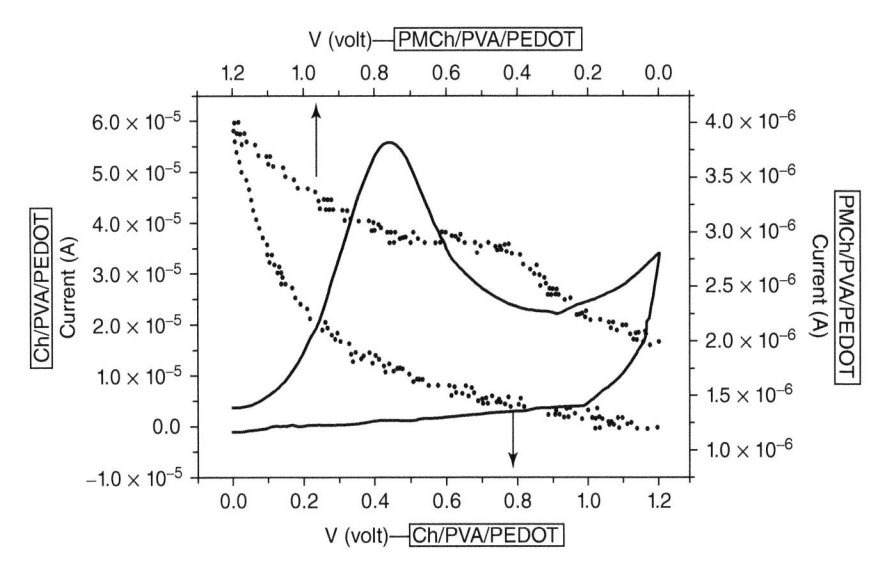

Figure 5.14 Properties of the chitosan/PVA/PEDOT blend fibers. Source: Kiristi et al. 2013 [62]. Reproduced with permission from Elsevier.

5.5 Regenerated Chitosan Fibers

In addition to the natural micro- and electrospun fibers, researchers have also produced regenerated chitosan fibers. For this, chitosan was dissolved in a binary ionic liquid and regenerated using a wet and dry spinning technique [63]. Chitosan (5%) was dissolved and heated at 80 °C to have a viscosity of 5 Pa. Solution was extruded through a 60 μm diameter spinneret at a pressure of 0.4 MPa and coagulated first using ethanol and later using 5% sodium hydroxide solution. The fibers obtained had a strength of 2.1 cN/dtex and modulus of 83.5 cN/dtex, much stronger than the common chitosan fibers [63].

5.6 Conclusions

Chitosan is one of the few biopolymers that has been extensively studied and used for medical, environmental, energy, textile, and other applications. Inherent antimicrobial activity, easy solubility, and ability to be made into various forms are some of the advantages of chitosan. However, stability in aqueous environments and mechanical properties of chitosan-based products are concerns for many applications. Despite considerable efforts on chemical and physical modifications including cross-linking, achieving the desired level of properties in chitosan-based products is a challenge. In addition to conventional solution spinning, chitosan has been made into electrospun fibers with properties suitable for skin and bone tissue engineering. Several studies have demonstrated that chitosan fibers can be used for controlled release of drugs and sorption of pollutants from wastewater. Chitosan has also been blended with various biopolymer and synthetic polymers to develop fibers for specific applications. Similarly, a few studies have been done on developing regenerated chitosan fibers. Recent studies have focused on using additives such as graphene oxide, carbon nanotubes, and carbon black to improve the applicability of chitosan. New technologies such as microwave heating that substantially reduces the extraction time and hence cost and use of unconventional sources of chitin such as plants and fungi provide an opportunity to explore new applications for chitosan fibers.

Acknowledgments

Authors thank the Centre for Incubation, Innovation, Research and Consultancy and Jyothy Institute of Technology for their support to complete this work. Narendra Reddy thanks the Department of Biotechnology, Ministry of Science and Technology, Government of India for support through the Ramalingaswami Reentry Fellowship. M. S. Santosh thanks Karnataka Council for Technological Upgradation for funding part of this work.

References

1 Rinaudo, M. (2006). Chitin and chitosan: properties and applications. *Progress in Polymer Science* 31: 603–632.
2 Boamah, P.O., Huang, Y., Hua, M. et al. (2015). Sorption of heavy metal ions onto carboxylate chitosan derivatives – a mini-review. *Ecotoxicology and Environment Safety* 116: 113–120.
3 Islam, S., Bhuiyan, M.A.R., and Islam, M.N. (2016). Chitin and chitosan: structure, properties and applications in biomedical engineering. *Journal of Polymers and the Environment* 25: 854–866.
4 Chen, G., Mi, J., Wu, X. et al. (2011). Structural features and bioactivities of the chitosan. *Journal of Biological Macromolecules* 49: 543–547.
5 Younes, I. and Rinaudo, M. (2015). Chitin and chitosan preparation from marine sources. Structure properties and applications. *Marine Drugs* 13: 1133–1174.
6 Liman, Z., Selmi, S., Sadok, S., and Abed, A.E. (2011). Extraction and characterization of chitin and chitosan from crustacean byproducts: biological and physiochemical properties. *African Journal of Biotechnology* 40 (4): 640–647.
7 Teli, M.D. and Sheikh, J. (2012). Extraction of chitosan from shrimp shells waste and application in antibacterial finishing of bamboo rayon. *International Journal of Biological Macromolecules* 50: 1195–1200.
8 Kucukgulmez, A., Celik, M., Yanar, Y. et al. (2011). Physiochemical characterization of chitosan extracted from *Metapenaeus stebbingi* shells. *Food Chemistry* 126: 1144–1148.
9 Kannan, M., Nesakumari, M., Rajarathnam, K., and Singh, R.A.J.A. (2010). Production and characterization of mushroom chitosan under solid-state fermentation conditions. *Advance Biological Research* 4 (1): 10–13.

10 Tayel, A.A., Ibrahim, L.A.S., Al-Saman, M.A., and Moussa, S.H. (2014). Production of fungal chitosan from date wastes and its application as a biopreservative for minced meat. *International Journal of Biological Macromolecules* 69: 471–475.

11 Knidri, H.E., Khalfaouy, R.E., Laajeb, A. et al. (2016). Eco-friendly extraction and characterization of chitin and chitosan from the shrimp shell waste via microwave irradiation. *Process Safety and Environmental Protection* 104: 395–435.

12 Sagheer, F.A.A., Al-Sughayer, M.A., Muslim, S., and Elsabee, M.Z. (2009). Extraction and characterization of chitin and chitosan from marine sources in Arabian Gulf. *Carbohydrate Polymers* 77: 410–419.

13 Cai, J., Yang, J., Du, Y. et al. (2006). Enzymatic preparation of chitosan from the waste *Aspergillus niger* mycelium of citric acid production plant. *Carbohydrate Polymers* 64: 151–157.

14 Upadhyaya, L., Singh, J., Agarwal, V., and Tewari, R.P. (2013). Biomedical applications of carboxymethyl chitosans. *Carbohydrate Polymers* 91: 452–466.

15 Demina, T.S., Gilman, A.B., Akopova, T.A., and Zelenetskii, A.N. (2014). Modification of the chitosan structure and properties using high energy chemistry methods. *High Energy Chemistry* 48 (5): 293–302.

16 East, G.C. and Qin, Y. (1993). Wet spinning of chitosan and the acetylation of chitosan fibers. *Journal of Applied Polymer Science* 50: 1773–1779.

17 Li, L., Yuan, B., Liu, S. et al. (2012). Preparation of high strength chitosan fibers by using ionic liquid as spinning solution. *Journal of Materials Chemistry* 22: 8585–8594.

18 Hirano, S., Nagamura, K., Zhang, M. et al. (1999). Chitosan staple fibers and their chemical modification with some aldehydes. *Carbohydrate Polymers* 38: 293–298.

19 Lee, S., Park, S., and Kim, Y. (2007). Effect of the concentration of sodium acetate (SA) on crosslinking of chitosan fiber by epichlorohydrin (ECH) in a wet spinning system. *Carbohydrate Polymers* 70: 53–60.

20 Knaul, J.Z., Hudson, S.M., and Creber, K.A.M. (1999). Crosslinking of chitosan fibers with dialdehydes: proposal of a new reaction mechanism. *Journal of Polymer Science Part B: Polymer Physics* 37: 1079–1094.

21 Yang, Q., Dou, F., Liang, B., and Shen, Q. (2005). Studies of crosslinking reaction on chitosan fiber with glycerol. *Carbohydrate Polymers* 59: 205–210.

22 Knaul, J.Z., Hudson, S.M., and Creber, K.A.M. (1999). Improved mechanical properties of chitosan fibers. *Journal of Applied Polymer Science* 72: 1721–1732.

23 Pati, F., Adhikari, B., and Dhara, S. (2011). Development of ultrafine chitosan fibers through modified wet spinning technique. *Journal of Applied Polymer Science* 121: 1550–1557.

24 Lee, C.K., Kim, S.J., Kim, S.I. et al. (2006). Preparation of chitosan microfibers using electro-wet-spinning and their electroactuation properties. *Smart Materials and Structures* 15: 607–611.

25 Ozarkar, S., Jassal, M., and Agrawal, A.K. (2008). pH and electrical actuation of single walled carbon nanotube/chitosan composite fibers. *Smart Materials and Structures* 17: 055016–055025.

26 Yan, W., Shen, L., Ji, Y. et al. (2014). Chitin nanocrystal reinforced wet-spun chitosan fibers. *Journal of Applied Polymer Science* 131: 40852–40858.

27 Zhao, J., Luo, C., Han, W. et al. (2012). Fabrication and properties of poly(L-lactide)/hydroxyapatite/chitosan fiber ternary composite scaffolds for bone tissue engineering. *Journal of Polymer Engineering* 32: 283–289.

28 Funakoshi, T., Majima, T., Iwasaki, N. et al. (2005). Novel chitosan based hyaluronan hybrid polymer fibers as a scaffold in ligament tissue engineering. *Journal of Biomedical Materials Research* 74A: 338–346.

29 Xu, X., Zhuang, X., Cheng, B. et al. (2010). Manufacture and properties of cellulose/O-hydroxyethyl chitosan blend fibers. *Carbohydrate Polymers* 81: 541–544.

30 Zhang, H., Wu, C., Zhang, Y. et al. (2010). Elaboration, characterization and study of a novel affinity membrane made from electrospun hybrid chitosan/nylon-6 nanofibers for papain purification. *Journal of Materials Science* 45: 2296–2304.

31 Schueren, L.V., Steyaert, I., Schoenmaker, B.D., and Clerck, K.D. (2012). Polycaprolactone/chitosan blend nanofibers electrospun from an acetic acid/formic acid solvent system. *Carbohydrate Polymers* 88: 1221–1226.

32 Nguyen, T.T.T., Chung, O.H., and Park, J.S. (2011). Coaxial electrospun poly(lactic acid)/chitosan (core/shell) composite nanofibers and their antibacterial activity. *Carbohydrate Polymers* 86: 1799–1806.

33 Wawro, D. and Pighinelli, L. (2011). Chitosan fibers modified with Hap/β-TCP nanoparticles. *International Journal of Molecular Sciences* 12: 7286–7300.

34 Li, L., Yuan, B., Liu, S. et al. (2013). N-Acyl chitosan and its fiber with excellent moisture absorbability and retentivity: preparation in a novel [Gly]Cl/water homogenous system. *Journal of Applied Polymer Science* 129: 3282–3289.

35 Shim, I.K., Lee, S.Y., ParK, Y.J. et al. (2008). Homogenous chitosan-PLGA composite fibrous scaffolds for tissue engineering. *Journal of Biomedial Materials Research* 247: 247–255.

36 Qu, L.J., Guo, X.Q., Tian, M.W., and Lu, A. (2014). Antimicrobial fibers based on chitosan and polyvinyl-alcohol. *Fibers and Polymers* 15 (7): 1357–1363.

37 Haaparanta, A.M., Järvinen, E., Cengiz, I.F. et al. (2014). Preparation and characterization of collagen/PLA, chitosan/PLA, and collagen/chitosan/PLA hybrid scaffolds for cartilage tissue engineering. *Journal of Materials Science: Materials in Medicine* 25 (4): 1129–1136.

38 Peirano, F., Vincent, T., and Guibal, V. (2008). Diffusion of biological molecules through hollow chitosan fibers. *Journal of Applied Polymer Science* 107: 3568–3578.

39 Tasselli, F., Mirmohseni, A., Dorraji, M.S.S., and Figoli, A. (2013). Mechanical, swelling and adsorptive properties of dry-wet spun chitosan hollow fibers crosslinked with glutaraldehyde. *Reactive and Functional Polymers* 73: 218–223.

40 Lee, R., Lee, K.H., Kang, E. et al. Microfluidic wet spinning of chitosan-alginate microfibers and encapsulation of HepG2 cells in fibers. *Biomicrofluidics* 5: 0222208.

41 Hirano, S. and Hayashi, H. (2003). Some fragrant fibres and yarns based on chitosan. *Carbohydrate Polymers* 54: 131–136.
42 Qin, Y., Zhu, C., Chen, J. et al. (2006). The absorption and release of silver and zinc ions by chitosan fibers. *Journal of Applied Polymer Science* 101: 776–771.
43 Albanna, M.Z., Bou-Akl, T.H., Walters, H.L., and Matthew, H.W.T. (2012). Improving the mechanical properties of chitosan-based heart value scaffolds using chitosan fibers. *Journal of the Mechanical Behavior of Biomedical Materials* 5: 171–180.
44 Qin, Y., Hu, H., Luo, A. et al. (2006). Effect of carboxymethylation on the absorption and chelating properties of chitosan fibers. *Journal of Applied Polymer Science* 99: 3110–3115.
45 Heinemann, C., Heinemann, S., Bernhardt, A. et al. (2008). Novel textile chitosan scaffolds promote spreading, proliferation and differentiation of osetoblasts. *Biomacromolecules* 9: 2913–2920.
46 Wawro, D., Steplewski, W., Sobczak, E. et al. (2012). Antibacterial chitosan fibers containing silver nanoparticles. *Fibers and Textiles in Eastern Europe* 68 (96): 24–31.
47 Zhou, Y., Yang, H., Liu, X. et al. (2013). Electrospinning of carboxyethyl chitosan/poly (vinyl alcohol)/silk fibroin nanoparticles for wound dressings. *International Journal of Biological Macromolecules* 53: 88–92.
48 Tuzlakoglu, K., Alves, C.M., Mano, J.F., and Reis, R.L. (2004). Production and characterization of chitosan fibers and 3-D fiber mesh scaffolds for tissue engineering applications. *Macromolecular Bioscience* 4: 811–819.
49 Sousa, F., Guebitz, G.M., and Kokol, V. (2009). Antimicrobial and antioxidant properties of chitosan enzymatically functionalized with flavonoids. *Process Biochemistry* 44: 749–756.
50 Sun, K. and Li, Z.H. (2011). Preparations, properties and applications of chitosan-based nanofibers fabricated by electrospinning. *Express Polymer Letters* 5 (4): 342–361.
51 Areias, A.C., Gomez-Tejedor, J.A., Sencadas, V. et al. (2012). Assessment of parameters influencing fiber characteristics of chitosan nanofiber membrane to optimize fiber mat production. *Polymer Engineering and Science* 52 (6): 1293–1300.
52 Skotak, M., Leonov, A.P., Larsen, G. et al. (2008). Biocompatible and biodegradable ultrafine fibrillary scaffold materials for tissue engineering by facile grafting of L-Lactide onto chitosan. *Biomacromolecules* 9: 1902–1908.
53 Horzum, N., Boyaci, E., Eroglu, A.E. et al. (2010). Sorption efficiency of chitosan nanofibers toward metal ions at low concentrations. *Biomacromolecules* 11: 3301–3308.
54 Schiffman, J.D. and Schauer, C.L. (2007). Crosslinking chitosan nanofibers. *Biomacromolecules (a)* 8: 594–600.
55 Schiffman, J.D. and Schauer, C.L. (2007). One-step crosslinking of crosslinked chitosan fibers. *Biomacromolecules (b)* 8: 2665–2667.
56 Edmondson, D., Cooper, A., Jana, S. et al. (2012). Centrifugal electrospinning of highly aligned polymer nanofibers over a large area. *Journal of Materials Chemistry* 22: 18646–18652.

57 Schiffman, J.D., Blackford, A.C., Wegst, U.G.K., and Schauer, C.L. (2011). Carbon black immobilized in electrospun chitosan membranes. *Carbohydrate Polymers* 84: 1252–1257.

58 Frohbergh, M.E., Katsman, A., Botta, G.P. et al. (2012). Electro spun hydroxyapatite-containing chitosan nanofibers crosslinked with genipin for bone tissue engineering. *Biomaterials* 33: 9167–9178.

59 Kiechel, M.A. and Schauer, C.L. (2013). Non-covalent crosslinkers for electro spun chitosan fibers. *Carbohydrate Polymers* 95 (1): 123–133.

60 Hussain, A., Collins, G., Yip, D., and Cho, C.H. (2013). Functional 3-D cardiac co-culture model using bioactive chitosan nanofiber scaffolds. *Biotechnology and Bioengineering* 110: 637–647.

61 Pangon, A., Saesoo, S., Saengkrit, N. et al. (2016). Hydroxyapatite-hybridized chitosan/chitin whisker bionanocomposite fibers for bone tissue engineering applications. *Carbohydrate Polymers* 144: 419–427.

62 Kristi, M., Oksuz, A.U., Oksuz, L., and Ulusoy, S. (2013). Electrospun chitosan/PEDOT nanofibers. *Materials Science and Engineering C* 33: 3845–3850.

63 Ma, B., Qin, A., Li, X., and He, C. (2013). High tenacity regenerated chitosan fibers prepared by using the binary ionic liquid solvent (Gly-HCl)-[Bmim]Cl. *Carbohydrate Polymers* 97: 300–305.

6

Collagen Fibers

Jinlian Hu and Yanting Han

Institute of Textiles and Clothing, The Hong Kong Polytechnic University, Hung Hom, Kowloon, Hong Kong, China

6.1 Introduction

Type I collagen fiber (CF) is one of the most abundant structural proteins in vertebrates. It exists in animal bones, tendons, cartilage, interosseous membrane, skin, ligaments, and other connective tissue, accounting for 25% or more (up to 35%) of total protein [1]. As the basic unit of CFs, individual collagen triple helices further assemble in a complex and hierarchical manner, which results in the macroscopic fibers and networks. Collagen fibril normally has a diameter of 10–500 nm. It is further gathered into small fibril bundle with a diameter of 1–2 μm by glycoproteins and proteoglycans [2]. Through covalent cross-linking, collagen fibrils can be formed into CFs [3]. Due to its short length, natural CF is different from the conventional fibers, which can be easily applied in textile fields. Thus, to adapt to human demands, regenerated CFs with specific properties are obtained by spinning methods with collagen protein extracted from substrates. Because of its excellent biocompatibility and biodegradability, and weak antigenicity, collage is regarded as suitable material for bio-applications. Besides, CFs possess affinity to chemicals and mental ions, which expand its application to various fields.

6.2 Source and Structure of CF

CF comes from most tissues of animal in different arrangements to play different functions (Figure 6.1). These tissues included tendons and ligaments. Tendons consist of bundles of collagenous fiber arranged in parallel, forming cords that have great tensile strength, thus to anchor or bind muscles to bones. The distinctive structural feature of fibrillar collagens is 300 nm uninterrupted triple helical structure, aggregating themselves into highly ordered fibrils with a typical quarter-staggered fibril array. In tissue, CF influences the mechanical properties based on fibril arrangement.

Handbook of Fibrous Materials, First Edition. Edited by Jinlian Hu, Bipin Kumar, and Jing Lu.
© 2020 Wiley-VCH Verlag GmbH & Co. KGaA. Published 2020 by Wiley-VCH Verlag GmbH & Co. KGaA.

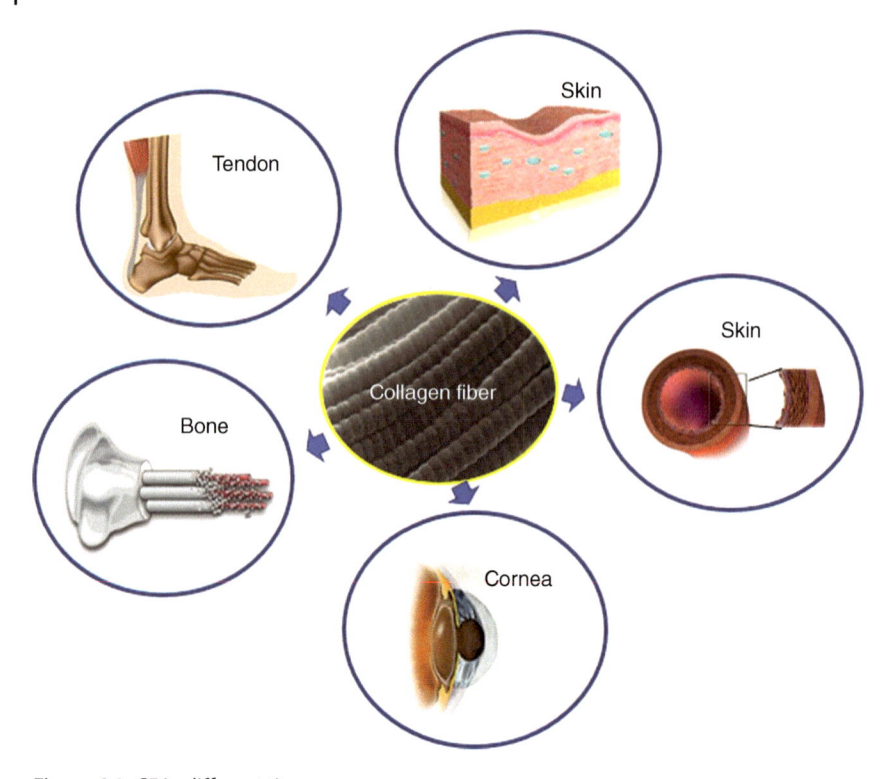

Figure 6.1 CF in different tissues.

The defining feature of collagen is a right-handed triple helix in which three parallel peptides in a left-handed triple helix are held together in a helical conformation by hydrogen bonds. At present, nineteen types of collagen have been identified, among which seven types of collagen have fibrous structure. The collagen used nowadays is mainly derived from the skin of vertebrates, which contains mainly type I collagen. In this chapter, we only focus on fibrils and fibers composed of type I collagen.

As shown in Figure 6.2, CFs have the hierarchical structure. In fibrillar collagen, collagen triple helix, known as tropocollagen (TC), assemble into fibrils, which agglomerate to form fibers. These fibril-forming collagens have large sections of homologous sequences. Individual TC monomers are arranged into microfibril, and individual microfibrils are packaged into collagen fibril, namely, the fibrillogenesis of collagen. Due to the enhanced effect of fibrillogenesis, collagen fibrils show good thermal stably. The strong macromolecular structure assembled in a complex, hierarchical manner provides CF and its network the ability to support stress multidimensionally.

The primary structure of collagen is the sequence structure of peptide chain. Collagen has a most unusual repetitious tripeptide sequence of Gly-X-Y, in which X can be any amino acid but is frequently a proline while Y is frequently a hydroxyproline or hydroxylysine. Collagen also contains some alanine, serine, arginine, and a small amount of methionine. Glycine is located inside the triple

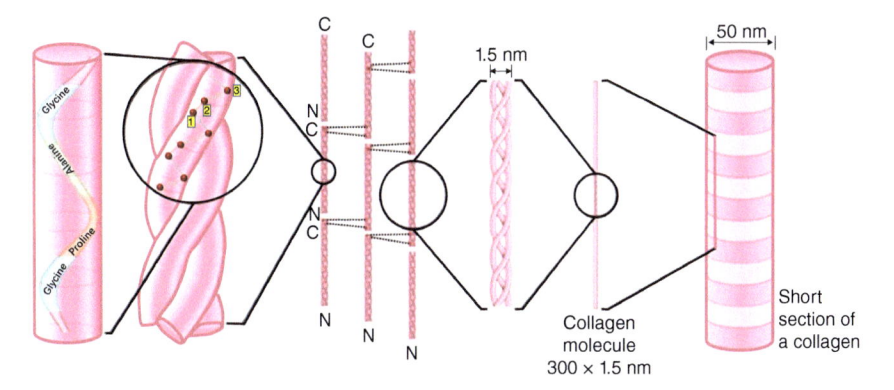

Figure 6.2 Hierarchical structure of CF.

helix bundle, providing a rotatable flexibility for the molecular chain, while proline increases the rigidity of the polypeptide chain, which results in the formation of a nonspherical conformation. Hydroxyproline is present in the triple helix bundle domain of collagen proteins. It provides additional hydrogen bonding capability that stabilizes the mature protein. Within the triple helix domains of nonfibrillar collagens, the repeated sequence is disrupted at certain location.

The secondary structure of the protein involves the folding ways of the polypeptide chain. Three left-handed helical (non-α spiral) polypeptide chains are intertwined to form a tropocollagen with a right-handed triple helix structure. Each peptide chain has about 1000 amino acid residues with a relative molecular mass between 92 000 and 100 000, so the relative molecular mass of the entire collagen molecule is about 300 000. In the formation of single-stranded spiral, hydrogen bonds cannot be generated within each peptide chain residues. However, when the peptides are stranded into three strands of helical chain, hydrogen bonds are formed between each peptide chain, resulting in a stable triple helix structure.

The tertiary structure of collagen refers to the three-dimensional (3D) position of the tropocollagen when forming the collagen microfibrils. Based on the secondary structure, the tropocollagen is further curly folded by the secondary bond to form a specific conformation. The tertiary structure of collagen reveals the role of secondary bonds between peptide chains in protein molecules, such as hydrogen bonds, ionic bonds, van der Waals forces, and hydrophobic bonds generated by nonpolar and polar groups on the side chain of amino acid residues. In addition to the secondary bonds, there are aldehydes condensation, aldol condensation, and aldol-histidine cross-linking between collagen molecules, which allows the individual polypeptides to be firmly connected, thus to imbue the fiber with exceptional strength and rigidity [4, 5].

The quaternary structure of collagen refers to that collagen microfibrils, which are arranged in parallel into a bundle through the covalent cross-linking to form the collagen fibrils with good stability, flexibility, and toughness [1]. By end-to-end aggregation, collagen fibrils will further gather into a bundle of CF [6]. It is observed that dark and white stripes are distributed in the entire CF.

The intermolecular distance was 67 nm, and the longitudinal distance between adjacent molecules was 40 nm. CFs generally covalently cross-linked by histidine and lysine between the C-terminus and the N-terminus. Due to the isolable and insolubility of individual natural microfibrils and fibrils, it is difficult to use most standard structure-determination techniques to identify the accurate features of collagen structure.

6.3 Isolation of Natural CF

Different from the cocoon silk that is continuous long independent fiber that can be reel off after degumming and collected by roller, CF in tissue is either discontinuous or usually intertwined with each other, leading to the difficulty to collect the CF using traditional method in textile fields. Research work has been carried out in development of newer methods to obtain CF from animal skin. However, as the protein, CF goes through denaturation and hydrolysis into gelatin under harsh conditions such as high temperature and strong acid and enzyme. Thus, physical treatment under moderate condition is preferred to obtain CF. Liao et al. [7] prepared the CF from animal skin through a series of procedures of leather processing including liming, splitting, deliming, bating, and pickling and then the grinding and sieving processes. Obtained CF has a size with 1–4 μm in diameter and 0.5–1 mm in length, which was used for metal ion absorption. J.D. Ambrósio et al. [8] prepared short leather fiber with size between 0.25 and 1.68 mm by smash and sieves as filler as enhancer. Joseph et al. [9] reported that leather fiber with average diameter of 4.5 μm and length of 258.5 μm is available from tanneries. However, strictly speaking, leather fiber is the chemically modified CF because tanning agents are reacted with collagen for stability. To prepare the pure long CF, spinning methods are adopted.

6.4 Spinning of CF

6.4.1 Extraction of Collagen

Natural collagen is usually extracted from tissues such as skin, tendons, and bones. The dissolution of collagen is, however, impeded by the low solubility of natural collagen due to its covalent cross-links including Schiff base [10], which formed a stably fiber network. Therefore, chemical/biological agents are used to cleave the covalent bonds and get the CF dissolved. To date, the noncovalent-bonded collagen can be obtained by mainly three widely used methods: acid extraction, alkali extraction, and enzyme extraction. The most common solvent systems include a neutral salt solution (0.15–0.20 M NaCl) and dilute acid solution (0.5 M acetic acid):

(1) *Enzyme*: For insoluble collagen, pepsin is able to cleave the cross-linked regions at the telopeptide without damaging the integrity of the triple helix, and then the collagen can be extracted by further treatment. Briefly, raw

materials are suspended in 0.5 mol/l acetic acid in the presence of pepsin. The extraction of collagen with limited pepsin digestion contributes to a higher yield. Due to the high activity of enzyme, enzyme extraction is highly efficient and nonpolluting. As enzyme is sensitive to temperature and pH, the control of the reaction process is essential.

(2) *Acid extraction*: Acid acetic acid, sulfuric acid, citric acid, and hydrochloric acid are commonly used for acid extraction. Acid will make partial hydrolysis of collagen to get small molecules of collagen [11]. Generally, collagen is extracted using 0.5 M acetic acid solution, which proved to be effective without too much damage to triple helix structure of collagen. The extraction normally is conducted for 16 hours with intermittent stirring. The collagen solution after centrifugation is salt out by solution of NaCl. To further purify the collagen, it can be redissolved in acetic acid and reprecipitated as described above. To avoid the degradation of collagen, temperature should be as lower as possible [12], for example, 15 °C.

(3) *Alkali extraction*: Alkali sodium hydroxide, calcium hydroxide, sodium carbonate, and lime are commonly used. The steps for alkali extraction of CFs are similar to acid extraction, but large amount of alkali consumption will result in environment pollution. There is a standard method for alkali extraction of collagen. Briefly, raw materials are shaken in 0.5 M NaOH containing 10% NaCl at 4 °C [13]. After the CF is dissolved, the collagen is recovered by salting out, purified by redissolving in acid, dialyzed against 0.01 M acetic acid, and lyophilized. Practically, in the process of preparing the collagen, a variety of ways are combined and used to achieve the best effect. For example, acid extraction and alkali extraction can be combined together to improve the extraction efficient process.

Owing to the excellent biological features and physiochemical properties of collagen, it has been among the most widely used biomaterials for biomedical applications particularly when delivered in the form of gels, films, injectables, and coatings. To further extend the utility of collagen for use in medical devices and to address issues such as fixation in defect sites and stability for tissue engineering (TE) and guided bone regeneration (GBR), the ability to manufacture mechanically robust fibers and fabrics is important. To date, mainly three spinning methods including wet spinning, electrospinning, and microfluidic spinning (Figure 6.3) have been reported to prepare CFs.

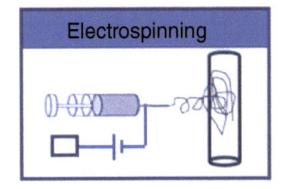

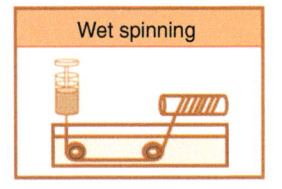

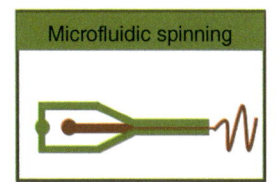

Figure 6.3 Spinning methods used to prepare CF.

6.4.2 Electrospinning of CF

Electrospinning is a process that uses an electric field to control the formation and deposition of polymers. This process is remarkably efficient, rapid, and inexpensive [14]. Generally, in electrospinning, the collagen solution is fed through a thin needle opposite to a collecting plate or target. When a high voltage is applied to the solution, a jet is formed as soon as the applied electric field strength overcomes the surface tension of the solution. When traveling toward the grounded collecting plate, the jet becomes thinner as a consequence of solvent evaporation, and fibers are formed. Splaying of the fibers because of splitting of the jet can occur as well. The fibers are deposited on the grounded target as a nonwoven highly porous mesh. Placing a rotating collector between the needle and the grounded plate allows deposition of the fibers on the collector with a certain degree of alignment. Mats with randomly oriented or aligned fibers can be formed by using a stationary or a rotating collector. Fibers with diameters in the range of 100 nm to few micrometers can be created depending on the polymer physical properties (e.g. viscosity, electrical conductivity, and surface tension), applied electric field, polymer flow rate, and the distance between the needle tip and the collector.

The possibility to produce meshes with a high surface area makes electrospinning an ideal solution for applications, which require seeding of cells, like vascular tissue engineering. Moreover, the architecture of the produced nonwoven meshes is similar to that found in most natural extracellular matrices, thus enhancing the applicability of these scaffolds [15]. Therefore, the electrospinning process has the potential to produce collagen fibrils that closely mimic, and at some points may even fully reproduce, the structural and biological properties of the natural polymer [16]. Matthews et al. [17] used hexafluoro-2-propanol (HFP) as solvent; the viscosity and conductivity of the solution, the spinning voltage, the flow rate, and the collecting distance were adjusted in preliminary experiments to optimize the electrospinning conditions. During electrospinning of an 8% w/v collagen solution in HFP, jet formation was first observed at 12 kV. Increasing the voltage to values between 19 and 21 kV offered stable operating conditions to electrospin continuously for more than three hours. Dry CFs could be collected at a distance of ~15 cm on a grounded plate with evaporation of HFP. It is noted that the CFs produced by this method using the mentioned parameters have similar dimensions as native collagen fibrils and range from 100 to 600 nm (Figure 6.4). Although strong organic solvent such as HFP may influence the structure of electrospinning of collagen, it is proved that only minor differences between structures of collagen electrospun from HFP and structures of collagen electrospun from phosphate buffered saline (PBS)-based solutions. The bending moduli of the electrospun fibers ranged from 1.3 to 7.8 GPa at ambient conditions and ranged from 0.07 to 0.26 MPa when immersed in PBS buffer.

6.4.3 Wet Spinning of CF

Among the various fiber manufacturing processes available, wet spinning has the potential to convert biomolecules into fibers without need for high voltage

Figure 6.4 Calfskin type I collagen electrospun onto a static, cylindrical mandrel. Source: Matthews et al. 2002 [17]. Reproduced with permission from American Chemical Society.

during manufacture and is less likely to be associated with denaturation. Wet spinning is conventional methods to produce man-made fibers such as viscose rayon. This fiber spinning technology is based on nonsolvent-induced phase separation, whereby polymer dope solutions are extruded through a spinneret into a nonsolvent coagulation bath in which it continuously polymerizes so as to form a solid long fiber.

The coagulation bath must be either a poor solvent or a nonsolvent with respect to the polymer. The diameter of fibers can be tuned by changing the needle(s) diameter, the polymer composition, and its volumetric flow rate [18]. Traditional wet and dry spinning processes have been most commonly used to produce protein fibers including CF.

Hirano et al. [19] described the production of chitosan–CFs by wet spinning from an aqueous 2% acetic acid–methanol solution spun into an aqueous 5% ammonia solution containing 40–43% ammonium sulfate. Continuous fibers with an average diameter of 36 μm were produced. In a method reported by Fofonoff and Bell [20], CFs with diameter more than 100 μm were prepared by wet spinning from an aqueous 0.05% acetic acid (0.5 M) solution into a coagulating bath consisting of alkaline alginic acid/boric acid (pH 8–10) with heating at 35±2°C. CF can be formed by polymerization when the acid in the collagen is neutralized upon contact with the neutralizing solution [21]. The obtained fibers are subsequently dehydrated in acetone and ethanol baths. The morphology

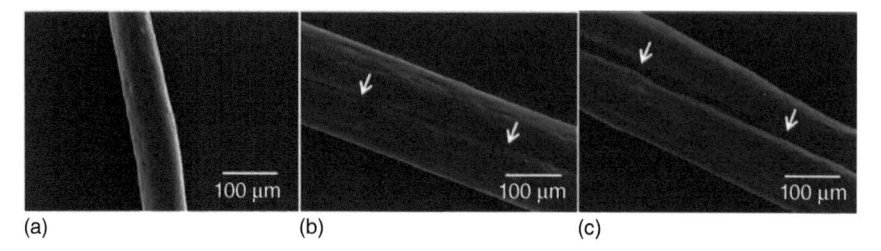

Figure 6.5 SEM images of fibers wet spun from suspensions with varied collagen concentration: (a) 1.2%, (b) 1.4%, and (c) 1.6% collagen (w/v) suspensions. Surface striations (marked with arrow) are observed in fibers derived from wet spinning suspensions with increased collagen concentrations. Source: Arafat et al. 2015 [21]. Reproduced with permission from Elsevier.

of thus obtained fibers depends on the collagen concentration (Figure 6.5). An additional example is provided by Furukawa et al. [22] in which solubilized collagen is spun through a spinneret into a coagulating bath comprising an aqueous solution of an inorganic salt, such as sodium sulfate, sodium chloride, ammonium sulfate, magnesium chloride, or aluminum sulfate. To enhance the CF properties, different strategies were successful as cross-linking and stretching of the filaments. By controlling the spinning conditions, different physical parameters can be achieved. Therefore, the quality of the fibers prepared from collagen dispersion is up to the composition and concentration of the coagulation bath and the spinneret design. In addition, post cross-linking treatment is proved to be effective method to improve the stability of CF against trypsin digestion. For example, using chemical agent such as glutaraldehyde, polyethylene glycol (PEG) in a solution for cross-linking acceptable low proliferation inhibition was observed interpreted as low cytotoxic effect [21, 23].

6.4.4 Microfluidic Spinning of CF

Microfluidic spinning is defined as the formation of fibers in a microchannel using a coaxial flow of a prepolymer and a cross-linking agent. In general, microfluidics deals with the processing of small amounts of fluids in channels with dimensions of tens to hundreds of micrometers, which are mostly embedded in small chips regularly consisting of polydimethylsiloxane (PDMS). This is similar to wet pinning, except that the cross-linking agent is supplied by the coaxial flow directly instead of the bath. To prevent polymerization inside of the system, the streams need to be coaxed close to the nozzle, which can be achieved by using microfluidic systems made by microfabrication techniques. The short distance of coaxial flow facilitates the formation of cell-laden fibers as the cells are only briefly exposed to a high shear stress [24]. Therefore, microfluidic spinning method is a promising platform for continuous fabrication of fibers with tunable morphological, structural, and chemical features. The diameters of the fabricated fibers can vary from ten to several hundreds of micrometers, depending on parameters such as flow rates of core and sheath solutions, channel dimensions, and solutions viscosities. The rotation speed of the collecting roller may also affect the fiber diameter as it imposes an external elongation stress on the fiber. In general, the size of the fibers fabricated with microfluidic systems is smaller than wet spinning but larger than electrospun fibers.

Microfluidic fiber spinning methods have been used to incorporate CFs. In comparison with blended fibers, the endless, plain CFs would be highly desirable due to a better performance including an expected higher mechanical stability, better biocompatibility, and less complexity in processing. Enomoto et al. [25] prepared the CF with a diameter from 8 to 30 µm by microfluidic spinning system with phosphate buffer. The obtained CF showed good bioactivity. Haynl et al. [26] reported that microfibers produced by microfluidic methods exhibited tensile strength and Young's modulus exceeding that of fibers produced in classical wet spinning devices and even that of natural tendon, and they showed lower diameters (Figure 6.6). It can be assumed that the smaller diameters of our microfibers led to increased mechanical stability due to better longitudinal orientation of collagen fibrils and less defect structures like voids, entanglements, free chain ends, and foreign particles in contrast to the situation in bigger diameter fibers.

6.4.5 Collagen Composite Fiber

CF in an electrospun form usually lacks mechanical integrity upon hydration. Previously established methods of cross-linking with glutaraldehyde, successful in increasing the strength of the electrospun structures, also carry an enhanced risk of cytotoxicity and calcification when used in vivo under inadequate removing of unreacted glutaraldehyde molecules. Thus, in addition to avoid denaturation of the native triple helical structure and the instability of regenerated collagen materials in the hydrated state, polymer component with stability is complexed with collagen to generate enhanced composite fibers while maintaining a high level of bioactivity through the presence of collagen. On the other hand, blending of collagen with synthetic polymers can provide methods to tailor mechanical and bioactive properties of the spinning scaffolds. The commonly used polymers for preparing collagen composite fiber include poly(lactic acid) (PLA) [27], polyurethane (PU) [28], polycaprolactone (PCL) [29], polydioxanone (PDO) [30], hydroxyapatite (HAP) [31], chitosan [32], poly(N-isopropyl acrylamide) [33], etc.

6.5 Application of CF

In addition to its high affinity and biodegradable, CF shows outstanding strength and flexibility. At present, the application CF is divided into two categories: one is based on the biological properties of collagen including its good biocompatibility and low antigenicity and biodegradability, which is suitable for cosmetics, biomedical materials, tissue engineering materials, and drug release fields, and the other is based on the unique fiber performance, such as good mechanical properties and highly chemical reactivity, making the CF preferable material for paper, textile, adsorbing materials, etc.

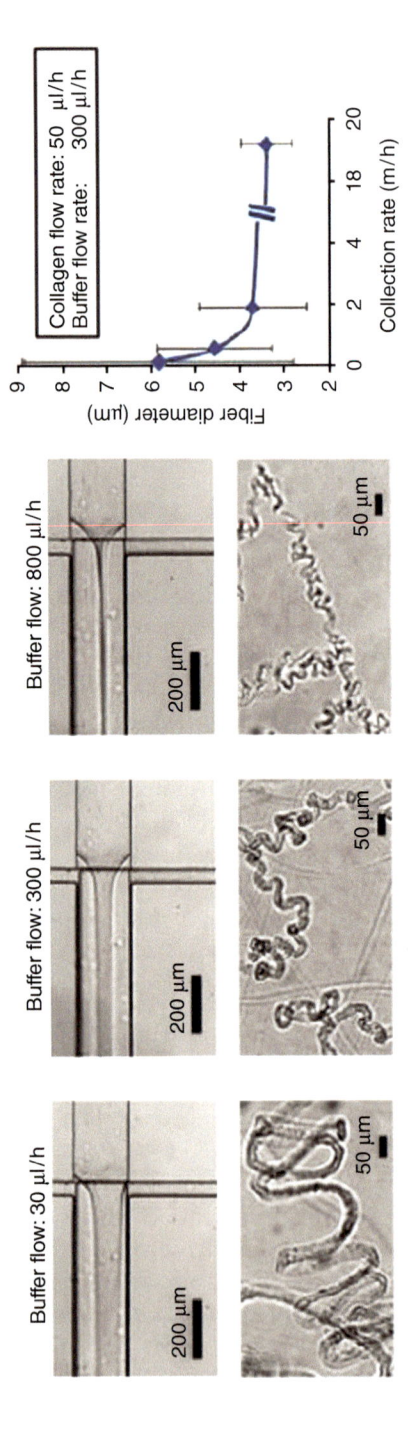

Figure 6.6 Using a constant collagen flow (here, 50 μl/h) and varying buffer flow rates enables the extrusion of microfibers into water with adjustable diameters. Source: Haynl et al. 2016 [26]. Reproduced with permission from American Chemical Society.

(1) *Bio-application*: Because of collagen's good mechanical stability, biocompatibility, biodegradability, low immunogenicity, and ability to promote cellular attachment and growth, collagenous materials, especially fibers, are used in medical and biomedical applications such as drug delivery [34], wound healing [35], and tissue engineering [36]. Rapid expansion of understanding of the excretal cellular matrix and progress toward the generation of synthetic matrix components suggests that synthetic CFs offer compelling advantages over traditional polymers for soft tissue repair and replacement. The CFs obtained by mechanical or chemical treatment still retain the structural characteristics and properties of natural CFs. CFs are regarded as a "green" fiber, which has been widely used in various fields. The CF extracted from the natural materials by acid, alkali, and enzymatic methods is a natural linear polymer whose amino acid composition is very close to that of the human tissues, so it can be applied on field of medical and healthcare such as burns and orthopedics, beauty, trauma treatment, tissue repair, and wound hemostasis.

(2) *Fiber as enhancer*: One alternative for natural CF, especially leather fiber, is to incorporate the fiber as filler in thermosetting and thermoplastic polymers. J.D. Ambrósio et al. [8] prepared the composites with recycled poly(vinyl butyral) (PVB) and wet blue leather fiber. PVB showed good interfacial adhesion with leather fibers. Increasing the leather fibers led to a significant increase in the tensile modulus while a reduction in the tensile strain in breaks. In the work of Joseph et al. [9], the waste leather fiber-filled polycaprolactone (PCL) composites were prepared by twin-screw extrusion. Addition of leather fibers resulted in improvement of tensile modulus of neat PCL and reduction in percentage crystallinity of PCL matrix. Leather fibers also increased the water uptake. They suggested that these biodegradable composites could be used to develop low-cost materials suitable for applications in wearing products and bags.

(3) *Metal adsorption application*: Calfskin that shows highly reactivity to tannins is utilized to prepare natural CF membrane for immobilization of tannins for adsorption metal ions in solution [37, 38] as shown in Figure 6.7. Vegetable tannins have multiple adjacent phenolic hydroxyls and exhibit specific affinity to metal ions. Thus, they promise to be a versatile agent for the treatment of metal ions containing wastewater. However, vegetable tannins are water-soluble compounds, which restrict their practical application. Due to the highly reactive activity to tannins, animal skin including calfskin is used to immobilize tannins. Attempts had been made to immobilize tannins onto various water-insoluble matrices.

Liao et al. [7] prepared a novel adsorption membrane by immobilizing condensed vegetable tannins onto CF membrane; these tannin-immobilized membranes possess proper physical properties and excellent adsorption and desorption characteristics for UO^{2+}. The effectiveness and practicability of tannin-immobilized membrane in the recovery of UO^{2+} from aqueous solution are obvious. Later, they found that tannin-immobilized CF matrixes are effective adsorbents in the recovery of Au(III) from aqueous solutions. Because the B ring of bayberry tannins is a pyrogallol structure, whereas the

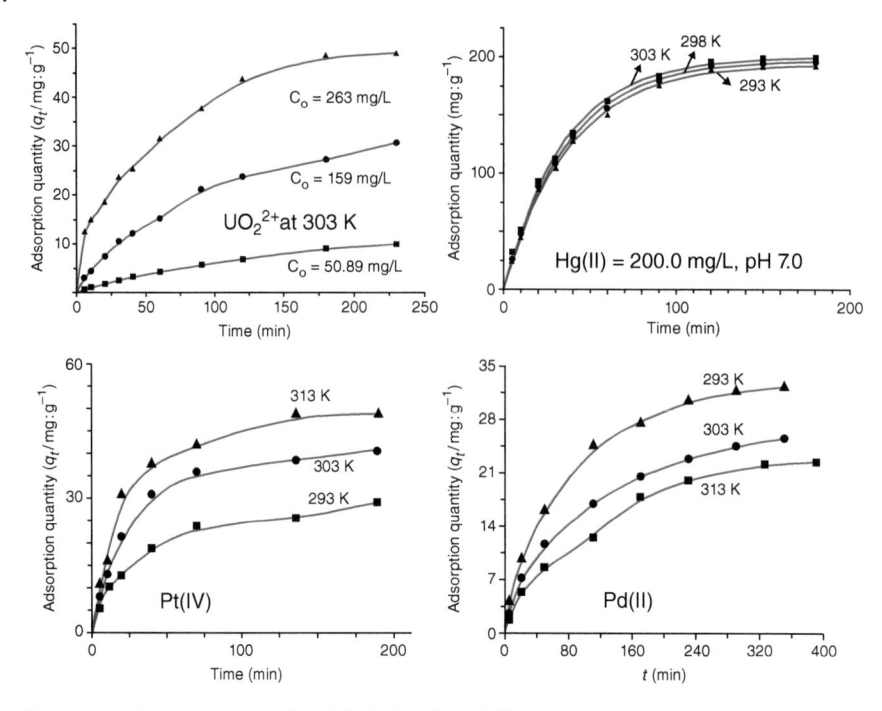

Figure 6.7 Absorption rate of modified skin CF to different metal ions ([7], [37], [38]).

B ring of larch tannins is a catechol structure, bayberry tannins have a higher reaction activity with Au(III). Therefore, immobilized bayberry tannins have a higher adsorption capacity for Au(III) than immobilized larch tannins [39]. Besides, the adsorption of other metal ions including platinum(IV) and palladium(II) on bayberry tannin-immobilized CF (BTICF) membrane was investigated [37]. Results showed that the membrane exhibits ability of selective adsorption to Pt(IV) and Pd(II) in the mixture solutions of metal ions.

(4) *Template for synthetic metal fiber*: According to the principles of leather processing, CF that contains abundant functional groups, like –OH, –COOH, and –NH$_2$, is ready to react with some metal ions, such as Cr(III), Al(III), Zr(IV), Ti(IV), and so forth. Hence, the metal fibers are supposed to generate via the reaction between metal ion and skin CF. On the basis of this idea, Deng et al. [40] synthesized the alumina fiber by reaction between Al(III) and CF from animal skin that was first stabilized by vegetable tannins to enhance its fixing capacity to Al(III). In their research, hierarchical alumina fiber (Figure 6.8) with ordered mesoporous distribution was successfully prepared. This metal fiber showed their improve abilities to be used as catalyst or catalyst support with high activity and selectivity owing to its high surface area and shape-selective properties [40, 41, 43]. Base on this method, hierarchical mesoporous ZrO$_2$ fiber and TiO$_2$ fiber were also successfully synthesized by using CF as template as shown in Figure 6.8 [41, 42]. Therefore, animal skin including calfskin can be used to prepare CF for general template to synthesize porous metal oxide fibers.

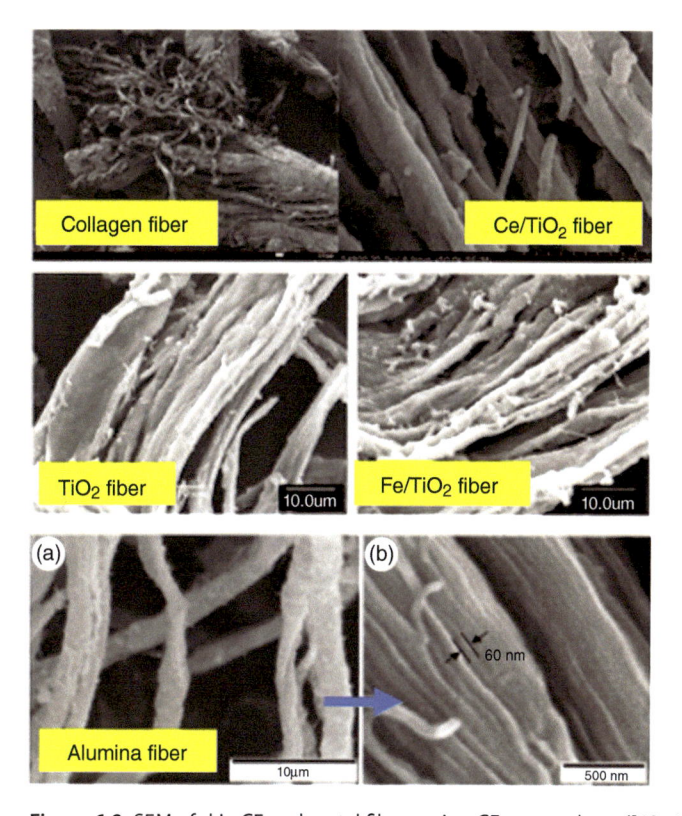

Figure 6.8 SEM of skin CF and metal fibers using CF as templates ([40–42]).

(5) *Metal nanoparticle supporter*: Due to the affinity to metal ions, water insolubility, hydrophilicity, flexibility, and softness, CF from animal skin is promising candidate of biopolymers for preparing supported metal nanoparticles. Wu et al. [44] use CF as a natural polymeric support to synthesize a novel palladium (Pd) nanoparticle catalyst. To achieve a stable immobilization of Pd on CF support, they used epigallocatechin-3-gallate (EGCG) to graft onto CF surface, acting both as dispersing and stabilizing agent for Pd nanoparticles. This EGCG-CF nanocatalyst presented high activity and selectivity for the liquid hydrogenation of allyl alcohol and retained the catalytic efficiency after recycling for five times. Later, Wu et al. [45] further used this method to prepare size-controlled gold nanoparticles (AuNPs) supported on CF, which was successfully applied as catalyst for 4-nitrophenol reduction. In addition to catalyst, metal nanoparticle-loaded skin CFs can also be used as high-performance microwave absorption. Guo et al. [46] designed and fabricated the controllable Ag@Ni core–shell NPs assembled in the hierarchical intertexture of skin CF stabled with bayberry tannin (SCF-BT) by a simple one-step route (Figure 6.9). This modified CF exhibited excellent microwave absorption performance in the whole X-band and C-band and some part of the S-band due to its multiple defective site polarization and interfacial polarization, as well as significant eddy current effect and anisotropic energy for the microwave energy dissipation.

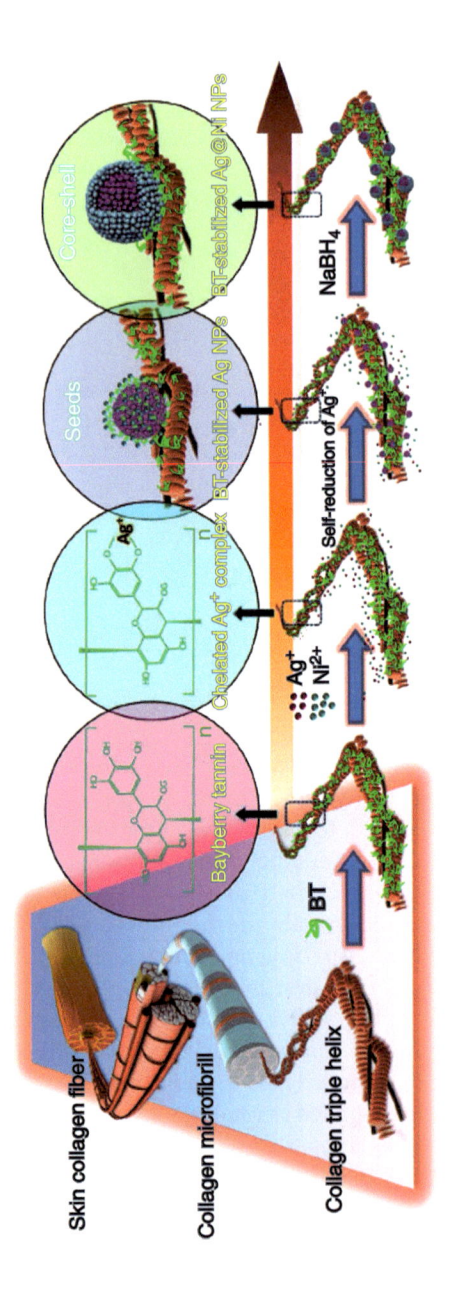

Figure 6.9 Schematic diagram showing the preparation mechanism of the one-step seeding growth of Ag@Ni core–shell NPs on skin CF. Source: Guo et al. 2012 [46]. Reproduced with permission from Royal Society of Chemistry.

6.6 Perspectives

Due to its unique structure, CF exhibits outstanding mechanical and biological properties, which has significantly benefited the development of functional materials, which is incomparable to synthetic polymers. According to the current development and based on our own understanding, perspectives are given for future development of CF:

(1) *Structure functions*: Although the molecule structure has been widely studied, the functions, especially its intrinsic smart behavior in materials' point of view, are rarely explored. Hence, more properties and smart function of CF regarding to its hierarchical structure from micro to macro can be investigated. For example, environmental (e.g. pH, thermo, magnetic, etc.) stimulus response of CF should be intensively studied for their potential application.

(2) *Preparation methods*: To date, CF with controllable length and diameter can be only achieved by spinning methods. Although the spinning method provides facile path to prepare CF, the regenerated structure of spun CF can be different from natural CF, which would weak its natural properties. Therefore, more efforts can be paid on how to obtain CF without breaking its natural structure and thus how to utilize the CF to prepare materials in various formats.

(3) *Multifunctions and high performance*: To meet the ever-increasing demand for novel materials, multifunctions and high performance should be realized on CF-based materials. For example, by chemical modification and/or compositing technology, the novel CF or its composites would exhibit enhanced property due to the synergistic effect resulting from natural CF and polymers.

(4) *Extended applications*: Generally, the application of CF, especially by spinning method, is focused on biological application such as wound dressing and scaffold. Some other potential application including metal and microwave absorption has been reported; however, the complex and high-cost preparation process limited its practical development. As major component of skin, CF should have advantages when applied in files such as artificial skin. Besides, the environmental stimuli-responsive behavior of CF may also contribute to its application as robotic sensors. On the other hand, back to its fiber form, 3D fibrous network can be achieved by CF, which could have broad range of application as covering materials to provide strength, flexibility, protectiveness, permeability, etc.

References

1 Buehler, M.J. (2006). Nature designs tough collagen: explaining the nanostructure of collagen fibrils. *Proceedings of the National Academy of Sciences of the United States of America* 103: 12285–12290.
2 Giudici, C., Viola, M., Tira, M.E. et al. (2003). Molecular stability of chemically modified collagen triple helices. *FEBS Letters* 547: 170–176.

3 Ushiki, T. (2002). Collagen fibers, reticular fibers and elastic fibers. A comprehensive understanding from a morphological viewpoint. *Archives of Histology and Cytology* 65: 109–126.
4 Drach, B., Kuksenko, D., and Sevostianov, I. (2016). Effect of a curved fiber on the overall material stiffness. *International Journal of Solids and Structures* 100–101: 211–222.
5 Eppell, S.J., Smith, B.N., Kahn, H., and Ballarini, R. (2006). Nano measurements with micro-devices: mechanical properties of hydrated collagen fibrils. *Journal of the Royal Society, Interface* 3: 117–121.
6 Raspanti, M., Viola, M., Sonaggere, M. et al. (2007). Collagen fibril structure is affected by collagen concentration and decorin. *Biomacromolecules* 8: 2087–2091.
7 Liao, X., Ma, H., Wang, R., and Shi, B. (2004). Adsorption of UO_2^{2+} on tannins immobilized collagen fiber membrane. *Journal of Membrane Science* 243: 235–241.
8 Ambrósio, J.D., Lucas, A.A., Otaguro, H., and Costa, L.C. (2011). Preparation and characterization of poly(vinyl butyral)-leather fiber composites. *Polymer Composites* 32: 776–785.
9 Joseph, S., Ambone, T.S., Salvekar, A.V. et al. (2015). Processing and characterization of waste leather based polycaprolactone biocomposites. *Polymer Composites* 38: 2889–2897.
10 Eyre, D.R., Paz, M.A., and Gallop, P.M. (1984). Cross-linking in collagen and elastin. *Annual Review of Biochemistry* 53: 717–748.
11 Kittiphattanabawon, P., Benjakul, S., Visessanguan, W. et al. (2005). Characterisation of acid-soluble collagen from skin and bone of bigeye snapper (*Priacanthus tayenus*). *Food Chemistry* 89: 363–372.
12 Nalinanon, S., Benjakul, S., and Kishimura, H. (2010). Collagens from the skin of arabesque greenling (*Pleurogrammus azonus*) solubilized with the aid of acetic acid and pepsin from albacore tuna (*Thunnus alalunga*) stomach. *Journal of the Science of Food and Agriculture* 90: 1492–1500.
13 Sadowska, M., Kołodziejska, I., and Niecikowska, C. (2003). Isolation of collagen from the skins of Baltic cod (*Gadus morhua*). *Food Chemistry* 81: 257–262.
14 Huang, Z.-M., Zhang, Y.-Z., Kotaki, M., and Ramakrishna, S. (2003). A review on polymer nanofibers by electrospinning and their applications in nanocomposites. *Composites Science and Technology* 63: 2223–2253.
15 Yang, L., Fitie, C.F., van der Werf, K.O. et al. (2008). Mechanical properties of single electrospun collagen type I fibers. *Biomaterials* 29: 955–962.
16 Sell, S.A., McClure, M.J., Garg, K. et al. (2009). Electrospinning of collagen/biopolymers for regenerative medicine and cardiovascular tissue engineering. *Advanced Drug Delivery Reviews* 61: 1007–1019.
17 Matthews, J.A., Wnek, G.E., Simpson, D.G., and Bowlin, G.L. (2002). Electrospinning of collagen nanofibers. *Biomacromolecules* 3: 232–238.
18 Yaari, A., Schilt, Y., Tamburu, C. et al. (2016). Wet spinning and drawing of human recombinant collagen. *ACS Biomaterials Science & Engineering* 2: 349–360.

19 Hirano, S., Zhang, M., Nakagawa, M., and Miyata, T. (2000). Wet spun chitosan–collagen fibers, their chemical N-modifications, and blood compatibility. *Biomaterials* 21: 997–1003.
20 Fofonoff, T.W. and Bell, E. (1999). Method for spinning and processing collagen fiber. Google Patents.
21 Arafat, M.T., Tronci, G., Yin, J. et al. (2015). Biomimetic wet-stable fibres via wet spinning and diacid-based crosslinking of collagen triple helices. *Polymer* 77: 102–112.
22 Furukawa, M., Takada, M., Murata, S., and Sasayama, A. (1994). Process for producing regenerated collagen fiber. Google Patents.
23 Tronci, G., Kanuparti, R.S., Arafat, M.T. et al. (2015). Wet-spinnability and crosslinked fibre properties of two collagen polypeptides with varied molecular weight. *International Journal of Biological Macromolecules* 81: 112–120.
24 Lee, P., Lin, R., Moon, J., and Lee, L.P. (2006). Microfluidic alignment of collagen fibers for in vitro cell culture. *Biomedical Microdevices* 8: 35–41.
25 Enomoto, S., Yajima, Y., Watabe, Y. et al. (2015). One-step microfluidic spinning of collagen microfibers and their application to cell cultivation. In: *2015 International Symposium on Micro-NanoMechatronics and Human Science (MHS)*, 1–4. IEEE.
26 Haynl, C., Hofmann, E., Pawar, K. et al. (2016). Microfluidics-produced collagen fibers show extraordinary mechanical properties. *Nano Letters* 16: 5917–5922.
27 Yang, F., Murugan, R., Wang, S., and Ramakrishna, S. (2005). Electrospinning of nano/micro scale poly(L-lactic acid) aligned fibers and their potential in neural tissue engineering. *Biomaterials* 26: 2603–2610.
28 Chen, R., Huang, C., Ke, Q. et al. (2010). Preparation and characterization of coaxial electrospun thermoplastic polyurethane/collagen compound nanofibers for tissue engineering applications. *Colloids and Surfaces B: Biointerfaces* 79: 315–325.
29 Phipps, M.C., Clem, W.C., Grunda, J.M. et al. (2012). Increasing the pore sizes of bone-mimetic electrospun scaffolds comprised of polycaprolactone, collagen I and hydroxyapatite to enhance cell infiltration. *Biomaterials* 33: 524–534.
30 McClure, M.J., Sell, S.A., Simpson, D.G., and Bowlin, G.L. (2009). Electrospun polydioxanone, elastin, and collagen vascular scaffolds: uniaxial cyclic distension. *Journal of Engineered Fibers and Fabrics* 4: 18–25.
31 Teng, S.-H., Lee, E.-J., Wang, P., and Kim, H.-E. (2008). Collagen/hydroxyapatite composite nanofibers by electrospinning. *Materials Letters* 62: 3055–3058.
32 Chen, Z., Wang, P., Wei, B. et al. (2010). Electrospun collagen–chitosan nanofiber: a biomimetic extracellular matrix for endothelial cell and smooth muscle cell. *Acta Biomaterialia* 6: 372–382.
33 Chen, M., Dong, M., Havelund, R. et al. (2010). Thermo-responsive core–sheath electrospun nanofibers from poly(N-isopropylacrylamide)/polycaprolactone blends. *Chemistry of Materials* 22: 4214–4221.

34 Zeng, J., Xu, X., Chen, X. et al. (2003). Biodegradable electrospun fibers for drug delivery. *Journal of Controlled Release* 92: 227–231.

35 Chen, J.-P., Chang, G.-Y., and Chen, J.-K. (2008). Electrospun collagen/chitosan nanofibrous membrane as wound dressing. *Colloids and Surfaces A: Physicochemical and Engineering Aspects* 313: 183–188.

36 Li, M., Mondrinos, M.J., Gandhi, M.R. et al. (2005). Electrospun protein fibers as matrices for tissue engineering. *Biomaterials* 26: 5999–6008.

37 Ma, H.-w., Liao, X.-P., Liu, X., and Shi, B. (2006). Recovery of platinum(IV) and palladium(II) by bayberry tannin immobilized collagen fiber membrane from water solution. *Journal of Membrane Science* 278: 373–380.

38 Huang, X., Liao, X., and Shi, B. (2009). Hg(II) removal from aqueous solution by bayberry tannin-immobilized collagen fiber. *Journal of Hazardous Materials* 170: 1141–1148.

39 Liao, X., Zhang, M., and Shi, B. (2004). Collagen-fiber-immobilized tannins and their adsorption of Au(III). *Industrial and Engineering Chemistry Research* 43: 2222–2227.

40 Deng, D., Tang, R., Liao, X., and Shi, B. (2008). Using collagen fiber as a template to synthesize hierarchical mesoporous alumina fiber. *Langmuir* 24: 368–370.

41 Xiao, G., Huang, X., Liao, X., and Shi, B. (2013). One-pot facile synthesis of cerium-doped TiO_2 mesoporous nanofibers using collagen fiber as the biotemplate and its application in visible light photocatalysis. *The Journal of Physical Chemistry C* 117: 9739–9746.

42 Cai, L., Liao, X., and Shi, B. (2010). Using collagen fiber as a template to synthesize TiO_2 and Fe_x/TiO_2 nanofibers and their catalytic behaviors on the visible light-assisted degradation of orange II. *Industrial and Engineering Chemistry Research* 49: 3194–3199.

43 Xiao, G., Zhou, J., Huang, X. et al. (2014). Facile synthesis of mesoporous sulfated Ce/TiO_2 nanofiber solid superacid with nanocrystalline frameworks by using collagen fibers as a biotemplate and its application in esterification. *RSC Advances* 4: 4010–4019.

44 Wu, H., Wu, C., He, Q. et al. (2010). Collagen fiber with surface-grafted polyphenol as a novel support for Pd(0) nanoparticles: synthesis, characterization and catalytic application. *Materials Science and Engineering C* 30: 770–776.

45 Wu, H., Huang, X., Gao, M. et al. (2011). Polyphenol-grafted collagen fiber as reductant and stabilizer for one-step synthesis of size-controlled gold nanoparticles and their catalytic application to 4-nitrophenol reduction. *Green Chemistry* 13: 651–658.

46 Guo, J., Wang, X., Miao, P. et al. (2012). One-step seeding growth of controllable Ag@Ni core–shell nanoparticles on skin collagen fiber with introduction of plant tannin and their application in high-performance microwave absorption. *Journal of Materials Chemistry* 22: 11933–11942.

7

Electrospun Fibers for Filtration

Xia Yin[1], Jianyong Yu[2], and Bin Ding[1,2]

[1] *Key Laboratory of Textile Science & Technology, Ministry of Education, College of Textiles, Donghua University, Shanghai 201620, China*
[2] *Innovation Center for Textile Science and Technology, Donghua University, Shanghai 200051, China*

7.1 Introduction

With the rapid growth of industrialization, urbanization, and modern agricultural development, environmental pollution is becoming more and more serious [1–3]. Among them, air pollution is a major health hazard that is responsible for respiratory diseases, tuberculosis, allergies, and asthma [4–6]. Besides the harmful gas, the suspended dusts are generally considered to contribute to air pollution, the size of which are 0.01–100 μm, and the smaller particle, the deeper it can enter into respiratory tract. In particular, the fine particles (PM with an aerodynamic diameter of 2.5 μm or less [PM2.5]) in the air are the most important air pollutants, having a negative impact on both industrial production and the health of human beings [7–10]. Therefore, finding effect way to achieve the protection and control of air pollution, especially the PM2.5 pollution, is an urgent issue to solve.

As for the sources of the fine particles, they are emitted mainly by anthropogenic activities, for example, vehicle exhaust, industrial production, biomass burning, secondary nitrate and secondary sulfates, etc., which can be grouped as normal and medium-high temperature particles according to the conditions that the particles are produced [11]. Due to the recognized threats of fine particles to the production facilities and human bodies, many prevention measures are taken to solve this problem: source control, usage of clean energy, air filtration technologies, and so on [12]. Among them, air filtration technology is the most widely used method for its high performance, low energy consumption, wide application range, and low cost. Traditional air filtration materials made of gauze can trap particles with bigger size and have very good air permeability, but they have almost no effect for particles with less than the order of magnitude of submicron [13]; nonwoven fibrous membranes (based on different fibers like glass, polypropylene [PP], polyester, and polyethylene [PE], etc.), which bear disordered arrangement of fibers, have relatively higher filtration efficiency [14], but their structure characteristics of still thicker fiber and larger pore size lead to

Handbook of Fibrous Materials, First Edition. Edited by Jinlian Hu, Bipin Kumar, and Jing Lu.
© 2020 Wiley-VCH Verlag GmbH & Co. KGaA. Published 2020 by Wiley-VCH Verlag GmbH & Co. KGaA.

limited filtration performance for ultrafine particles [15]. We can enhance the filtration efficiency of nonwoven fibrous membranes by increasing the fiber density of the membranes, but this way, the air permeability would be sacrificed [16, 17]. Therefore, achieving air filtration materials with higher filtration efficiency and at the same time with the lower pressure drop is always a difficult problem faced.

For fibrous filter media, it is easy to understand that the smaller the fiber diameter, the better filtration efficiency it will achieve. Meanwhile, the theory called "slip effect" tells us that when the diameter of fiber is small, the airflow would bypass the fiber so that cause less resistance against airflow, consequently leading to a smaller pressure drop across the fibrous media [18–20]. Thus, nanofibers become very promising to form filter media with high filtration performance for ultrafine particles. So far, a variety of techniques such as drawing, template synthesis, phase separation, islands in a sea, self-assembly, electrospinning, and solvothermal synthesis have been developed to fabricate nanofibers [21–26]. Comprehensively considering the complexity of the equipment, processing controllability, spinnable fiber range, structural controllability, productivity, and cost, electrospinning is often regarded as the preferred method to prepare uniform nanofibers in a continuous process and at a long length scales [27–29]. Electrospun nanofibrous membranes possess several intriguing features such as smaller pore size, higher porosity, remarkable specific surface area (SSA), and interconnected porous structure that make them very attractive in air filtration area including disposable respirators, automotive cabin air filters, industrial filter paper, cleanroom air purification systems, indoor air purifiers, and so on [14, 30–32].

Taking advantage of abovementioned fascinating features of electrospun nanofibrous membranes, the electrospinning technique is of huge interest for use in air filtration fields. Actually, there is another technique so-called electro-netting; strictly speaking, it can be regarded as the derived nanofabrication method from the technology of electrospinning. The diameter of nanofibers made by electro-netting can fall below 50 nm, further reduced than electrospun fibers (average diameter 100–500 nm) [33]. Thus, this technique of electro-netting attracted more attentions of researchers in recent years for their application in filtration area. In this chapter, two nanofabrication technologies, electrospinning and electro-netting, will be introduced, including the structures and properties of nanofibrous membranes that they can fabricate, the principle and formation mechanisms of each method, and their application in air filtration area for trapping both normal temperature and medium-high temperature ultrafine particles.

7.2 Fabrication Technologies

7.2.1 Electrospinning

Electrospinning, also named as electrostatic spinning, is a simple but effective and highly versatile technique that has been widely used to fabricate micro- and nanoscale fibers. This is a kind of technology that micro/nanofibers are formed

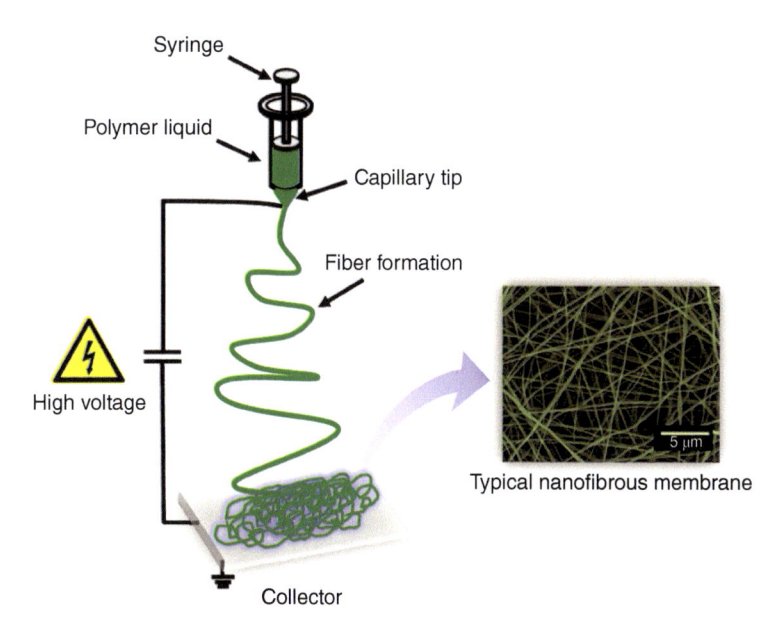

Figure 7.1 Schematic diagram of the basic setup for electrospinning.

from polymeric solutions or melts under the application of an electric field [34]. The schematic diagram to demonstrate the process of electrospinning is shown in Figure 7.1. During the electrospinning, the polymer liquid (a solution or melt) is put in the syringe that will push the liquid out from the tip with a certain speed, and a high voltage electric field (typically 10–50 kV) is applied to the polymer liquid. Gradually increasing the intensity of the electric field, a jet is drawn from the spinneret under a constant flow rate flying toward the collector; between the syringe tip and the collector, the solvent gradually evaporates and the polymer solidifies, and finally the solid micro/nanofiber is collected as an interconnected mats (as shown in Figure 7.1) on the oppositely charged grounded collector [35]. The diameters of the formed polymer fibers are from micrometers to few tens nanometers depending on different polymer type and technological conditions.

The process parameters that would influence the electrospinning include polymer solution properties (polymer type, solution concentration, solution viscosity, solution conductivity, etc.), spinning parameters (electric voltage, injection speed, collecting distance, etc.), and environmental parameters (mainly ambient temperature and ambient humidity).

7.2.2 Electro-netting

Since always accompanied with electrospinning, electro-netting is also called electrospinning/netting (ESN) process. As a variant of electrospinning process, this technique involves more higher voltage to induce the formation of not only a liquid jet but also the phase separation-induced splitting of small charged droplets between the syringe tip and the grounded collector, as shown in Figure 7.2a. Through this technique of electro-netting, a two-dimensional (2D)

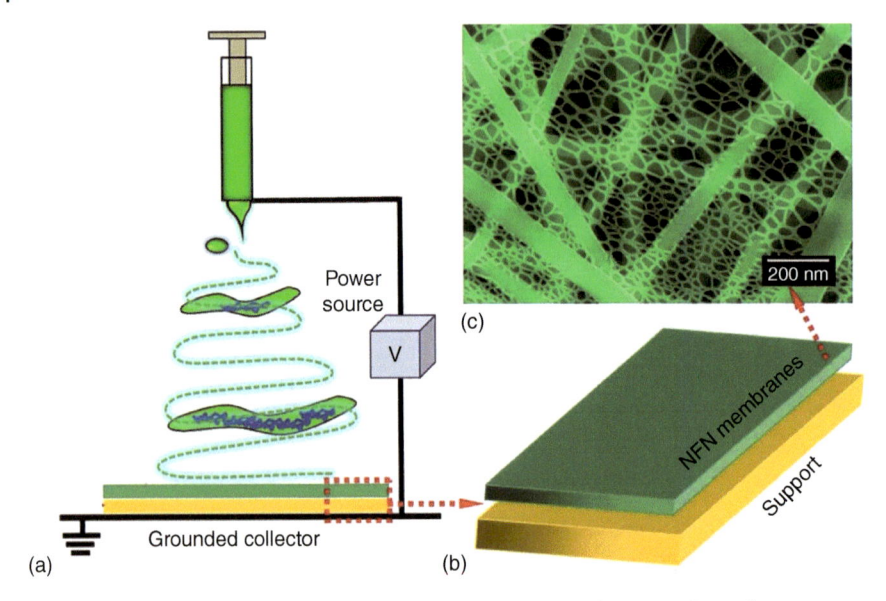

Figure 7.2 Setup for electro-netting and its formed NF/N membrane. (a) Setup for electro-netting. (b) NF/N membranes on support material. (c) Typical Field Emission Scanning Electron Microscope (FE-SEM) image of NF/N membranes. Source: Wang et al. 2012 [94]. Reproduced with permission of Royal Society of Chemistry.

spider-web-like-structured nanonets are fabricated between the primary electrospun fibers to construct a kind of aggregate structure termed as "electrospun nanofiber/net (NF/N) membranes" [33, 94] (shown in Figure 7.2c). This membrane possesses the unique "nanonet" structure comprising interlinked ultrathin nanowires with diameter of only 5–40 nm, which break the limitation of common electrospun fibers usually with large average diameter (100–500 nm) [36].

The origins of this novel one-step ESN process can be traced to the year of 2006 through a research article published in the journal of "Nanotechnology" by Ding et al. [37]. Compared with electrospinning, electro-netting needs more rigorous technological requirements, thus leading to the controllable preparation even harder. The formed inimitable NF/N structure exhibits many fascinating characteristics, such as extremely small diameter, super large SSA, and high porosity, which make it an attractive candidate in filtration, sensing, body protection, and many other environment, energy, and healthcare fields.

7.3 Principles and Theories

7.3.1 Fundamental Theory of Electrospinning

Although the setup for electrospinning is quite simple, the fundamental theory behind is very complex involving electric field, fluid rheology, polymer physics and chemistry, electrostatics, etc. [38, 39]. At first, the formation of micro/nanofibers by electrospinning was often thought to be caused by splitting

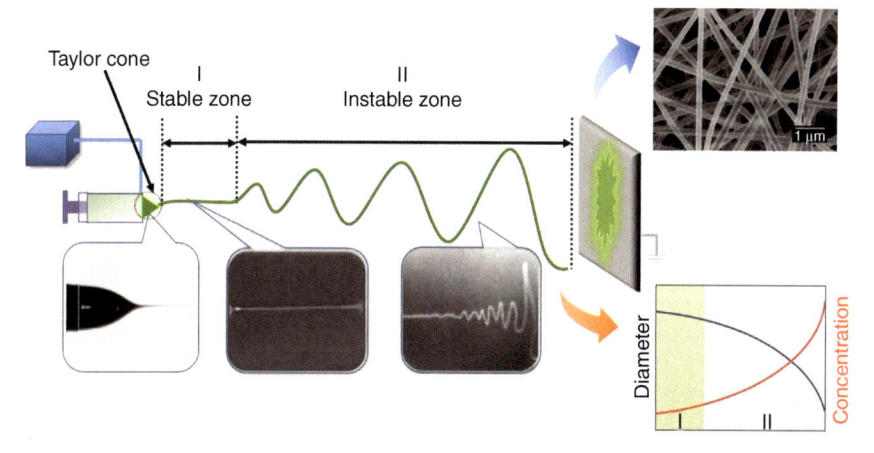

Figure 7.3 Fundamental theory and dynamic evolving process of electrospinning. Source: Wang et al. 2006 [46]. Reproduced with permission of American Chemical Society.

or splaying of the electrified jet as a result of repulsion between surface charges. With further research, it is indicated that the formation process actually is a dynamic evolving process, mainly controlled by the interaction of several physically unstable processes in electric field [40, 41]. The fundamental theories on electrospinning mainly can be described by three zones: (i) Taylor cone zone, (ii) jet stable zone, and (iii) jet in flight-instability zone, as shown in Figure 7.3.

During the electrospinning, the syringe is filled with a melt or blend polymer solution. When high voltage is applied, the solution becomes highly charged; the solution droplet at the syringe tip then will suffer two major types of forces: the surface tension and electrostatic repulsion force. When the voltage is increased to the critical value and any further increase will destroy the equilibrium, the hemispherical surface of the liquid at the tip elongates to form a conical shape, usually with a half angle of 49.3°, referred to as the Taylor cone [42, 43] (see insert in Figure 7.3). Just after this moment, the repulsive force of the charged polymer overcomes the surface tension of the solution, and as a result, a charged jet erupts from the tip of the Taylor cone and moves toward the collector.

After the charged jet ejecting from the Taylor cone, the jet first follows a straight path in the direction of the counter electrode. In this stable zone, the jet diameter decreases typically due to evaporation of the solvents and longitudinal deformations induced by electric forces [44, 45]. Following, the jet will bend and turn sideways and begin to perform spiraling (see insert in Figure 7.3); the electric forces continue to stretch and thin the jet by large ratios. It is revealed by researchers that as the diameter of the jet decreases, the path of the jet again becomes unstable and a new, smaller diameter electrical bending instability occurred [40, 46]. A succession of three or more smaller diameter bending instabilities is often observed before the jet solidified. Thanks to these instabilities, finally the length of the fiber increases enormously, and their cross-sectional diameter decreases to a fraction of a micron. Thus, during the whole process of fabrication, from the syringe tip to the collector, the solvent gradually evaporates so that the polymer concentration increases, and the fiber diameter decreases to the final microscale or even nanoscale (the corresponding change curve is shown in Figure 7.3.).

7.3.2 Formation Mechanism of the Nanofiber/Net (NF/N) Membranes

When the fantastic NF/N membrane with special spider-web-like structure was first demonstrated by Ding et al., its potential application value triggered enormous researchers to explore the spinnable polymers, influencing parameters, morphology, and so on; they also wanted to reveal what happened to the charged polymer jets and droplets during such small distance between the syringe tip and grounded collector and how to control the final collected structure [47]. However, the formation mechanism of NF/N structures is complex, and the consensus on it has not been reached yet.

In order to enrich the polymer species, develop more functional features, and broaden the applied fields for the polymeric nanonets, the systematic formation mechanism of the NF/N structure is eager to be created to achieve the controllable fabrication. Up to now, there are mainly four possible mechanisms proposed to explain the formation of the nanonet structure: (i) ions initiated splitting up of the electrospun fibers, (ii) intermolecular hydrogen bonding, (iii) intertwining among branching jets, and (iv) phase separation of charged droplets. The first one was proposed by Kim et al. [48], believing that nanonets are formed by the joints between many fibers and the possible joints occur at the apex of Taylor cone; the second one focuses on the hydrogen bonding, which possibly offers the link force for the interconnected fibers [49]. These two mechanisms do provide some qualitative and possible explanation for the nanonet formation, but the first one is just suitable for salt addition system, while the second one is only for PA-6/oligomer system [33]. The third plausible mechanism of nanonets is "intertwining among branching jets," pointed out by Tsou et al., who attributed the nanonet formation to the enormous tiny subsidiary jets [50], but it ignores that the tiny jets may not be strong enough to withstand the huge electric traction force during flying process. The last mechanism based on phase separation of charged droplets is regarded as the most comprehensive one so far, which systematically analyzes the forces exerted on the charged liquid and droplets to reveal the complex formation process for the nanonets [37, 51], and thus will be described next.

The mechanism of "phase separation of charged droplets" was first proposed by Ding et al. in 2006 after the observation of the defect poly(acrylic acid) (PAA) films and partly split PAA nanonets [37]. In a simple word, this mechanism states that the formation of nanonets is due to the phase separation of charged droplets generated between the syringe tip and grounded collector. They suggested that during the electro-netting process, besides the formation of polymer solution jet, it also sprayed some small charged droplets under a higher electric field [52] (see the schematic diagram illustrating the possible formation procedure of nanonets shown in Figure 7.4): having experienced instability and elongation process, the polymer solution jets become very long and thin and eventually form the polymer nanofibers; while, for small charged droplet, it deforms significantly to a thin liquid film, then the rapid phase separation leads the solvent part into pores, and the spider-web-like nanonet structure is gradually formed and collected.

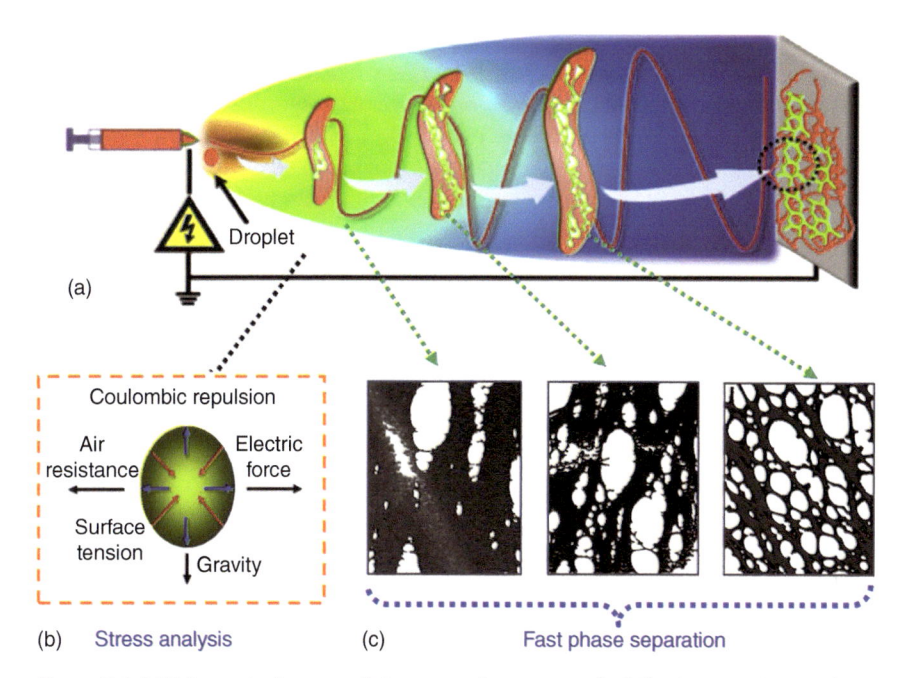

Figure 7.4 (a) Schematic diagram of electro-netting process. (b,c) The forces acting on the charged droplet and the possible process of nanonet formation during ESN process. Source: Wang et al. 2011 [54]. Reproduced with permission of The Royal Society of Chemistry.

After the charged droplets with microscale generated accompanying with the polymer solution jets, there are various forces exerted on the droplets under the electric field: electrostatic force, gravity, surface tension, drag force (air resistance), coulombic repulsion force, and viscoelastic force [53, 54], as shown in Figure 7.4b. The electrostatic force aims to carry the charged droplet from syringe tip to the grounded collector with a certain accelerated speed; the drag force between the surrounding air and the charged droplet would deform the droplets into films; the coulombic repulsion force acts an opposite function with surface tension and viscoelastic forces of the droplet. On the one hand, the coulombic repulsion force would like to expand the droplet; on the other hand, the droplet's surface tension and viscoelastic forces will contract it. During the electro-netting process, with the increasing moving speed of the charged droplet, the exerted drag force and the coulombic repulsion forces will be reinforced, so that leads to further expansion and distortion of the thin film. And at the same time, the fast phase separation causes the solvent part into pores (shown in Figure 7.4c); thus, the spider-web-like nanonet structure is finally constructed and collected.

7.4 Structure and Properties

7.4.1 Types and Structures of the Nanofiber Membranes

As a nanofabrication technique, electrospinning has been widely used due to not only simplest setup but also large number of spinnable material types and versatile structures and properties of nanofiber membranes. A variety of materials such as polymers (natural polymers like silk, chitin, chitosan, etc. and synthetic polymers like nylon 6, polystyrene, polyurethane, etc.) [55–58], ceramics (e.g. TiO_2, SiO_2, ZnO, etc.) [59, 60], carbon (including PAN based, polybenzoxazine based, polybenzimidazole based, and so on) [61, 62], and even metals [63] have been electrospun into uniform fibers with well-controlled sizes, compositions, and morphologies. In most situations, electrospun nanofibers are usually smooth solid fibers, and they are collected as nonwoven membranes with randomly arranged structures (as shown in Figure 7.6a); this kind of membrane structure is superior for filtration application because of thin fiber diameter, smaller pore size, and higher porosity. In order to explore the application of electrospun nanofibers, more and more researchers tried to develop nanofibers with versatile structures and morphologies through controlling processing parameters. So far, many kinds of novel structures have been successfully fabricated, including single fiber structure and aggregate structure.

The single fiber structure is important because it will directly influence the structure and property of the aggregated nanofiber membranes. For example, the cross-sectional shape and size, the pore distribution, and the internal structure of the single fiber will determine the porosity, SSA, and even physical and chemical properties of nanofiber membranes. So far, by regulating processing parameters including (i) the solution properties such as solution concentration, viscosity, elasticity, conductivity, and surface tension; (ii) spinning parameters like electric voltage, injection speed, collecting distance between syringe tip and ground collector, etc.; and (iii) the ambient parameters such as ambient temperature, ambient humidity, and air velocity, researchers can obtain single nanofibers other than smooth solid structure, but with bead-on-string, ribbon-like [64], helical [65], porous [66], core–shell [67], multichannel tubular [68], tube-in-tube [69], necklace-like [70], rice-grain shape [71], multicore cable-like [72], nanowire-in-microtube [73], and hollow [74] structures, as shown in the Figure 7.5a–l. These different structures can be used in various application fields: for example, the porous structure is suitable in application of adsorption field due to its high SSA; the multichannel tubular and hollow structures, because of the inner still air, can be used for thermal productive garments.

In addition to rich structures of single fiber, different nanofiber assembly morphologies were also reported via control of the processing conditions or collecting methods to produce fibrous nonwovens with versatile structures besides random oriented structure; the main patterns of them are shown in Figure 7.6. For instance, by using the rotating roller with high speed as collector, nanofibers were strongly stretched by such high speed to be aligned among the rotating orientation. Thus the highly aligned nanofibers were successfully fabricated, as shown in the Figure 7.6b; the uniform mesh-like patterning nanofibers can be fabricated by using electroconductive collector with woven structure, which is displayed in Figure 7.6c [75], and the fantastic spider-web-like nanofiber/net (shown in Figure 7.6d) can be prepared by the process of electro-netting. Therefore, we can obtain nanofibrous membranes with rich polymer type, various single fiber structures, and assembly morphologies to be highly attractive to numerous applications including filters, textiles, composites, sensors, biotechnology, and so on.

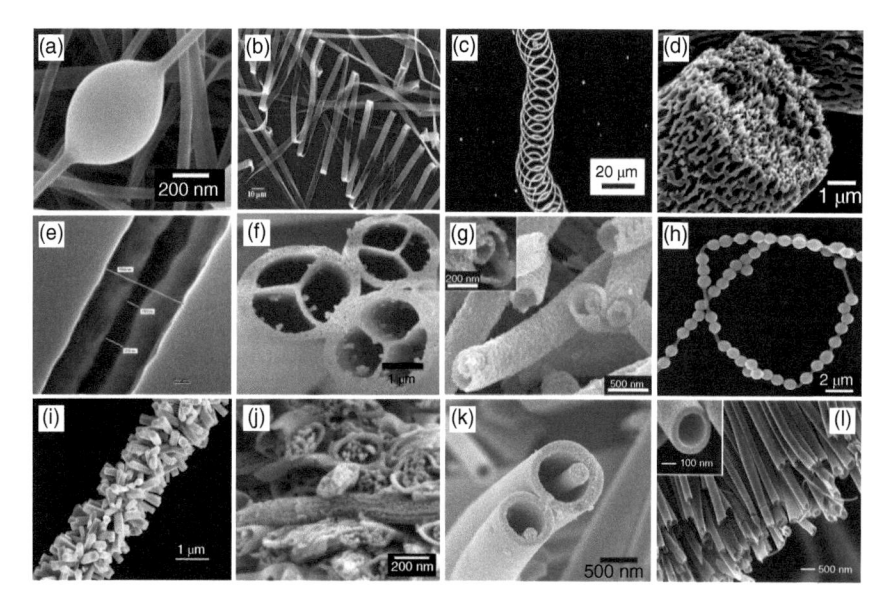

Figure 7.5 Various single fiber structures obtained by electrospinning. (a) Bead on string. (b) Ribbon-like. Source: Koombhongse et al. [64]. Reprinted with permission of John Wiley & Sons, Inc. (c) Helical. Source: Kessick and Tepper 2004 [65]. Reprinted with permission of American Institute of Physics. (d) Porous [67]. Source: Adapted from Ding et al. 2008 [66]. (e) Core-shell. Source: Sun et al. 2003 [67]. Reprinted with permission of John Wiley & Sons, Inc. (f) Multichannel tubular. Source: Zhao et al. 2007 [68]. Reprinted with permission of American Chemical Society. (g) Tube in tube. Source: Mou et al. 2010 [69]. Reprinted with permission of American Chemical Society. (h) Necklace-like. Source: Jin et al. 2009 [70]. Reprinted with permission of American Chemical Society. (i) Rice grain-shaped. Source: Shengyuan et al. 2011 [71]. Reprinted with permission of Royal Society of Chemistry. (j) Multicore cable-like. Source: Kokubo et al. 2007 [72]. Reprinted with permission of IOP Publishing Ltd. (k) Nanowire in microtube. Source: Chen et al. 2010 [73]. Reprinted with permission of American Chemical Society. (l) Hollow structures. Source: Li and Xia 2004 [74]. Reprinted with permission of American Chemical Society.

7.4.2 Structures and Species of the Nanofiber/Net (NF/N) Membranes

As mentioned previously, the technology of electro-netting needs higher requirements in processing control compared with electrospinning. Although a very broad range of polymers have been electrospun into micro- or nanofibers by electrospinning, it is founded that not all of those spinnable polymers can be fabricated into spider-web-like NF/N-structured membranes. However, with continuous efforts that have been applied to NF/N membranes, many kinds of polymer systems, such as polymers soluble and spinnable from water, biocompatible and biodegradable polymers, and polymer blends, can be successfully obtained with NF/N structure [33]. Some typical morphologies of NF/N membranes based on different polymer systems are shown in Figure 7.7.

Similar to the technology of electrospinning, many parameters including solution parameters, process parameters, and ambient parameters will influence the phase separation process during the electro-netting and the gained morphologies of NF/N membranes. Although a general knowledge of formation mechanism

184 | *7 Electrospun Fibers for Filtration*

Figure 7.6 Different aggregate structure of electrospun nanofiber. (a) Random oriented. (b) Aligned. Source: Mathew et al. 2006 [116]. Reprinted with permission of John Wiley & Sons, Inc. (c) Patterned. Source: Reprinted with permission of Zhang and Chang 2007 [75]. Copyright 2007 Wiley-VCH Verlag GmbH & Co. (d) Spider-web-like nanofiber/net structures. Source: Reprinted with permission of Wang et al. 2009 [117]. Copyright 2010 IOP Publishing Ltd.

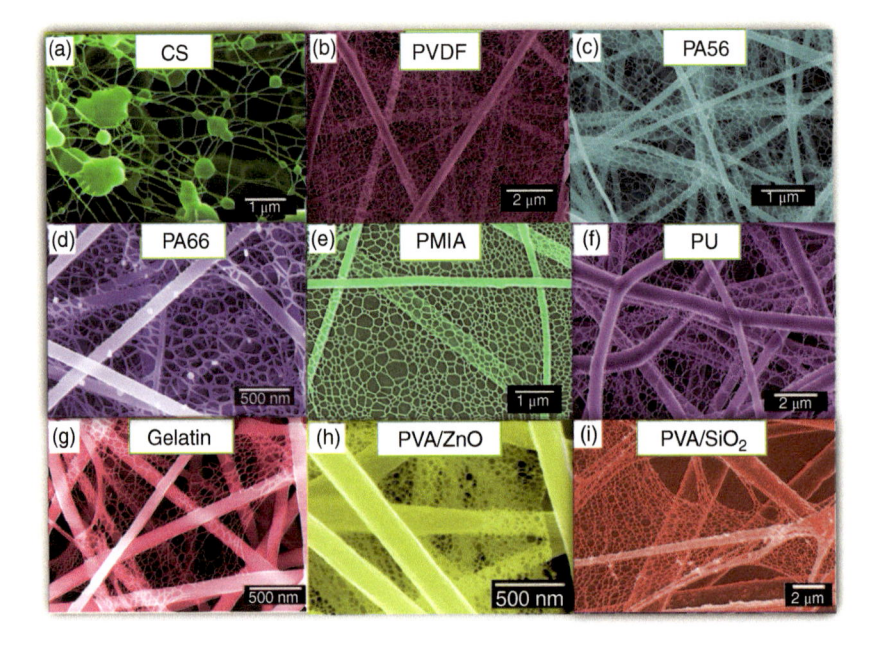

Figure 7.7 Several typical NF/N membranes based on different polymer systems. (a) CS. (b) PVDF. (c) PA-56. (d) PA-66. (e) PMIA. (f) polyurethane (PU). Source: Reprinted with permission of Hu et al. 2011 [54]. Copyright 2011 Wiley-VCH Verlag GmbH & Co. (g) Gelatin. Source: Reprinted with permission of Wang et al. 2011 [118]. Copyright 2011 Elsevier Ltd. (h) polyvinyl acetate (PVA)/ZnO. Source: Reprinted with permission of Ding et al. 2008 [119]. 2008 Elsevier B.V. (i) PVA/SiO_2. Source: Ding et al. 2010 [120]. Reproduced with permission of Elsevier.

and influencing factors of NF/N membranes has been known, the relationship between spinnable polymers and NF/N membranes is yet to be revealed. Recently, Ding's laboratory successfully fabricated the first pure-inorganic NF/N membranes (SiO_2, TiO_2, etc.) and even the first carbon NF/N membranes in the world, which will intensively broaden the application field of NF/N membranes. In addition to develop new types of NF/N membranes, researchers also engaged in controllable fabrication and structure optimization of the NF/N membranes, such as regulating the fiber and nanowire diameter, pore width, coverage rate of nanonets, and so on.

7.5 Application of Nanofibrous Membranes in Air Filtration

Due to many outstanding structural advantages of nanofibrous membranes, such as finer fiber diameter, smaller pore size, higher surface area to volume ratio, higher porosity, and vast possibilities for surface functionalization, nanofibrous membrane-based filter media (usually with random oriented nonwoven structure) are considered to be much promising in air filtration fields. According to the temperature of particle source, current electrospun nanofibrous filters can be divided into two groups: normal temperature filter and medium-high temperature filter.

7.5.1 Normal Temperature Filter

Normal temperature filtration media are those filters that can be conveniently used under 150 °C for a long time. They are usually the core filtration part of indoor air purifier, vehicle cabin filters, building filters, and personal respiratory protection products. The materials used for these filtration applications (no need to resist high temperature) can be natural polymers, synthetic polymers (mostly used like PA, PAN, PU, PVA, PEO, etc.) [76–80], polymer blends [81, 82], and polymer incorporated with functional materials [83, 84]. Among those normal temperature filters made by various polymer systems, if we classify them according to structure, there are mainly electrospun nanofiber membranes and electrospun NF/N membranes.

7.5.1.1 Electrospun Nanofiber Membranes

Nanofiber Filters with High Efficiency and Low Pressure Drop Compared with traditional filtration materials made of cotton cloth or gauze, nanofibrous membranes have more thinner fiber diameter and smaller pore size, so that they can trap even finer pollutant particles in the air. Gibson et al. [85] firstly fabricated PA-66 nanofibrous membranes for filtration use and revealed that they were extremely efficient at trapping airborne particles with diameter between 0.5 and 200 μm. Li et al. [86] prepared PA-6 electrospun nanofibers with diameter in the range of 120–700 nm; this fabricated nanofibrous membrane has filtration efficiency of above 80% for the airborne particles of 0.5 μm, which is much higher than that of traditional filtration materials. By further adjusting processing parameters, Ahn et al. [87] obtained PA-6 thinner fiber (average diameter of 80 nm) membrane with improved filtration efficiency of 99.99%, but its air pressure drop was too high (larger than 500 Pa). It is found that raising filtration efficiency and reducing pressure drop are always a pair of contradictory body when trying to improve the performance of nanofibrous filtration membranes. Thus, the achievement of air filters with both high efficiency and low pressure drop is always a challenge that researchers face these years.

According to the "slip effect," it is known that the thinner the fiber, the higher filtration efficiency, and meanwhile, if the diameter of fiber decreases close to the mean free path of air molecules (65.3 nm), the air will stride over the fiber so that the air pressure drop of the filtration materials can be dramatically reduced [88]. However, if the fiber assembly pack is too dense, it will destroy this "slip effect." Therefore, a good balance should be made between the fiber structure and assembly structure. Through careful controlling over the diameters of electrospun polyacrylonitrile fibers and aperture size of fiber assembly, Zhao et al. [89] reported novel slip-effect functional nanofibrous membranes with decreased air resistance (reduction rate of 40%) due to the slip flow of air molecules and indicated the optimal fiber diameter and pore size that are validated both theoretically and experimentally. In order to obtain the fibers with diameter in the vicinity of mean free path of air molecules, authors in this chapter selected polyacrylonitrile (PAN) with low molecular weight as electrospun polymer and lithium chloride (LiCl) to enhance the solution conductivity to obtain the bead-free nanofibers. According to the drag theory, the relationship between pore structure and slip effect was intensively investigated (shown in Figure 7.8): with decreasing of fiber diameter, the average pore size of the membrane decreases, and the drag force reduces as well, but if the fiber diameter and pore size continuously go down (e.g. fiber diameter lower than 70 nm), the drag force will then increase. That means too close

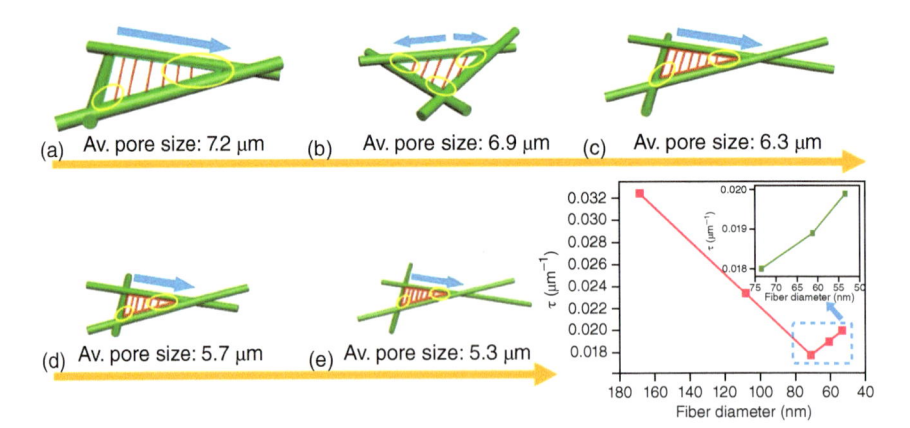

Figure 7.8 The relationship between pore structure and slip effect. The average pore size reduces with the decreasing of fiber diameter: (a) 168, (b) 108, (c) 71, (d) 60, and (e) 53 nm. (f) Equilibrium factor of PAN fibrous membranes with different fiber diameter. Source: Zhao et al. 2016 [89]. https://www.nature.com/articles/srep35472#rightslink; http://creativecommons.org/licenses/by/4.0/. Licensed under CCBY 3.0.

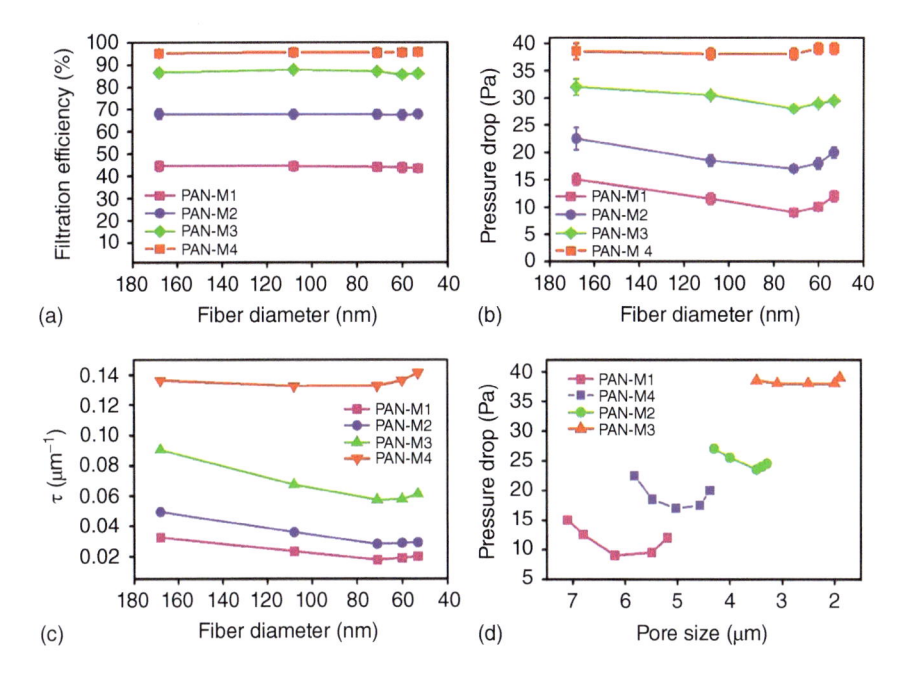

Figure 7.9 The influence of fiber diameter and pore size of the several PAN membranes on the filtration properties. (a) Filtration efficiency for several PAN membranes. (b) Pressure drop change with different fiber diameter. (c) Equilibrium factor change of PAN fibrous membranes with different fiber diameter. (d) Structure–function relationship between pressure drop and pore size. Source: Zhao et al. 2016 [89]. https://www.nature.com/articles/srep35472#rightslink; http://creativecommons.org/licenses/by/4.0/. Licensed under CCBY 3.0.

distance between two adjacent fibers will restrain the airflow and thus result in the increasing of the air pressure drop.

To further investigate the critical value of the fiber diameter and pore size of PAN nanofibrous membranes, Zhao et al. [89] then fabricated several PAN fibrous membranes with nearly identical filtration efficiencies by controlling the electrospinning time, as illustrated in Figure 7.9a (the points with the same color have nearly identical filtration efficiencies but with different fiber diameters). It is found that the fiber diameters between 60 and 100 nm are most effective for slip flow (achieving lowest pressure drop and drag force; see Figure 7.9b,c); if the pore size is too small (less than 3.5 μm), the air pressure drop of the nanofibrous membrane will be the same no matter how the fiber diameter changes (see Figure 7.9d). Thus the pore size should be larger than 3.5 μm for constructing slip-effect functional nanofibrous membranes. Ultimately, the fabricated PAN membranes displayed low air resistance of only 29.5 Pa, high purification efficiency of 99.09%, good transmittance of 77%, and long service life.

Instead of further reducing the fiber diameter and at the same time to keep higher porosity of the membrane to improve the filtration performance, some other researchers focused on fabricating electreted nanofibrous membranes to improve its filtration efficiency by electrostatic effect without deteriorating the pressure drop. Yeom et al. [84] prepared PA-6/boehmite nanofibrous membranes; this membrane had better filtration efficiency than pure PA-6 membrane when both membranes had almost the same air pressure drop. Cho et al. [83] added TiO_2 electreted nanoparticles when electrospinning PAN nanofibrous membranes and successfully fabricated PAN/TiO_2 nanofibrous membrane with higher filtration efficiency and lower air pressure drop. Recently, Wang et al. [90] reported a novel electret nanofibrous membrane with rich charges and desirable charge stability by using polyvinylidene fluoride (PVDF) as the matrix polymer and polytetrafluoroethylene nanoparticles (PTFE NPs) as the charge enhancer through the in situ charging technology of electrospinning. By adjusting the concentration of PTFE NPs (22 wt% PVDF polymer solution containing 0, 0.01, 0.05, and 0.1 wt% of PTFE NPs was prepared) and the applied voltage (four injection energy level as 20, 30, 40, and 50 kV, respectively) during electrospinning, it was found that the optimal employment of PTFE NPs should be 0.05 wt% and 40 kV was the best applied voltage. The electreted PVDF/PTFE nanofibrous membranes they fabricated were endowed with elevated surface potentials from 0.42 to 3.63 kV and had better charge stability (reduced decrement of charges from 75.4% to 17.5%).

Meanwhile, they also proposed the electret mechanism of this PVDF/PTFE nanofibrous membrane [90]. They thought there were four kinds of charge format: volume charges, surface charges, polarized dipole, and interfacial polarization between PVDF nanofibers and PTFE nanoparticles, as shown in Figure 7.10a–d. The good electreted effect of PVDF/PTFE nanofibrous membranes was attributed to, on the one hand, the strong polarity of fluorine-containing polymer (polarized dipole produced is much stable than volume charge and surface charge) and, on the other hand, the extra interfacial polarization between PVDF nanofibers and PTFE nanoparticles. With the increase of the concentration of PTFE NPs and injection energy, the electreted effect of PVDF/PTFE nanofibrous membranes was significantly improved. Finally, the resultant PVDF/PTFE nanofibrous membrane reached a

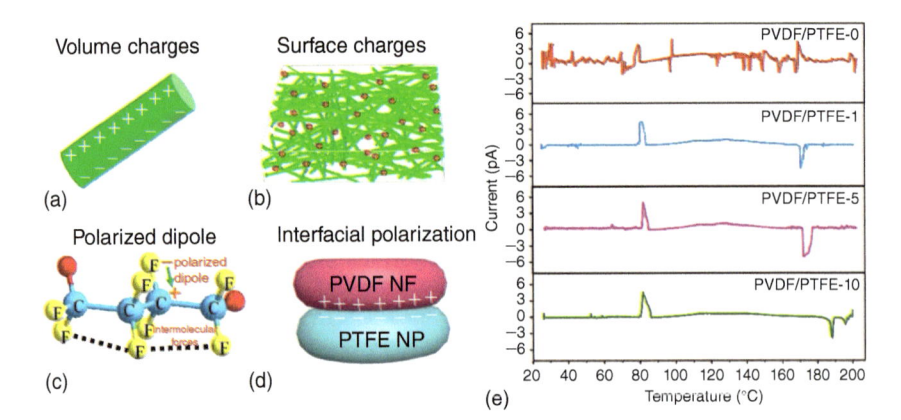

Figure 7.10 Schematic description of electric charge categories existing in PVDF nanofibrous membranes containing various PTFE NP concentrations. (a) Volume charges. (b) Surface charges. (c) Polarized dipole. (d) Interfacial polarization. (e) Thermally stimulated current spectra of PVDF fibrous membranes containing various PTFE NP concentrations. Source: Wang et al. 2016 [90]. Reprinted with permission of American Chemical Society.

high filtration efficiency of 99.972%, a low pressure drop of 57 Pa, and superior long-term service performance.

Nanofiber Filters with Functional Properties Besides the efforts made in the direction of achieving filtration materials with high efficiency and low pressure drop, researchers also did many work on developing air filters with functional properties.

One of the most important functions we concern about the filtration materials is mechanical property during the practical application; thus many mechanically enhanced nanofibrous membranes were continuously reported. Liu et al. [91] prepared polyacrylonitrile/poly(acrylic acid) (PAN/PAA) nanofibrous membranes as filter medium by electrospinning; the tensile strength of this blended nanofibrous mat improved as increasing PAA content, from 3.8 to 6.6 MPa, higher than pristine PAN. Wang et al. [82] fabricated tortuously structured polyvinyl chloride (PVC)/PU fibrous filtration medium with robust mechanical performances. They prepared blended solutions with different weight ratios of PVC/PU (10/0, 9/1, 8/2, and 7/3) and found that the increasing PU contents not only increased the average fiber diameter but also showed gradually increased adhesion among the adjacent fibers (as shown in Figure 7.11, indicated by dotted circle). When the weight ratio of PVC/PU was 8/2, the interfacial compatibility of PVC/PU reached to the optimum state, and the PVC/PU membranes showed the maximum tensile strength of 9.9 MPa. The as-prepared membranes showed filtration efficiency (99.5%) and pressure drop (144 Pa) performance for 300–500 nm sodium chloride aerosol particles.

In order to obtain mechanically enhanced air filters without sacrificing filtration performance, recently, Zhang et al. [92] combined multi-jet electrospinning and physical bonding process to create anti-deformed poly(ethylene oxide)@polyacrylonitrile/polysulfone (PEO@PAN/PSU) composite membranes with binary structures for effective air filtration. This membrane allows the ambiguous fiber framework including thin PAN nanofibers and fluffy PSU microfibers, and the mixed PEO component will be melted as bonding points between fibers after heat treatment (see the fabrication procedure in Figure 7.12). With increasing PEO concentration, the number of point-shaped bonding structures

Figure 7.11 FE-SEM images of PVC/PU membranes fabricated with varied weight ratios of PVC/PU: (a) 10/0, (b) 9/1, (c) 8/2, and (d) 7/3. Source: Wang et al. 2013 [82]. Reprinted with permission of Elsevier.

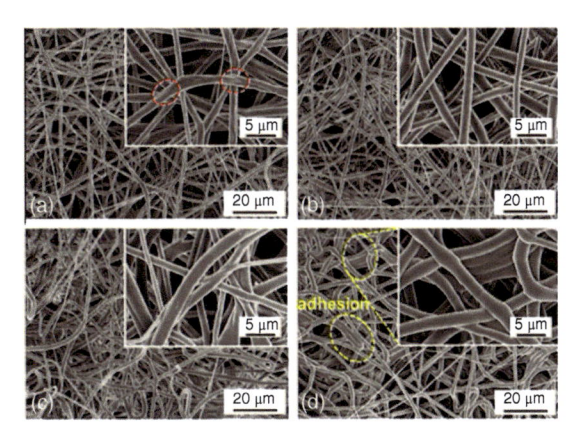

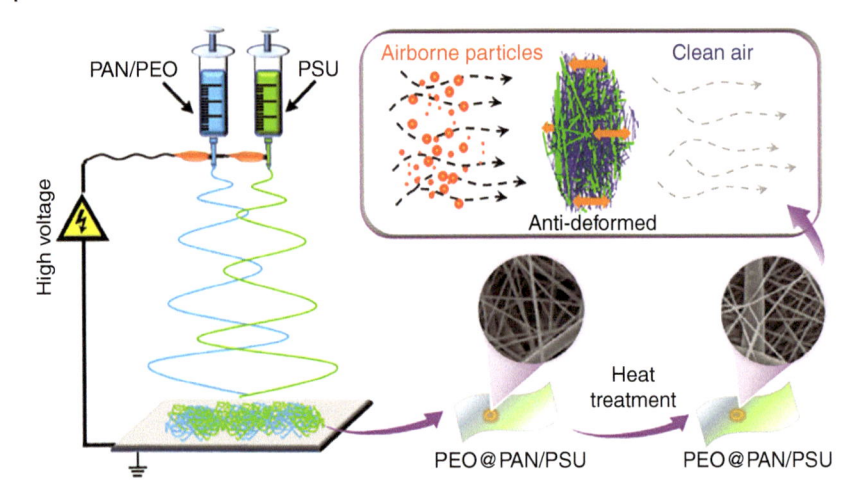

Figure 7.12 Schematic illustration of the fabrication procedure of the anti-deformed PEO@PAN/PSU composite membranes, including the multi-jet electrospinning device and physical bonding (heat treatment) process. Source: Zhang et al. [92]. Reprinted with permission of American Chemical Society.

increased dramatically; thus the mechanical properties of the membrane were strengthened. It was found that when the PEO concentration was of 1.5 wt%, the filtration efficiency of the PEO@PAN/PSU membrane would keep the same before and after the compacting treatment; in other words, the membrane's structural deformation can be neglected even under some external force. With the integrated features of small pore size, high porosity, and point-shaped bonding structure, the resultant composite membrane exhibited robust mechanical properties (8.2 MPa), high filtration efficiency of 99.992%, low pressure drop of 95 Pa, and desirable quality factor of 0.1 Pa^{-1}, and most importantly, it can avoid the potential safety hazards caused by unexpected structural collapsing under external forces during the practical application.

Besides the improvement of mechanical properties, air filtration materials can be endowed with other functions. For example, Wang et al. [93] report a novel superamphiphobic nanofibrous membrane by the combination of electrospun PAN/PU nanofibers and a new synthesized fluorinated polyurethane (FPU) containing terminal perfluoroalkane segment to achieve oil and non-oil aerosol particle filtration (see Figure 7.13). The introduction of low surface energy FPU enabled the PAN/PU composite membranes to have both superhydrophobicity and superoleophobicity. By tuning the solution composition, hierarchical structure, and the FPU content, the as-prepared membrane obtained excellent superamphiphobic properties of a water contact angle of 154° and an oil contact angle of 151°. Meanwhile, this kind of membrane can capture a range of different oil aerosol particles in a single-unit operation with filtration efficiency larger than 99.9%. Therefore, it is very promising in the filtration performance toward oil and non-oil aerosol particles and can be used in industry, respiratory protection equipment, and numerous air filtration fields.

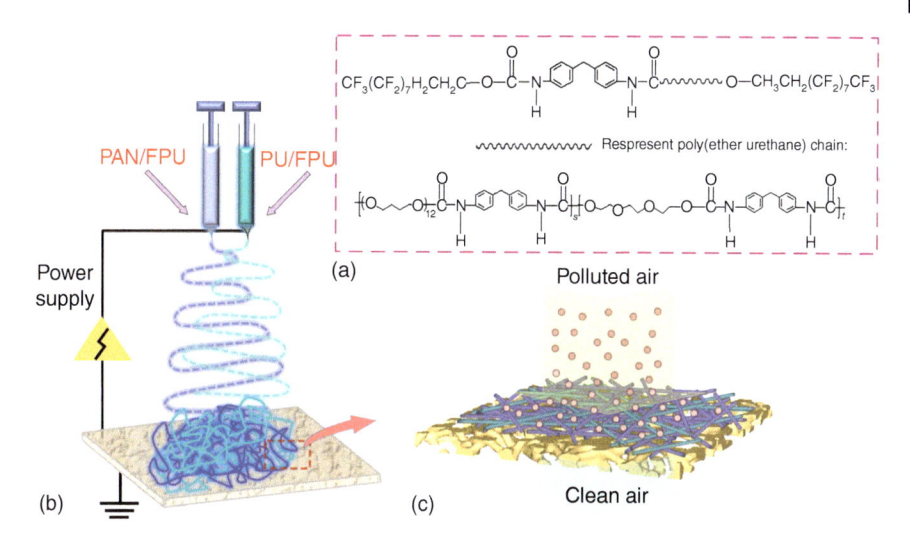

Figure 7.13 Schematic illustration of the fabrication procedure of the superamphiphobic nanofibrous membrane. (a) Illustration of the chemical structure of as-synthesized FPU. (b) Electrospinning device for the preparation of the filter membranes. (c) Filtration process of the superamphiphobic membranes for aerosol particles. Source: Wang et al. 2014 [93]. Reprinted with permission of Elsevier.

7.5.1.2 Electrospun Nanofiber/Net (NF/N) Membranes

In addition to the above electrospun nanofiber membranes fabricated by electrospinning, as mentioned previously, another promising technology of electro-netting is also widely used for air filter construction due to finer fiber diameter, smaller pore size, and higher porosity of the NF/N membranes.

For instance, Wang et al. [94] firstly created a two-tier composite structure as a novel airborne particulate filtration medium: the polyamide-66 (PA-66) NF/N membrane composed of traditional electrospun nanofibers and 2D spider-web-like nanonets as the top layer and the conventional nonwoven PP structure as the support (as shown in Figure 7.14). By regulating the solution properties (mainly the kinds of salt and salt concentration) and several electro-netting process parameters (such as applied voltage, ambient temperature, and humidity), the optimal nanonets with a high coverage rate (over 95%) and a layer-by-layer packing structure were finally achieved under a voltage of 30 kV, ambient temperature of 25 °C, and relative humidity (RH) of 25%. Owing to several fascinating features such as extremely small diameter, high porosity, and controllable coverage rate, the resultant two-tier composite filtration medium showed high filtration efficiency (up to 99.9%) toward NaCl aerosols, low pressure drop, facile filter cleaning, and more lightweight.

In order to further improve the filtration efficiency, Liu et al. [95] fabricated a bio-based polyamide-56 nanofiber/net (PA-56 NF/N) membrane with bimodal structures for effective air filtration via one-step ESN. This filtration medium contained completely covered 2D ultrathin (20 nm) nanonets and bonded scaffold fibers to improve the mechanical properties of the membrane.

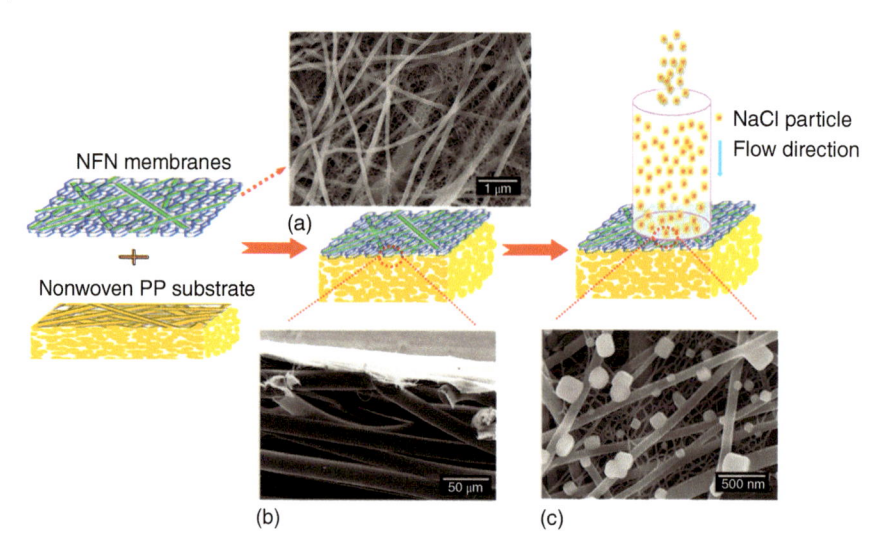

Figure 7.14 Schematic diagrams illustrating the fabrication of an NF/N layer deposited on the nonwoven PP substrate and the filtration process of the composite membrane. (a) FE-SEM image of PA-66 NF/N membranes deposited on the PP scaffold. FE-SEM images of the composite membrane. (b) Cross section before filtration. (c) Top view after filtration. Source: Wang et al. 2012 [94]. Reproduced with permission of Royal Society of Chemistry.

Through tuning the solution concentration and regulating the weight ratio of $HCOOH/CH_3COOH$, the optimal structure with high coverage of nanonets and stable bonded scaffold fibers was finally constructed. Taking advantages of extremely thin diameter, high porosity, and bonded scaffold, the resultant PA-56 NF/N membranes exhibit robust mechanical strength of 11.02 MPa, high filtration efficiency of 99.995%, pressure drop of 111 Pa, large dust holding capacity of 49 g/m^2, and long service life (dust-cleaning regeneration ability) for filtrating ultrafine airborne particles.

Although NF/N membranes with ultrathin fiber diameter can achieve higher filtration efficiency, the tiny pore size usually results in relatively high pressure drop. How to fabricate filter medium having higher filtration efficiency but at the same time with lower air pressure drop is always a challenging problem. Thus, many researchers tried ways to construct NF/N membrane that avoid dense packing structure to improve the filtration performance. For instance, Wang et al. [96] prepared an ultra-lightweight (2.94 gm^{-2}) nylon 6–polyacrylonitrile nanofiber/net binary (N6–PAN NNB)-structured membrane via multi-jet spinning (see Figure 7.15). The combination of high coverage N6 nanonets and low packing density of PAN nanofibers enabled the resultant NNB composite membrane to act as an efficient filtration medium. It is found that when the concentration of N6 solutions was 15 wt%, and the fiber spinning jet ratio of N6/PAN reached 2/2, the composite membrane exhibited high filtration efficiency (99.99%) and desirable quality factor (0.1163 Pa^{-1}) even under a high flow rate (90 l/min), better than the commercial glass fiber and melt-blown PP fiber-based media. Meanwhile, this as-prepared composite membrane showed

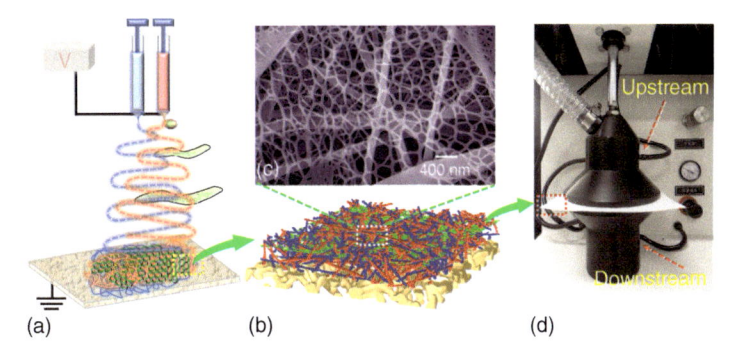

Figure 7.15 (a) The fabrication process of N6–PAN NNB composite membranes on a nonwoven substrate by multi-jet spinning. (b) Representation of the formed N6–PAN NNB composite membrane (N6 nanofibers are in red; N6 nanonets are in green; and PAN nanofibers are in blue) and nonwoven substrate (yellow part). (c) Typical FE-SEM image of the membrane. (d) Illustration of the filtration process of N6–PAN NNB composite filter medium. Source: Wang et al. [96]. Reprinted with permission of Royal Society of Chemistry.

a deep bed filtration pattern, rather than the surface filtration pattern of the N6 NF/N membrane, which was investigated by the way of three-dimensional computer simulation.

Different from incorporating low packing density of nanofibers into the NF/N membranes, Yang et al. [97] created a sandwich-structured polyamide-6/polyacrylonitrile/polyamide-6 (PA-6/PAN/PA-6) composite membrane for effective air filtration via sequential spinning. This composite membrane had three layers (see Figure 7.16): the upper and bottom PA-6 NF/N layers with ultrathin (~20 nm) nanonets and the middle PAN bead-on-string fiber layer as cavity structure. The existence of the middle PAN layer cannot only reduce the packing density of the whole composite membrane but also

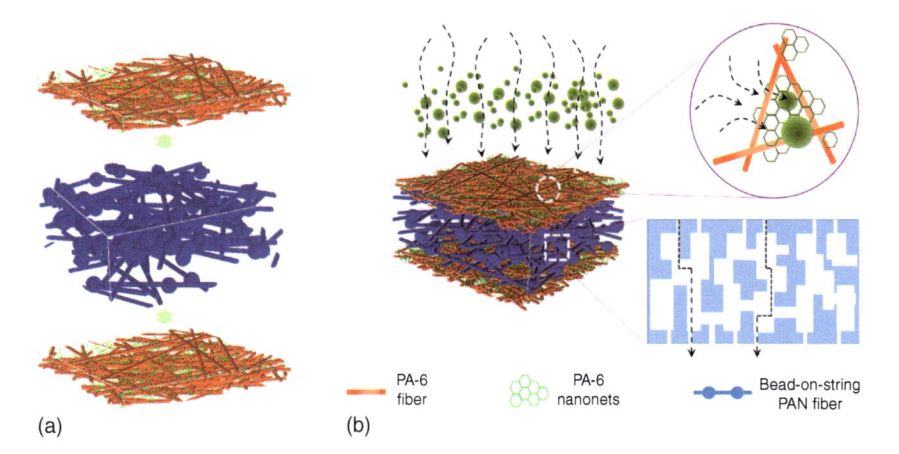

Figure 7.16 (a) Illustration of the sandwich-structured PA-6/PAN/PA-6 composite membrane. (b) Schematic diagram showing the filtration process of PA-6/PAN/PA-6 composite media for 300–500 nm NaCl airborne particles. Source: Yang et al. [97]. Reprinted with permission of Elsevier.

function as a support to strengthen the filter medium. By tuning the solution concentration and voltage applied, the number and size of the beads for the PAN layer were finely controlled, while by varying the weight ratio of PA-6 NF/N layers and PAN layer, the sandwich-structured composite filtration medium was finally optimized. Attributed to the integrated properties of ultrathin diameter offered by PA-6 NF/N layers, and stable cavity structure provided by PAN bead-on-string fiber layer, the resultant sandwich-structured PA-6/PAN/PA-6 composite membranes exhibit robust mechanical properties (tensile strength of 7.28 MPa, toughness of 1.77 MJ/m^3, and Young's modulus of 130.4 MPa) and much high filtration efficiency of 99.9998% with pressure drop of 117.5 Pa for 300–500 nm airborne particles. These novel composite membranes were regarded very promising in wide application fields of indoor air clean, engine intake, respirator, and heating, ventilation, and air conditioning, etc.

Zhang et al. [98] then tried a new strategy to create microwave-structured polyamide-6/poly(*m*-phenylene isophthalamide) nanofiber/net (PA-6/PMIA NF/N) membranes by combining the electro-netting and thicker staple fiber intercalating process. The embedded PMIA staple fibers, because of more thicker fiber diameter, constructed a cavity structure (microwave fluctuation and high-loft structures due to the supporting effect of the staple fibers) for the NF/N membranes so as to reduce the packing density. By optimizing the PA-6 concentrations and PMIA staple fiber arrangement, the resulting PA-6/PMIA NF/N filtration medium can efficiently filter ultrafine airborne particles with high filtration efficiency of 99.995%, low pressure drop of 101 Pa, and large dust-holding capacity of >50 g/m^2. Thus, it is proved that the way of embedding thicker fiber into NF/N membranes to build cavity structure is creative and effective. However, the staple fibers are short and randomly distributed, which somewhat limit the controllable fabrication and stability of the membranes. In order to construct more stable cavity structure, recently, the authors introduced filament into the NF/N membranes [99]; they allowed polyethylene terephthalate (PET) filaments to orderly embed into 2D PA-6 nanonet layer (see Figure 7.17) to construct a ripple-like composite membrane for effective air filtration. The pleat span provided by single filament enlarged with the increasing of filament diameter (see Figure 7.17h). By adjusting the filament diameter and interval gap, the membrane structure for the optimized filtration performance was obtained. It was found that the as-prepared ripple-like PA-6 NF/N filtration medium can filter the ultrafine airborne particles (300–500 nm) with high efficiency of 99.996% and lower air pressure drop of only 95 Pa due to its extremely small pore size ($\approx$0.35 μm) provided by PA nanonets and high porosity (92.5%) supplied by the built stable cavity structures.

In addition to construct stable cavity structure for NF/N filtration medium to achieve low air pressure drop, Zhang et al. [100] reported a highly integrated polysulfone/polyacrylonitrile/polyamide-6 (PSU/PAN/PA-6) filtration medium for multilevel physical sieving airborne particles via sequential spinning. This integrated air filter had three layers: the PSU microfiber (diameter of ~1 μm) layer, middle PAN nanofiber (diameter of ~200 nm) layer, and PA-6 nanonets

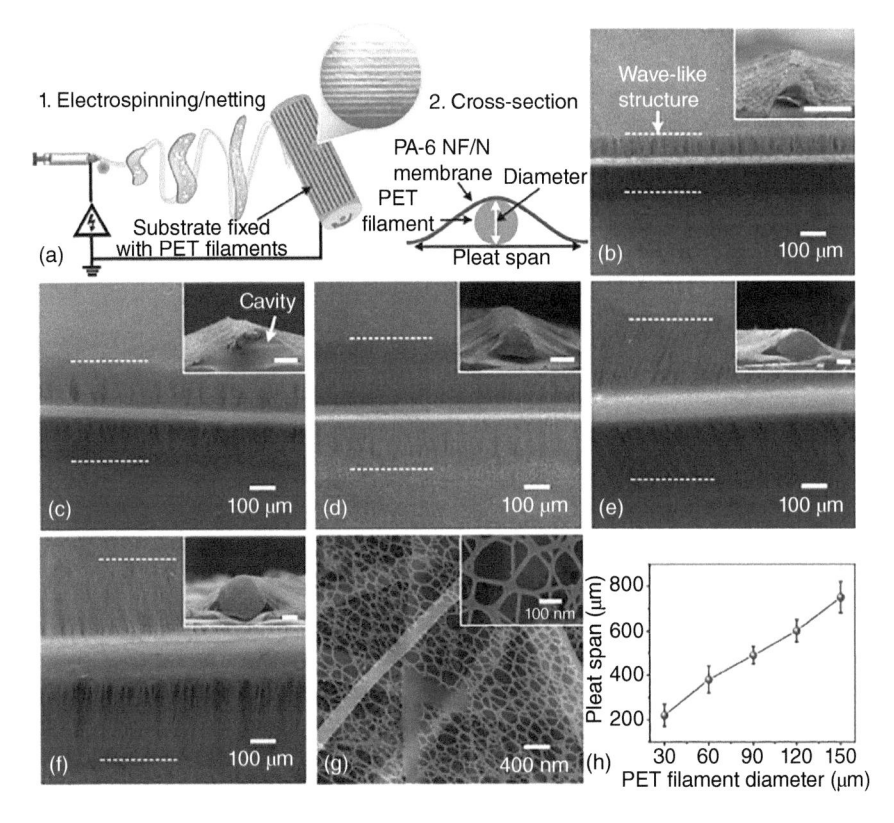

Figure 7.17 (a) Schematic showing (1) the fabrication process of the ripple-like PA-6 NF/N membrane and (2) the cross section of the PET embedded PA-6 NF/N filter. SEM images of PA-6 NF/N membranes embedded with PET filaments with various diameters of (b) 30, (c) 60, (d) 90, (e) 120, and (f) 150 μm and certain framework gap of 1 mm. Insets are the corresponding images of their cross-sectional views (the scale bar is 50 μm). (g) FE-SEM image of PA-6 nanonets. (h) The pleat span provided by single PET filament with various diameters. Source: Zhang et al. 2017 [99]. Reprinted with permission of John Wiley & Sons, Inc.

(diameter of ~20 nm) layer (see Figure 7.18). Different layers have varied pore structures and porosity so that the combined multilayer structure enables the filter to work efficiently to cut off penetration of particles with a relatively wide range without sacrificing the air pressure drop. Owing to the elaborate gradient structure and excellent hydrophobicity (Water contact angle of ~130°), such as-prepared highly integrated filter can have a high efficiency of 99.992%, a low pressure drop of 118 Pa, and a robust Quality factor value of 0.08 Pa^{-1} for filtrating the 300 nm NaCl particles. Moreover, it is worthy to mention that the filtration performance was exhibited after treatment of removing electric charges of the membrane. That means this integrated filtration medium traps airborne particles only in the way of physical sieving manner; it can avoid the potential safety hazards caused by unexpected electret failure. Thus, this kind of integrated air filters will be more stable and can be used even in the special filtration environment of high airflow speed and high humidity.

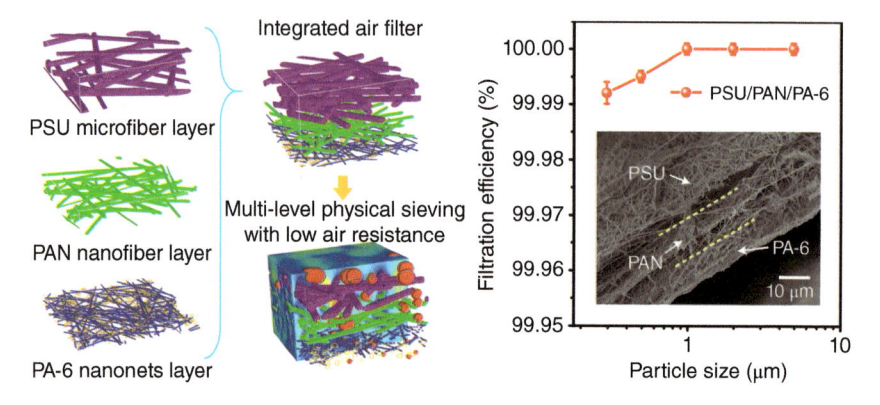

Figure 7.18 The left part: Illustration of the integrated air filter containing three layers (the upper PSU microfiber layer, middle PAN nanofiber layer, and the bottom PA-6 nanofiber/nets layer). The right part: The filtration efficiency of the integrated filter for trapping airborne particles with different size. Source: Zhang et al. 2016 [100]. Reprinted with permission of American Chemical Society.

7.5.2 Medium-High Temperature Filter

Besides dealing with airborne articles with normal temperature (usually for personal protecting area like indoor air cleaner, respirator, window gauze, etc.), filters used for the source control (e.g. automobile exhaust and most industrial dedusting processes such as coal fired boilers, municipal waste incinerators, etc.) are often required to operate within a range between 150 and 260 °C or even higher under some extreme environments. Polymers that are mentioned above for normal temperature filtration such as PA-6, PAN, and PU nanofibrous membranes cannot fulfill the requirements of medium-high temperature filtration. Only those polymers with excellent thermal stability (higher than 200 °C) such as polyethersulfone (PES) [101], polyetherimide [102], polyimide [103], and polymeta-phenylene isophthalamide [104] have feasibility to achieve medium-high temperature filtration.

Nakata et al. [101] prepared PES (glass transition temperature up to 220 °C) nanofibrous membranes as the high heat-resistant air filters. They found that the diameter profile of electrospun PES membrane was obviously affected by PES concentration, feeding rate, and needle-collector distance. After evaluating the filtration properties of PES nanofibrous membranes made by various electrospinning processes, it was found that both the filtration efficiency and pressure drop were significantly decreased over the pore size of 3.0 μm. When the average pore size was 3.2 μm, the PES membranes exhibited the best filtration performance with the high filtration efficiency (99.9998%) and low pressure drop (215 Pa) for removing 300 nm particles, which can meet the High efficiency particulate air Filter (HEPA) requirement. Li et al. [105] reported a novel hierarchical-structured polyetherimide–silica (PEI-SiO_2) nanofibrous membrane by a single-step strategy. The introduction of SiO_2 particles provided the membrane with not only electreted effects but also promising superhydrophobicity. Moreover, unlike other filtration media, such prepared membrane can be treated at 200 °C for 30 minutes without sacrificing filtration efficiency and pressure drop, which thus expanded the scope of application in air filtration fields.

However, the thermal stability of polymers is still limited so that they cannot be directly used as filtration media working under high temperature over 1000 °C. Thus, in the field of high temperature filtration, more effective and stable heat-resistant materials are needed to deal with high temperature waste from process and power industries and other extreme environments. We all know that ceramic materials have excellent resistance against corrosion, chemical erosion, and heat treatment. Therefore, using electrospun ceramic nanofibrous membranes with extremely high thermal stability and nanoscale fiber as the high temperature filtration media attracted particular attentions these years. So far, many different kinds of ceramic nanofibers such as SiO_2, Al_2O_3, ZrO_2, etc. [106–108] had been fabricated, but the common problem they all faced is the brittleness of the ceramic nanofibers. Then, considerable efforts have been made to fabricate flexible ceramic nanofibers, and some good achievements have been obtained [109–113], which actually shed the light on their successful application for high temperature filtration.

For instance, Mao et al. [114] fabricated a novel silica nanofibrous (SNF) membranes with remarkable flexibility and thermal stability as high temperature filtration media by the combination of electrospinning and sol–gel methods (the schematic illustration of the fabrication process; see Figure 7.19). Through regulating the precursor fiber composition and calcination temperature, the optimal SNF membrane was obtained with good flexibility (0.0156 gf cm), relatively high tensile strength (5.5 MPa), and tensile modulus (114 MPa), which also showed excellent filtration performances toward 300–500 nm sodium chloride aerosols in terms of a high filtration efficiency (99.99%) and relatively low pressure drop (163 Pa). Most importantly, this kind of filtration medium

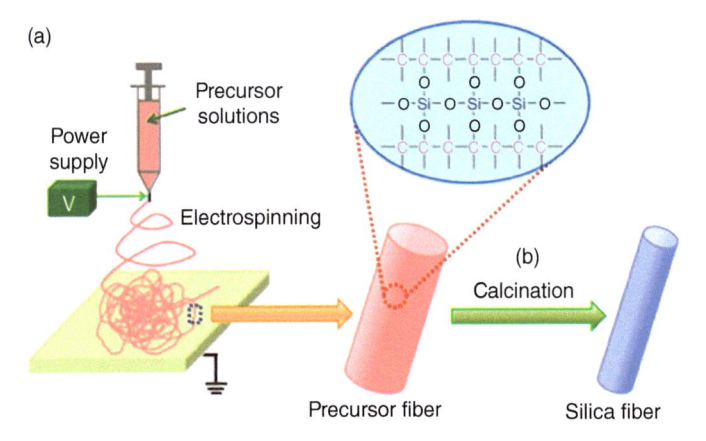

Figure 7.19 Schematic illustration of the fabrication process of silica nanofibrous membranes by the first step of (a) electrospinning and then (b) the calcination process. Source: Mao et al. 2012 [114]. Reproduced with permission of Royal Society of Chemistry.

exhibited excellent thermal stability up to 1000 °C; thus it has great potential for applications as high temperature filtration media to trap ultrafine particles. In order to further improve the application performance, recently, the authors successfully prepared flexible and highly temperature-resistant polynanocrystalline yttria-stabilized zirconia (YSZ) nanofibrous membranes by the combination of electrospinning and heat treatment process [115]. They used zirconium acetate (Zr: 15–16 wt%), yttrium nitrate hexahydrate, and poly(vinylpyrrolidone) (PVP) (molecular weight = 1 300 000) as a precursor solution. The mechanical properties of the YSZ membranes could be controlled by tuning the morphology and crystallite size of the fiber, from brittleness to flexibility. The filtration performances of the YSZ nanofibrous membranes with different fiber diameter were also carefully investigated (see Figure 7.20), and it was found that when the average fiber diameter of the YSZ membranes was 382 nm, the prepared membrane exhibited excellent performance: bending resistance of 400 times, heat resistance of 1473 K, and high filtration efficiency of 99.996% for 300–500 nm NaCl particles. Thus, this reported inorganic flexible nanofibrous membranes were regarded by authors as promising high temperature filtration membranes for a wide range of fields such as metallurgy, steel, electric power, and cement industries.

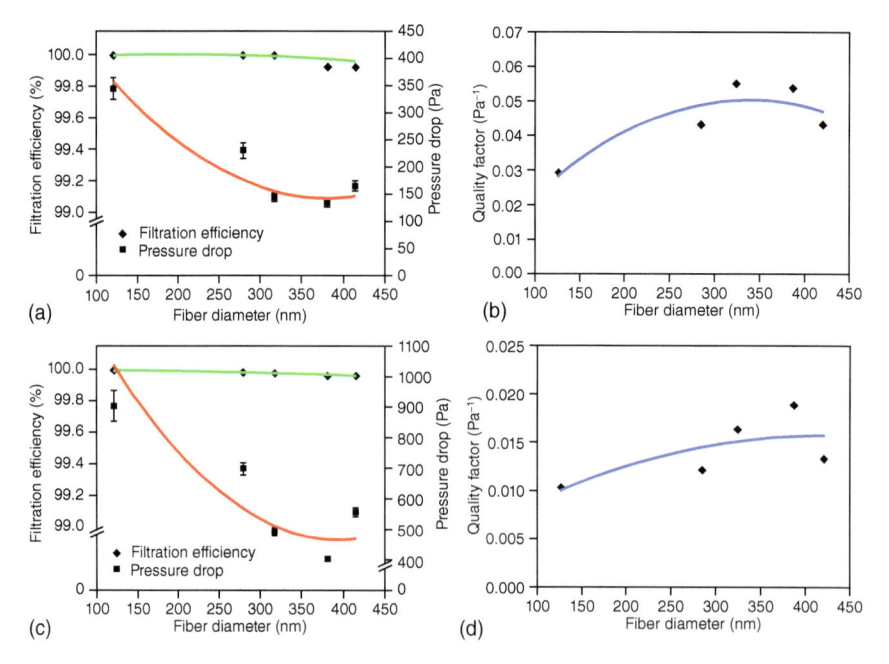

Figure 7.20 Filtration performance of the YSZ nanofibrous membranes with different fiber diameter. (a) Filtration efficiency and pressure drop. (b) Quality factor of the YSZ nanofibrous membranes with different fiber diameter under a face velocity of 32 l/min, respectively. (c) Filtration efficiency and pressure drop. (d) Quality factor of the relevant membranes with different fiber diameter under a face velocity of 85 l/min, respectively. Source: Mao et al. 2012 [115]. Reprinted with permission of American Ceramic Society.

7.6 Future Trends

Nanofibrous membranes fabricated by electrospinning and electro-netting, due to the integrated good properties of fine fiber diameter, tiny pore size, high porosity, and SSA, have been widely regarded as an attractive option for ultrafine particle filtration. In the past ten years, researchers made great efforts on structure design, controllable fabrication, and performance optimization of the nanofibrous filtration membranes and have already achieved many developments in fabricating membranes with both higher filtration efficiency and lower air pressure drop. However, further improving the filtration performance and realizing functionalization of the nanofibrous filters are still challenging, and many practical problems are waiting to be solved. For instance, the mechanical properties (tensile, compression, and abrasion properties) of nanofibrous membranes are relatively poor for the practical use; thus mechanical property enhanced nanofibrous membranes by creating special structure, or chemical modification must be a direction of development in the future. The processes of electrospinning and electro-netting actually can endow nanofibrous membranes with electret effects, and the resultant electrostatic effect will improve the filtration efficiency without sacrificing air pressure drop, so constructing stable and efficient electreted nanofibrous membranes will shred the light on the achievement of air filtration materials with further high performance. In addition, the functionalization of the air filters should be another direction of development, such as air filters with moisture resistant, harmful gas removing, superamphiphobic, and so on.

References

1. Macpherson, C.C. (2013). Climate change matters. *Journal of Medical Ethics* 40 (4): 288–290.
2. Holmes, C.D. (2014). Air pollution and forest water use. *Nature* 507 (7491): E1–E2.
3. Kjellstrom, T., Butler, A.J., Lucas, R.M. et al. (2010). Public health impact of global heating due to climate change: potential effects on chronic non-communicable diseases. *International Journal of Public Health* 55 (2): 97–103.
4. Fajersztajn, L., Veras, M., Barrozo, L.V. et al. (2013). Air pollution: a potentially modifiable risk factor for lung cancer. *Nature Reviews Cancer* 13 (9): 674–678.
5. World Health Organization. (2017). 7 million premature deaths annually linked to air pollution (March 25, 2014) https://www.who.int/mediacentre/news/releases/2014/air-pollution/en/
6. Lave, L.B. and Seskin, E.P. (2013). *Air Pollution and Human Health*. Taylor & Francis London, UK.
7. Querol, X., Alastuey, A., Rodriguez, S. et al. (2001). Monitoring of PM10 and PM2.5 around primary particulate anthropogenic emission sources. *Atmospheric Environment* 35 (5): 845–858.

8 Rodrıguez, S., Querol, X., Alastuey, A. et al. (2004). Comparative PM10–PM2. 5 source contribution study at rural, urban and industrial sites during PM episodes in Eastern Spain. *Science of the Total Environment* 328 (1): 95–113.

9 Oberdörster, G., Ferin, J., Gelein, R. et al. (1992). Role of the alveolar macrophage in lung injury: studies with ultrafine particles. *Environmental Health Perspectives* 97: 193.

10 Yim, S.H. and Barrett, S.R. (2012). Public health impacts of combustion emissions in the United Kingdom. *Environmental Science & Technology* 46 (8): 4291–4296.

11 Peukert, W. (1998). High temperature filtration in the process industry. *Filtration & Separation* 35 (5): 461–464.

12 Song, Y., Zhang, Y., Xie, S. et al. (2006). Source apportionment of PM2. 5 in Beijing by positive matrix factorization. *Atmospheric Environment* 40 (8): 1526–1537.

13 Donovan, R.P. (1985). *Fabric Filtration for Combustion Sources*, 1e. New York and Basel: Marcel Dekker, Inc.

14 Wang, C.-S. and Otani, Y. (2012). Removal of nanoparticles from gas streams by fibrous filters: a review. *Industrial & Engineering Chemistry Research* 52 (1): 5–17.

15 Adiletta, J.G. (1999) Fibrous nonwoven web. US Patent 5,954,962.

16 Shim, W.S. and Lee, D.W. (2013). Quality variables of meltblown submicron filter materials. *Indian Journal of Fiber &Textile Research* 38: 132–137.

17 Liu, Y., Cheng, B., Wang, N. et al. (2012). Development and performance study of polypropylene/polyester bicomponent melt-blowns for filtration. *Journal of Applied Polymer Science* 124 (1): 296–301.

18 Hung, C.-H. and Leung, W.W.-F. (2011). Filtration of nano-aerosol using nanofiber filter under low Peclet number and transitional flow regime. *Separation and purification technology* 79 (1): 34–42.

19 Yoon, K., Hsiao, B.S., and Chu, B. (2008). Functional nanofibers for environmental applications. *Journal of Materials Chemistry* 18 (44): 5326–5334.

20 Ma, H., Hsiao, B.S., and Chu, B. (2011). Thin-film nanofibrous composite membranes containing cellulose or chitin barrier layers fabricated by ionic liquids. *Polymer* 52 (12): 2594–2599.

21 Liu, X. and Ma, P.X. (2009). Phase separation, pore structure, and properties of nanofibrous gelatin scaffolds. *Biomaterials* 30 (25): 4094–4103.

22 Suzuki, A. and Arino, K. (2012). Polypropylene nanofiber sheets prepared by CO_2 laser supersonic multi-drawing. *European Polymer Journal* 48 (7): 1169–1176.

23 Tanaka, S., Doi, A., Nakatani, N. et al. (2009). Synthesis of ordered mesoporous carbon films, powders, and fibers by direct triblock-copolymer-templating method using an ethanol/water system. *Carbon* 47 (11): 2688–2698.

24 Wang, D., Sun, G., and Chiou, B.S. (2007). A high-throughput, controllable, and environmentally benign fabrication process of thermoplastic nanofibers. *Macromolecular Materials and Engineering* 292 (4): 407–414.

25 Qiu, P. and Mao, C. (2010). Biomimetic branched hollow fibers templated by self-assembled fibrous polyvinylpyrrolidone structures in aqueous solution. *ACS Nano* 4 (3): 1573–1579.

26 Wang, X. and Li, Y. (2002). Selected-control hydrothermal synthesis of α-and β-MnO_2 single crystal nanowires. *Journal of the American Chemical Society* 124 (12): 2880–2881.

27 Fong, H., Chun, I., and Reneker, D. (1999). Beaded nanofibers formed during electrospinning. *Polymer* 40 (16): 4585–4592.

28 Ding, B., Kimura, E., Sato, T. et al. (2004). Fabrication of blend biodegradable nanofibrous nonwoven mats via multi-jet electrospinning. *Polymer* 45 (6): 1895–1902.

29 Wu, J., Wang, N., Zhao, Y. et al. (2013). Electrospinning of multilevel structured functional micro-/nanofibers and their applications. *Journal of Materials Chemistry A* 1 (25): 7290–7305.

30 Thavasi, V., Singh, G., and Ramakrishna, S. (2008). Electrospun nanofibers in energy and environmental applications. *Energy & Environmental Science* 1 (2): 205–221.

31 Sahay, R., Kumar, P.S., Sridhar, R. et al. (2012). Electrospun composite nanofibers and their multifaceted applications. *Journal of Materials Chemistry* 22 (26): 12953–12971.

32 Zhou, C., Chu, R., Wu, R. et al. (2011). Electrospun polyethylene oxide/cellulose nanocrystal composite nanofibrous mats with homogeneous and heterogeneous microstructures. *Biomacromolecules* 12 (7): 2617–2625.

33 Wang, X., Ding, B., Sun, G. et al. (2013). Electro-spinning/netting: a strategy for the fabrication of three-dimensional polymer nano-fiber/nets. *Progress in Materials Science* 58 (8): 1173–1243.

34 Reneker, D.H. and Chun, I. (1996). Nanometre diameter fibres of polymer, produced by electrospinning. *Nanotechnology* 7 (3): 216.

35 Bhardwaj, N. and Kundu, S.C. (2010). Electrospinning: a fascinating fiber fabrication technique. *Biotechnology advances* 28 (3): 325–347.

36 Lin, K., Chua, K.-N., Christopherson, G.T. et al. (2007). Reducing electrospun nanofiber diameter and variability using cationic amphiphiles. *Polymer* 48 (21): 6384–6394.

37 Ding, B., Li, C., Miyauchi, Y. et al. (2006). Formation of novel 2D polymer nanowebs via electrospinning. *Nanotechnology* 17 (15): 3685.

38 Greiner, A. and Wendorff, J.H. (2007). Electrospinning: a fascinating method for the preparation of ultrathin fibers. *Angewandte Chemie International Edition* 46 (30): 5670–5703.

39 Li, Z. and Wang, C. (2013). *Effects of Working Parameters on Electrospinning*. Heidelberg: Springer.

40 Wendorff, J.H., Agarwal, S., and Greiner, A. (2012). *Electrospinning: Materials, Processing, and Applications*. Chichester: Wiley.

41 Teo, W.E. and Ramakrishna, S. (2006). A review on electrospinning design and nanofibre assemblies. *Nanotechnology* 17 (14): R89–R106.

42 Higuera, F.J. (2003). Flow rate and electric current emitted by a Taylor cone. *Journal of Fluid Mechanics* 484 (484): 303–327.

43 Xie, J. and Wang, C.H. (2007). Encapsulation of proteins in biodegradable polymeric microparticles using electrospray in the Taylor cone-jet mode. *Biotechnology and Bioengineering* 97 (5): 1278–1290.

44 Han, T., Reneker, D.H., and Yarin, A.L. (2007). Buckling of jets in electrospinning. *Polymer* 48 (20): 6064–6076.

45 Reneker, D.H. and Yarin, A.L. (2008). Electrospinning jets and polymer nanofibers. *Polymer* 49 (10): 2387–2425.

46 Wang, C., Hsu, C., Lin, J. (2006). Scaling laws in electrospinning of polystyrene solutions. *Macromolecules* 39 (22): 7662–7672.

47 Ding, B., Wang, M., Wang, X. et al. (2010). Electrospun nanomaterials for ultrasensitive sensors. *Materials Today* 13 (11): 16–27.

48 Barakat, N.A.M., Kanjwal, M.A., Sheikh, F.A. et al. (2009). Spider-net within the N6, PVA and PU electrospun nanofiber mats using salt addition: novel strategy in the electrospinning process. *Polymer* 50 (18): 4389–4396.

49 Pant, H.R., Bajgai, M.P., Nam, K.T. et al. (2010). Formation of electrospun nylon-6/methoxy poly (ethylene glycol) oligomer spider-wave nanofibers. *Materials letters* 64 (19): 2087–2090.

50 Tsou, S.Y., Lin, H.S., and Wang, C. (2011). Studies on the electrospun Nylon 6 nanofibers from polyelectrolyte solutions: 1. Effects of solution concentration and temperature. *Polymer* 52 (14): 3127–3136.

51 Chun, J.Y., Hong, G.P., Surassmo, S. et al. (2014). Study of the phase separation behaviour of native or preheated WPI with polysaccharides. *Polymer* 55 (16): 4379–4384.

52 Mit-uppatham, C., Nithitanakul, M., and Supaphol, P. (2004). Ultrafine electrospun polyamide-6 fibers: effect of solution conditions on morphology and average fiber diameter. *Macromolecular Chemistry and Physics* 205 (17): 2327–2338.

53 Grimm, R.L. and Beauchamp, J. (2005). Dynamics of field-induced droplet ionization: time-resolved studies of distortion, jetting, and progeny formation from charged and neutral methanol droplets exposed to strong electric fields. *The Journal of Physical Chemistry B* 109 (16): 8244–8250.

54 Wang, X., Ding, B., Yu, J., et al. (2011). Electro-netting: fabrication of two-dimensional nano-nets for highly sensitive trimethylamine sensing. *Nanoscale* 3 (3): 911–915.

55 Ding, B., Kim, H.Y., Lee, S.C. et al. (2002). Preparation and characterization of nanoscaled poly (vinyl alcohol) fibers via electrospinning. *Fibers and Polymers* 3 (2): 73–79.

56 Kim, B., Park, H., Lee, S.-H. et al. (2005). Poly (acrylic acid) nanofibers by electrospinning. *Materials Letters* 59 (7): 829–832.

57 Li, X., Lin, L., Zhu, Y. et al. (2013). Preparation of ultrafine fast-dissolving cholecalciferol-loaded poly (vinyl pyrrolidone) fiber mats via electrospinning. *Polymer Composites* 34 (2): 282–287.

58 Chen, C., Wang, L., and Huang, Y. (2007). Electrospinning of thermo-regulating ultrafine fibers based on polyethylene glycol/cellulose acetate composite. *Polymer* 48 (18): 5202–5207.

59 Li, D. and Xia, Y. (2003). Fabrication of titania nanofibers by electrospinning. *Nano Letters* 3 (4): 555–560.

60 Fan, M., Hui, W., Li, Z. et al. (2012). Fabrication and piezoresponse of electrospun ultra-fine Pb ($Zr_{0.3}$, $Ti_{0.7}$) O_3 nanofibers. *Microelectronic Engineering* 98: 371–373.
61 Deng, L., Young, R.J., Kinloch, I.A. et al. (2013). Carbon nanofibres produced from electrospun cellulose nanofibres. *Carbon* 58: 66–75.
62 Yan, H., Mahanta, N.K., Wang, B. et al. (2014). Structural evolution in graphitization of nanofibers and mats from electrospun polyimide–mesophase pitch blends. *Carbon* 71: 303–318.
63 Khalil, A., Hashaikeh, R., and Jouiad, M. (2014). Synthesis and morphology analysis of electrospun copper nanowires. *Journal of Materials Science* 49 (8): 3052–3065.
64 Koombhongse, S., Liu, W., and Reneker, D.H. (2001). Flat polymer ribbons and other shapes by electrospinning. *Journal of Polymer Science Part B* 39 (21): 2598–2606.
65 Kessick, R. and Tepper, G. (2004). Microscale polymeric helical structures produced by electrospinning. *Applied Physics Letters* 84 (23): 4807–4809.
66 Ding, B., Li, C., Du, J., and Shiratori, S. (2008). *Nanotechnology Research: New Nanostructures, Nanotubes and Nanofibers* (ed. X. Huang). New York: Nova Science Publishers, Inc.
67 Sun, Z., Zussman, E., Yarin, A.L. et al. (2003). Compound core–shell polymer nanofibers by co-electrospinning. *Advanced Materials* 15 (22): 1929–1932.
68 Zhao, Y., Cao, X., and Jiang, L. (2007). Bio-mimic multichannel microtubes by a facile method. *Journal of the American Chemical Society* 129 (4): 764–765.
69 Mou, F., Guan, J.-G., Shi, W. et al. (2010). Oriented contraction: a facile nonequilibrium heat-treatment approach for fabrication of maghemite fiber-in-tube and tube-in-tube nanostructures. *Langmuir* 26 (19): 15580–15585.
70 Jin, Y., Yang, D., Kang, D. et al. (2009). Fabrication of necklace-like structures via electrospinning. *Langmuir* 26 (2): 1186–1190.
71 Shengyuan, Y., Peining, Z., Nair, A.S. et al. (2011). Rice grain-shaped TiO_2 mesostructures-synthesis, characterization and applications in dye-sensitized solar cells and photocatalysis. *Journal of Materials Chemistry* 21 (18): 6541–6548.
72 Kokubo, H., Ding, B., Naka, T. et al. (2007). Multi-core cable-like TiO_2 nanofibrous membranes for dye-sensitized solar cells. *Nanotechnology* 18 (16): 165604.
73 Chen, H., Wang, N., Di, J. et al. (2010). Nanowire-in-microtube structured core/shell fibers via multifluidic coaxial electrospinning. *Langmuir* 26 (13): 11291–11296.
74 Li, D. and Xia, Y. (2004). Direct fabrication of composite and ceramic hollow nanofibers by electrospinning. *Nano Letters* 4 (5): 933–938.
75 Zhang, D. and Chang, J. (2007). Patterning of electrospun fibers using electroconductive templates. *Advanced Materials* 19 (21): 3664–3667.

76 Kim, G.T., Ahn, Y.C., and Lee, J.K. (2008). Characteristics of Nylon 6 nanofilter for removing ultra fine particles. *Korean Journal of Chemical Engineering* 25 (2): 368–372.

77 Mei, Y., Wang, Z., and Li, X. (2013). Improving filtration performance of electrospun nanofiber mats by a bimodal method. *Journal of Applied Polymer Science* 128 (2): 1089–1094.

78 Sambaer, W., Zatloukal, M., and Kimmer, D. (2011). 3D modeling of filtration process via polyurethane nanofiber based nonwoven filters prepared by electrospinning process. *Chemical Engineering Science* 66 (4): 613–623.

79 Qin, X.H. and Wang, S.Y. (2008). Electrospun nanofibers from crosslinked poly (vinyl alcohol) and its filtration efficiency. *Journal of Applied Polymer Science* 109 (2): 951–956.

80 Patanaik, A., Jacobs, V., and Anandjiwala, R.D. (2010). Performance evaluation of electrospun nanofibrous membrane. *Journal of Membrane Science* 352 (1): 136–142.

81 Yun, K.M., Suryamas, A.B., Iskandar, F. et al. (2010). Morphology optimization of polymer nanofiber for applications in aerosol particle filtration. *Separation and Purification Technology* 75 (3): 340–345.

82 Wang, N., Raza, A., Si, Y. et al. (2013). Tortuously structured polyvinyl chloride/polyurethane fibrous membranes for high-efficiency fine particulate filtration. *Journal of Colloid and Interface Science* 398: 240–246.

83 Cho, D., Naydich, A., Frey, M.W. et al. (2013). Further improvement of air filtration efficiency of cellulose filters coated with nanofibers via inclusion of electrostatically active nanoparticles. *Polymer* 54 (9): 2364–2372.

84 Yeom, B.Y., Shim, E., and Pourdeyhimi, B. (2010). Boehmite nanoparticles incorporated electrospun nylon-6 nanofiber web for new electret filter media. *Macromolecular Research* 18 (9): 884–890.

85 Gibson, P., Schreuder-Gibson, H., and Rivin, D. (2001). Transport properties of porous membranes based on electrospun nanofibers. *Colloids and Surfaces A* (187): 469–481.

86 Li, L., Frey, M.W., and Green, T.B. (2006). Modification of air filter media with nylon-6 nanofibers. *Journal of Engineered Fibers and Fabrics.* 1: 1–22.

87 Ahn, Y., Park, S., Kim, G. et al. (2006). Development of high efficiency nanofilters made of nanofibers. *Current Applied Physics* 6 (6): 1030–1035.

88 Przekop, R. and Gradoń, L. (2008). Deposition and filtration of nanoparticles in the composites of nano-and microsized fibers. *Aerosol Science and Technology* 42 (6): 483–493.

89 Zhao, X., Wang, S., Yin, X. et al. (2016). Slip-Effect functional air filter for efficient purification of $PM_{2.5}$. *Scientific Reports* 6: 35472.

90 Wang, S., Zhao, X., Yin, X. et al. (2016). Electret polyvinylidene fluoride nanofibers hybridized by polytetrafluoroethylene nanoparticles for high-efficiency air filtration. *ACS Applied Materials & Interfaces* 8 (36): 23985–23994.

91 Liu, Y., Park, M., Ding, B. et al. (2015). Facile electrospun Polyacrylonitrile/poly (acrylic acid) nanofibrous membranes for high efficiency particulate air filtration. *Fibers and Polymers* 16 (3): 629–633.

92 Zhang, S., Liu, H., Yin, X. et al. (2016). Anti-deformed polyacrylonitrile/polysulfone composite membrane with binary structures for effective air filtration. *ACS Applied Materials & Interfaces* 8 (12): 8086–8095.

93 Wang, N., Zhu, Z., Sheng, J. et al. (2014). Superamphiphobic nanofibrous membranes for effective filtration of fine particles. *Journal of Colloid and Interface Science* 428: 41–48.

94 Wang, N., Wang, X., Ding, B. et al. (2012). Tunable fabrication of three-dimensional polyamide-66 nano-fiber/nets for high efficiency fine particulate filtration. *Journal of Materials Chemistry* 22 (4): 1445–1452.

95 Liu, B., Zhang, S., Wang, X. et al. (2015). Efficient and reusable polyamide-56 nanofiber/nets membrane with bimodal structures for air filtration. *Journal of Colloid and Interface Science* 457: 203–211.

96 Wang, N., Yang, Y., Al-Deyab, S.S. et al. (2015). Ultra-light 3D nanofibre-nets binary structured nylon 6–polyacrylonitrile membranes for efficient filtration of fine particulate matter. *Journal of Materials Chemistry A* 3 (47): 23946–23954.

97 Yang, Y., Zhang, S., Zhao, X. et al. (2015). Sandwich structured polyamide-6/polyacrylonitrile nanonets/bead-on-string composite membrane for effective air filtration. *Separation and Purification Technology* 152: 14–22.

98 Zhang, S., Liu, H., Yu, J. et al. (2016). Microwave structured polyamide-6 nanofiber/net membrane with embedded poly (m-phenylene isophthalamide) staple fibers for effective ultrafine particle filtration. *Journal of Materials Chemistry A* 4 (16): 6149–6157.

99 Zhang, S., Liu, H., Zuo, F. et al. (2017). A controlled design of ripple-like polyamide-6 nanofiber/nets membrane for high-efficiency air filter. *Small*: 1603151.

100 Zhang, S., Tang, N., Cao, L. et al. (2016). Highly integrated polysulfone/polyacrylonitrile/polyamide-6 air filter for multilevel physical sieving airborne particles. *ACS Applied Materials & Interfaces* 8 (42): 29062–29072.

101 Nakata, K., Kim, S.H., Ohkoshi, Y. et al. (2007). Electrospinning of poly (ether sulfone) and evaluation of the filtration efficiency. *Sen'i Gakkaishi* 63 (12): 307–312.

102 Moon, S., Choi, J., and Farris, R.J. (2008). Preparation of aligned polyetherimide fiber by electrospinning. *Journal of Applied Polymer Science* 109 (2): 691–694.

103 Fukushima, S., Karube, Y., and Kawakami, H. (2010). Preparation of ultrafine uniform electrospun polyimide nanofiber. *Polymer Journal* 42 (6): 514–518.

104 Huang, C., Chen, S., Reneker, D.H. et al. (2006). High-strength mats from electrospun poly (p-Phenylene Biphenyltetracarboximide) nanofibers. *Advanced Materials* 18 (5): 668–671.

105 Li, X., Wang, N., Fan, G. et al. (2015). Electreted polyetherimide–silica fibrous membranes for enhanced filtration of fine particles. *Journal of Colloid and Interface Science* 439: 12–20.

106 Shao, C., Kim, H., Gong, J. et al. (2002). A novel method for making silica nanofibres by using electrospun fibres of polyvinylalcohol/silica composite as precursor. *Nanotechnology* 13 (5): 635.

107 Yu, H., Guo, J., Zhu, S. et al. (2012). Preparation of continuous alumina nanofibers via electrospinning of PAN/DMF solution. *Materials Letters* 74: 247–249.

108 Zhao, Y., Tang, Y., Guo, Y. et al. (2010). Studies of electrospinning process of zirconia nanofibers. *Fibers and Polymers* 11 (8): 1119–1122.

109 Guo, M., Ding, B., Li, X. et al. (2009). Amphiphobic nanofibrous silica mats with flexible and high-heat-resistant properties. *The Journal of Physical Chemistry C* 114 (2): 916–921.

110 Zhao, F., Wang, X., Ding, B. et al. (2011). Nanoparticle decorated fibrous silica membranes exhibiting biomimetic superhydrophobicity and highly flexible properties. *RSC Advances* 1 (8): 1482–1488.

111 Wang, X., Cui, F., Lin, J. et al. (2012). Functionalized nanoporous TiO_2 fibers on quartz crystal microbalance platform for formaldehyde sensor. *Sensors and Actuators B* (171): 658–665.

112 Yang, L., Raza, A., Si, Y. et al. (2012). Synthesis of superhydrophobic silica nanofibrous membranes with robust thermal stability and flexibility via in situ polymerization. *Nanoscale* 4 (20): 6581–6587.

113 Han, W., Ding, B., Park, M. et al. (2015). Facile synthesis of luminescent and amorphous $La_2O_3-ZrO_2$: Eu^{3+} nanofibrous membranes with robust softness. *Nanoscale* 7 (34): 14248–14253.

114 Mao, X., Si, Y., Chen, Y. et al. (2012). Silica nanofibrous membranes with robust flexibility and thermal stability for high-efficiency fine particulate filtration. *RSC Advances* 2 (32): 12216–12223.

115 Mao, X., Bai, Y., Yu, J. et al. (2016). Flexible and highly temperature resistant polynanocrystalline zirconia nanofibrous membranes designed for air filtration. *Journal of the American Ceramic Society* 99 (8): 2760–2768.

116 Mathew, G., Hong, J., Rhee, J. et al. (2006). Preparation and anisotropic mechanical behavior of highly-oriented electrospun poly (butylene terephthalate) fibers. *Journal of Applied Polymer Science* 101 (3): 2017–2021.

117 Wang, X., Ding, B., Yu, J. et al. (2009). A highly sensitive humidity sensor based on a nanofibrous membrane coated quartz crystal microbalance. *Nanotechnology* 21 (5): 055502.

118 Wang, X., Ding, B., Yu, J. et al. (2011). Large-scale fabrication of two-dimensional spider-web-like gelatin nano-nets via electro-netting. *Colloids and Surfaces B* 86 (2): 345–352.

119 Ding, B., Ogawa, T., Kim, J. et al. (2008). Fabrication of a super-hydrophobic nanofibrous zinc oxide film surface by electrospinning. *Thin Solid Films* 516 (9): 2495–2501.

120 Ding, B., Li, C., Wang, D. et al. (2010). *Fabrication and Application of Novel Two-Dimensional Nanowebs via Electrospinning*, Nanotechnology: Nanofabrication, Patterning, and Self Assembly, 51–69. New York: Nova Science Publishers, Inc.

8

Aramid Fibers

Manjeet Jassal, Ashwini K. Agrawal, Deepika Gupta, and Kamlesh Panwar

Indian Institute of Technology, Department of Textile Technology, New Delhi, 110016 Delhi, India

8.1 Introduction

As per the US Federal Trade Commission, aramid fibers, or aromatic polyamides, are manufactured fibers that have at least 85% of the amide linkages attached directly to two aromatic rings. These were the first organic fibers to be used for reinforcement in advanced composites. Due to their outstanding thermal and mechanical properties, these fibers have been able to displace inorganic fibers and metal wires from the market of high-performance applications. These fibers have been used widely in areas such as protective clothing for hostile environment that are exposed to heat, chemicals, and radiations.

It was due to their inability to dissolve in a solvent and no melting point that the production of first aramid fiber took place long after the commercialization of nylon in 1938. DuPont launched the very first aramid fiber under the trade name Nomex® in 1960. It was the first *m*-aramid fiber with very high temperature resistance. In order to achieve further superior properties, the attention was drawn toward *p*-aramid fibers leading to the development of Kevlar® in 1971. A similar *p*-aramid, Twaron, was launched by Teijin toward the end of 1980s.

DuPont Co. (USA) is world's leading producer of *m*-aramid fibers (under the trade name Nomex). Teijin (Japan) is the only other producer of *m*-aramid fibers (under the trade name Teijinconex). DuPont Co. (USA) is world's leading producer of *p*-aramid fibers also, followed by Teijin (Japan), which is the second largest producer. Other major producers include Rhône-Poulenc (France), Toray Industries Inc. (Japan), Acordis, and Akzo Nobel (NV, Netherlands). The history of the development of the aramid fibers is summarized in Figure 8.1. Several reviews [1–10] are available on the synthesis, spinning, properties, and applications of aramid fibers.

8.2 Preparation of Aromatic Polyamides

Due to their characteristic features such as rigid structures, no melting point, and inability to dissolve in a solvent, aramids could not be prepared by melt spinning

Handbook of Fibrous Materials, First Edition. Edited by Jinlian Hu, Bipin Kumar, and Jing Lu.
© 2020 Wiley-VCH Verlag GmbH & Co. KGaA. Published 2020 by Wiley-VCH Verlag GmbH & Co. KGaA.

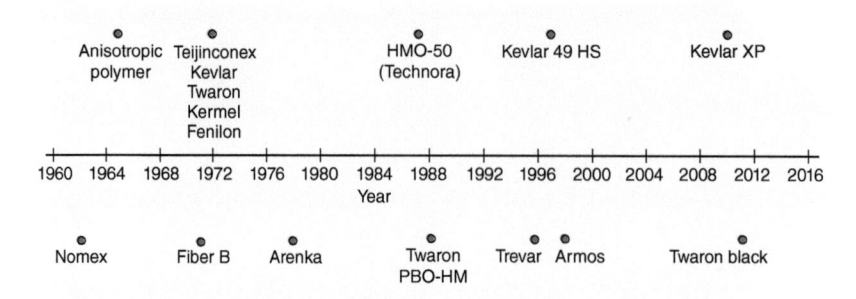

Figure 8.1 Timeline of important historical developments of aramid fibers.

and conventional bulk polymerization. Instead, the aromatic polyamides are mainly synthesized by low temperature polycondensation and direct polycondensation in solution using phosphites [11–13]. Aramids could be synthesized in the form of AB homopolymers with reaction between an amine group and a carboxylic acid halide group or AABB aramids by reaction between aromatic diamines and diacids or diacid chlorides.

8.2.1 Low Temperature Polycondensation

Due to low reactivity of aromatic diamines toward aromatic dicarboxylic acids, aromatic diacid chlorides are generally used for the synthesis of aramids. Low temperature polycondensation is carried out at temperatures below 50 °C to avoid degradation, side reaction, and cross-linking.

The basic reaction involved in the synthesis of aramids by low temperature polycondensation is shown in Figure 8.2.

Based on these reactions, the synthesis of Nomex (m-aramid) and Kevlar (p-aramid) [14] could be shown in Figures 8.3 and 8.4, respectively.

Generally, 5–20 wt% initial monomer concentration is used with equimolar ratios of diamines and diacid chloride. Solvent is selected either from

$$n\, H_2N\text{—Ar—COCl} \longrightarrow [\text{HN—Ar—CO}]_n + n\, HCl$$

$$\quad\quad\quad\quad\quad\quad\quad\quad\quad\quad\quad\quad\quad A \quad\quad\quad\quad B$$

Figure 8.2 Reaction mechanism for the synthesis of aramids by low temperature polycondensation.

Figure 8.3 Reaction mechanism for the synthesis of Nomex (m-aramid).

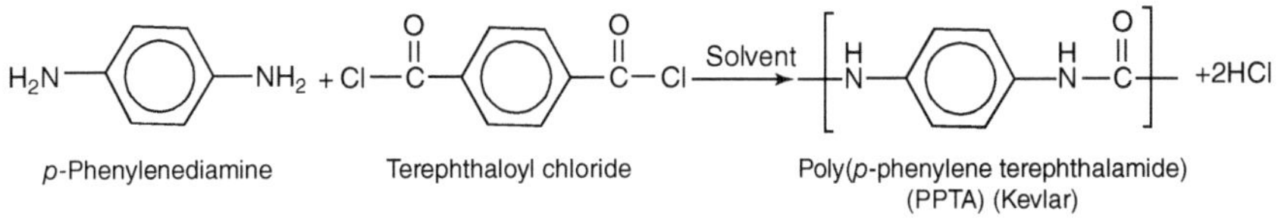

8.2 Preparation of Aromatic Polyamides

Figure 8.4 Reaction mechanism for the synthesis of Kevlar (p-aramid).

halogenated nonaromatic hydrocarbons (such as methyl ethyl ketone, chloroform, methylene chloride, dimethylcyanamide, and acetonitrile) or amide (such as N-methyl-2-pyrrolidone, dimethylacetamide, tetramethyl urea, and hexamethylphosphoramide). The incorporation of aryl-ether linkages in aramid structure has been reported [15–17] to increase its solubility and processability without significantly altering their thermal properties. Similarly, various other approaches such as incorporation of other flexible functionalities [18–20], N-alkylation of the backbone, and introduction of bulky pendant groups [21–23] have been explored for improving the solubility of the aramids.

Low temperature polycondensation could be carried out by two different methods: interfacial polycondensation and solution polycondensation. In the case of interfacial polycondensation, the monomers are dissolved in two different immiscible solvents, and the reaction is carried out at the interface. However, in solution polycondensation, reaction is carried out in an inert organic solvent. As the interfacial polycondensation is diffusion dependent, high molecular weight polymers with a broad molecular weight distribution are synthesized. Due to this reason, solution polycondensation is preferred to interfacial polycondensation.

The properties of the synthesized polymer depend upon various factors such as the solubility–concentration–temperature relationship and the salt concentration. The molecular weight of the final polymer is dependent upon the stoichiometry of the solvent and reactant mixture. The inherent viscosity is highest at an optimum reactant concentration, although it decreases on both decreasing and increasing concentration from the optimum value. The viscosity drop at low reactant concentration is due to the occurrence of side reactions. However, at high reactant concentration, the viscosity drop is due to reduced mobility of reactant as a result of gelation much before high inherent viscosity is achieved. A suitable solvent could be useful in delaying the gelation until high molecular weight is attained. Also, small amounts of various alkaline and alkali earth metal salts such as lithium hydroxide, lithium chloride, calcium hydroxide, and calcium chloride are used to increase the solvating power of organic solvents [23, 24]. As the polymerization reaction is an exothermic process, so at high reactant concentration, the large amount of heat released causes an increase in side reactions, thus reducing the inherent viscosity. So, the temperature of the reaction is generally kept very low (0 to −15 °C), thus reducing side reactions and increasing the inherent viscosity.

As a large amount of salt is generated during low temperature polycondensation reaction that further neutralizes HCl, the process requires expensive corrosion-resistant materials adding significantly to the cost of the process. The acid dichloride, being highly active, must be stored in refrigerator under dry conditions to avoid their hydrolysis.

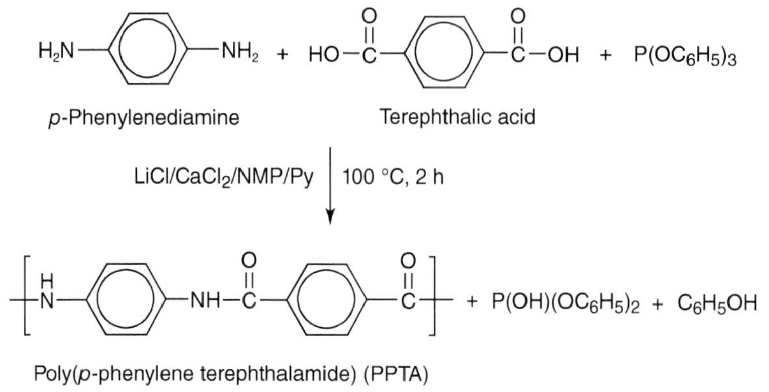

Figure 8.5 Reaction mechanism for the synthesis of poly(p-phenylene terephthalamide) (PPTA) through direct polycondensation.

8.2.2 Direct Polymerization in Solution Using Phosphites

Synthesis of aramids using aromatic diacids and aromatic diamines is a straightforward and direct synthesis method. Diacids are reasonably cheap materials but show lesser activity toward diamines. As a result, their reaction with diamines is carried out in the presence of certain catalysts (generally phosphorus- or sulfur-based compounds) [25–27]. These catalysts activate the diacids forming an N-P type intermediate and thus completing the polycondensation. The reaction mechanism for the synthesis of aramids by direct polycondensation is shown in Figure 8.5.

Shoji et al. [28] reported the synthesis of aramids using diacids and diamines with ether linkage. The presence of ether linkages led to an increase in the polymer mobility that, on the other hand, reduced the glass transition temperature as well as melting point of the synthesized aramid. Polycondensation of isophthalic acid with 3,4′-oxydianiline led to the formation of high molecular weight aramids. The group further synthesized high molecular weight aramid using aromatic diacid and 4,4′-oxydianiline by bulk polycondensation.

Besides polycondensation, aramids are also synthesized by low temperature hydrogen transfer reactions. This reaction occurs in between a diacid and diisocyanate forming an intermediate polymer, which on removal of carbon dioxide on subsequent heating leads to the formation of aromatic polyamide. The reaction mechanism for hydrogen transfer reaction is shown in Figure 8.6. However, this is not the commonly used reaction for the synthesis of aramids because of the higher cost of diisocyanates and the difficulty in eliminating all the carbon dioxide.

8.2.3 Copolyaramids

Aramids possess very high melting point and show inability to dissolve in a solvent, which make their processing very difficult. In order to improve their processability, copolymers of aramids are synthesized [29–31]. The incorporation of diamines with significantly distant amino groups leads to lowered spatial

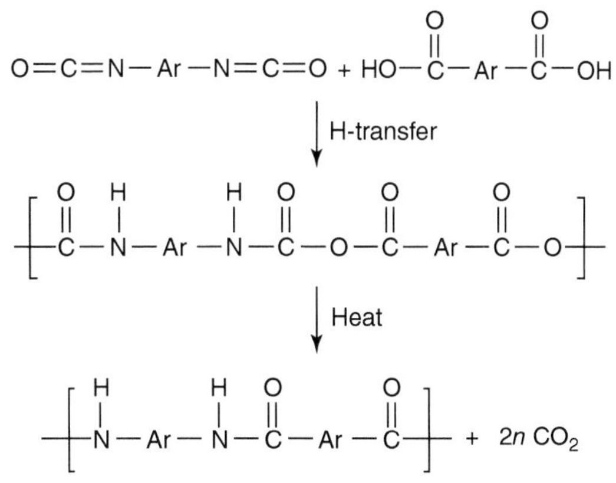

Figure 8.6 Reaction mechanism for the synthesis of aramids by hydrogen transfer reaction.

$$O=C=N-Ar-N=C=O + HO-\overset{O}{\overset{\|}{C}}-Ar-\overset{O}{\overset{\|}{C}}-OH$$

↓ H-transfer

$$\left[\overset{O}{\overset{\|}{C}}-\overset{H}{\overset{|}{N}}-Ar-\overset{H}{\overset{|}{N}}-\overset{O}{\overset{\|}{C}}-O-\overset{O}{\overset{\|}{C}}-Ar-\overset{O}{\overset{\|}{C}}-O\right]$$

↓ Heat

$$\left[-\overset{H}{\overset{|}{N}}-Ar-\overset{H}{\overset{|}{N}}-\overset{O}{\overset{\|}{C}}-Ar-\overset{O}{\overset{\|}{C}}-\right] + 2n\ CO_2$$

Table 8.1 Polymerization conditions for certain copolyaramids.

Monomers	Solvent	Polymer	η_{inh}
1. p-Phenylenediamine Terephthaloyl chloride Isophthaloyl chloride	HMPA/NMP	95/5 Copoly(p-phenylene terephthalamide/isophthalamide)	4.48, 3.9
2. p-Phenylenediamine p-Phenylene diisocyanate Terephthaloyl chloride	HMPA/NMPLiCl HMPA/NMPLiCl HMPA/NMPLiCl	90/10 Copoly(p-phenylene p-phenylene diisocyano terephthalamide) 95/5 Copolymer 80/20 Copolymer	1.83 2.87 2.82
3. p-Phenylenediamine 2,5-Bis(p-aminophenyl)-1,3,4-oxadiazole Terephthaloyl chloride	HMPA/NMPLiCl	Copoly(p-phenylene)/ 2,5-bis-(p-aminophenyl)-1,3,4-oxadiazo terephthalamide	4.48
4. p-Phenylenediamine 2-Chloro-p-phenylenediamine Terephthaloyl chloride	DMAC/Li$_2$CO$_3$	50/50 Copoly(p-phenylene/ 2-chloro-p-phenylene terephthalamide)	3.85

HMPA, hexamethyl phosphoric triamide; DMAC, dimethyl acetamide; NMP, N-methyl-2-pyrrolidone.
Source: With prior permission from Indian Journal of Fibre & Textile Research.

density of hydrogen bonding and rate of crystallization. On the other hand, the fiber drawability could be improved by using another comonomer with two phenylene groups connected with a single atom. The polymerization conditions for copolyaramids are summarized in Table 8.1.

8.3 Aramid Solutions

Aramid polymers exhibit high melting point, and often their decomposition temperature is higher than their melting temperature. This makes the fiber processing

Figure 8.7 Directional orientation of rodlike molecules of PPTA.

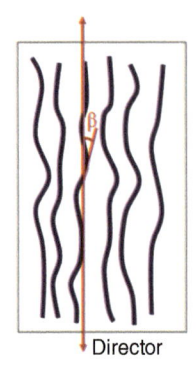

$\beta = 9-12°$

High molecular orientation

Director

of aramids via melt spinning unfeasible. Therefore aramid fibers are made by solution spinning. However, dissolution of aramids is also challenging because of their rigid structure. It is generally achieved by the use of highly polar solvents without or with inorganic salts or strong acids. Isotropic solutions of aramids can be produced by dissolving homopolymers (e.g. poly(*m*-phenylene isophthalamide) (*m*-DPI)) in solvents such as *N*-methyl-2-pyrrolidone (NMP) and DMAc. In the case of *p*-aramids, solubility can be increased by copolymerization as explained previously. As mentioned earlier, *p*-aramids dissolve in highly polar solvents containing inorganic salts and strong acids. They form isotropic solutions only at low polymer concentrations. For example, poly(*p*-phenylene terephthalamide) (PPTA) forms isotropic solutions at concentrations lower than 10 wt% and ambient temperature.

One of the distinctive features of semirigid polymer is that they exhibit molecular orientation when subjected to external forces such as electrical, magnetic, shear, and elongational flow. The above characteristics are shown in lyotropic solutions such as poly(*p*-benzamide) and PPTA [32]. Figure 8.7 shows the molecular orientation of rodlike molecules of PPTA along the direction or axis of orientation, with degree of orientation described by β that ranges from 9° to 12°.

As mentioned earlier factors such as concentration of solution, solvent, molecular weight of polymer, and temperature affect the orientation and phase behavior in lyotropic nematic polymer solutions. Example of a lyotropic solution is that of PPTA in 100% sulfuric acid where the viscosity and molecular arrangement are the function of concentration as shown in Figure 8.8. With increasing concentration the molecular orientation increases, leading to the formation of smectic and nematic phases in liquid crystals (Figure 8.8a). From Figure 8.8b it can be seen that initially the viscosity increases with increase in the concentration and solution shows little or no orientation of crystal domains. On further increasing the concentration (18–23% PPTA), the viscosity decreases, since in this region the crystal domains perfectly orient showing anisotropic behavior. This particular concentration region where the viscosity is lowest (i.e. formation of nematic liquid crystal) and the polymer is highly oriented is of interest for spinning. Unlike the case of isotropic polymers where an increase in the viscosity is expected with concentration, lyotropic solutions show decrease in the viscosity as a result of formation of nematic liquid crystal phase.

For solution spinning, low viscosity at high polymer concentration is required. Figure 8.9 shows a phase diagram for PPTA in 100% sulfuric acid. As mentioned

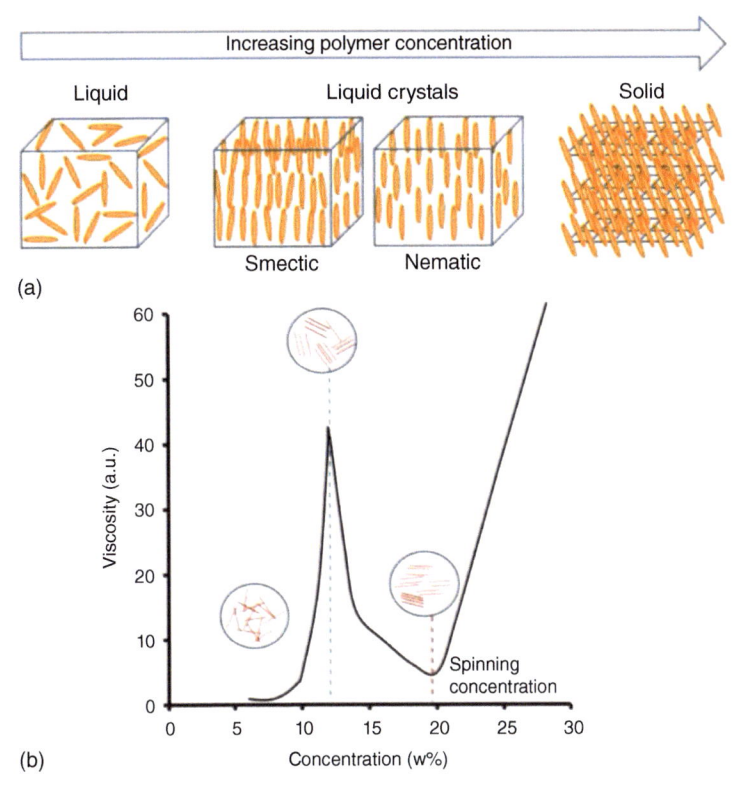

Figure 8.8 (a) Structure of lyotropic nematic phase with respect to concentration. (b) A typical viscosity vs. concentration curve for lyotropic solutions.

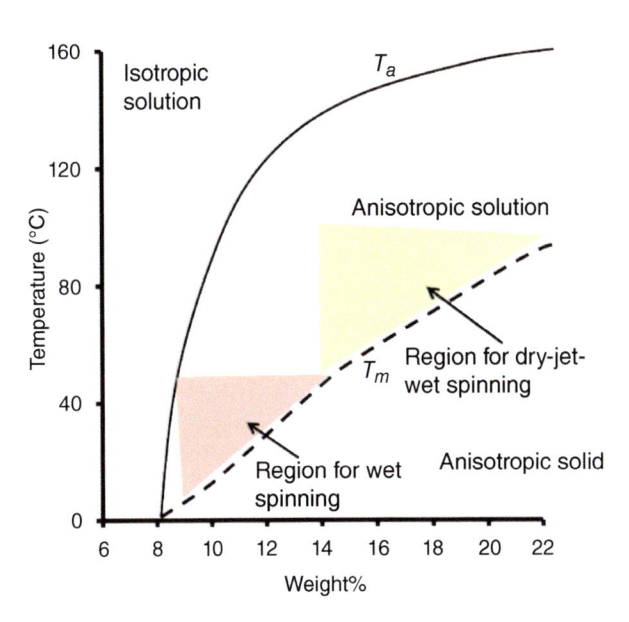

Figure 8.9 Phase diagram of anisotropic solution of PPTA in 100% H_2SO_4.

earlier, critical concentration is dependent on the molecular weight and temperature. These dopes at high concentration of 18–23% are solid up to about 80 °C. Therefore, they are required to be melted prior to spinning. On the other hand, for coagulation at low temperatures, dry-jet-wet spinning is done. In this process of spinning of PPTA in nematic phase, fiber precipitation occurs in the air gap, and the acid is removed in the coagulation bath. It is also evident from the phase diagram (Figure 8.9) that up to concentrations of 10%, dope viscosity remains low and such dopes can be wet-spun.

8.4 Spinning of Aramid Fibers

The spinning of aramid fibers can be done using dry spinning, wet spinning, or dry-jet-wet spinning process.

8.4.1 Dry Spinning

The dry spinning of aramid fibers is done by extruding the polymer solution through a spinneret mounted above a heated column. The polymer solution is extruded in the presence of hot inert gas (or air), and solvent gets evaporated from the emergent fiber. The solidified fiber is collected at the bottom of the column.

The process for dry spinning of Nomex fibers from dimethylformamide (DMF), and DMAc solutions was developed by DuPont. Nomex, derived from the m-aramid polymer solution, is disoriented when in solution, unlike p-aramid (e.g. Kevlar). The fibers get roughly orientated during the extrusion of the solution; however it exhibits poorer strength as compared with its p-aramid counterpart. The extent of fiber orientation can be increased by increasing the shear rate through the spinneret capillary. However, inhomogeneity is developed due to solvent diffusion and formation of skin due to fast solvent evaporation from surface than from bulk. The as-spun fibers are drawn to enhance their physical properties. Typical mechanical properties of fibers are on the order of 0.6 GPa with a 30% elongation at break.

8.4.2 Wet Spinning

In the wet spinning process, spinneret is submerged in the coagulation bath containing solvent and non-solvent. Polymer–solvent and non-solvent interaction affects the structure and leads to the ultimate properties. Important variables that control this process are polymer concentration, solution composition and its temperature, composition and temperature of coagulating solution, the rate of extrusion, and the residence time in the coagulating bath. The isotropic solution, spun by wet spinning, is exposed to the coagulant as soon as it exits the spinneret nozzle, preventing the solution from complete attenuation and hence resulting in low tenacity, intermediate modulus, and high elongation. Aramids such as Kevlar and Teijinconex have been spun using wet spinning process.

8.4.3 Dry-Jet-Wet Spinning

In a typical dry-jet-wet spinning (Figure 8.10), the anisotropic polymer dope of ~20 wt% concentration (in ~80 wt% of concentrated sulfuric acid) is extruded through spinnerets maintained at 100 °C through an air gap of about 1 cm into a cold extrusion water bath kept at 0–4 °C. The fiber precipitates and undergoes strain, and the solvent (acid) gets eliminated in the coagulation bath. The spinneret capillary and air gap causes orientation and alignment of crystal domains, resulting in highly oriented and crystalline as-spun fibers. When the anisotropic solution passes through the capillary hole, the shear due to the capillary causes some of the liquid crystal domains to orient along the direction of shear; however, at the capillary exit, some disorientation of liquid crystal domain occurs because of solution viscoelasticity. The disorientation is quickly overcome by sudden quenching of the filament in cooling bath. This orientation in filament is particularly important as it results in highly crystalline, high strength, and high modulus fiber. The fibers produced by this process show tenacity and initial modulus of two to four times that of a fiber prepared by a conventional wet spinning process.

If the same process is used to produce Kevlar filaments of large diameter, then the core of the filament might lose its orientation because the time required to quench the core will increase with the square of the filament radius. However, the outer part of the filament, i.e. the skin, will have a high degree of orientation due to quick quenching. This creates a "skin-core" effect, because the overall properties of the filaments, such as tensile strength per unit cross-sectional

Figure 8.10 Schematic diagram showing dry-jet-wet spinning of lyotropic solution.

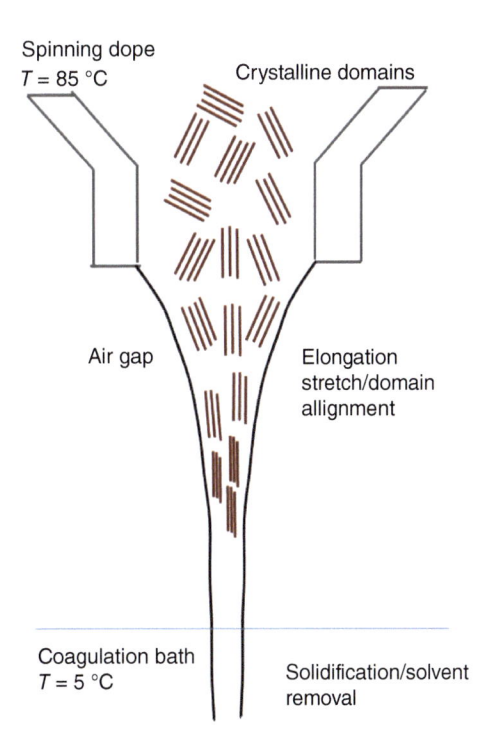

area, will diminish owing to a decrease in average orientation. The dry-jet-wet spinning process allows using a low temperature coagulant without concern for freezing the spinning dope. The extruded solution experiences more strain in the air gap and get fully stretched, thereby developing a higher degree of molecular orientation.

8.4.4 Aramid Nanofibers

Traditional methods of synthesizing aramid fibers constitute bottom-up approach where macrofibers are produced from monomers. In recent times, aramid macroscopic yarns are broken down into stable nanofiber dispersion by a top-down approach where deprotonation and dissolution in strong solvents such as dimethyl sulfoxide (DMSO) and potassium hydroxide (KOH) have been employed to produce nanoscale fibers of Kevlar [33, 34]. This could be a result of equilibrium between the antagonistic forces, namely, electrostatic repulsion, $\pi-\pi$ stacking, and van der Waals attraction. The other method of synthesizing nanoscale fibers of N-substituted Kevlar by electrospinning was demonstrated by Yeager et al. [35]. The N-substituted Kevlar showed better dissolution that leads to uniform aramid nanofibers (ANFs) as compared with nonuniform nanofibers of unsubstituted Kevlar dispersions in strong sulfuric acid [36]. However, electrospun substituted Kevlar nanofibers showed unoriented crystalline domains and lacked microvoids generally present in amorphous regions. Similarly, m-aramid nanofibers have been synthesized simply by dissolving pristine m-aramid polymer in DMAc at 80 °C followed by electrospinning [37]. The mechanical and chemical stability of ANFs has been improved by various methods such as post-treatment by microwave irradiation [37], washing, and heat treatment [38]. Schematic of different approaches of producing ANFs has been shown in Figure 8.11.

Nanoscale aramid fibers could be expected to possess high or better strength, rigidity, toughness, and many other properties. Further, fibers in nano-dimension would facilitate proper distribution of stress over multiple polymer chains and bonds. ANFs can be useful for nanoscale reinforcement for better mechanical properties [39], composite epoxy adhesives [37], ultrafiltration of water and hemodialysis [40], separation of oil and water [41], and battery separators [42]. Detailed applications of ANFs have been given in Section 8.6.

8.4.5 Fiber Heat Treatment

The physical and tensile properties of as-spun aramid fibers can be further enhanced by heat treatment under tension. In wet-spun yarns both tenacity and modulus increase exponentially with increasing temperature (and draw ratio). The effective increase begins at about 360 °C (T_g of polymer), and the properties attain a maximum at nearly 550 °C (T_m of polymer). During the process of heat treatment, wet-spun fibers show maximum increase in crystallinity, orientation, and structure perfection. In dry-jet-wet spun yarns, a jump in the modulus is observed, which becomes independent of temperature greater than 200 °C. The final modulus is a function of precursor modulus (or orientation), with tenacity

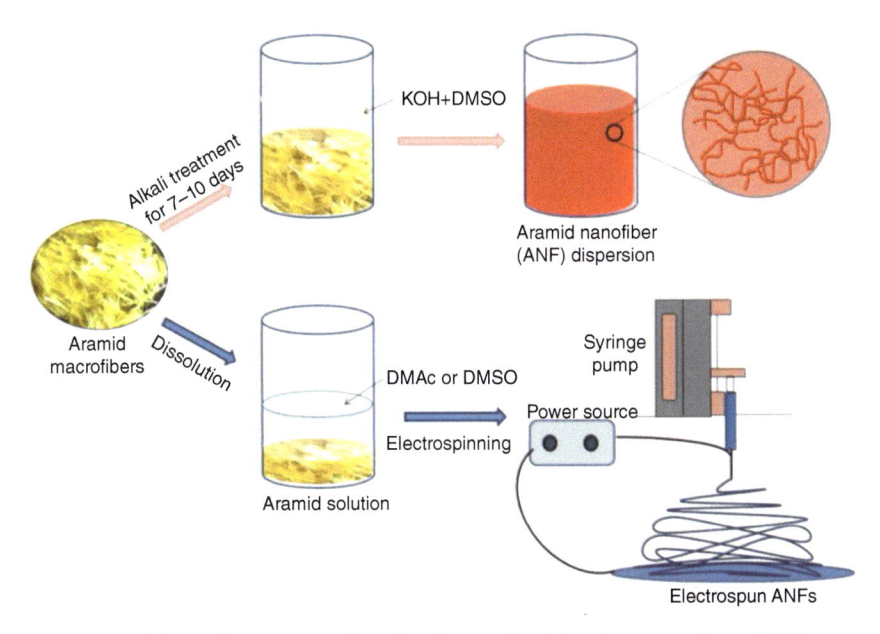

Figure 8.11 Schematic representation of various methods for the preparation of ANFs.

remaining constant. Structural perfections also occur that is indicated by density increase.

8.5 Influence of Structure on Properties

The important properties of aramids, particularly tensile properties, are associated with the fiber structure at molecular level. As mentioned earlier, when the solution is subjected to shear or elongational flow, the liquid crystalline domains tend to align in the direction of shear or flow and achieve a high molecular orientation. Fiber formation from a nematic lyotropic solution is best achieved at a critical concentration showing minimum viscosity and at a temperature close to its anisotropic transition temperature. These conditions are important as they facilitate solution orientation prior to spinning.

8.5.1 Fiber Structure

Different types of Kevlar fibers such as Kevlar 119, Kevlar 29, Kevlar 49, and Kevlar 149 show slight systematic variations in structure. PPTA is essentially a highly crystalline polymer. Morphologically, Kevlar 29 has a single-phase crystalline structure with imperfections in the lattice and a crystallinity of about 68%. The lattice arrangement of pseudo-orthorhombic crystals of PPTA (Figure 8.12a) has been given by Northolt and Aartsen [43]. Unit cell dimensions determined by Northolt are $a = 7.85$, $b = 5.15$, and $c = 12.8$. Panar and coworkers [44] proposed a paracrystalline distorted crystal structure. Intermolecular hydrogen bonds between the C=O and the N–H groups of the amide unit lead

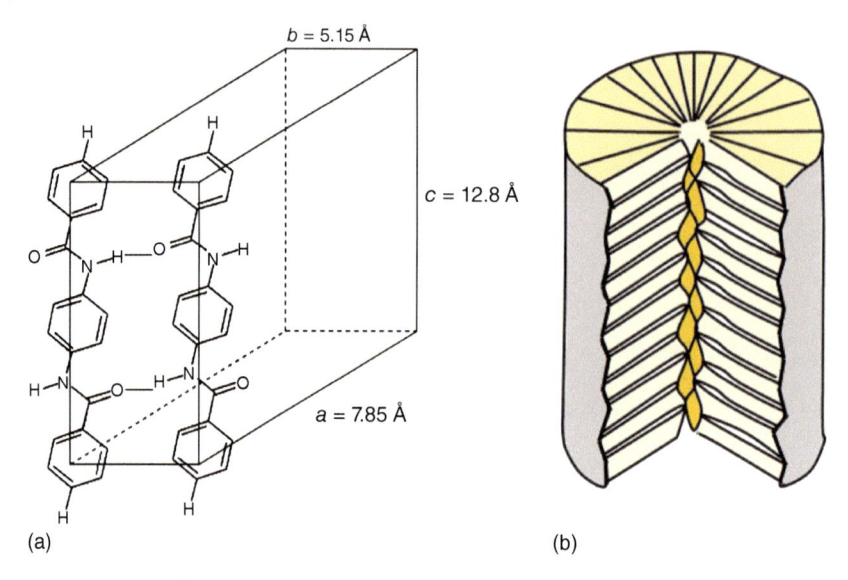

Figure 8.12 (a) Pleated sheet structure in Kevlar 49 fibers. (b) Unit cell of PPTA crystal.

to the formation of sheets linked with hydrogen bonds within the b–c plane. Using polarized optical microscopy, it has been shown that these imperfect crystals pack to form pleated sheets (Figure 8.12b) and the pleats are in a direction perpendicular to the fiber axis. Transmission electron microscopy and micro-X-ray diffraction measurements have shown that pleated sheets are formed in the plane of hydrogen bond and are arranged radially, forming a fibrillar structure.

The highly oriented polymer chains in the direction of fiber axis impart the necessary high longitudinal modulus of elasticity to the aramid fiber. The presence of aromatic rings and the conjugation of the electrons give the aramids their chemical stability and stiffness toward mechanical strain. The transverse hydrogen bonds and the covalent bonds in the direction of fiber axis lead to a unique mechanical property, anisotropy. Hydrogen bonds are important for the stability of the amide groups; however, they are much weaker than the existing covalent bonds along the fiber direction. Also, there is absence of any bonding between stacked sheets of polymer or pleats. The differential nature of bonding in each direction results in a high longitudinal strength and a low transverse strength, which is an important property for applications such as ballistics. The anisotropy of mechanical properties suggests that such materials should be used meticulously in engineering design and product development.

A skin-core morphology is also very common due to the processing of fibers as mentioned earlier. According to Li et al. [45], the high strength of PPTA fibers may be attributed to their fibrillar structure, the intertwining arrangement of the fibrils, and the presence of a highly oriented skin layer. On the other hand, appearance of defects that may be due to disorientation, at the ends of molecular crystallites in the core of the fiber, may lead to overall mechanical vulnerability.

Emergence of different morphological patterns from liquid crystalline solutions, due to various spinning processes, has been shown by Dayal et al. [46]. The liquid crystal polymer system in the isotropic state undergoes slow evaporation and exhibits a rough surface with no internal texture at the core. In the isotropic metastable region where the solvent evaporation is at a fast rate, the fiber develops a smooth skin within which isolated domains of microfibrillar structures are formed. These types of microfibrils are normally seen during the spinning process. In the unstable nematic region, the spinodal decomposition is dominant, generating the broken spiral or concentric ring patterns that proliferate toward the core of the fiber. Thus, the evolved morphology depends strongly on the initial thermodynamic states governed by the phase diagram, kinetics of phase separation, and the rate of solvent evaporation.

8.5.2 Fiber Properties

Aramid fibers possess medium to ultra-high tensile strength, medium to low extension, and moderate to ultra-high modulus. The densities for these crystalline fibers range from $1.35\,g/cm^3$ in the case of low oriented m-DPI (e.g. Nomex) to $1.45\,g/cm^3$ for fibers from PPTA (e.g. Kevlar). Meta-oriented phenylene rings are the major constituents of heat- and flame-resistant aramid fibers, whereas para-oriented phenylene rings are mainly present in ultra-high strength high modulus aramid fibers.

8.5.2.1 Chemical Properties

The PPTA fibers are resistant to many organic solvents and salts, but there can be a considerable loss of strength on using strong acids. Due to high crystallinity, aramid fibers have high glass transition temperature (T_g) and therefore are difficult to dye. By using recommended cationic dyes [47, 48] and pyridine for solvent dyeing, deep shades and excellent color fastness have been reported for Nomex and polyisophthalamides of both 4,4'-diaminodiphenylmethane and 2,2-bis(4-aminophenyl)propane.

The aromatic nature (i.e. electron conjugation) of p-aramids is responsible for oxidative reactions in the presence of ultraviolet (UV) light. This leads to change in color and loss of some mechanical strength. Kevlar 49 has a major drawback in that it strongly absorbs radiation in UV region around 250 nm and a low broad absorption is observed around 330 nm. As a consequence these fibers cannot be directly exposed to sunlight and should not be used for applications where there is presence of UV light. Light from the sun has wavelength in the range of 300–400 nm, and Kevlar, after absorption of this radiation in air, shows severe degradation of its mechanical properties.

Further, the amide linkages in aramids render these fibers hydrophilic, and they absorb moisture reversibly. Water molecules get absorbed by the amide linkages present at the chain ends, intramicrofibrillar spaces, crystal defects, and microvoids, which tend to reduce the mechanical properties [49].

Table 8.2 Properties of some aramids.

Fiber	Characteristics	Crystalline modulus, GPa	Tensile modulus, GPa (gpd)	Tensile strength, GPa (gpd)	Extension to break (%)	Moisture regain (%)
Nomex			17 (140)	0.6 (5)	22	4.5
Fiber B			128 (1000)	— (22)	5	—
Kevlar 29	Regular	153	70 (550)	2.9 (23)	4	5–7
Kevlar 49	High modulus	156	135 (950)	2.9 (23)	2.8	3–4
Kevlar 100	Colored	156	60 (475)	— (23)	3.9	—
Kevlar 119	High durability	156	55 (430)	3.1 (24)	4.4	5–7
Kevlar 129	High strength	156	99 (780)	3.4 (26.5)	3.3	4–6
Kevlar 149	Ultrahigh modulus	156	143 (1100)	2.3 (18)	1.5	1–1.2
Twaron	Regular	156	79 (622)	3	3.3	7
Twaron HM	High modulus	156	123 (960)	(22.1)	2	3.5
Technora	Regular	91	70 (570)	3.3 (27)	4.3	3.5–5
Trevar V106			77 (657)	—	3.7	—
Ekonol			136 (1100)	3.8 (31)	2.6	—
Vectran			91 (700)	3.2 (25)	—	<0.1
Armos			140–145	4.5	3	—

Source: With prior permission from Indian Journal of Fibre & Textile Research.

8.5.2.2 Mechanical Properties

Mechanical or tensile properties of the aramid fibers are controlled by the fiber structure as discussed earlier. The fiber modulus is governed by overall molecular orientation, crystallinity, formation of H-bonds between the polar amide groups on adjacent chain, and highly symmetrical internal fiber structure. Orientation of polymer chains promotes efficient load sharing between molecular chains. The properties of various aramids are summarized in Table 8.2.

Young et al. [50] have shown that fibers produced by different processing conditions develop different molecular structures, thereby imparting different mechanical properties. According to Dobb and Johnson [51], high molecular orientation, low skin-to-core ratio (more chains per unit cross-sectional area), and large crystallite size promote the development of high modulus in Kevlar. Fiber strength is critically affected by morphological defects, such as microvoids and the presence of cracks in the high modulus variant Kevlar 149. Typical stress–strain curve of Kevlar aramid yarn as compared with other high-performance fibers is shown in Figure 8.13. The figure shows that Kevlar yarn has a break tenacity of 18–26 gpd, which is twice as that of nylon or E-glass fibers and more than five times that of steel wire. Kevlar exhibits an unusually high initial modulus of 475–1100 gpd, which is two times that of steel wire and fiberglass and elongation of only 1.5–4.4%. As a spun fiber, Kevlar 29 has a modulus of 62 GPa. Crystalline orientation can be increased by heat treatment under tension as described in Section 8.4.4. The resulting Kevlar 29 fiber has a modulus of 131 GPa. Kevlar fibers, being thermally stable at high temperature,

Figure 8.13 Stress–strain behavior of Kevlar compared with other high-performance fibers.

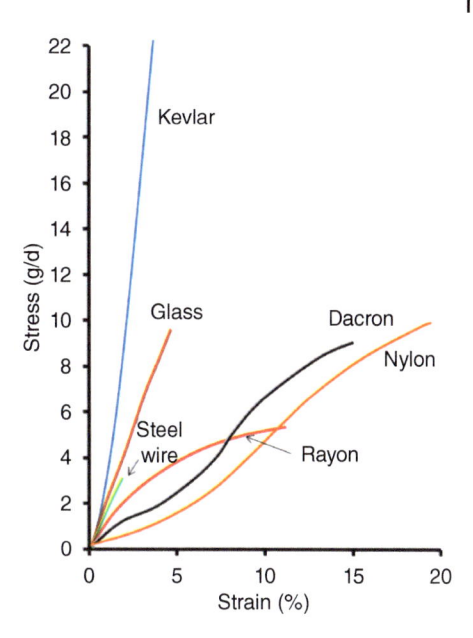

retain their strength and modulus even at temperatures as high as 300 °C. Kevlar shows creep at low load due to high crystallinity, but under high strain it significantly creeps.

8.5.2.3 Thermal Properties

There is no melting point of aramid fibers; however decomposition occurs at temperature as high as 550 °C. Aramids very easily take up moisture, reversibly. The differential scanning calorimetry studies of PPTA reveal moisture evolution at about 100 °C and a second-order transition at 360 °C, followed by melting and decomposition at 500–575 °C in nitrogen and until about 362 °C in air. Glass transition temperature (T_g) ranges from 250 °C to >400 °C. Some aramid fibers such as Nomex show shrinkage in a high heat source or a flame. Aramid fibers typically burn only with difficulty because of high limiting oxygen index (LOI) values. Burning produces a thick char covering the surface and acts as a thermal barrier. Pyrolysis of PPTA fiber produces p-phenylenediamine, aniline, benzonitrile, etc., indicating a homolytic cleavage of amide bond.

Such high thermal stability of aramids can be attributed to higher bond dissociation energies of C—C and C—N bonds in the main polymer chain in aromatic compounds than those in aliphatic ones. Excellent thermostability of poly(p-phenylene benzobisthiazole) is due to its fully aromatic character and highly rigid molecule. Introduction of flexible groups such as –O– into main polymer chain (e.g. Technora by Tenjin) leads to more flexible chains, thereby lowering the thermal stability.

Aramids show excellent retention of mechanical properties at elevated temperatures. This is due to the presence of C–N linkage of the amide group in PPTA and the electron conjugation between the amide groups and the aromatic ring in p-aramids. This results in the yellow color of these fibers and increased chain rigidity. At 300 °C Kevlar retains about 50% of its strength of that at room temperature, while 70% of the modulus remains of this level. Aramids have a slight negative longitudinal coefficient of thermal expansion of about -2 to -4×10^{-6}/K and a positive transverse expansion of 60×10^{-6}/K. Its high crystallinity results in negligible shrinkage both at high temperature in air (<0.1% at 177 °C) and in hot water (<0.1% at 100 °C). They also have a low thermal conductivity that varies by about an order of magnitude in the longitudinal vs. transverse direction.

8.6 Applications

Due to their characteristic properties such as high strength, low density, compressive strength, low flammability, no melting point (direct degradation above 500 °C), and good resistance to impact, abrasion, and chemical and thermal degradation, aramids show a wide range of applications in the field of reinforcement as well as apparel-related applications [52]. Various applications of aramids are summarized in Table 8.3.

However, due to the difference in the properties of *m*-aramid and *p*-aramid, they find specific applications as per their properties. The *m*-aramid (Nomex) has excellent thermal resistance and good textile properties but inferior mechanical properties for high-performance fiber. Thus, it is generally used for protective clothing, hot gas filters, industrial-coated fabrics, felt scrims, and reinforcement of rubberized belts and hoses. The *p*-aramid (Kevlar), on the other hand, shows unique combinations of toughness, extra high tenacity and modulus, and exceptional thermal stability. Thus, *p*-aramid covers diverse end uses in areas such as industrial, aerospace, military, and civilian applications for various purposes such as cut, heat and bullet/fragment resistant (ballistic) apparel, hard armors, brake and transmission friction parts, reinforced tires and rubber goods, ropes and cables, various forms of composites, industrial gloves, circuit board reinforcements, etc. Various other copolymers containing combination of meta- and para-substituted monomers have also been developed for optimum properties. Various applications of aramids are discussed below in detail.

8.6.1 Composites with Soft Materials

Applications of high-performance composite entail reinforcement by fibrous materials with superior mechanical properties. Especially, PPTA fibers, i.e. Kevlar, are well suited for high-performance composite applications, as they possess a high tensile strength and modulus with a high thermal resistance and chemical inertness, and moreover, they exhibit low electrical conductivity compared with metallic or carbon fibers.

Since the advent of Kevlar in 1972, the availability of these aramid fibers has advanced in materials science, particularly in the areas of fiber-reinforced composites. Kevlar fibers have been readily composited with materials like epoxy matrix [53], nylon [54], carbon fibers, and glass fibers for improved mechanical properties and impact resistance.

Table 8.3 Various applications of aramids.

Contributing properties	Application area	Typical end uses	Aramids
Reinforcement applications			
High specific strength and stiffness	Composites with soft materials	Tyres, automotive products (belt, hose), conveyor belts	p-Aramid
Good adhesion to rubber, axial compressive strain below ~0.8%	Advanced composites	Air craft structural components, wound pressure vessel	
High fatigue resistance, weight reduction, and impact damage tolerance	Ropes and cables	Telecommunication cables, civil engineering construction	
No creep and outstanding dimensional stability, nonrusting	Short fiber application	Friction products (brakes and clutches), gaskets and thixotropes and additives	Kevlar
—	Communication	Reinforcement of delicate glass fiber for optical fiber	
—	Other leisure industry, transport industry	Sailing boat hulls, tennis rackets, light containers, air ducting in aircraft industry, moisture wicking fiber	
Apparel-related applications			
High specific strength and stiffness, excellent thermal resistance, good textile properties	Protective application	For workers, furnishing, fire proof doors, protective gloves	*m*-aramid (Nomex)
Excellent cut and puncture resistance	Ballistic protection	Aramid sewing threads, soft and hard body armor	
Dynamic energy absorption characteristics	Perm-selective use	Hollow fiber permeation separation membranes	
—	Electrical application	Motors and transformers	*m*-Aramid (Nomex)
High volume resistivities and high dielectric strengths			

8.6.2 Advanced Composites

Composites reinforced with macroscopic aramid fibers are significantly useful in areas where their stiffness-to-weight and strength-to-weight ratios render them more desirable than conventional materials such as steel. Structural components of aircrafts have been imparted with high properties using such types of engineered materials. High impact resistance of these advanced composites ensures effective endurance of stresses and strains incurred in flight, which remain structurally intact and withstand catastrophic impact. On the contrary, carbon fibers, which exhibit good stiffness and compressive strength, are unsuitable as they undergo brittle fracture because of their rigid coplanar ring structure.

Recently, composites of aramid-boehmite have been prepared via sol–gel technique, which are transparent and yellow in color. Due to the presence of chemical bonding between the inorganic and organic phases, there is a significant enhancement in the mechanical strength and a thermal stability up to 450 °C. Such composites can be suitable as matrix for advanced applications [55].

As mentioned earlier, nanoscale reinforcement in composite materials is desired as it provides greater interfacial area leading to excellent stress transfer amid the matrix and filler [56]. Due to their higher surface-to-volume ratio, ANFs offer better interaction between fibers and the matrix and hence improved reinforcement as compared with conventional fibers. Guan et al. [57] prepared PVA/ANF nanocomposite films by solution casting technique. The presence of 5 wt% ANF-3 in PVA matrix resulted in significant enhancement of strength and toughness due to formation of multiple hydrogen bonds between the two components. Lin et al. [39] integrated ANFs in epoxy resin resulting in nanocomposites with enhanced elastic modulus, fracture toughness, and strength comparable to that of CNT-reinforced composites.

8.6.3 Ultrafiltration and Hemodialysis

ANFs are insoluble in water but impart hydrophilicity as they possess high susceptibility to moisture. Because of this property, Nie et al. [40] have shown that ANF-modified polysulfone membranes can suitably be used for ultrafiltration of water, separation of proteins, and adsorption of small molecules like creatinine toxins during dialysis. ANF-modified polymeric membranes show improved antifouling properties. ANFs have also been shown to improve the filtration efficiency of glass fiber filter media used for separation of water from water-in-oil emulsion [41].

8.6.4 Ropes and Cables

Ropes and cables made of aramid fibers are employed in a wide range of applications such as crane pendants, boom suspension ropes in mining machine, deep sea mooring, offshore and marine applications, heavy lifting, bridge stay cables, guyed towers, and antennas. Properties such as high specific strength and specific modulus, lack of creep, and outstanding dimensional stability render aramid fibers used as ropes and cables that are stronger and lighter than steel or other materials. Unlike iron, cables prepared from Kevlar 29 do not rust and can be used in oil rigs to probe large depths because in air, the specific strength of Kevlar is 7 times that of steel, while in sea water, the strength increases to more than 20 times that of steel.

8.6.5 Industrial Protective Apparel

In industries, superior materials are required that facilitate protection against flame and fire [58], electric arc flash, cut hazards, chemicals, dry particles of lead and asbestos, and hazardous aerosols. The blends of Nomex and Kevlar ensure insulating protection by the material due to very low shrinkage of Kevlar in flames. The excellent mechanical and chemical properties of Kevlar result in protective clothing being cut, aberration, and chemical resistance with comfortable fit.

Breathable, waterproof, and thermally resistant materials have been prepared by Park et al. [59] by electrospinning of *m*-aramid in DMAc/LiCl solution. As compared with commercial polytetrafluoroethylene membranes, *m*-aramid nanofibrous membrane showed better water vapor permeability with good thermal (stable up to 450 °C) and mechanical resistance. Such materials can potentially be used for breathable, waterproof, and thermal-resistant clothing.

8.6.6 Ballistics

Ballistic protection is a category of protective clothing that function to protect individual from the bullets and metal fragments of handheld weapons and explosives [60, 61]. The basic requirements of body armor fit for such application should be that it must be lightweight, allow sufficient movement, be cost effective, and give maximum protection against trauma/impact. Technically, it is important that the material used for such application should absorb energy in a given time frame, accumulate strain energy, and dissipate energy in the form of heat. Kevlar has replaced nylon in ballistic applications. This is because it has high T_g and thermal stability that can withstand high temperatures due to ballistic collision. Their extremely crystalline and ordered nature gives rise to high dynamic modulus, which assists in rapid wave propagation and fast response to longitudinal deformation. Kevlar provides high toughness that is effective in assimilation of longitudinal strain energy and transverse kinetic energy [62]. A lightweight composite from Kevlar reinforcing fiber for anti-ballistic application having polybenzoxazine (PBA)/urethane prepolymer (PU) alloys as a matrix was investigated by Rimdusit et al. [63]. The appropriate thickness of Kevlar reinforced 80/20 PBA/PU composite panel was 30 plies and 50 plies to resist the penetration from the ballistic impact equivalent to levels II-A and III-A of NIJ standard.

8.6.7 Permselective Use

Due to their outstanding chemical, thermal, mechanical, and hydrolytic stability and permselective properties, especially in reverse osmosis, aramids find application as membrane polymers. Hollow fibers of Nomex, being excellent semipermeable membranes, are used commercially for desalination of seawater or brackish water. Carrera-Figueiras and Aguilar-Vega [64] studied the gas barrier properties of aramids such as poly(benzophenone isophthalamide) (DBF/ISO) and poly(benzophenone-5-*tert*-butylisophthalamide) (DBF/TERT) homopolymers and their copolyamides with different DBF/TERT ratios. The copolymerization was observed to control their gas barrier properties. The presence of bulky groups increases fractional free volume and interchain spacing, thus increasing the diffusion coefficients and gas permeability. Similarly, resin-coated aramid papers (both Nomex and Kevlar) have also been used to study the helium permeability [65]. Nomex, being dense in nature, was observed to show lesser permeability, while Kevlar, being porous in nature, showed higher permeability. Due to their high abrasion resistance and good dimensional stability, *m*-aramids are also used for filter bags in baghouses for the purpose of air cleaning [66, 67]. These *m*-aramid bags remain dimensionally stable at high temperature (200 °C) with growth and shrinkage of approximately 1% only.

8.6.8 Electrical Application

Aramids possess high Young's modulus and excellent heat resistance due to which they are used for electrical insulation in generators, motors, transformers, and other electrical equipments. Their characteristic dielectric strength, thermal stability, cryogenic capability, chemical compatibility, moisture insensitivity, mechanical toughness, and radiation resistance make *m*-aramid papers suitable for such applications. These *m*-aramid papers are prepared using two physically distinct forms, floc and fibrids [68]. Floc is yarn cut to short lengths that provide mechanical strength. Fibrids are thin microscopic filmlike particles that provide dielectric strength and maintain the sheet integrity by binding the floc. A sheet is formed by combining these two forms and is kept under temperature and pressure. These components and the processing conditions of the sheets impart mechanical strength and impermeability, making them suitable for high temperature insulation. The electrical and mechanical properties of *m*-aramid papers are unaffected at temperatures up to 200 °C and show high sustainability for at least 10 years of continuous exposure at 220 °C. By varying the thickness, degree of densification, and floc-to-fibrid ratio, these electrical properties could be varied to a large extent. Two different commercially available poly-*p*-substituted phenylene terephthalamide-based aramid films with biaxial orientation, i.e. Micron and Aramica, have also been explored for their electrical properties [69]. Biaxial orientation imparts very high Young's modulus, and high heat resistance is owed to the *p*-aromatic polyamide linkages, making these films suitable for electrical insulation. The presence of aramids on electrical cables minimizes cable sagging and prevents the cables from coming into close contact with neighboring electrical cables.

8.6.9 Communication

p-Aramids, due to their high modulus and strength, low density, and resistance to creep, find important application in communication cables. They provide lightweight flexible support structures for fiber-optic cables and ductile power conductors from excessive loading or axial strain. Untwisted aramid yarn that is present along the length of the cable provides very high modulus to resist stretching. Twisted aramid yarns, on the other hand, inserted as a ripcord, provide maximum strength for tearing the protective sheathing when installing or repairing the cable. Mower [70] reported low modulus aramid fibers with braided construction to be more tolerant of bending fatigue in communication cables. It was observed that bending-induced damage increased and residual strength decreased with increase in the modulus of aramid fiber.

8.7 Future Trends

Due to their characteristic properties such as low density, high strength, compressive strength, low flammability, no melting point (direct degradation before their melting), and good resistance to impact, abrasion, and chemical and thermal degradation, aramids have highly been explored for high-performance technical textiles and composites for aircraft, aerospace, automobile, marine, and other industries. *p*-Aramid fibers possess high tensile strength and modulus, low breakage elongation, and good resistance to chemicals but have poor compression strength and lack in fatigue strength. So, *p*-aramids are more suitable for various high-performance applications. *m*-Aramid fibers, on the other hand, show excellent thermal resistance properties and thus more suitable for electrical and flame-retardant applications. Aramid fibers also can be tailor-made to use for reinforcement of glass fibers for fiber optics and telecommunication, permselective application (hollow fiber), asbestos replacement in gaskets (Kevlar pulp), and various other new applications that are continuously being explored.

References

1 Jassal, M. and Ghosh, S. (2002). Aramid fibres: an overview. *Indian Journal of Fibre & Textile Research* 27: 290–306.
2 Ward, I.M. (1987). *Developments in Oriented Polymers-2*. London and New York: Elsevier Applied Science.
3 Lewin, M. and Preston, J. (eds.) (1985). *Handbook of Fiber Science and Technology Vol III: High technology fibers Part B*. New York and Basel: Marcel Dekker, Inc.
4 Jassal, M. (2000). Aramid fibers. In: *Man made Fibers-II: Solution Spun Fibers*, 1e (eds. A.K. Agrawal and M. Jassal). India: Nodal Centre for Upgradation of Textile Education.
5 Lewin, M. (2007). *Handbook of Fiber Chemistry*. Boca Raton, FL: Taylor & Francis.
6 Hearle, J.W.S. (2001). *High Performance Fibres*. Cambridge, England: Woodhead Publishing Ltd.
7 Salem, D.R. (2001). *Structure Formation in Polymeric Fibers*. Munich: Hanser Verlag.

8 Akato, K. and Bhat, G. (2017). High performance fibers from aramid polymers. In: *Structure and Properties of High-Performance Fibers* (ed. G. Bhat), 245–266. Cambridge, England: Woodhead Publishing.

9 Deopura, B.L. and Padaki, N.V. (2015). *Synthetic Textile Fibres: Polyamide, Polyester and Aramid Fibres in Textiles and Fashion*, 1e (ed. R. Sinclair), 97–114. Cambridge, England: Woodhead Publishing.

10 Li, G. (2017). High-performance rigid-rod polymer fibers. In: *Structure and Properties of High-Performance Fibers* (ed. G. Bhat), 141, 166. Oxford: Woodhead Publishing.

11 Mittal, V. (2015). Advances in synthesis and properties of engineering polymers. In: *Manufacturing of Nanocomposites with Engineering Plastics*, 1–13. Cambridge, England: Woodhead Publishing.

12 Wypych, G. (2016). PPTA poly(p-phenylene terephthalamide). In: *Handbook of Polymers*, 2e (ed. G. Wypych), 542–545. Toronto, Canada: ChemTec Publishing.

13 Silver, F.M. and Dobinson, F. (1978). Aromatic polyamides. II. Synthesis, wet-spinning, and fiber thermal characterization of poly (4,4′-oxanilideterephthalamide) and copolymers with 4,4′-oxydianiline or isophthalic acid. *Journal of Polymer Science Part A* 16 (9): 2141–2149.

14 Afshari, M., Sikkema, D.J., Lee, K., and Bogle, M. (2008). High performance fibers based on rigid and flexible polymers. *Polymer Reviews* 48 (2): 230–274.

15 Zulfiqar, S., Lieberwirth, I., and Sarwar, M.I. (2008). Soluble aramid containing ether linkages: synthesis, static and dynamic light scattering studies. *Chemical Physics* 344 (1): 202–208.

16 Spiliopoulos, I.K., Mikroyannidis, J.A., and Tsivgoulis, G.M. (1998). Rigid-rod polyamides and polyimides derived from 4,3-diamino-2,6-diphenyl or di(4-biphenylyl)-p-terphenyl and 4-Amino-4-carboxy-2,6-diphenyl-p-terphenyl. *Macromolecules* 31 (2): 522–529.

17 Liaw, D.J., Liaw, B.Y., and Tseng, J.M. (1999). Synthesis and characterization of novel poly (amide-imide)s containing hexafluoroisopropylidene linkage. *Journal of Polymer Science Part A* 37 (14): 2629–2635.

18 Shi, H., Zhao, Y., Zhang, X. et al. (2004). Packing mode and conformational transition of alkyl side chains in N-alkylated poly (p-benzamide) comb-like polymer. *Polymer* 45 (18): 6299–6307.

19 Abe, A., Saito, Y., Imaizumi, M. et al. (2005). Surface derivatization of poly (p-phenylene terephthalamide) fiber designed for novel separation and extraction media. *Journal of Separation Science* 28 (17): 2413–2418.

20 More, A.S., Pasale, S.K., and Wadgaonkar, P.P. (2010). Synthesis and characterization of polyamides containing pendant pentadecyl chains. *European Polymer Journal* 46 (3): 557–567.

21 Spiliopoulos, I.K. and Mikroyannidis, J.A. (1996). Soluble, rigid-rod polyamide, polyimides, and polyazomethine with phenyl pendent groups derived from 4,4-diamino-3,5,3,5-tetraphenyl-p-terphenyl. *Macromolecules* 29 (16): 5313–5319.

22 Liu, Y.L. and Tsai, S.H. (2002). Synthesis and properties of new organosoluble aromatic polyamides with cyclic bulky groups containing phosphorus. *Polymer* 43 (21): 5757–5762.

23 Zhou, S., Zhang, M., Wang, R. et al. (2017). Synthesis and characterization of new aramids based on *o*-(*m*-triphenyl)-terephthaloyl chloride and *m*-(*m*-triphenyl)-isophthaloyl chloride. *Polymer* 109: 49–57.

24 García, J.M., García, F.C., Serna, F., and José, L. (2010). High-performance aromatic polyamides. *Progress in Polymer Science* 35 (5): 623–686.

25 Silver, F.M. (1979). Aromatic polyamides. IV. A novel synthesis of sulfonated poly (para-phenyleneterephthalamide): polymerization of terephthalic acid and para-phenylenediamine in sulfur trioxide. *Journal of Polymer Science, Polymer Chemistry Edition* 17 (11): 3519–3533.

26 Silver, F.M. (1979). Aromatic polyamides. V. A general synthesis of polyamides from aromatic diacids and aromatic diamines in sulfur trioxide. *Journal of Polymer Science, Polymer Chemistry Edition* 17 (11): 3535–3542.

27 Mallakpour, S. and Kolahdoozan, M. (2008). Synthesis and properties of novel soluble aromatic polyamides derived from 5-(2-phthalimidyl-3-methyl butanoylamino) isophthalic acid and aromatic diamines. *Reactive and Functional Polymers* 68 (1): 91–96.

28 Shoji, Y., Mizoguchi, K., and Ueda, M. (2008). Synthesis of aramids by polycondensation of aromatic dicarboxylic acids with aromatic diamines containing ether linkages. *Polymer Journal* 40 (8): 680–681.

29 Silver, F.M., Dobinson, F., and Ridgway, J.S. (1978). Aromatic polyamides. III. Copoly (4, 4′-oxanilideterephthalamide-4, 4′-phenyleneterephthalamide): Synthesis, wet-spinning, fiber thermal characterization, and stability in sulfuric acid. *Journal of Polymer Science* 16 (9): 2151–2157.

30 Boivin, J. and Brisson, J. (2004). Synthesis of rod copolyaramids with short flexible spacers. *Journal of Polymer Science Part A* 42 (20): 5098–5112.

31 Springer, H., Abu Obaid, A., Prabawa, A., and Hinrichsen, G. (1998). Influence of hydrolytic and chemical treatment on the mechanical properties of aramid and copolyaramid fibers. *Textile Research Journal* 68 (8): 588–594.

32 Chae, H.G. and Kumar, S. (2006). Rigid-rod polymeric fibers. *Journal of Applied Polymer Science* 100 (1): 791–802.

33 Yang, M., Cao, K., Lang Sui, Y.Q. et al. (2011). Dispersions of aramid nanofibers: a new nanoscale building block. *ACS Nano* 5 (9): 6945.

34 Li, J., Fan, J., Liao, K. et al. (2016). Facile fabrication of a multifunctional aramid nanofiber-based composite paper. *RSC Advances* 6 (93): 90263–90272.

35 Yeager, M.P., Hoffman, C.M., Xia, Z., and Trexler, M.M. (2016). Method for the synthesis of para-aramid nanofibers. *Journal of Applied Polymer Science* 133 (42): 1–8.

36 Srinivasan, G. and Reneker, D.H. (1995). Structure and morphology of small diameter electrospun aramid fibers. *Polymer International* 36 (2): 195–201.

37 On, S.Y., Kim, M.S., and Kim, S.S. (2017). Effects of post-treatment of meta-aramid nanofiber mats on the adhesion strength of epoxy adhesive joints. *Composite Structures* 159: 636–645.

38 Ryu, S.Y., Chung, J.W., and Kwak, S.Y. (2017). Amphiphobic meta-aramid nanofiber mat with improved chemical stability and mechanical properties. *European Polymer Journal* 91: 111–120.

39 Lin, J., Bang, S.H., Malakooti, M.H., and Sodano, H.A. (2017). Isolation of aramid nanofibers for high strength and toughness polymer nanocomposites. *ACS Applied Materials & Interfaces* 9 (12): 11167–11175.

40 Nie, C., Yang, Y., Peng, Z. et al. (2017). Aramid nanofiber as an emerging nanofibrous modifier to enhance ultrafiltration and biological performances of polymeric membranes. *Journal of Membrane Science* 528: 251–263.

41 Shin, C. and Chase, G. (2006). Separation of water-in-oil emulsions using glass fiber media augmented with polymer nanofibers. *Journal of Dispersion Science and Technology* 27 (4): 517–522.

42 Jeon, K.S., Nirmala, R., Navamathavan, R. et al. (2014). The study of efficiency of Al_2O_3 drop coated electrospun meta-aramid nanofibers as separating membrane in lithium-ion secondary batteries. *Materials Letters* 132: 384–388.

43 Northolt, M.G. and van Aartsen, J.J. (1973). On the crystal and molecular structure of poly-(p-phenylene terephthalamide). *Journal of Polymer Science, Polymer Letters Edition* 11 (5): 333–337.

44 Panar, M., Avakian, P., Blume, R.C. et al. (1983). Morphology of poly(p-phenylene terephthalamide) fibers. *Journal of Polymer Science, Polymer Physics Edition* 21 (10): 1955–1969.

45 Li, S., McGhie, A., and Tang, S. (1993). Internal structure of Kevlar® fibres by atomic force microscopy. *Polymer* 34 (21): 4573–4575.

46 Dayal, P., Guenthner, A.J., and Kyu, T. (2007). Morphology development of main-chain liquid crystalline polymer fibers during solvent evaporation. *Journal of Polymer Science Part B* 45 (4): 429–435.

47 Silver, F.M. and Dobinson, F. (1977). Aromatic polyamides. I. Poly-p-xylyleneterephthalamide: synthesis, fiber spinning, and characterization. *Journal of Polymer Science* 15 (10): 2535–2538.

48 Higashi, F., Goto, M., and Kakinoki, H. (1980). Synthesis of polyamides by a new direct polycondensation reaction using triphenyl phosphite and lithium chloride. *Journal of Polymer Science* 18 (6): 1711–1717.

49 Yahaya, R., Sapuan, S., Jawaid, M., Leman, Z., Zainudin, E. (2015) Effect of moisture absorption on mechanical properties of natural fibre hybrid composite. *Proceedings of the 13th International Conference on Environment, Ecosystems and Development (EED '15)* (1–2 April 2015). Kuala Lumpur, Malaysia.

50 Young, R.J., Lu, D., Day, R.J. et al. (1992). Relationship between structure and mechanical properties for aramid fibres. *Journal of Materials Science* 27 (20): 5431–5440.

51 Dobb, M., Johnson, D., and Saville, B. (1977). Supramolecular structure of a high-modulus polyaromatic fiber (Kevlar 49). *Journal of Polymer Science* 15 (12): 2201–2211.

52 Hahn, C. (2000). Characteristics of p-aramid fibers in friction and sealing materials. *Journal of Industrial Textiles* 30 (2): 146–165.

53 Reis, P., Ferreira, J., Santos, P. et al. (2012). Impact response of Kevlar composites with filled epoxy matrix. *Composite Structures* 94 (12): 3520–3528.

54 Yu, Z., Ait-Kadi, A., and Brisson, J. (1991). Nylon/Kevlar composites. I: mechanical properties. *Polymer Engineering & Science* 31 (16): 1222–1227.

55 Ahmad, Z., Sarwar, M.I., Krug, H., and Schmidt, H. (1997). Preparation and properties of composites of Kevlar-Nomex copolymer and boehmite. *Die Angewandte Makromolekulare Chemie* 248: 139–151.

56 Wagner, H.D. (2007). Nanocomposites: paving the way to stronger materials. *Nature Nanotechnology* 2 (12): 742–744.

57 Guan, Y., Li, W., Zhang, Y. et al. (2017). Aramid nanofibers and poly(vinyl alcohol) nanocomposites for ideal combination of strength and toughness via hydrogen bonding interactions. *Composites Science and Technology* 144: 193–201.

58 Horrocks, A.R. (2016). Technical fibres for heat and flame protection. In: *Handbook of Technical Textiles*, 2e (eds. A.R. Horrocks and S.C. Anand), 237–270. Cambridge, UK: Woodhead Publishing.

59 Park, Y.S., Lee, J.W., Nam, Y.S., and Park, W.H. (2015). Breathable properties of m-Aramid nanofibrous membrane with high thermal resistance. *Journal of Applied Polymer Science* 132 (8): 1–6.

60 Rebouillat, S. (2016). ARAMIDS: 'Disruptive', open andcontinuous innovation. In: *Advanced Fibrous Composite Materials for Ballistic Protection*, 2e (ed. X. Chen), 11–70. Cambridge, England: Woodhead Publishing.

61 Chen, X. and Zhou, Y. (2016). Technical textiles for ballistic protection. In: *Handbook of Technical Textiles*, 2e (ed. X. Chen), 169–192. Cambridge, UK: Woodhead Publishing.

62 Bajaj, P. (1997). Ballistic protective clothing: an overview. *Indian Journal of Fibre & Textile Research* 22 (4): 274–291.

63 Rimdusit, S., Pathomsap, S., Kasemsiri, P. et al. (2011). Kevlar^{TM} fiber-reinforced polybenzoxazine alloys for ballistic impact application. *Engineering Journal* 15 (4): 23–40.

64 Carrera-Figueiras, C. and Aguilar-Vega, M. (2007). Gas permeability and selectivity of benzophenone aromatic isophthalic copolyamides. *Journal of Polymer Science Part B* 45 (15): 2083–2096.

65 Bubacz, M., Beyle, A., Hui, D., and Ibeh, C.C. (2008). Helium permeability of coated aramid papers. *Composites Part B Engineering* 39 (1): 50–56.

66 Kohli, A., Forsten, H. H., and Wyss, K. H. (2008) Bag filter comprising meta-aramid and acrylic fiber. US Patent 7,456,120, filed 13 Septembre 2006 and issued 25 November 2008.

67 Greatorex, A. T. (1998) Filter unit and process for manufacturing a filter unit. US Patent 5,753,001, filed 2 August 1995 and issued 19 May 1998.

68 Bhatia, A. (1995) Aramid papers with improved dimensional stability. *Proceedings: Electrical Electronics Insulation Conference and Electrical Manufacturing & Coil Winding Conference*, (18–21 September 1995). Rosemont, USA: IEEE.

69 Yasufuku, S. (1995). Application of aramid film to electrical and electronic uses in Japan. *IEEE Electrical Insulation Magazine* 11 (6): 27–33.

70 Mower, T.M. (2000). Sheave-bending and tensile fatigue of aramid-fiber strength members for communications cables. *International Journal of Fatigue* 22 (2): 121–135.

9

Conductive Fibers

Tung Pham and Thomas Bechtold

University Innsbruck, Research Institute of Textile Chemistry and Textile Physics, Hoechsterstrasse 73, Dornbirn 6850, Austria

9.1 Introduction

Future smart textiles will exhibit a high level of integration of electronic components; thus miniaturization and appropriate strategies for integration of electric functionalities are essential.

The realization of conductive textiles is the basic requirement for the integration of miniaturized (e.g. nanosized) sensors in a textile structure, e.g. humidity, temperature, pressure, gas sensors, etc., as well as for connection concepts of sensors in large-area textile sensor networks. Instead of separated manufacturing of electronic devices, which then are mounted on a textile product, the buildup of a device is integrated into the different steps of the textile production. Functional elements are formed during yarn formation, mounted through weaving or knitting, and placed through embroidery and printing. Standardized textile processed products will be replaced by combination of individually designed steps. Thus, in the near future, integrated electronic devices will be produced through an appropriate combination of materials and specialized textile techniques. Finding the balance between electrical conductivity, flexibility, and comfort of the textile is a challenge.

9.2 Production of Conductive Fibers: Principles and Technologies

9.2.1 Conductivity

Fibers and their wide variety can be used for many purposes from insulation to conducting applications. Early examples in textiles consisting of fibers with electrical conductivity could be found not only in antistatic and shielding but also heating elements.

The conductivity is defined by Ohm's law as follows:

$$U = R \times I$$

Handbook of Fibrous Materials, First Edition. Edited by Jinlian Hu, Bipin Kumar, and Jing Lu.
© 2020 Wiley-VCH Verlag GmbH & Co. KGaA. Published 2020 by Wiley-VCH Verlag GmbH & Co. KGaA.

9 Conductive Fibers

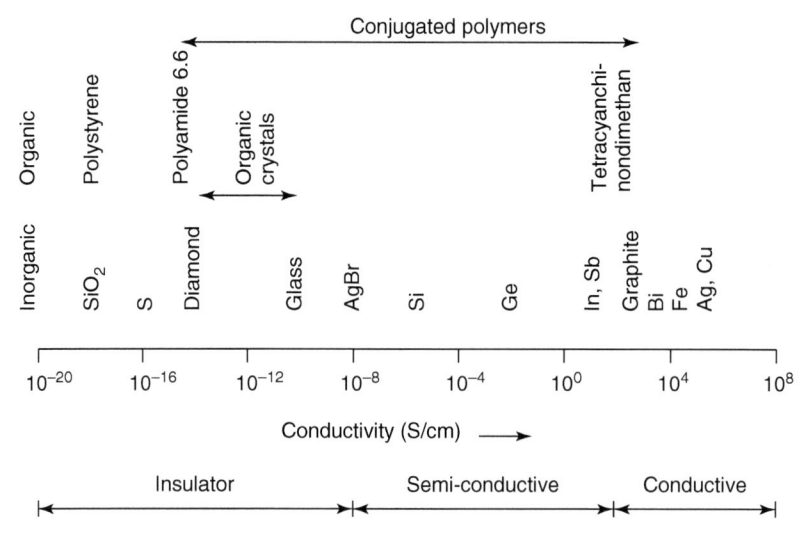

Figure 9.1 Conductivity range of typical material classes (according to [1]).

where I is the current through a resistor and U is the drop in potential across it.

The constant R is called the resistance and is proportional to the length l of the sample and inversely proportional to the sample cross section A:

$$R = \rho \frac{l}{A}$$

where ρ is defined as the bulk resistivity (Ω cm). Its inverse $\sigma = 1/\rho$ is the conductivity (S/cm).

Depending on the conductivity range, materials can be divided in three different classes, namely, insulator, semiconductive, and conductive materials (Figure 9.1) [1].

The typical values of resistivity and/or conductivity of some common materials are summarized in Table 9.1.

Depending on applications, different materials and approaches and materials are selected in order to fit the application-specific requirements. Examples on conductive fibers are given in Table 9.2. However, the values of the resistivity in different original publications are often given as linear resistance in Ω/cm, which indicates resistance per length unit.

9.2.2 Metal Fibers: Coating and Deposition

In very early stages of the development of conductive networks and/or connection in textile structures, insulated wires were fixed on a fabric with a very low degree of integration. While the problem of insulation could be handled, the electrical connection to the functional element required manual welding. The cables were stiff and did not exhibit sufficient textile character. Earlier works in producing metal yarns from metal filaments were dated back to the 1930s where the use of iron, nickel, chromium, or stainless metal wires is reported [7–9].

Table 9.1 Conductivity of different materials from conductive to insulation.

Material	Strength (cN/dtex)	Resistivity (Ω cm)	Conductivity (S/cm)	References
Ag	—	—	6.8×10^5	[2]
Cu	—	—	6.5×10^5	[2]
Stainless steel	—	—	1.8×10^4	[2]
Al	—	—	4.0×10^5	[2]
Graphite	—	—	5.0×10^2	[2]
PPy	—	—	1–10	[3]
PAn	—	—	1–5	[4]
Cellulose (Lyocell)	45–50	10^{14}–10^{16}	—	[5]
Cellulose (Viscose)	25–45	10^6–10^9	—	[1]
Cellulose (Cotton)	35–70	10^6–10^8	—	[1]
PA6,6	40–100	10^9–10^{12}	4×10^{-12}	[6]
PET	35–130	10^{11}–10^{14}	—	[1]
PP	22–55	$>10^{13}$	—	[1]
Carbon	200–300	10^{-3}	—	[1]
Glass	100–350	10^{12}–10^{15}	—	[1]

Today's approaches to integrating cables and wires during weaving or knitting already have reached technical scale. Full-scale prototypes with integration of different conductive materials such as wires and stainless steel yarn in knitted fabric were produced. A wide range of conductive metallic materials can be positioned on textile structures by embroidery techniques, which successfully could be used to manufacture conductive structures from stainless steel yarn, aluminum, or copper wires [10, 11].

Due to the nature of metals having lower elasticity, spinning techniques have been investigated to produce conductive polymer/metal wire composite yarns [12]. Wires of different types and diameters were obtained using core spinning and wrap spinning techniques to form a complex yarn. In order to introduce and/or to improve yarn elasticity, the concept of hybrid yarn was developed, where a conductive yarn (Bekinox BK50) winds around an elastic core yarn (Polyamide/Lycra) in a direct twisting device used in knitted fabrics as strain sensors [13]. Test results show that the hybrid yarns can effectively enhance the electromechanical properties of the knitted strain sensors without compromising processability and comfortability [13]. Another approach is to coat textile fibers with metals [14–16]. For example, Kalanyan et al. demonstrated that the deposition of wolfram on polyamide 6 fiber nonwoven mats using low temperature atomic layer deposition (ALD) leads to a conductivity of about 1000 S/cm [16]. The same technique was also applied to form platinum conductive layer via (methylcyclopentadienyl)trimethyl platinum and ozone on polyamide 6 nonwoven, having effective conductivities as high as ~5500–6000 S/cm [17]. Also metals like gold can be deposited as coating on yarns [14, 15].

Table 9.2 Examples on conductive fibers/yarns/fabrics with best achieved conductivity.

Composition	Fineness	Tensile strength/tenacity/modulus/elongation	Resistivity[a]	Conductivity (S/cm)	References
Metal based					
Stainless steel/polyester yarn	200 tex	5.97 cN 0.3 cN/tex — 57.6%	100 Ω/cm	—	[13]
Stainless steel/polyester yarn	400 tex	10.87 cN 0.27 cN/tex — 63.2%	50 Ω/cm	—	[13]
Nonwoven polyamide/tungsten deposition	—	—	—	1000	[16]
Nonwoven polyamide/platinum deposition	—	—	—	5700	[17]
Poly(p-phenylene terephthalamide)/Poly(styrene-butadiene-styrene) nanosilver coating	—	—	0.15 Ω/cm	—	[25]
Polyurethane/silver nanowire ink coating	—	—	7.3×10^{-5} Ω/cm	—	[27]
Stainless steel	130–820 tex	—	<75–10 Ω/m	—	[66]
polyamide/silver plated	4–125 tex	—	<2000–5 Ω/m	—	[65]
Ag-coated monofilament	2.2 tex	—	68 Ω/cm	—	[67]
Ag-coated multifilament	47 tex	—	1 Ω/cm	—	[67]
Ag-coated Cu/polyester	61 tex	—	0.0925 Ω/cm	—	[68]
Intrinsically conductive polymer based					
Cotton/poly(3-hexylthiophene) coating	—	—	5.9×10^7 Ω/cm	—	[49]
Polypyrrole nano fibers	4–5 μm	—	—	0.5	[69]

Polyaniline/gelatin blend	100–650 nm	6.56–10.49 MPa — 0.49–0.09%	0.021	[70]
Polyaniline	200–1000 nm	—	10^{-3}–10^{2}	[71]
Polyaniline/polyamide blend	150–500 nm	—	6.2×10^{-7}	[72]
Polyaniline/polyamide coat	800–2000 nm	—	0.4–2	[73]
Polyaniline/silica	1–5 μm	—	5×10^{-7}–1.07	[74]
Polyaniline/Poly(D,L-lactide)	0.94 μm	—	0.044	[75]
Poly-3-hexylthiophene/polyethylene blend	380–1200 nm	—	0.16–0.3	[76]
Cotton/polypyrrole coating	—	—	5.8	[77]
Carbon based				
Cotton/graphene coating		374 Ω/cm	—	[52]
Polyurethane/carbon black coating	—	—	140	[55]
Polypropylene/carbon black/graphite melt spinning	—	—	0.32	[59]
Polyester/polypropylene/MWCNT coating	—	—	0.52	[60]
Cotton/MWCNT coating	—	20 Ω/cm	—	[78]
Carbon/polyethylene	26 dtex	—	0.18–1.39	[79]
		24 cN/tex 126 cN/tex 105%		
Cellulose/carbon	—	— 8–14 cN/tex 300–700 cN/tex 4–12%	0.0024–0.18	[80]

a) The values of the resistivity in different original publications are often given as linear resistance in Ω/cm, which indicates resistance per length unit.

Figure 9.2 Laser scanning image of a cellulose knitted fabric coated with nanosized silver particles by reduction from silver salt $AgNO_3$ (a) 0.004 mol, (b) 0.005 mol, and (c) 0.009 mol.

Printing techniques such as screen printing and ink-jet printing can be used to form conductive layers from metals on textile fabrics with the distinct advantage of being able to build conductive areas independent of fabric structure and to position the conductive areas [18–22]. Techniques for making inks conductive include the use of silver flakes in polyurethane (PU) [19] or acrylic binder [22]. In Ref. [23], a three functional layer design was used to fabricate the durable conductive tracks consisting of an interface layer providing a smooth surface followed by a conductive silver layer and finally an encapsulation layer, which was screen-printed on top to provide upper-side protection and electrical insulation [23].

Also the in situ synthesis of nanosized silver particles to form conductive tracks was reported [24–26]. The improvement of the conductivity was achieved by a further sintering step at 200 °C for 30 minutes of the formed silver particles on the textile surfaces [24]. Additional elastic performance was achieved, along with electrical conductivity, by coating elastic rubber fibers with silver nanoparticles [25]. The conductive fibers exhibited outstanding stability against repeated external deformations with the help of the elastic rubber materials that act as elastic scaffolds and can be used to fabricate highly sensitive textile-based pressure sensor [25]. In Ref. [26], the polyamide 6 fabric was first treated with potassium permanganate to oxidize fabric surface and thus to increase number of assembled particles on the fabric surface. In another study, core–shell conductive fibers with high stretchability are developed by writing silver nanowires ink on pre-strained polyurethane fibers. The core–shell elastic fibers are twisted, forming flexible piezo-resistive fiber construction that shows desirable sensitivity to pressure and bending deformations [27].

Another approach to forming nanosized silver particles on fibers is the reduction from silver salt, e.g. $AgNO_3$. Using different $AgNO_3$ concentrations, different surface topology, and resistance could be achieved. Examples are given for 0.004, 0.005, and 0.009 mol $AgNO_3$ (Figure 9.2a–c).

By initiating the radical graft polymerization of glycidyl methacrylate monomer on plain woven cotton fabric using 10 kGy dose of electron beam irradiation, followed by the hydrazination of the epoxy ring to introduce reducing agents into the fiber structure, the electroless deposition of silver from silver salt solution could also be realized. The deposited silver nanoparticles make a conductive pathway through contact network in the cotton fabric [28].

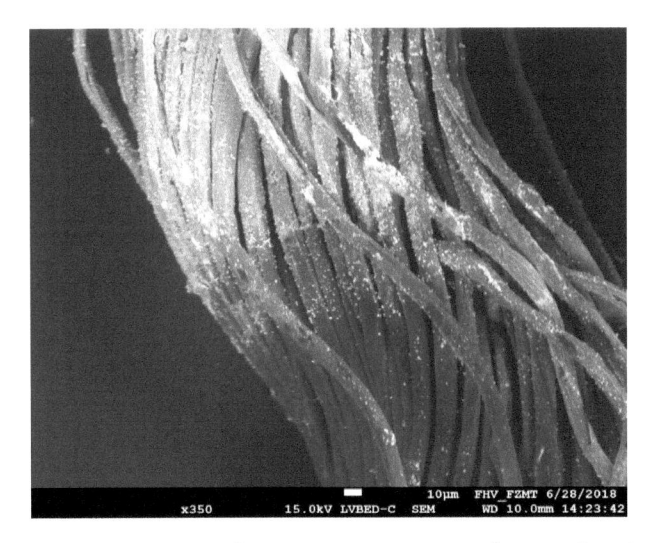

Figure 9.3 Scanning Electron Microscopy image of copper-deposited cellulose fiber by electroless copper deposition using nanosized silver seeds (bright areas: copper layers).

In our own work, further electroless deposition was initiated using nanosized silver seeds as catalysts. The techniques enable the localized formation of a thin copper layer on cellulose fibers (Figure 9.3).

Another approach was developed by Liu et al. to coat cotton fabric with copper through reduction using immobilized silver ions as catalyst [29].

9.2.3 Intrinsically Conductive Polymers and Coating

As described according to IUPAC Compendium of Chemical Terminology, intrinsically conducting polymer (ICP) composed of macromolecules having fully conjugated sequences of double bonds along the chains. The bulk electrical conductivity of an ICP is comparable to that of some metals and results from its macromolecules acquiring positive or negative charges through oxidation or reduction by an electron acceptor or electron donor (charge transfer agent), termed a dopant. Examples of ICPs are polyacetylene (PA), polypyrrole (PPy), polyaniline (PAn or PANI), polythiophene (PT or PTh), poly(p-phenylene vinylene) (PPV), poly(p-phenylene) (PPP), and polyfluorene (PF) (Figure 9.4) [30].

Unlike polymeric electrolytes, in which charge is transported by dissolved ions, charge in ICPs is transported along and between polymer molecules via generated charge carriers, e.g. holes and electrons [31]. The unique characteristic of conducting polymers is given by backbones of contiguous sp^2-hybridized carbon centers where the π-electrons delocalize over the whole polymer chain. The conductive characteristic of conducting polymers is transformed from insulating to conducting through a doping process. In the p-doped state (electron accepting), the main chain of the conducting polymer is oxidized with counteranion doping for keeping the electron neutrality of the whole molecule. There are holes in the main chains (lost electrons) that make the conducting polymer p-type conducting. In the n-doped state (electron donating), the main chain of the conducting

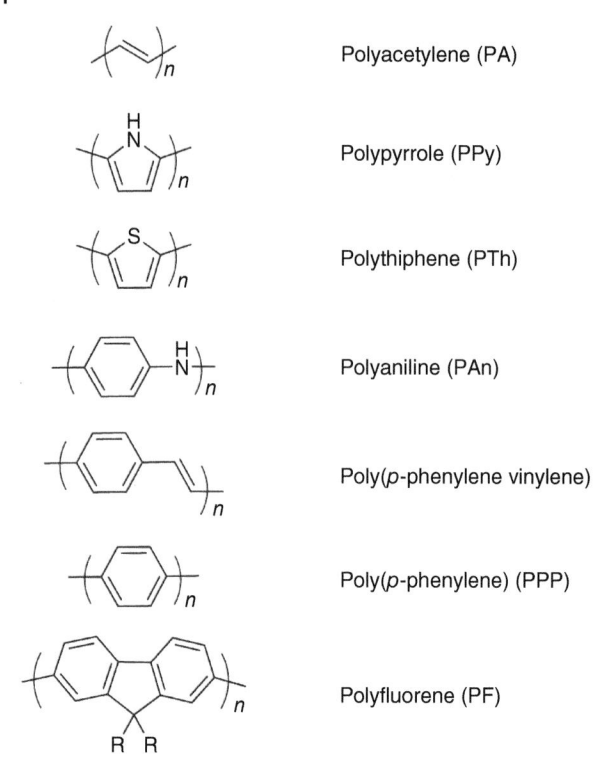

Figure 9.4 Main chain structures of several representative conjugated polymers (according to [30]).

polymer is reduced with counteraction doping for keeping the electron neutrality of the whole molecule. There are electrons in the main chains that make the conducting polymer n-type conductive [30]. In its oxidized form, the conductivity of PANI, for example, increases about 10 orders of magnitude.

Similar to the fiber coating approach with metals, synthetic fibers can also be functionalized through chemical oxidative deposition, in which conducting electroactive polymers such as PANI, PPy, and PT are used to coat textiles for improving tensile strength and thermal stability [32–37, 77]. PPy has high mechanical strength and is electroactive in organic and aqueous solutions. PT and its derivatives in p- or n-type forms can be used for application in field-effect transistors in flexible logic circuits. Furthermore, surface deposition of electroactive polymers increases conductivity of the fibers by 1 order of magnitude [38–40]. Cotton fabric coated with fibrillary PPy reached conductivity of 5.8 S/cm and can be used for electromagnetic interference shielding [77]. Application of PPy in electrically triggered color-changing fabric was realized by selective deposition of PPy on the cotton side of a polyester-covered cotton fabric, while the thermochromic ink was painted on the polyester side. The fabric showed remarkable resistive heating and color-changing properties under an applied current [34]. Also coating of three-dimensional fibrous scaffolds of polylactide

(PLA) with PPy in combination of chondroitin sulfate (CS) was investigated with respect to deliver electrical stimulation to cells in PLA scaffolds [41].

For application as biosensor, polypropylene (PP) fibers were coated with conductive PPy coating using iron(III) chloride as an oxidant, water as a solvent, and 5-sulfosalicylic acid as a dopant. The biological attachment of avidin onto the fiber surface was achieved through the inclusion of a carboxyl functional group via 3-thiopheneacetic acid in the monomer. The immobilized avidin was then successfully used to capture biotin [42].

In order to improve the adhesion between the coating layer and fiber surface, low temperature (oxygen and argon) plasma treatments were applied for polyester fabrics leading to the activation of the fiber surface and generation of a microscopically rough surface, prior to the deposition of PPy by in situ polymerization. The enhanced adhesion was confirmed by abrasion tests and subsequent surface resistivity measurements [43].

In another work, oxidative in situ polymerization of poly(3,4-ethylenedioxythiophene)–p-toluenesulfonic acid (PEDOT:PTSA) directly on a textile polyester fleece was realized. The finished textiles can be used as high-performance textile heating elements, e.g. in carpets, electric blankets, or automotive seat heaters [44]. A combined approach of coating of gold nanoparticles and deposition of thin layers of the conductive polymer poly(3,4-ethylenedioxithiophene) (PEDOT) was applied to increase the conductivity of plain cotton yarns. The application of the modified cotton yarns ranged from passive devices such as resistors to active devices, e.g. organic electrochemical transistors (OECTs) and organic field-effect transistors (OFETs) [45]. Oxidative chemical vapor deposition (OCVD) of PEDOT was also used to coat viscose fibers, leading to conductivity of 14.2 S/cm [46].

Using a direct coating method, a polyurethane-based formulation with poly(3,4-ethylenedioxythiophene)–polystyrene sulfonate (PEDOT:PSS) dispersion and ethylene glycol was applied to poly(ethylene terephthalate) (PET) substrates, and resistivity drop by a factor of 10 was registered [47]. With similar system, however, in poly(styrene-r-Bu acrylate) P(St-BA) aqueous solution, Wu et al. [48] created a conductive composite with outstanding stretchability. The elastic conductive composite is then formed on the PET nonwoven fabric substrate by dip coating, leading to conductive soft fabric structure.

Dip coating technique was also applied to coat poly(3-methylthiophene) (P3MT) polyester fabric [56] and poly(3-hexylthiophene) (P3HT) on cotton fabric. In the case of P3HT-coated cotton, a decrease by several orders of magnitude in both surface and volume resistivity was observed compared with untreated cotton [49].

Another approach in utilizing intrinsically conductive polymers in synthetic fibers was melt blending. In Ref. [50], a melt processable PANI complex was blended with polypropylene under different mixing conditions and melt-spun into fiber filaments under different draw ratios. The continuous conductive pathways (interpenetrating network [IPN]) of PANI depended strongly on two factors: the size of the initial dispersed conductive phase, which depended on the melt blending conditions, and the stress applied to orient this phase to a fibril-like morphology, which was controlled by the draw ratio of the fiber [50].

9.2.4 Carbon-Based Fibers (Carbon Black, Carbon Nanotubes, Graphene)

Graphene woven fabrics were synthesized through chemical vapor deposition (CVD) using Cu meshes consisting of wires with ~60 μm in diameter as substrates. The fabrication of the textiles involved growing graphene on the substrate, removing the Cu mesh wires, and subsequently collapsing the graphene to form double-layer microribbons. Such polymers could be also embedded in polydimethylsiloxane (PDMS) or PET films [51]. Conductive textiles could also be produced by immobilizing graphene via reduction from graphene oxide on cotton fabric by using a conventional dip and dry method. The electrical conductivity of the fabric is enhanced 3 orders of magnitude as the number of coating cycles was increased from 1 to 20. The surface conductivity of the resulting graphene depended on the reducing agent type and concentration [52]. Also infusion of pristine few-layer graphene (FLG) from natural bulk graphite onto fabrics using an interfacial trapping method led to higher electrical conductivities [53].

Conductive fibers can be realized by spinning carbon nanotubes (CNTs) containing polymers, for example, by coating techniques using functionalized multi-walled carbon nanotubes (MWCNTs). However the conductivity of these yarns and fibers is decreased by the drawing process during fiber fabrication [60, 78]. Another approach involved spray coating technique to coat electrospun polyurethane nanofiber webs with MWCNT [54].

Wet spinning technique has been investigated by different research groups to generate conductive fibers with carbon black (CB) and CNT. Seyedin et al. [55] reported on wet-spun PU fibers a range of conducting fillers including CB, single-walled carbon nanotubes (SWCNTs), and chemically converted graphene. They found that the electrical and mechanical properties of the conductive fibers were strongly dependent on the aspect ratio of the filler and the interaction between the filler and the elastomer. The high aspect ratio SWCNT filler resulted in fibers with the highest electrical properties and reinforcement, while fibers produced from the low aspect ratio CB had the highest stretchability. Furthermore, PU/SWCNT fibers presented the largest sensing range (up to 60% applied strain) and the most consistent and stable cyclic sensing behavior [55]. In another work [56], alkali hydrolyzed polyester fabric was treated with SWCNT, MWCNT, and CB in combination with nanosized TiO_2 and citric acid as a cross-linking agent through exhaustion method and post-curing. Alternatively, Jin et al. developed a new dissolving coating method to impregnate CB into the polyester fiber surface to enhance the fiber conductivity [57].

To produce sheath–core structured fibers, a CNT solution was used to form the core portion of the fibers, while the sheath was made from polymer such as polyvinyl alcohol (PVA) in a wet spinning process. It was found that stretching fibers in the coagulation bath was a significant step in the formation of a structured core. The drawn flexible fibers could be woven into a fabric for potential use as a pressure sensor [58].

Another approach to preparing conductive polypropylene composite is to mix CB and graphite nanoplatelets (GNP) into a polypropylene matrix during a melt mixing process [45]. Conductive fibers with a sheath–core structure were then

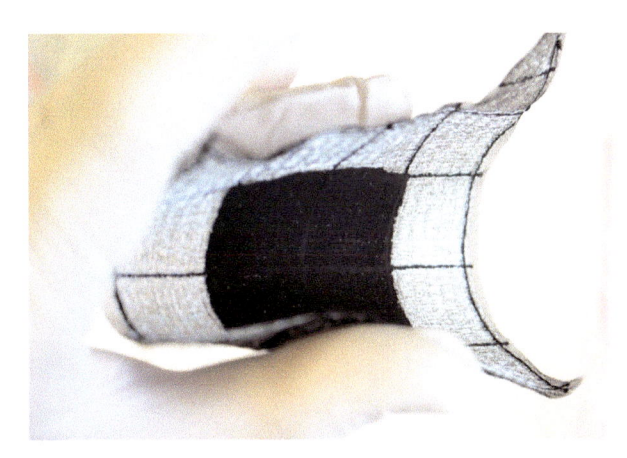

Figure 9.5 Coated textile electrode using active carbon black conductive paste.

produced on a bicomponent melt spinning line with the core materials from hybridized GNP/CB/PP composite and the sheath from polyamide [59]. Very similar concept was investigated by Lin et al. [60], however with the PP/MWCNT as conductive coating layer on PET yarns. The test results indicate that tensile strength of the conductive yarns increases with an increase in the coiling speed that contributes to a more orientated MWCNT arrangement in the yarn direction as well as a higher adhesion between PP/MWCNT and PET yarns [60]. Melt mixing technique was also applied for polycarbonate/MWCNT, subsequently melt-spun to monofilament fiber [61].

Another possibility to obtaining flexible conductive yarns is demonstrated by laser structuring [62], but these materials seem to be challenging regarding their cost efficiency for the use in large-area textile applications. Metal sputtering and CVD techniques are currently studied in basic research approaches; however at present the technology is still far from being on a cost level for commercialization.

Utilizing the conductive nature of active CB, conductive paste can be used to coat fabrics to form conductive areas and pattern in textiles (Figure 9.5) as alternatives to conductive polymers.

Another example is the use of carbon fibers as electrode in redox flow batteries, whereas the carbon fiber rowing is placed on a textile substrate using tailored fiber placement (technical embroidery) technique (Figure 9.6).

9.2.5 Combination of Different Techniques

Some research groups combined different concepts in producing conductive textile fibers. Hu et al. [63] applied the technique of dip coating for highly transparent polyethylene terephthalate mesh fabrics with CNT, followed by ink-jet printing with PEDOT:PSS to develop textile-based CNT/PEDOT:PSS composite conductors. These coated fabrics show a combination of conductivity transparency and flexibility suitable for use in flexible displays [63]. Similar approach was also developed in Ref. [64], where polyethylene terephthalate fabric was first dip-coated with reduced graphene oxide (RGO) sheets and

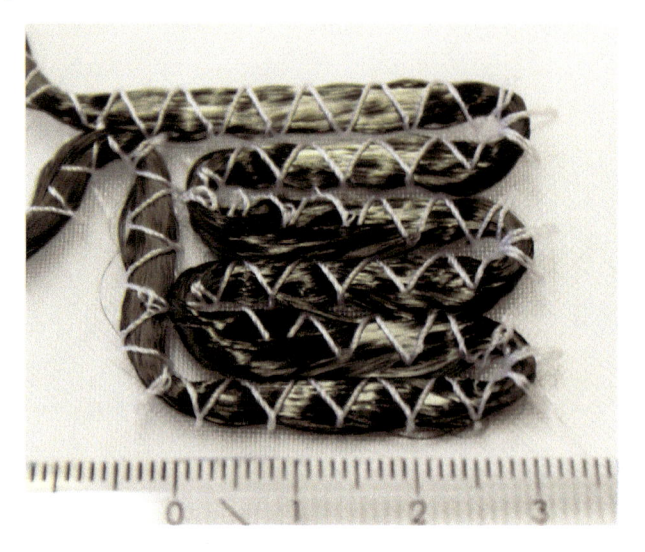

Figure 9.6 Embroidered carbon fibers as electrode in redox flow batteries.

then coated with PPy by in situ polymerization. The RGO–PPy coated samples exhibited much lower surface resistivity values than samples coated only with PPy and RGO, respectively [64].

Table 9.2 gives some examples on conductive fibers, yarns, or fabrics, with their typical material parameters such as fiber fineness, fiber strength, tenacity, Young's modulus, and best achieved electrical conductivity. Please note that the values and units are taken as published in the original publications.

9.3 Integration of Conductivity into Textile Structures

In the traditional concept of smart textiles, these were textile structures (garment, protective clothes, interior) that were able to carry electronic devices. The electronic devices were manufactured separately and integrated into the textile product during the procedures of assembly. Such concepts required complex manufacturing operations and often considerable amounts of manual work, e.g. for positioning and welding. The final products often could be defined as textile-coated electronic device [81]. The possible miniaturization of devices to the structural dimensions of textiles, e.g. the diameter of a thread, offers new strategies for integration of electronic functionality into textile products [82].

The integration of electronics, switches, actuators, displays, control elements, and panels has to occur on the level of the textile structure. Instead of an assembly of electronic components and a textile product as two components, which are merged at a late stage of production, the installation of an electronic device will be integrated into the textile production step.

The desired functionality is provided by the garment/textile while the electronics disappear and function as part of the textile structure. The integration of such *smart* textile sensors are a key component of a future suitable Internet of Everything.

Table 9.3 Stages of integration of conductivity in textile material and their characteristics.

Production stage	Product	Examples	Production volume required	Characteristics	Flexibility in design
Spinning	Yarn, ply	Conductive yarn	High	—	High
Weaving	Fabric	Fabric with electrical lines as weft	High	Rectangular crossing of threads	Medium
Knitting	Fabric, garment	Heating textiles using Ag-coated yarn	Medium	Loops in direction of production	—
Embroidery	Motifs	Sensor elements on fabric	Low	Free from	—
Printing	Print	Heating elements	Low	Free form	—
Garment production	Garment	Cloths for monitoring vital parameters in sports	Low	Flexible	—

As a result of a high level of integration, the borders between textile manufacturing and production of electronic devices disappear.

The textile processes have to be adapted with regard to their ability to process conductive materials and to form conductive structures; at the same time the performance of conductive elements and the production of a device have to be developed with consideration of the processing requirements of a certain textile technique. The overall concept of a textile product, both garment and technical product, thus will determine fiber material, processing techniques, and final performance, which at the same time also will limit the degree of freedom in choice of conductive material and concepts of integration (Table 9.3).

9.3.1 Fiber Material Selection/Yarn Production

Natural fibers such as cotton, flax, hemp, wool, and silk are staple fibers that exhibit limited length and require a spinning process to form threads. Man-made fibers, e.g. polyester, polyamide, and cellulosics, are initially produced and also available in form of endless fibers and filaments.

The selection of a fiber material follows the planned application; thus the development of conductive structures on the level of yarn formation has to follow the requirements defined by the product and by the manufacturing techniques. The conductive elements have to be processed on machinery that has been designed to handle textile threads, which are flexible, bendable, and elastic. Relevant strategies to integrating conductive elements include the following.

Production of threads with conductive properties can be achieved through a number of different techniques:

- Electroless deposition and electrochemical coating [83].
- CVD and sputtering [46].

- Coating with conductive polymers [84–86].
- Integration of metal wires (core–shell yarns).
- Spinning of metal staple fibers or conductive fibers to yarn.
- Wrapping of yarns with metal fillets.
- Ply formation with conductive threads.

Each of the techniques offers specific advantages and weaknesses. Using deposition and coating techniques, a layer of conductive material is formed on a fibrous substrate. For example, a very thin copper layer can be formed on a polyester fabric using electroless deposition technique (Figure 9.7). As a result a bendable conductive yarn structure is obtained, which exhibits textile characteristics, as long as thickness and mechanical properties of the conductive layer do not alter the yarn properties negatively. The adhesion between conductive layer and basic yarn fibers must be sufficient in height to withstand the mechanical and abrasive wear during formation of textile fabrics.

For technical applications and functional products, e.g. sensors that are built in a product technical aspect, e.g. corrosion resistance, mechanical stability, and overall lifetime, are of relevance, and the optical appearance of the elements is of lower. A substantial drawback for use of coated threads in garment applications results from the color of the coating or metal fibers used, which has to be considered at an early stage of the design.

Independent of the used coating and deposition technique, the layer formation will lead to interlinkages between neighboring fibers and filaments, which alter the mechanical properties of the threads to increased stiffness, lowered elasticity, and reduced flexibility. The overall conductivity of the structure will be dependent on the presence of a sufficient number of interlinkages between the fibers; thus tension and bending will influence conductivity negatively. The use of knitted fabric with strain-dependent electrical resistance allows production of textile-based electromechanical elongation sensors [13].

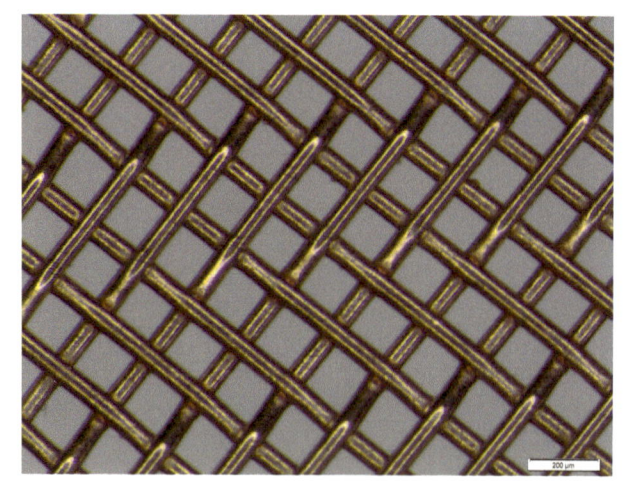

Figure 9.7 Polyester fabric after electroless coating with copper.

Use of conductive fibers, e.g. conductive polymers and metal fibers, allows direct production of conductive yarn; thus the mechanical properties and the optical appearance of the structures will be dependent on the material used. Electrical current is transferred between individual fibers through direct physical contact, which remains intact during bending, stretching, and compression, while individual fibers still can move in the yarn to adapt for length changes during bending. Through use of conductive polymer fibers or metal fibers with diameters in the range of 10–20 μm, the ratio between diameter and bending radius can be kept at low level; thus higher durability in bending tests can be achieved in comparison to metal wire-based concepts.

Another technique to form threads with conductive elements is based on the production of core–shell yarns, e.g. through friction spinning, which however limits the fineness of thread structures to rather coarse yarns. Such a technique allows production of conductive yarns with textile appearance as the yarn surface is built from conventional textile fibers, which can be dyed and finished with the basic textile material, which makes the conductive structure invisible. In many cases the textile characteristics of the core–shell yarns permit processing on standard textile machinery.

A drawback can result from the different strength elongation behavior of the conductive metal core compared with the more flexible and elastic fiber-based shell. Under mechanical stress, breakage of the conductive core will cause risk of interruption and loss of function. This property will require more attention in the case of woven structures and fabrics where the conductive element has been positioned in straight form, while the loops in knitted structures will permit higher deformation without development of high loads in the direction of the yarns. An improvement of yarn flexibility can be realized through use of ply yarns, which combine a textile fiber-based yarn with a metal fillet or wire (Figure 9.8). The screwed structure of the threads permits better adaptation on stress during

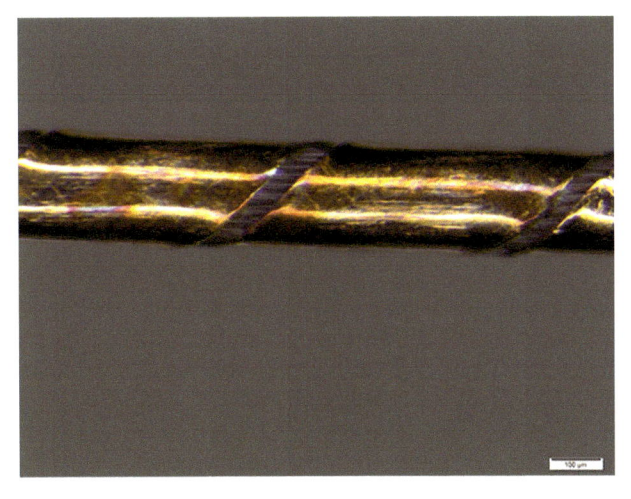

Figure 9.8 Synthetic filament yarn wrapped with copper fillet.

bending and elongation in yarn direction, which positively influences mechanical resistance.

As an important difference between core–shell structures and the other concepts mentioned above, the conductive structure in core–shell yarns is covered by an insulated coating of nonconductive standard textile fibers. This leads to an inherent protection against short circuits; however similar to the use of plastic insulated wires, later contacting requires local access through the fiber shell wound around the electrical line. Reversely the use of uncoated electrical lines permits simple connection to external devices; however dependent on the planned application, attention has to be paid to prevention of short circuits through direct contact between the open lines in the fabric.

9.3.2 Fabric Production

The integration of conductive elements by means of threads allows generation of well-defined pattern, and the design of an electrical circuit however is dependent on the manufacturing technique used.

Weaving technologies form fabric through defined crossings of threads. The direction and position of a conductive element thus has to follow the direction of the threads either in weft or warp direction. When conductive yarns are used as weft, in the weaving step the conductive element is integrated over the full width of the fabric. Mechanical stress and abrasive wear for weft yarns are substantially lower when compared with warp yarn, where abrasive forces result from the repetitive and rapid movement of the reed.

Special weaving techniques also allow the formation of localized patterns, however on the cost of substantial losses in conductive material, which have to be removed at the end of the weaving process.

Knitting processes including warp knitting can form a conductive structure in direction longitudinal and cross direction. The fabric production is based on conjunction of loops and highly elastic as well as rigid structures being formed, which contain the conductive elements. The process of loop formation requires flexible and bendable threads with appropriate frictional behavior to be processed by a series of needles working the needle bed of the knitting machine. Such concepts already are widely in use for heating textiles, e.g. by integration of silver-coated threads. The integration as weft or warp can be achieved more easily; however elastic properties have to be considered with care as the elongation of a loop-based structure is different compared with a straight laid weft.

9.3.3 Embroidery/Sewing

Originally designed as technique to form decorative textile motifs on a basic fabric, embroidery techniques have been adapted to process a wide range of materials including wires, carbon rovings, and coated yarns. The principle of embroidery technique is based on the formation of patterns through crossing of two yarns, one or both of which can be made from conductive fibers. The major advantage of embroidery techniques lies in the high flexibility of patterns available as no specific direction is determined by the technology. Localized circuits can be formed, with high geometric accuracy (<0.1 mm).

Restrictions with regard to yarn diameter and yarn properties result from the principle used in embroidery machines, where the front yarn is pushed through the basic fabric by means of a needle and then has to form a loop on the backside of the fabric. The backing yarn has to be transported through this loop, which thus is an essential requirement to form the embroidery. Limitations also arise from limited textile characteristics of wires and stiff material, which exhibit low tendency to build the required loop at the backside of the fabric, which requires adaptations of the equipment used. Soutache techniques allow positioning of thick yarns and fiber rovings as well as stiff non-textile elements on one side of the fabric.

In a special technique the basic fabric is removed, e.g. through dissolution of PVA fleece in hot water or through organic acid hydrolysis of cellulose fabric, thus releasing the embroidered structure as motif. The chemical stability of the conductive fibers used in such an application thus has to be selected with regard to the later processing conditions to avoid corrosion or destruction of the conductive fibers or films.

A distinct advantage of embroidery lies in the possibility to apply such a technique at a final stage of the manufacturing, e.g. to form localized connections to the conductive structures in the fabric. This allows to build up electrical lines to external devices as well as to form position sensors and circuits at the appropriate position.

Similar requirements have to be considered when sewing techniques are applied to process conductive structures and to connect external devices to circuits inside the fabric.

9.3.4 Printing

Printing techniques permit formation of localized patterns of conductive structures on textile fibers. Through repeated printing, multilayered structures can be implemented and combined with conductive elements already present in the fabric; similarly to production of integrated circuits, contacting of printed structures also can be achieved by embroidery techniques, which allows mechanical contact formation without welding.

As a general disadvantage the formation of the circuits will always be on the surface of the material, which is of lower relevance for technical products that are designed to operate inside a device.

At present two major printing techniques are in use: screen printing and ink-jet printing. Both techniques have been used to form conductive structures on textile material.

In screen printing the conductive structure is placed on the textile in the form of a printing paste, which consists of conductive polymers, finely divided metal powders (e.g. silver), or finely dispersed carbon. To achieve conductivity a critical minimal content of conductive solids has to be achieved to permit sufficient probability for direct contact formation between individual particles. At the same a binder system is applied in the paste, which forms the elastic matrix between the particles and binds the printing to the fibrous textile structure. Printing of pastes containing conductive polymers with positive temperature coefficient can be used to from flexible heating elements, which protect against overheating through increase of electrical resistance with temperature.

Ink-jet printing is more sensitive with regard to the presence of particulate material, which deposits inside the printing head or leads to abrasive effects during operation. Techniques to using ink-jet printing to form conductive patterns based on silver thus use solution of silver salts or silver complexes, which are printed on the fabric and then transferred to release silver upon heating.

Printing techniques allow formation of complex multilayered architectures and insulating layer can be printed as separator.

When polymer pastes, e.g. silicon-based pastes, are deposited on the conductive structures, protective layers can be applied locally and thus be useful to protect integrated electronic structures against corrosive environmental conditions during use (sweat, rain) and maintenance (washing).

9.4 Applications/Examples

Traditionally textile and garment production were organized as a more or less independent series of production steps that finally deliver the product, e.g. fiber production, yarn production, fabric production, textile chemical treatments (dyeing, finishing), and assembly of different components to obtain the final product, e.g. garment, home textile, and technical product.

The integration of wearable electronics requires a reconsideration of a product with regard to the desired functionality.

9.4.1 Material Selection

Selection and combination of conductive material is dependent on the conditions applied during use and maintenance, e.g. mechanical stress, bending cycles, elongation, compression, and abrasive effects. The selection of conductive materials has to consider the conditions applied during washing, e.g. temperature, mechanics, and chemical present during laundry and tumbler drying. Even when individual components are resistant to the conditions in use, corrosion can appear through galvanic element formation when incompatible materials have been connected, e.g. copper/steel/aluminum.

The final position of sensor elements in the garment defines the presence and position of integrated electrical lines in the fabric, which has to be produced and positioned during cutting of pieces in such an order that the lines will appear in right form at the appropriate part of the garment only. In the same way application of embroidery techniques and printing techniques has to be arranged into the production steps to form the final functional element through localized processing.

The overall concept of a smart textile depends on the function to be delivered. Relevant aspects that define the design are:

- *Energy consumption*: Devices that consume low amounts of energy, e.g. sensors, or switches or identification devices can be designed to operate with minimum energy consumption, which permits power generation through energy harvesting using textile-based antennas or piezoelectric effects. In applications where larger amounts of power have to be transferred in textile structures, e.g. electrical heat generation, the flexible connections have to be powerful and mechanically resistant to avoid interrupts, which could cause the risk of hot spots and short circuits. Devices for high power application usually will use metal wires or metal fiber-based yarns for energy transfer, while low energy applications permit use of coatings, conductive films, and prints.
- *Dimension of electronic components*: The functionality defines the dimension of the electronic elements, which can be fully integrated and cased or has to be built as separated unit that is disconnected before laundry. RFID devices that harvest sufficient energy to deliver a signal for identification can be built in the form of a fully coated tag that is integrated as washable device into the textile. Devices that contain sensors or switches connected to microcomputer often are built as separable units, which contain the electronic device as removable component, which are disconnected before use. As an example miniaturized electronic elements were placed on narrow stripes, which then were integrated into textile structures, e.g. through wrapping, weaving, or embroidery [87].
- *Functionality and transfer*: While Radio Frequency Identification (RFID) tags can be miniaturized and operated with minimal energy requirements, other devices, e.g. monitoring of electrocardiogram (ECG), will require conductive elements that form connections to the skin as well as to the evaluating electronic device. Thus combinations of conductive materials will be assembled to build the final product, e.g. skin electrodes, conductive lines inside the textile structure, and connections to the data acquisition unit [88].

The examples given must be understood as representatives to demonstrate the wide range of material concepts as well as applications.

9.4.2 Washable Sensors (Example: Moisture Bed Sensor)

For support of caregivers in care homes, a washable incontinence detector has been built up for detection of conductive liquids (urine) during sleep. For corrosion resistance and to achieve washability at 80–90 °C, the sensor textile was produced from stainless steel yarns using embroidery techniques (Figure 9.9). The connection to the electronic device is achieved through press buttons, which allow disconnection of the sensor textile for washing, while the electronic remains fixed to the bed of a patient.

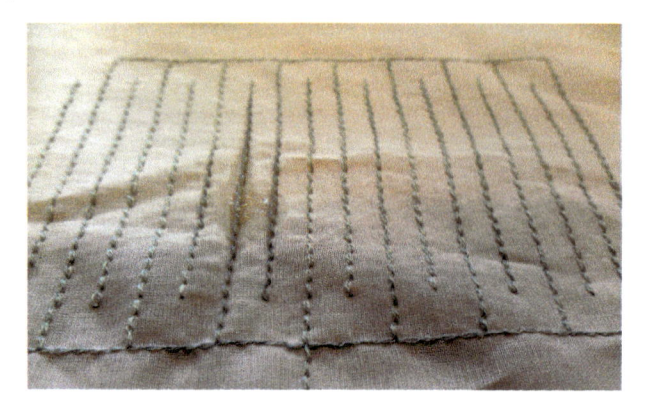

Figure 9.9 Washable wetness sensor for bedding (embroidered stainless steel).

9.4.3 Textile Electrodes

3D structures can replace metal film electrodes in batteries, which allows production of thicker coatings on electrodes. Figure 9.10 shows a current collector made of copper wires, coated with carbon, designed to serve as cathode for lithium-iron phosphate batteries.

The application of carbon felt and other nonwoven structures for application as current collectors in energy storage has been studied in detail. The use of such devices for flexible batteries and also in supercapacitors has been investigated in detail [89–94].

Another example on textile based conductive structure is demonstrated by the round-shaped flexible heating element which is produced by embedding an embroidered metal yarn heating structure into a silicone layer. As a result a flexible pan was obtained, which could reach temperatures above 150 °C (Figure 9.11).

Figure 9.10 Embroidered textile current collector (copper wire) coated with carbon for Li-FePO$_4$ batteries.

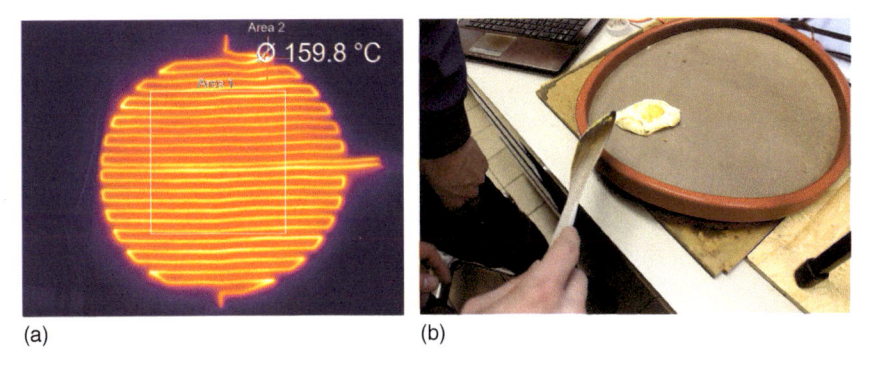

Figure 9.11 Thermographic photo of a flexible heating pan (a) and in use for frying and egg (b).

9.4.4 Flexible Devices

An example for combination of several technological approaches is given by Dias and Monaragla, who could prepare electroluminescent threads by a combination of several approaches. A core–shell construction of a yarn was coated with a dielectric paste and a phosphor ink and finally wrapped with a thin copper wire. The system is operated with alternating current and permits integration of luminescent devices, e.g. into car interior. Through its complex structure the material will exhibit sensitivity to strain and abrasion; thus coverage through coating will be required to improve mechanical stability [95].

9.4.5 Wearable Electronics

Medical applications and therapy form a central field of application for wearable electronics as the added value of such devices can be demonstrated directly. Representative examples are integration of ECG electrodes in clothing in long-term observation [96, 97] and integrated stimulation electrodes for rehabilitation and therapy [98, 99].

9.4.6 Summary and Future Trends

During the last 20 years, a lot of different concepts for conductive materials has been developed by scientists and partly scaled up to commercial application. Technicians today may choose from organic polymers, semiconductors, and metal-based material. Dependent on the production scheme, the implementation of a conductive element can be undertaken during the production of a textile fabric, e.g. in the form of conductive yarns or metal wires, or applied at a late stage, e.g. in the form of prints with conductive inks or embroidery elements. This allows deep integration of smart electronic devices in textiles through combination of different techniques and materials. As a result important textile characteristics such as flexibility and elasticity can be maintained.

The toolkit of conductive fibers available can be understood as the first and highly relevant step for the future production of smart textiles for a wide range of applications.

Future trends in the development of conductive fibers for smart textiles will include the following aspects:

- Miniaturization of conductive lines to the dimension of fibers will be important to reducing dimensions of devices further.
- At present many functional elements, sensors, actuators, and other devices are still understood as technically designed functional element, which offers limited potential for miniaturization or integration. Size reduction to the dimension of threads and formation of flexible structures will facilitate the placement of devices during fabric production.
- Devices with minimum energy consumption and elements of energy harvesting will reduce dimensions of conductive elements, e.g. for local power supply and wireless signal transfer. Wireless communication and data reading further will contribute to the miniaturization of smart electronic elements in textiles.
- Conductive fibers will play a significant role in all applications that include transfer of electrical current and line-based information transfer.

In any case the further development of smart textile products will require the formation of development consortia including expertise from electronics, information technology, textiles, and materials sciences. Isolated developments will be rare, and successful concepts will require consortia built with complementary expertise. From the point of textile production, the formation of a functional electronic element will be based on a series of organized steps placed along the value added chain, e.g. spinning, fabric formation, printing, embroidery, and garment technology, which finally will lead to an electronic device with maximum level of integration.

References

1 Bobeth, W. (1993). *Textile Faserstoffe: Beschaffenheit und Eigenschaften*. Berlin, Heidelberg: Springer-Verlag.
2 Geetha, S., Satheesh Kumar, K.K., Rao, C.R.K. et al. (2009). EMI shielding: methods and materials—A review. *Journal of Applied Polymer Science* 112: 2073–2086. https://doi.org/10.1002/app.29812.
3 Nakata, M., Taga, M., and Kise, H. (1992). Synthesis of electrical conductive polypyrrole films by interphase oxidative polymerization—effects of polymerization temperature and oxidizing agents. *Polymer Journal* 24 (5): 437–441.
4 Stejskal, J., Riede, A., Hlavatá, D. et al. (1998). The effect of polymerization temperature on molecular weight, crystallinity, and electrical conductivity of polyaniline. *Synthetic Metals* 96 (1): 55–61. https://doi.org/10.1016/S0379-6779(98)00064-2.
5 Lenz, J., Schurz, J., and Eichinger, D. (1994). Properties and structure of Lyocell and viscose type fibres in the swollen state. *Lenzinger Berichte* 7: 19–25.

6 Zhang, K. and Ge, M. (2016). Preparation and characterization of PA6/titania conductive fibres. *Indian Journal of Fibre and Textile Research* 41 (2): 150–155.

7 James, E.S. (1936). Metal reducing method. US Patent 2, 050, 298.

8 Grisley, F. (1936). Improvements in blankets, pads, quilts, clothing, fabric, or the like, embodying electrical conductors. UK Patent GB445195.

9 Costanzo, R.J. (1936). Electrically heated socks. UK Patent GB1128224.

10 Aguiló-Aguayo, N. and Bechtold, T. (2014). Characterisation of embroidered 3D electrodes by use of anthraquinone-1,5-disulfonic acid as probe system. *Journal of Power Sources* 254C: 224–231.

11 Lenninger, M., Froeis, T., Scheiderbauer, M. et al. (2013). High current density 3D-electrodes manufactured by technical embroidery. *Journal of Solid State Electrochemistry* 17 (8): 2303–2309.

12 Celik Bedeloglu, A. and Sunter, N. (2013). Investigation of polyacrylic/metal wire composite yarn characteristics manufactured on fancy yarn machine. *Materials and Manufacturing Processes* 28 (6): 650–656. https://doi.org/10.1080/10426914.2012.727118.

13 Guo, L., Berglin, L., and Mattila, H. (2012). Improvement of electro-mechanical properties of strain sensors made of elastic-conductive hybrid yarns. *Textile Research Journal* 82 (19): 1937–1947. https://doi.org/10.1177/0040517512452931.

14 Schwarz, A., Hakuzimana, J., Gasana, E. et al. (2008). Gold coated polyester yarn. *Advances in Science and Technology* 60: 47–51.

15 Schwarz, A., Hukazimana, J., Kaczynska, A. et al. (2010). Gold coated yarns through electroless deposition. *Surface and Coatings Technology* 204: 1214–1418.

16 Kalanyan, B., Oldham, C.J., Sweet, W.J. III,, and Parsons, G.N. (2013). Highly conductive and flexible nylon-6 nonwoven fiber mats formed using tungsten atomic layer deposition. *ACS Applied Materials & Interfaces* 5 (11): 5253–5259. https://doi.org/10.1021/am401095r.

17 Mundy, J.Z., Shafiefarhood, A., Li, F. et al. (2016). Low temperature platinum atomic layer deposition on nylon-6 for highly conductive and catalytic fiber mats. *Journal of Vacuum Science an Technology, A: Vacuum, Surfaces, and Films* 34 (1): 01A152/1–01A152/6. https://doi.org/10.1116/1.4935448.

18 Karaguzel, B., Merritt, C.R., Kang, T. et al. (2008). Utility of nonwovens in the production of integrated electrical circuits via printing conductive inks. *Journal of The Textile Institute* 99: 37–45.

19 Araki, T., Nogi, M., Suganuma, K. et al. (2011). Printable and stretchable conductive wirings comprising silver flakes and elastomers. *IEEE Electron Device Letters* 32: 1424–1426.

20 Romaguera, V.S., Madec, M.B., and Yeates, S.G. (2009). Inkjet printing of conductive polymers for smart textiles and flexible electronics. *Materials Research Society Symposium Proceedings* 1192: 26–31.

21 Rai, P., Lee, J., Mathur, G.N., and Varadan, V.K. (2012). Carbon nanotubes polymer nanoparticle inks for healthcare textile. *Proceedings of SPIE* 8548 https://doi.org/10.1117/12.946253.

22 Rai, P., Kumar, P.S., Oh, S. et al. (2012). Smart healthcare textile sensor system for unhindered-pervasive health monitoring. *Proceedings of SPIE* 8344 https://doi.org/10.1117/12.921253.

23 Yang, K., Torah, R., Wei, Y. et al. (2013). Waterproof and durable screen printed silver conductive tracks on textiles. *Textile Research Journal* 83 (19): 2023–2031. https://doi.org/10.1177/0040517513490063.

24 Kardarian, K., Busani, T., Osorio, I. et al. (2014). Sintering of nanoscale silver coated textiles, a new approach to attain conductive fabrics for electromagnetic shielding. *Materials Chemistry and Physics* 147 (3): 815–822. https://doi.org/10.1016/j.matchemphys.2014.06.025.

25 Lee, J., Kwon, H., Seo, J. et al. (2015). Conductive fiber-based ultrasensitive textile pressure sensor for wearable electronics. *Advanced Materials (Weinheim, Germany)* 27 (15): 2433–2439. https://doi.org/10.1002/adma.201500009.

26 Montazer, M. and Komeily Nia, Z. (2015). Conductive nylon fabric through in situ synthesis of nano-silver: preparation and characterization. *Materials Science and Engineering, C: Materials for Biological Applications* 56: 341–347. https://doi.org/10.1016/j.msec.2015.06.044.

27 Wei, Y., Chen, S., Yuan, X. et al. (2016). Multiscale wrinkled microstructures for piezoresistive fibers. *Advanced Functional Materials* 26 (28): 5078–5085. https://doi.org/10.1002/adfm.201600580.

28 Krishnanand, K., Thite, A., and Mukhopadhyay, A.K. (2016). Electro-conductive cotton fabric prepared by electron beam induced graft polymerization and electroless deposition technology. *Journal of Applied Polymer Science* 133: 44576. https://doi.org/10.1002/app.44576.

29 Liu, S., Hu, M., and Yang, J. (2016). A facile way of fabricating a flexible and conductive cotton fabric. *Journal of Materials Chemistry C* 4: 1320–1325. https://doi.org/10.1039/C5TC03679H.

30 Li, Y. (ed.) (2015). *Organic Optoelectronic Materials*. Springer. ISBN: 978-3-319-16861-6.

31 IUPAC Gold Book. *Compendium of Chemical Terminology*. International Union of Pure and Applied Chemistry. http://publications.iupac.org/compendium/index.html.

32 Malinauskas, A. (2001). Chemical deposition of conducting polymers. *Polymer* 42: 3957–3972.

33 Li, H., Shi, G., Ye, W. et al. (1997). Polypyrrole-carbon fibre composite film prepared by chemical oxidative polymerization of pyrrole. *Journal of Applied Polymer Science* 64: 2149–2154.

34 Huang, G., Liu, L., Wang, R. et al. (2016). Smart color-changing textile with high contrast based on a single-sided conductive fabric. *Journal of Materials Chemistry C: Materials for Optical and Electronic Devices* 4 (32): 7589–7594. https://doi.org/10.1039/C6TC02051H.

35 Kim, B.C., Innis, P.C., Wallace, G.G. et al. (2013). Electrically conductive coatings of nickel and polypyrrole/poly(2-methoxyaniline-5-sulfonic acid) on nylon Lycra textiles. *Progress in Organic Coatings* 76 (10): 1296–1301. https://doi.org/10.1016/j.porgcoat.2013.04.004.

36 Firoz Babu, K., Siva Subramanian, S.P., and Anbu Kulandainathan, M. (2013). Functionalisation of fabrics with conducting polymer for tuning capacitance and fabrication of supercapacitor. *Carbohydrate Polymers* 94 (1): 487–495. https://doi.org/10.1016/j.carbpol.2013.01.021.

37 Molina, J., Esteves, M.F., Fernandez, J. et al. (2011). Polyaniline coated conducting fabrics. Chemical and electrochemical characterization. *European Polymer Journal* 47 (10): 2003–2015. https://doi.org/10.1016/j.eurpolymj.2011.07.021.

38 Anbarasan, R., Vasudevan, T., Kalaignan, G.P., and Gopalan, A. (1999). Chemical grafting of aniline and o-toluidine onto poly(ethylene terephthalate) fibre. *Journal of Applied Polymer Science* 73: 121–128.

39 Yin, X., Kobayashi, K., Yoshino, K. et al. (1995). Percolation conduction in polymer composites containing polypyrrole coated insulating polymer fibre and conducting polymer. *Synthetic Metals* 69: 367–368.

40 Bhadani, S.N., Sen Gupta, S.K., Sahu, G.C., and Kumari, M. (1996). Electrochemical formation of some conducting fibres. *Journal of Applied Polymer Science* 61: 207–212.

41 Hiltunen, M., Pelto, J., Ellae, V., and Kellomaeki, M. (2016). Uniform and electrically conductive biopolymer-doped polypyrrole coating for fibrous PLA. *Journal of Biomedical Materials Research, Part B: Applied Biomaterials* 104 (8): 1721–1729. https://doi.org/10.1002/jbm.b.33514.

42 McGraw, S.K., Alocilja, E., Senecal, A., and Senecal, K. (2012). Synthesis of a functionalized polypyrrole coated electrotextile for use in biosensors. *Biosensors* 2: 465–478. https://doi.org/10.3390/bios2040465.

43 Montarsolo, A., Varesano, A., Mossotti, R. et al. (2012). Enhanced adhesion of conductive coating on plasma-treated polyester fabric: a study on the ageing effect. *Journal of Applied Polymer Science* 126 (4): 1385–1393. https://doi.org/10.1002/app.36762.

44 Opwis, K., Knittel, D., and Gutmann, J.S. (2012). Oxidative in situ deposition of conductive PEDOT:PTSA on textile substrates and their application as textile heating element. *Synthetic Metals* 162 (21–22): 1912–1918. https://doi.org/10.1016/j.synthmet.2012.08.007.

45 Mattana, G., Cosseddu, P., Fraboni, B. et al. (2011). Organic electronics on natural cotton fibres. *Organic Electronics* 12 (12): 2033–2039. https://doi.org/10.1016/j.orgel.2011.09.001.

46 Bashir, T., Skrifvars, M., and Persson, N.-K. (2011). Production of highly conductive textile viscose yarns by chemical vapor deposition technique: a route to continuous process. *Polymers for Advanced Technologies* 22 (12): 2214–2221. https://doi.org/10.1002/pat.1748.

47 Aakerfeldt, M., Straaaat, M., and Walkenstroem, P. (2013). Electrically conductive textile coating with a PEDOT-PSS dispersion and a polyurethane binder. *Textile Research Journal* 83 (6): 618–627. https://doi.org/10.1177/0040517512444330.

48 Wu, C.-H., Shen, H.-P., Don, T.-M., and Chiu, W.-Y. (2013). Fabrication of flexible conductive films derived from poly(3,4-ethylenedioxythiophene)-poly(styrenesulfonic acid) (PEDOT:PSS) on the nonwoven fabrics substrate. *Materials Chemistry and Physics* 143 (1): 143–148. https://doi.org/10.1016/j.matchemphys.2013.08.037.

49 Cohen David, N., David, Y., Katz, N. et al. (2016). Electro-conductive fabrics based on dip coating of cotton in poly(3-hexylthiophene). *Polymers for Advanced Technologies* https://doi.org/10.1002/pat.3857.

50 Soroudi, A., Skrifvars, M., and Liu, H. (2011). Polyaniline-polypropylene melt-spun fiber filaments: the collaborative effects of blending conditions and fiber draw ratios on the electrical properties of fiber filaments. *Journal of Applied Polymer Science* 119 (1): 558–564. https://doi.org/10.1002/app.32655.

51 Li, X., Sun, P., Fan, L. et al. (2012). Multifunctional graphene woven fabrics. *Scientific Reports* 2: 395.

52 Shateri-Khalilabad, M. and Yazdanshenas, M.E. (2013). Fabricating electroconductive cotton textiles using graphene. *Carbohydrate Polymers* 96: 190–195.

53 Woltornist, S.J., Alamer, F.A., McDannald, A. et al. (2015). Preparation of conductive graphene/graphite infused fabrics using an interface trapping method. *Carbon* 81: 38–42. https://doi.org/10.1016/j.carbon.2014.09.020.

54 Yoo, H.J., Kim, H.H., Cho, J.W., and Kim, Y.H. (2012). Surface morphology and electrical properties of polyurethane nanofiber webs spray-coated with carbon nanotubes. *Surface and Interface Analysis* 44 (4): 405–411. https://doi.org/10.1002/sia.3817.

55 Seyedin, S., Razal, J.M., Innis, P.C., and Wallace, G.G. (2016). A facile approach to spinning multifunctional conductive elastomer fibres with nanocarbon fillers. *Smart Materials and Structures* 25 (3): 035015/1–035015/9. https://doi.org/10.1088/0964-1726/25/3/035015.

56 Ebrahimbeiki Chimeh, A. and Montazer, M. (2016). Fabrication of nano-TiO_2/carbon nanotubes and nano-TiO_2/nanocarbon black on alkali hydrolyzed polyester producing photoactive conductive fabric. *Journal of the Textile Institute* 107 (1): 95–106. https://doi.org/10.1080/00405000.2015.1012881.

57 Jin, X., Xiao, C., An, S. et al. (2006). Preparation and properties of a new coating method for preparing conductive polyester fibers with permanent conductivity. *Journal of Applied Polymer Science* 102: 2685–2691. https://doi.org/10.1002/app.24095.

58 Park, G., Jung, Y., Lee, G.-W. et al. (2012). Carbon nanotube/poly(vinyl alcohol) fibers with a sheath-core structure prepared by wet spinning. *Fibers and Polymers* 13 (7): 874–879. https://doi.org/10.1007/s12221-012-0874-5.

59 Nilsson, E., Oxfall, H., Wandelt, W. et al. (2013). Melt spinning of conductive textile fibers with hybridized graphite nanoplatelets and carbon black filler. *Journal of Applied Polymer Science* 130 (4): 2579–2587. https://doi.org/10.1002/app.39480.

60 Lin, J.-H., Lin, Z.-I., Pan, Y.-J. et al. (2016). Manufacturing techniques and property evaluations of conductive composite yarns coated with polypropylene and multi-walled carbon nanotubes. *Composites Part A: Applied Science*

and Manufacturing 84: 354–363. https://doi.org/10.1016/j.compositesa.2016.02.004.

61 Bautista-Quijano, J.R., Poetschke, P., Bruenig, H., and Heinrich, G. (2016). Strain sensing, electrical and mechanical properties of polycarbonate/multiwall carbon nanotube monofilament fibers fabricated by melt spinning. *Polymer* 82: 181–189. https://doi.org/10.1016/j.polymer.2015.11.030.

62 Thüringen-GreenTec (2012). Project-Nr. 2012 FE 9001., Verbund-Nr.: 2012 VF 0001: Erzeugung hochelastischer metallisierter Fasern als Grundbaustein für textilbasierte elektronische Komponenten durch kontinuierliche Laser- und elektrochemische Strukturierung von metallisierten Monofilamenten. Teilprojekt: Technologieentwicklung zur Erzeugung von metallisierten Monofilamenten auf der Basis von Spezialpolymeren – Lasertex.

63 Hu, B., Li, D., Manandharm, P. et al. (2012). CNT/conducting polymer composite conductors impart high flexibility to textile electroluminescent devices. *Journal of Materials Chemistry* 22 (4): 1598–1605. https://doi.org/10.1039/C1JM14121J.

64 Berendjchi, A., Khajavi, R., Yousefi, A.A., and Yazdanshenas, M.E. (2016). Improved continuity of reduced graphene oxide on polyester fabric by use of polypyrrole to achieve a highly electro-conductive and flexible substrate. *Applied Surface Science* 363: 264–272. https://doi.org/10.1016/j.apsusc.2015.12.030.

65 Bae, J. and Hong, K.H. (2012). Electrical properties of conductive fabrics for operating capacitive touch screen displays. *Textile Research Journal* 83 (4): 329–336. https://doi.org/10.1177/0040517512464298.

66 Kursun Bahadir, S. and Jevšnik, S. (2016). Optimization of hot air welding process parameters for manufacturing textile transmission lines for e-textiles applications: Part I: Electro-conductive properties. *Textile Research Journal* 87 (2): 232–243. https://doi.org/10.1177/0040517516629140.

67 Ding, J.T.F., Tao, X., Au, W.M., and Li, L. (2014). Temperature effect on the conductivity of knitted fabrics embedded with conducting yarns. *Textile Research Journal* 84 (17): 1849–1857. https://doi.org/10.1177/0040517514530026.

68 Jeong, M.J., Yun, T.-I., Baek, J.J., and Kim, Y.T. (2015). Wireless power transmission using a resonant coil consisting of conductive yarn for wearable devices. *Textile Research Journal* 86 (14): 1543–1548. https://doi.org/10.1177/0040517515586163.

69 Kang, T.S., Lee, S.W., and Joo, J. (2005). Electrically conducting polypyrrole fibers spun by electrospinning. *Synthetic Metals* 153: 61–64.

70 Li, M., Guo, Y., and Wei, Y. (2006). Electrospinning polyaniline-contained gelatin nanofibers for tissue engineering applications. *Biomaterials* 27: 2705–2715.

71 Cardenas, J.R., Franca, M.G.O., and Vasconcelos, E.A. (2007). Growth of sub-micron fibres of pure polyaniline using the electrospinning technique. *Journal of Physics D: Applied Physics* 40: 1068–1071.

72 Hong, K.H. and Kang, T.J. (2006). Polyaniline–nylon 6 composite nanowires prepared by emulsion polymerization and electrospinning process. *Journal of Applied Polymer Science* 99: 1277–1286.

73 Hong, K.H., Oh, K.W., and Kang, T.J. (2005). Preparation of conducting nylon-6 electrospun fiber webs by the in situ polymerization of polyaniline. *Journal of Applied Polymer Science* 96: 983–991.

74 Choi, S.S., Chu, B.Y., and Hwang, D.S. (2005). Preparation and characterization of polyaniline nanofiber webs by template reaction with electrospun silica nanofibers. *Thin Solid Films* 477: 233–239.

75 McKeon, K.D., Lewis, A., and Freeman, J.W. (2010). Electrospun poly(D,L-lactide) and polyaniline scaffold characterization. *Journal of Applied Polymer Science* 115: 1566–1572.

76 Laforgue, A. and Robitaille, L. (2008). Fabrication of poly-3-hexylthiophene/polyethylene oxide nanofibers using electrospinning. *Synthetic Metals* 158: 577–584.

77 Lu, M., Xie, R., Liu, Z. et al. (2016). Enhancement in electrical conductive property of polypyrrole-coated cotton fabrics using cationic surfactant. *Journal of Applied Polymer Science* 133 (32) https://doi.org/10.1002/app.43601.

78 Shim, B.S., Chen, W., Doty, C. et al. (2008). Smart electronic yarns and wearable fabrics for human biomonitoring made by carbon nanotube coating with polyelectrolytes. *Nano Letters* 8: 4151–4157.

79 Nilsson, E., Rigdahl, M., and Hagström, B. (2015). Electrically conductive polymeric bi-component fibers containing a high load of low-structured carbon black. *Journal of Applied Polymer Science* 132: 42255. https://doi.org/10.1002/app.42255.

80 Härdelin, L. and Hagström, B. (2015). Wet spun fibers from solutions of cellulose in an ionic liquid with suspended carbon nanoparticles. *Journal of Applied Polymer Science* 132: 41417. https://doi.org/10.1002/app.41417.

81 Maiellaro, G., Ragonese, E., Gwoziecki, R. et al. (2013). Ambient light organic sensor in a printed complementary organic TFT technology on flexible plastic foil. *IEEE Transactions on Circuits and Systems I* 61: 1036–1043.

82 Lee, J. (2015). Conductive fiber-bases ultrasensitive textile pressure for wearable electronics. *Advanced Materials* 27: 2433–2439.

83 Kim, S.M., Kim, I.Y., and Kim, H.R. (2017). Production of electromagnetic shielding fabrics by optimization of electroless silver plating conditions for PET fabrics. *Journal of the Textile Institute* 108 (6): 1065–1073. https://doi.org/10.1080/00405000.2016.1219449.

84 Schwarz, A., Hakuzimana, J., Westbroek, P. et al. (2012). A study on the morphology of thin copper films on para-aramid yarns and their influence on the yarn's electro-conductive and mechanical properties. *Textile Research Journal* 82 (15): 1587–1596, 10 pp. DOI:https://doi.org/10.1177/0040517511431291.

85 Mohammadian, M., Afzali, A., Mottaghitalab, V., and Haghi, A.K. (2012). Washing and rubbing fastness of electroless plated polyester conductive fabric. *Materiale Plastice (Bucharest, Romania)* 49 (3): 182–185.

86 Lee, H.M., Choi, S., Jung, A., and Ko, S. (2013). Highly conductive aluminum textile and paper for flexible and wearable electronics. *Angewandte Chemie International Edition* 52 (30): 7718–7723. https://doi.org/10.1002/anie.201301941.

87 Zysset, C., Kinkeldei, T., Muenzenrieder, N. et al. (2013). Combining electronics on flexible plastic strips with textiles. *Textile Research Journal* 83 (11): 1130–1142. https://doi.org/10.1177/0040517512468813.

88 Kannaian, T., Neelaveni, R., and Thilagavathi, G. (2013). Design and development of embroidered textile electrodes for continuous measurement of electrocardiogram signals. *Textile Research Journal* 42 (3): 303–318. https://doi.org/10.1177/1528083712438069.

89 Fan, T., Zhao, C., Xiao, Z. et al. (2016). Fabricating of high-performance functional graphene fibers for micro-capacitive energy storage. *Scientific Reports* 6: 29534. https://doi.org/10.1038/srep29534.

90 Xie, K., Zhang, K., Han, Y. et al. (2016). A novel TiO_2-wrapped activated carbon fiber/sulfur hybrid cathode for high performance lithium sulfur batteries. *Electrochimica Acta* 210: 415–421. https://doi.org/10.1016/j.electacta.2016.05.172.

91 Cheng, X., Zhang, J., Ren, J. et al. (2016). Design of a hierarchical ternary hybrid for a fiber-shaped asymmetric supercapacitor with high volumetric energy density. *Journal of Physical Chemistry C* 120 (18): 9685–9691. https://doi.org/10.1021/acs.jpcc.6b02794.

92 Ma, K., Cheng, J.P., Liu, F., and Zhang, X. (2016). Co-Fe layered double hydroxides nanosheets vertically grown on carbon fiber cloth for electrochemical capacitors. *Journal of Alloys and Compounds* 679: 277–284. https://doi.org/10.1016/j.jallcom.2016.04.059.

93 Zhang, Y., Hu, Z., An, Y. et al. (2016). High-performance symmetric supercapacitor based on manganese oxyhydroxide nanosheets on carbon cloth as binder-free electrodes. *Journal of Power Sources* 311: 121–129. https://doi.org/10.1016/j.jpowsour.2016.02.017.

94 Dong, L., Xu, C., Li, Y. et al. (2016). Simultaneous production of high-performance flexible textile electrodes and fiber electrodes for wearable energy storage. *Advanced Materials (Weinheim, Germany)* 28 (8): 1675–1681. https://doi.org/10.1002/adma.201504747.

95 Dias, T. and Monaragla, R. (2012). Development and analysis of novel electroluminescent yarns and fabrics for localized automotive interior illumination. *Textile Research Journal* 82 (11): 1164–1176. https://doi.org/10.1177/0040517511420763.

96 Cho, H., Lim, H., Cho, S., and Lee, J. (2015). Development of textile electrode for electrocardiogram measurement based on conductive electrode configuration. *Fibers and Polymers* 16 (10): 2148–2157. https://doi.org/10.1007/s12221-015-5317-7.

97 Trindade, I.G., Martins, F., and Baptista, P. (2015). High electrical conductance poly(3,4-ethylenedioxythiophene) coatings on textile for electrocardiogram monitoring. *Synthetic Metals* 210 (Part_B): 179–185. https://doi.org/10.1016/j.synthmet.2015.09.024.

98 Aakerfeldt, M., Lund, A., and Walkenstroem, P. (2015). Textile sensing glove with piezoelectric PVDF fibers and printed electrodes of PEDOT:PSS. *Textile Research Journal* 85 (17): 1789–1799. https://doi.org/10.1177/0040517515578333.

99 Zieba, J., Frydrysiak, M., and Tokarska, M. (2011). Research of textile electrodes for electrotherapy. *Fibres & Textiles in Eastern Europe* 19 (5): 70–74.

10

Phase Change Fibers

Subrata Mondal

National Institute of Technical Teachers' Training and Research, Department of Mechanical Engineering, FC Block, Sector III, Salt Lake City, Kolkata, 700106, West Bengal, India

10.1 Introduction

In the last few decades, there are tremendous research activities to search efficient systems for the energy storage. Among the various energy storage systems (ESS), thermal energy storage (TES) is an attractive way to reduce the imbalance between energy supply and demand. Thermal ESS could be based on latent heat storage (LHS), sensible heat storage (SHS), and thermochemical heat storage processes. In the SHS materials, storage of energy is associated with the rise in temperature of TES materials either in liquid or solid form. Thermal energy would store/release in the LHS system, when it undergoes phase change from solid to liquid and vice versa. However, energy would store after reversible chemical reaction between the substances in the thermochemical ESS [1–4]. Phase change materials (PCMs) based on LHS system is an attractive way of storing thermal energy. PCM based on LHS system has two major advantages, viz. high energy storage density and isothermal nature of the storage process [5].

Nowadays, consumers are not only looking for the aesthetic look for their garments, but they are also looking for the added functionality to their garments specially comfort and ergonomic. Functional textiles are end users' requirement specific and can be tailored to achieve performance requirement of users' specific purposes. These textiles are specially tailored to perform a predefined functionality above its normal functions. Functional or interactive textiles induce shift from the passive to the active functionality [6].

The recent advances of various smart materials have opened up opportunities for many areas in the advanced material fields. PCM is one such type of smart materials gaining momentum for the applications in temperature regulating and thermal balance of advanced material applications [6]. This chapter aims to provide an overview for the principle of PCMs, various types of PCMs, PCMs for thermal management, and various techniques to prepare phase change fibrous structures and their applications in advanced material fields with a special focus on thermoregulating textiles.

Handbook of Fibrous Materials, First Edition. Edited by Jinlian Hu, Bipin Kumar, and Jing Lu.
© 2020 Wiley-VCH Verlag GmbH & Co. KGaA. Published 2020 by Wiley-VCH Verlag GmbH & Co. KGaA.

10.2 Phase Change Materials (PCMs)

PCMs are high heat of fusion and heat of solidification substances which melt or solidify at a certain temperature range. Therefore, PCMs are capable of storing and releasing large amounts of heat energy upon its phase changes. When the material changes from solid to liquid and vice versa, the heat is absorbed or released. Therefore, PCMs are classified as LHS materials. LHS can be achieved through (i) liquid–solid, (ii) solid–liquid, (iii) solid–gas, (iv) liquid–gas, and (v) solid–solid phase changes. However, only solid–liquid and vice versa phase changes of PCMs are suitable for practical applications [6].

10.2.1 Principle of Phase Change Materials

Solid to liquid or vice versa PCMs are used practically for the development of smart materials. Liquid to gas PCMs are not used for practical applications due to the large volume changes or high pressure required to store the materials in gas phase. However, solid to solid PCMs are not practical due to too low heat storage or release during the phase change processes. PCMs can absorb (and store) and release heat while the materials change from solid to liquid or vice versa. The amount of energy absorbed or released depends on the value of heat of fusion (ΔH_f) or heat of solidification (or heat of crystallization) (ΔH_c), respectively. Heat of fusion and heat of solidification or heat of crystallization can be measured by using differential scanning calorimetry (DSC) in heating and cooling cycles, respectively (Figure 10.1). The temperature of a PCM remains constant during the phase change processes (Figure 10.2) [6].

PCMs have enormous potential for the development of thermoregulating textiles. The heat exchange of textiles with the human body parts and the environment plays a major role for the thermal state of the human body. In general, thermal comfort is defined as the perception of mind that expresses satisfaction with the thermal environment, whereas dissatisfaction may be caused by warm/cold discomfort for the body [3, 7]. However, thermal dissatisfaction may

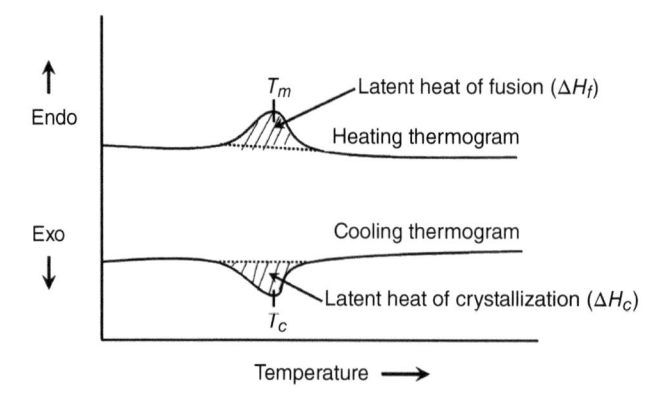

Figure 10.1 Schematic of DSC heating and cooling thermograms for PCMs (T_m, crystal melting temperature; T_c, crystallization temperature).

Figure 10.2 Thermal energy vs. temperature of solid–liquid phase changes.

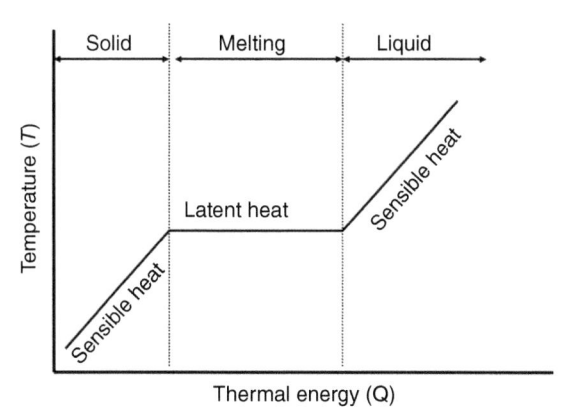

also be caused by an unwanted heating or cooling of particular part of the body (local discomfort). Transfer of dry heat from a textile is complex process, which involves conduction, convection, and radiation mode of heat transfer [8]:

$$H_{dry} = M - W - C_{res} - E_{res} - E - S$$

where H_{dry} is the dry heat loss, M is the metabolic rate, W is the external work, C_{res} is the convective respiratory heat loss, E_{res} is the evaporative respiratory heat loss, E is the evaporative heat loss from the skin, and S is the change in body heat content, in which all values are expressed in W/m².

The movement of thermal energy through a textile is an important factor for the comfort. The human body would attempt to maintain a core body temperature around 37 °C. The balance between heat productions by the body and heat loss to the environment is the comfort factor. The body would be in a state of comfort when the body temperature is slightly below the core body temperature, and there is no moisture on the skin [9]. The loss of heat through perspiration and heat flux through fabrics are very important for the comfort of the wearer [10, 11].

TES is a process for high or low temperature ESS for later use; hence, TES bridges the time gap between energy requirements and energy use. Therefore, apart from the advanced textile fields, PCM could also be applicable in other advanced materials applications. Thermoregulating effect of PCM incorporated material would result from either heat absorption or release of the PCMs, which could be incorporated into the fibrous structure. An active thermal barrier effect may result from either heat absorption or heat emission of the PCMs of PCM incorporated materials, which regulates with the surrounding environment and, subsequently, adopts it to the thermal needs (i.e. activity level and ambient temperature) [12].

10.2.2 Various Types of Phase Change Materials

Simple and classical example of a PCM is water, and its various phases are ice (solid), liquid water (liquid), and steam (gas). PCMs could be classified into three categories such as (i) organic PCMs, (ii) inorganic PCMs, and (iii) eutectic PCMs of organic–organic, organic–inorganic, and inorganic–inorganic mixtures.

Table 10.1 Melting point temperature and latent heat of absorption for various PCMs.

Materials	Melting point (°C)	Latent heat of absorption, ΔH (J/g)	References
Organic PCMs			
n-Octadecane	35.6	233.8	[3]
n-Hexadecane	22.0	185.3	[3]
Capric acid	30.2	142.7	[13]
Tetradecanol	37.0	207.3	[13]
Lauric acid	43.2	177.7	[13]
Butyl stearate	19	200	[14]
Inorganic PCMs			
Glauber salt	32.4	254	[15]
$CaCl_2 \cdot 6H_2O$	29	190.8	[16]
Eutectic PCMs			
Lauric and stearic acid (75.5 : 24.5%)	37	182.7	[17]
Tetradecanol and capric acid (38 : 62%)	22.74	153.4	[13]
Tetradecanol and lauric acid (53.6 : 46.4%)	26.59	162.7	[13]
Tetradecanol and myristic acid (71.84 : 28.16%)	36.46	208	[13]

10.2.2.1 Organic PCMs

Organic PCMs commonly belong to two groups, viz. paraffin and fatty acids. Some of the organic PCMs with their melting point temperature and heat of absorption are presented in Table 10.1. Paraffin is linear alkanes with chemical formula of C_nH_{2n+2}. PCM based on paraffin has good TES capability per unit mass with little or no subcooling. Paraffinic PCMs are widely used to prepare the thermoregulated advanced materials due to their several outstanding properties such as noncorrosiveness, chemical and thermal stability, and low subcooling [18]. Polyethylene glycol (PEG) with appropriate molecular weight could also be used as PCMs for manufacturing thermoregulated fibrous structures for advanced material applications [6]. By selecting appropriate carbon atoms of hydrocarbons, the phase transition temperature of PCMs could be tailored for specific applications [6]. Various fatty acids, viz. capric, lauric, palmitic, and stearic acids (SAs), and their binary mixture are widely used as PCMs for thermal balance. Organic PCMs are relatively expensive; however, they have high latent heat per unit volume and low density. Key advantages and disadvantages of organic PCMs are tabulated in Table 10.2.

10.2.2.2 Inorganic PCMs

Inorganic PCMs generally belong to the hydrated salts that have number of hydrate molecules [6]. For specific applications, heat absorbing, and releasing temperature interval of PCMs should be in the specific range, for example, for textile purpose it should be around 10–35 °C. Couple of inorganic PCMs with their melting point temperature and heat of absorption are tabulated in

Table 10.2 Key advantages and disadvantages of organic PCMs.

Advantages	Disadvantages
○ Organic PCMs are generally noncorrosive ○ Chemically stable on repeated phase change cycles ○ High specific latent heat of fusion and solidification ○ Low vapor pressure	○ Low thermal conductivity ○ High change in volume as compared with inorganic PCMs on phase change ○ Organic PCMs are flammable

Source: Farid et al. 2004 [15].

Table 10.3 Major advantages and disadvantages of inorganic PCMs.

Advantages	Disadvantages
○ High specific heat ○ High thermal conductivity ○ Nonflammable ○ Cheap as compared with the organic PCMs	○ Corrosive to metals ○ Suffer from decomposition on repeated uses

Source: Farid et al. 2004 [15].

Table 10.1. Hydrated salts absorb high energy at its melting point while release energy when it solidifies [19]:

$$Na_2SO_4 \cdot 10H_2O \rightarrow Na_2SO_4 + 10H_2O + \Delta H$$

$$Na_2SO_4 + 10H_2O \rightarrow Na_2SO_4 \cdot 10H_2O - \Delta H$$

Inorganic PCMs have several major advantages and disadvantages as tabulated in Table 10.3.

10.2.2.3 Eutectic PCMs

Eutectic PCMs are mixture of two or more components in such a proportion that their combined melting point and solidification point temperature is single, and lower than the individual component. Eutectic mixture of PCMs could be classified as organic–organic, organic–inorganic, and inorganic–inorganic types. Some of the eutectic PCMs with their melting point temperature and latent heat of fusion are tabulated in Table 10.1.

10.3 Phase Change Fibers

10.3.1 How PCM Works with Fibrous Structures

Here author explains the working principles of PCM in fibrous structure, taking an example of specific filed such as advanced textile applications. The PCMs that change phases from 10 to 35 °C would be suitable for the development of fibrous structures for smart textiles. Textile with PCM-containing fibrous structure will

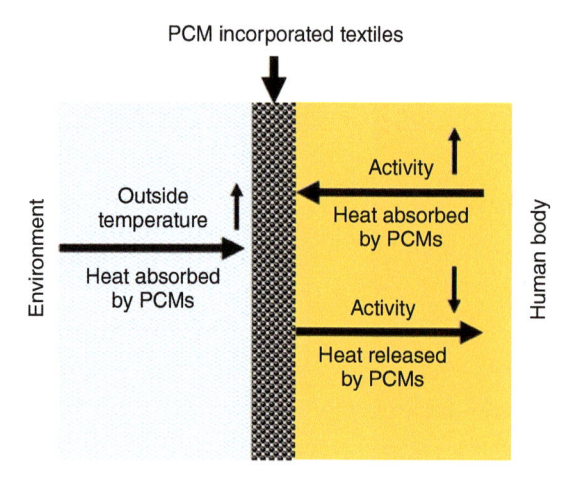

Figure 10.3 Schematic representation of interactions for the PCM incorporated textiles with environment and human body.

react with the changes of environmental temperature and the temperatures in different areas of body. PCM in fibrous structure absorbs body heat when the body temperature increases and PCM melts. However, melted or semi-melted PCM gives off heat when body temperature decreases. In this way, PCM incorporated textiles could keep the body in a comfortable state and are used for the thermal management of garments [3, 6, 20]. The process of phase change is a dynamic; hence, the PCMs inside the fibrous structures are constantly changing from solid to liquid and vice versa depending upon the physical activity of person and external environmental temperature (Figure 10.3).

10.3.2 Microencapsulation of PCMs

Though PCMs are used for the absorption, storage, or release of thermal energy, however, to effectively use the PCMs, they need to be packed in an inert and durable micro-container. Microencapsulation is a process of incorporating micro-sized solid or liquid droplet or gases inside the compatible thin solid wall structure. Tiny particles or droplets of PCMs are surrounded by a coating with a uniform wall, which is generally made with polymer. Therefore, microencapsulated PCM consists of two major parts, such as the PCM inside microcapsules known as core and the surrounding wall known as shell. The diameter of microcapsules varies from 100 μm to 1 mm. Scanning electron microscopic image for the microcapsule of paraffin core and polystyrene shell is shown in Figure 10.4 [21]. Microencapsulation of solid, liquid, or gases can be prepared by using various techniques such as interfacial polymerization, in situ polymerization, chemical coacervation, solvent evaporation, spray drying, fluidized bed methods, etc. [3, 22]. The core–shell structure of microencapsulated PCM structure formation process by single-stage emulsion technique is schematically illustrated in Figure 10.5. The selection of microencapsulation methods depends on desired characteristics and end applications in the fibrous structure. The properties of microcapsules, their size, shapes, shell materials, method of applications, and compatibility with the formulation of polymer depend on the type of applications. To preserve the functional properties of microencapsulated

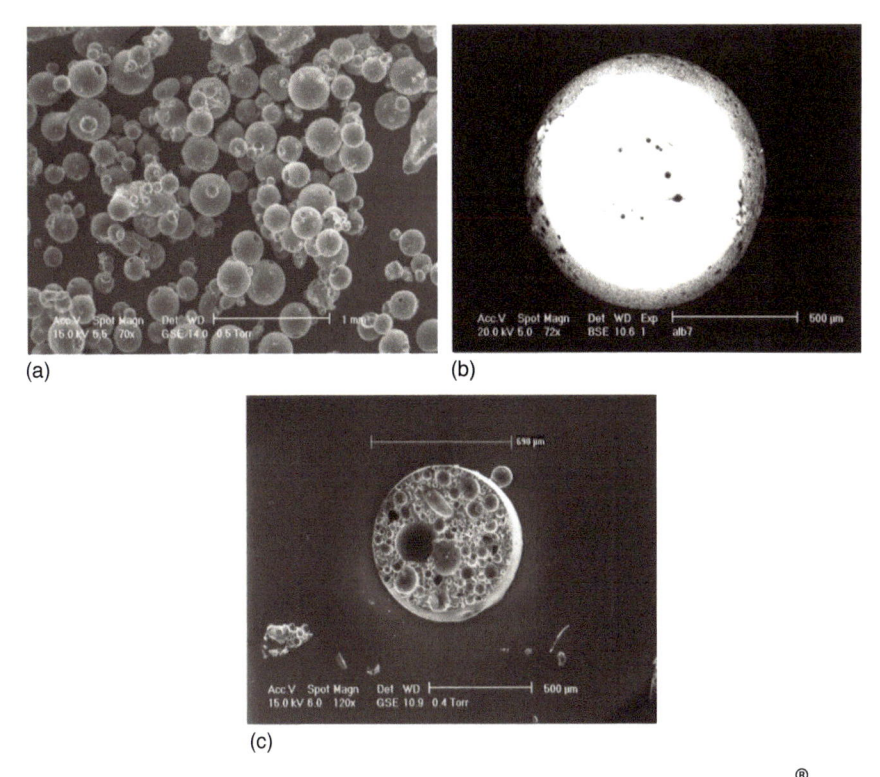

Figure 10.4 Scanning electron microscopic images of microcapsules containing PRS® paraffin wax as core materials and polystyrene as shell material. (a): Overall view of a batch of micro particles. (b) and (c) Cross sections of single microparticles. Source: Sanchez et al. 2007 [21]. Reproduced with permission of Springer Nature.

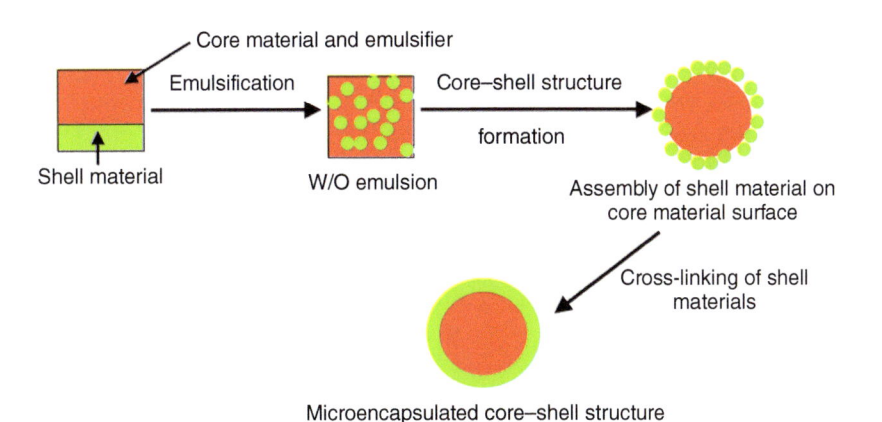

Figure 10.5 Schematic of process flow for the core–shell structure formation of microencapsulated PCMs by single-stage emulsion technique.

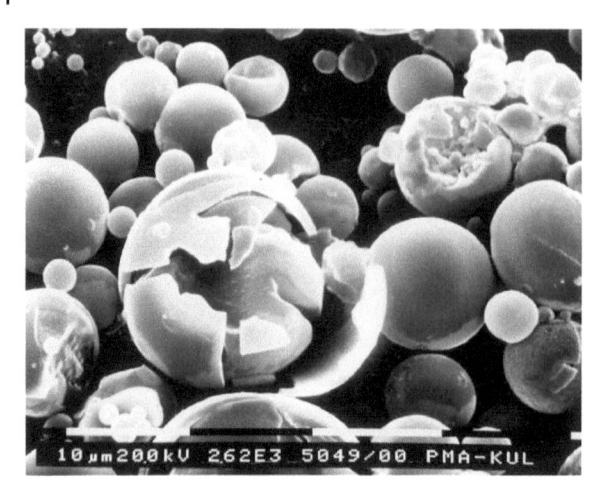

Figure 10.6 Scanning electron microscope image showing the thickness of urethane shell material. PCMs were cryogenically fractured. Source: Stappers et al. 2005 [24]. Reproduced with permission of Electrochemical Society.

PCMs over repeated phase transition cycles, PCM should remain within the shell over the entire product life cycles and should increase heat transfer area. Further, PCM microcapsules should be resistant to mechanical and thermal stress [15, 23]. Therefore, durability and thickness of shell (Figure 10.6) materials in core–shell structure is very important for the thermal balance and mechanical property of microencapsulated PCMs.

If PCMs are applied directly on the fibrous structure without microencapsulation, following problems may arises [25]:

(a) Leaking of PCM from the surface in its liquid or semi liquid forms.
(b) Changes for the numbers of hydrates in the salts specially in humid condition.
(c) Diffusion of low viscous melted PCMs from the textile substrate.

10.3.3 Techniques to Prepare Phase Change Fibrous Structures

Principally, PCM incorporated fibers could be made through two different approaches: (i) PCM microcapsules can be dispersed into the polymer solution or melt prior to the extrusion of fiber, and (ii) PCM microcapsules could be attached on the fiber surface by coating, finishing, etc. [6]. Various proposed PCM incorporated fibrous structures are schematically shown in Figure 10.7.

10.3.3.1 Coating of Encapsulated PCM on Fibrous Structure

Microencapsulated PCMs could be attached on the surface of fibrous structure of synthetic fiber during fiber finishing process, while for natural fiber during the preparation stage prior to the weaving or during the fabric finishing process by using a suitable binder (Figure 10.7a). In a typical coating process of encapsulated PCM on fiber, the microsphere PCMs are dispersed in a water solution, which contains a surfactant, dispersant, antifoaming agent, a thickener, and polymer binder [6]. Durability of microencapsulated PCM on the fiber surface depends on the interaction of microencapsulated PCM with the fiber surface. In general, softness and smoothness of the fibers are decreases after coating of PCMs on the fiber surfaces.

Figure 10.7 Schematic of PCM incorporated fibrous structures. (a) Encapsulated PCM-coated fiber. (b) Encapsulated PCM-entrapped fiber. (c) Core–sheath structure of PCM/polymer hollow fibers.

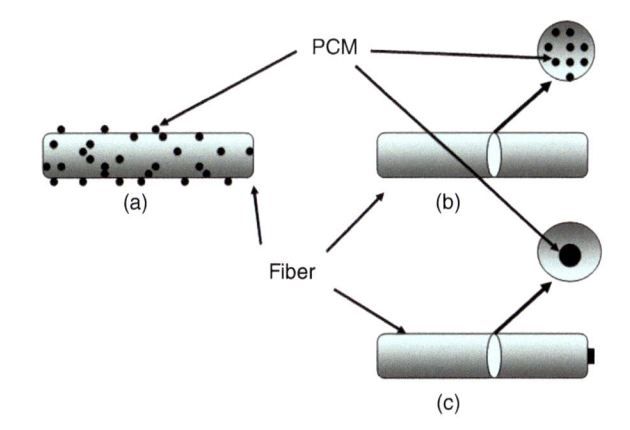

10.3.3.2 PCM Incorporated Fibrous Structure by Conventional Spinning

In this method, microencapsulated PCMs are added in the polymer solution or melt, and then PCM incorporated fibers are spun according to the methods of conventional melt spinning, dry or wet spinning, etc. Microencapsulated PCM incorporated fibers store or release heat energy with the changes of surrounding environment or human body temperature [6, 26]. Polypropylene monofilament fibers incorporated with PCM can be prepared by melt spinning with a PCM loading up to 12% [27]. Li et al. reported melt-spun polyethylene fibers embedded with microencapsulated PCMs and alginate fiber that contained PCM microcapsules spun by using wet spinning method [28]. Form-stable polyacrylonitrile (PAN) and SA fiber can be prepared by solution blending. Phase change transition temperature and phase transition enthalpy of PCM incorporated fiber increase with increasing stearic acid (SA) content in the fiber matrix [29]. SA could be attached with PAN matrix by intermolecular forces that ensure homogeneous distribution of PCMs in the PAN matrix [29, 30]. PCM incorporated composite fibre can be prepared by electrospinning technique [31–33]. Heat storage capacity for some of the phase change fibers has been tabulated in Table 10.4. However,

Table 10.4 Heat storage capacity of PCM incorporated fibers.

Material	PCMs	Enthalpy of melting (J/g)	Fiber preparation method	References
Polyacrylonitrile (PAN)	Eutactic mixture of fatty acids[a]	120.1	Electrospinning	[32]
Carbon fibre mats	Capric acid	110.3	Electrospinning of PAN followed by carbonization	[32]
Polyethylene terephthalate	Lauric acid	70.76	Electrospinning	[31]
Polyacrylonitrile	Capric acid	165.4	Electrospinning	[32]
Polyethylene terephthalate	Adipic acid dioctadecyl esters (75%)	66.5	Electrospinning	[33]

a) Capric acid and lauric acid.

Table 10.5 Some of the key advantages and disadvantages of melt spinning and solution spinning of phase change fiber.

Spinning methods	Advantages	Disadvantages
Melt spinning	(i) PCMs are permanently locked within the fiber matrix. Therefore, durable phase change fiber as compared with coating or finishing of microencapsulated PCM on fiber surface (ii) Thermal properties of polymer of the fiber matrix will affect the resultant thermoregulating properties of phase change fiber (iii) Environmental friendly method due to the elimination of toxic solvent (as compared with solution spinning)	(i) Shell materials of microcapsules of PCM should be resistant to temperature of polymer melt (ii) Particle size and size distribution of microsphere of PCM affects the spin ability and mechanical properties of fibers (iii) Uniform distribution and dispersion of microencapsulated PCM in fiber matrix is difficult to achieve. (iv) Poor process ability and poor properties of fibers with high concentration of microencapsulated PCMs (v) Microcapsules of PCMs may not have the structural integrity to withstand the force exerted on the microcapsule during extrusion of fibers
Solution spinning (wet/dry spinning)	(i) PCMs are permanently incorporated locked within the fiber matrix. Therefore, durable phase change fiber as compared with coating or finishing of microencapsulated PCM on fiber surface (ii) Thermal properties of polymer of the fiber matrix will affect the resultant thermoregulating properties of phase change fiber (iii) Better mechanical properties of fiber in the case of wet solution spinning (as compared with melt spinning)	(i) Shell materials of microcapsules should be resistant to the dope solution (ii) Particle size and size distribution of microsphere of PCM affects the spin ability and mechanical properties of fibers (iii) In the case of wet spinning, PCM should not leach out from the microencapsulated PCM by solution of wet spinning bath (iv) Uniform distribution and dispersion of microencapsulated PCM in fiber matrix may be difficult to achieve (v) Poor process ability and poor properties of fibers with high concentration of microencapsulated PCMs (vi) Microcapsules of PCMs do not have the structural integrity to withstand the force exerted on the microcapsule during extrusion of fibers

advantages and disadvantages of conventional spinning processes for the preparation of phase change fibers have been summarized in Table 10.5.

10.3.3.3 Phase Change Fibers by Electrospinning

Electrospinning is an efficient technique to prepare ultrafine nanofibers and produces continuous nanofibers from submicron diameters down to nanometer diameters in a high potential electric field by using a conductive polymer solution. Application of high voltage creates electrically charged jet of polymer solution or melt. Before reaching the jet to the target surface, solvent evaporates, and liquid polymer solidifies, and electrospun nanofibrous mats are collected on the conductive plate connected with ground. Electrospun fibers of lauric acid and polyethylene terephthalate (PET) show smooth cylindrical surface with a latent heat of fusion around 70.76 J/g while the diameter of the fiber ranges from several tens to several hundred nanometers [31]. Sheath–core structure of polymer and PCM nanofiber can be prepared by coaxial electrospinning technique. Wan et al. reported nanofiber of PAN as sheath and isopropyl palmitate and paraffin oil as PCM in the core material [34]. PCMs are well encapsulated in the hollow PAN fiber to ensure thermal balance with the sudden environmental temperature changes due to the homothermal effect of PCM incorporate fiber [34]. Zdraveva et al. reported poly(vinyl alcohol) nanofiber incorporated with PCM by using emulsion electrospinning technique. Nanofibers maintained morphological integrity during the phase change and heat-regulating performances over the cyclic phase changes [35]. Phase change electrospun fibers based on PEG as PCM and cellulose acetate as matrix have been reported by Chen et al. Morphology of composite fibers shows cylindrical and smooth surface (Figure 10.8). PCM such as PEG has been distributed both on the surface and in the core of the fiber matrix [36]. Mechanical properties

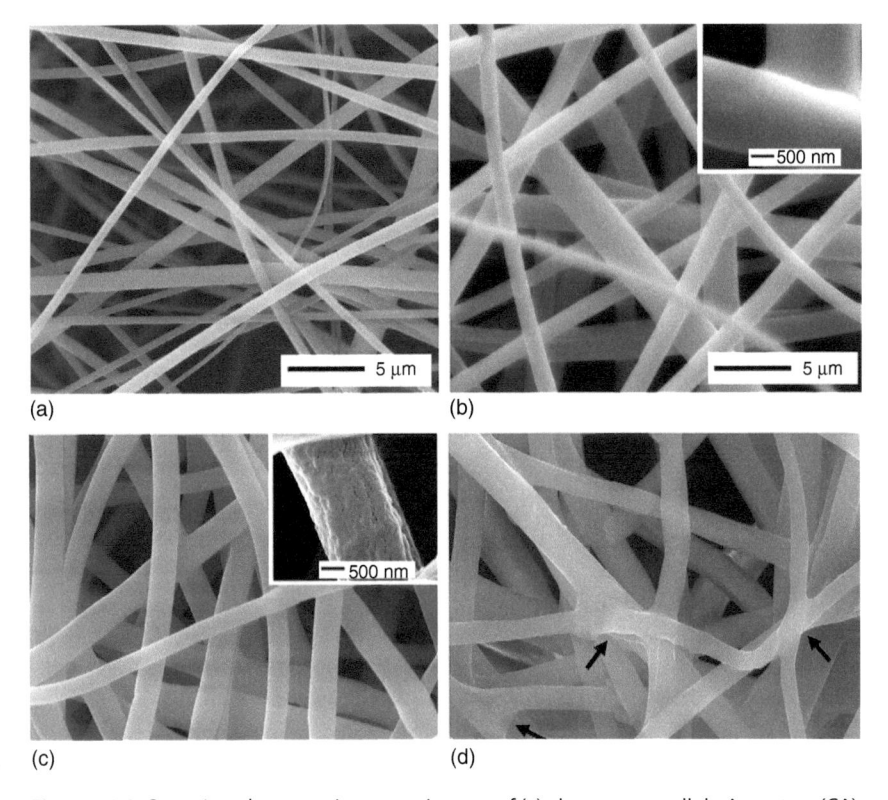

Figure 10.8 Scanning electron microscopy images of (a) electrospun cellulosic acetate (CA) fibers, (b) electrospun phase change fiber of polyethylene glycol (PEG)/CA, (c) washed phase change fiber of PEG/CA, and (d) electrospun PEG/CA composite fibers after 100 heating/cooling cycles treated. Source: Chen et al. 2007 [36]. Reproduced with permission of Elsevier.

of the PCM incorporated fibers generally decrease. The extent of decrease for tensile properties depends on the distribution and/or dispersion of PCM phases in the matrix phase and the intermolecular forces between the two phases of PCM and the matrix of fiber. Chen et al. shows that the tensile properties of the electrospun composite phase change PET fibers were lower than that of the electrospun pure PET fibers [31].

10.4 Phase Change Fibers for Advanced Material Applications

PCMs could adapt to the thermal regulating functional performance of the materials made with PCM incorporated fibrous structures [37, 38] by altering their state phases in a defined temperature range. Applications of phase change fibrous materials include smart textile, shoes/accessories, automotive, electrical, civil engineering, and many other advanced material applications. The following is a brief summary for the application of PCM incorporated fibers for the advanced material applications.

10.4.1 Phase Change Fibers for Advanced Textile Applications

10.4.1.1 Sportswear

Active wear required a thermal balance between the heat generated by the body and the heat released into the environment while engaged in physical activity to reduce thermal stress situation. For the normal garment, heat generated by the human body during physical activity is often not released into the environment in the required amount and thus increases the thermal stress to the wearer. When sportswear was prepared by using PCM incorporated fibers, during physical activity, the wearer's body temperature increases, and the excess heat is absorbed by the encapsulated PCMs and released when necessary (for example, during the rest time) [6].

PCM incorporated fibrous materials could be used to manufacture garments for the sportswear in order to control the sport persons' body temperature dynamically [3]. PCM in the fibrous structure will absorb the body heat during the sporting activity, whereas absorbed heat by PCM will be released and keep the body warm during the resting. Varieties of sports apparel with PCM incorporated fibrous structure can be realized such as skiwear, hunting cloth, other outdoor and indoor sportswear, etc. [39]. Cold winter sports such as alpine climbing, ice climbing, and recreational skiing have further stimulated the demand of PCM incorporated fibrous textiles, which can keep the outdoor sport individuals dry, warm, and comfortable. Snowboard gloves, garments for ice climbing, and underwear for cycling and running are few more examples for the applications of PCM incorporated fibrous structures in sportswear [6].

10.4.1.2 Hospital Applications

PCM interacts with the microclimate around the human body and could respond to the fluctuations in surrounding temperature due to the changes in patient activity levels and external environment changes. Therefore, surgical apparel, bedding materials for patients, bandages, and other medical textile products prepared by using PCM incorporated fibrous structure could be applicable to the hospital applications. Microencapsulated PCM incorporated textiles could also be used for medical applications for hot and cold therapies [21].

10.4.1.3 Beddings and Accessories

PCM incorporated fibrous materials could be applicable in quilts, bed sheet, pillow cover, mattress, etc. to ensure dynamic temperature control in bed. Additional heat energy will be absorbed by PCM incorporated textiles when body temperature rises in order to absorb the body heat. However, in the case of drop of the body temperature, stored energy will be released and the body will keep warm [6].

10.4.1.4 Other Applications in Advanced Textiles

Textiles prepared with phase change fibers could be useful for the space suits, gloves, etc. to keep the astronauts in a comfortable state. Further, phase change fibers can be used for the manufacturing of thermal underwear, jackets, sports garments, skiwear, etc. Some of the other applications of phase change fibers in textile fields include textiles for the interior decorations, curtain, mattress, blankets, home furnishings, firefighter suits, etc.

10.4.2 Automotive Industries

Microencapsulated PCM incorporated fibrous structure can be applicable in automobile textiles, for example, seat covers. Paraffin-based PCM has several advantages such as high capacity of its heat storage capability, low toxicity, lack of hygroscopic properties, and low cost and obtains the targeted temperature range by blending, and all these excellent properties make paraffinic incorporated phase change fibers suitable for the automotive applications. Application of PCM incorporated fibrous structure in car interior and seat cover can provide superior thermal control to the passengers [6]. During the summer, temperature inside the automobile can increase significantly; in order to regulate the temperature inside, many cars are equipped with air conditioning systems, which is quite energy-intensive process. In order to regulate the interior temperature of car, the phase change fiber could be used in automotive interior, which could offer energy saving as well as improve the thermal comfort of car interior.

10.4.3 Electrical Applications

Phase change fibers can be used in electrical engineering as optical sensor fibers [40]. Phase change fibers act as passive elements and do not required any additional energy source; therefore, application of phase change fibers in electronic and computer is promising. Phase change fibers in thermal diode can be used to improve the effectiveness of heat sink [41]. Management of generated heat from the electronic devices is a very important issue as the size of electronic component decreases while its speed increases. Inappropriate thermal management of electronic components can reduce its performance, and it could damage the electronic components. Electrospun nanofiber mat can be used to protect electronic devices by dissipation of heat from the electronic devices, computers, circuit board, etc. [42].

10.4.4 Other Applications

Phase change fibers can also be used for shoes and accessories. Phase change fibers can regulate the temperature by absorbing, storing, redistributing, and releasing of heat to prevent dramatic change of wearer head and feet temperatures. Therefore, the liner textiles made with phase change fibers could be applicable for helmet and shoes in order to regulate the wearer temperature. In civil engineering, it can be used as pipe wrap, building insulation, wallpapers, tiles, etc. Some of the other applications of PCM fibers include containers to regulate the temperature of food and drinks, household appliance insulation, etc. [43, 44].

10.5 Summary

LSH PCMs are suitable to develop thermoregulated advanced fibrous materials because these materials are efficient to store thermal energy and provide higher TES density with smaller temperature difference between storing and releasing of heat. In this chapter author provided an overview of PCM incorporated fibrous materials for advanced material applications with an emphasis on thermoregulating textiles. Generally, microencapsulated PCMs are applied on the substrates, for example, garment by finishing, coating, lamination, etc., and coating/lamination would affect the other properties of garments such as moisture permeability, fabric aesthetic look, garment hand, etc. The textile substrate prepared by using PCM incorporated fibers would be more effective to prepare thermoregulating textiles. Apart from the advanced textile applications of phase change fibers, this chapter also summarized the applications in electrical and automotive industries and other areas of advanced material applications. Uniform dispersion and distribution of microencapsulated PCMs throughout the fibrous structure is a major challenge, and this subsequently would affect the thermoregulating properties of PCM incorporated fibrous materials. Other challenges for manufacturing fibrous structure with PCMs include incorporation of PCMs in a single fibrous matrix or surface, cost, etc. Future research and developments in these areas are required for the further development.

References

1 Gasia, J., Miró, L., de Gracia, A. et al. (2016). Experimental evaluation of a paraffin as phase change material for thermal energy storage in laboratory equipment and in a shell-and-tube heat exchanger. *Applied Sciences* 6 (4): 112.

References

2 He, B. and Setterwall, F. (2002). Technical grade paraffin waxes as phase change materials for cool thermal storage and cool storage systems capital cost estimation. *Energy Conversion and Management* 43 (13): 1709–1723.

3 Sarier, N. and Onder, E. (2007). The manufacture of microencapsulated phase change materials suitable for the design of thermally enhanced fabrics. *Thermochimica Acta* 452 (2): 149–160.

4 Hasnain, S.M. (1998). Review on sustainable thermal energy storage technologies, Part I: Heat storage materials and techniques. *Energy Conversion and Management* 39 (11): 1127–1138.

5 Sharma, A., Tyagi, V.V., Chen, C.R., and Buddhi, D. (2009). Review on thermal energy storage with phase change materials and applications. *Renewable and Sustainable Energy Reviews* 13 (2): 318–345.

6 Mondal, S. (2008). Phase change materials for smart textiles – an overview. *Applied Thermal Engineering* 28 (11–12): 1536–1550.

7 Budd, G.M. (1989). Ergonomic aspects of cold stress and cold adaptation. *Scandinavian Journal of Work, Environment & Health* 15 (Suppl 1): 15–26.

8 Nielsen, R., Olesen, B.W., and Fanger, P.O. (1985). Effect of physical-activity and air velocity on the thermal insulation of clothing. *Ergonomics* 28 (12): 1617–1631.

9 Shishoo, R.L. (1998). Technology for comfort. *Textile Asia* 19 (6): 93–110.

10 Sen, A.K. (2001). *Coated Textiles: Principle and Applications* (Tech ed. Damewood J), 133–154. Technomic Publishing Co.

11 Ying, B.-A., Kwok, Y.-L., Li, Y. et al. (2004). Assessing the performance of textiles incorporating phase change materials. *Polymer Testing* 23 (5): 541–549.

12 Lennox-Kerr, P. (1998). Comfort in clothing through thermal control. *Textile Month* (November), pp. 8–9.

13 Huang, J.Y., Lu, S.L., Kong, X.F. et al. (2013). Form-stable phase change materials based on eutectic mixture of tetradecanol and fatty acids for building energy storage: preparation and performance analysis. *Materials* 6: 4758–4775.

14 Feldman, D., Shapiro, M.M., and Banu, D. (1986). Organic phase change materials for thermal energy storage. *Solar Energy Materials* 13 (1): 1–10.

15 Farid, M.M., Khudhair, A.M., Razack, S.A.K., and Al-Hallaj, S. (2004). A review on phase change energy storage: materials and applications. *Energy Conversion and Management* 45 (9–10): 1597–1615.

16 Lane, G.A. (1980). Low temperature heat storage with phase change materials. *International Journal of Ambient Energy* 1: 155–168.

17 Sari, A. and Kaygusuz, K. (2002). Thermal performance of a eutectic mixture of lauric and stearic acids as PCM encapsulated in the annulus of two concentric pipes. *Solar Energy* 72 (6): 493–504.

18 Arjun, D. and Hayavadana, J. (2014). Thermal energy storage materials (PCMs) for textile applications. *Journal of Textile and Apparel, Technology and Management* 8 (4): 1–11.

19 Saito, A., Okawa, S., Shintani, T., and Iwamoto, R. (2001). On the heat removal characteristics and the analytical model of a thermal energy storage capsule using gelled Glauber's salt as the PCM. *International Journal of Heat and Mass Transfer* 44 (24): 4693–4701.

20 Wang, S.X., Li, Y., Hu, J.Y. et al. (2006). Effect of phase-change material on energy consumption of intelligent thermal-protective clothing. *Polymer Testing* 25 (5): 580–587.

21 Sanchez, L., Sanchez, P., de Lucas, A. et al. (2007). Micro encapsulation of PCMs with a polystyrene shell. *Colloid and Polymer Science* 285 (12): 1377–1385.

22 Onder, E., Sarier, N., and Cimen, E. (2008). Encapsulation of phase change materials by complex coacervation to improve thermal performances of woven fabrics. *Thermochimica Acta* 467 (1–2): 63–72.

23 Boh, B. and Knez, E. (2006). Microencapsulation of essential oils and phase change materials for applications in textile products. *Indian Journal of Fibre and Textile Research* 31: 72–82.

24 Stappers, L., Yuan, Y., and Fransaer, J. (2005). Novel composite coatings for heat sink applications. *Journal of the Electrochemical Society* 152 (7): C457–C461.

25 Mondal, S. (2011). Thermo-regulating textiles with phase-change materials. In: *Functional Textiles for Improved Performance, Protection and Health*, (eds. Pan, N. and Sun. G.) 163–183. Woodhead Publishing.

26 Tjonnas, M.S., Faerevik, H., Sandsund, M., and Reinertsen, R.E. (2015). The dry-heat loss effect of melt-spun phase change material fibres. *Ergonomics* 58 (3): 535–542.

27 Iqbal, K. and Sun, D.M. (2014). Development of thermo-regulating polypropylene fibre containing microencapsulated phase change materials. *Renewable Energy* 71: 473–479.

28 Li, W., Ma, Y.-J., Tang, X.-F. et al. (2014). Composition and characterization of thermoregulated fiber containing acrylic-based copolymer microencapsulated phase-change materials (MicroPCMs). *Industrial and Engineering Chemistry Research* 53 (13): 5413–5420.

29 Guo, J., Xiang, H.X., and Wang, Q.Q. (2011). Preparation and properties of form-stable phase change materials polyacrylonitrile fiber/stearic acid blends. In: *New and Advanced Materials, Parts 1 and 2*, Advanced Materials Research, vol. 197–198 (eds. H.Y. Zhou, T.L. Gu, D.G. Yang, et al.), 584–588. Stafa-Zurich: Trans Tech Publications Ltd.

30 Guo, J., Xiang, H.X., Wang, Q.Q., and Xu, D.Z. (2013). Preparation and properties of polyacrylonitrile fiber/binary of fatty acids composites as phase change materials. *Energy Sources, Part A: Recovery, Utilization, and Environmental Effects* 35 (11): 1064–1072.

31 Chen, C., Wang, L., and Huang, Y. (2008). A novel shape-stabilized PCM: electrospun ultrafine fibers based on lauric acid/polyethylene terephthalate composite. *Materials Letters* 62 (20): 3515–3517.

32 Cai, Y.B., Zong, X., Zhang, J.J. et al. (2013). Electrospun nanofibrous mats absorbed with fatty acid eutectics as an innovative type of form-stable phase change materials for storage and retrieval of thermal energy. *Solar Energy Materials and Solar Cells* 109: 160–168.

33 Chen, C.Z., Liu, S.S., Liu, W.M. et al. (2012). Synthesis of novel solid-liquid phase change materials and electrospinning of ultrafine phase change fibers. *Solar Energy Materials and Solar Cells* 96 (1): 202–209.

34 Wan, Y.F., Zhou, P.C., Liu, Y.D., and Chen, H.X. (2016). Novel wearable polyacrylonitrile/phase-change material sheath/core nano-fibers fabricated by coaxial electro-spinning. *RSC Advances* 6 (25): 21204–21209.

35 Zdraveva, E., Fang, J., Mijovic, B., and Lin, T. (2015). Electrospun poly(vinyl alcohol)/phase change material fibers: morphology, heat properties, and stability. *Industrial and Engineering Chemistry Research* 54 (35): 8706–8712.

36 Chen, C.Z., Wang, L., and Huang, Y. (2007). Electrospinning of thermo-regulating ultrafine fibers based on polyethylene glycol/cellulose acetate composite. *Polymer* 48 (18): 5202–5207.

37 Bendkowska, W., Tysiak, J., Grabowski, L., and Blejzyk, A. (2005). Determining temperature regulating factor for apparel fabrics containing phase change material. *International Journal of Clothing Science and Technology* 17 (3–4): 209–214.

38 Ying, Y.L.K., Li, Y., Yeung, C.-Y., and Song, Q.-W. (2004). Thermal regulating functional performance of PCM garments. *International Journal of Clothing Science and Technology* 16 (1/2): 84–96.

39 Shin, Y., Yoo, D.-I., and Son, K. (2005). Development of thermoregulating textile materials with microencapsulated phase change materials (PCM). II. Preparation and application of PCM microcapsules. *Journal of Applied Polymer Science* 96 (6): 2005–2010.

40 Priest, R. and Hughes, R. (1980). Thermally induced optical phase effects in fiber optic sensors. *Applied Optics* 19 (9): 1477–1483.

41 Janarthanan, B. and Sagadevan, S. (2015). Thermal energy storage using phase change materials and their applications: a review. *International Journal of ChemTech Research* 8 (6): 250–256.

42 Li, Y., Li, X.-W., He, J.-H., and Wang, P. (2014). Thermal protection of electronic devices with the Nylon 6/66-PEG nanofiber membranes. *Thermal Science* 18 (5): 1441–1446.

43 Chalco-Sandoval, W., Fabra, M.J., Lopez-Rubio, A., and Lagaron, J.M. (2017). Use of phase change materials to develop electrospun coatings of interest in food packaging applications. *Journal of Food Engineering* 192: 122–128.

44 Dugan, J. and Kuckhoff, E. (2007). Multicomponent fiber comprising a phase change material. US Patent WO 2007035483 A1.

11

Bicomponent Fibers

Rudolf Hufenus[1], Yurong Yan[2], Martin Dauner[3], Donggang Yao[4], and Takeshi Kikutani[5]

[1] Empa, Swiss Federal Laboratories for Materials Testing and Research, Laboratory for Advanced Fibers, Department of Functional Materials, Lerchenfeldstrasse 5, 9014 St. Gallen, Switzerland
[2] South China University of Technology, Department of Polymer Materials and Engineering, No. 381 Wushan Road, Tianhe, Guangzhou 510640, China
[3] Institute of Textile Technology and Process Engineering, German Institutes of Textile and Fiber Research, Department of Filament and Nonwoven Technologies, Körschtalstraße 26, 73770 Denkendorf, Germany
[4] Georgia Institute of Technology, School of Materials Science & Engineering, Atlanta, GA 30332-0295, USA
[5] Tokyo Institute of Technology, School of Materials and Chemical Technology, Department of Materials Science and Engineering, 2-12-1-S8-32, O-okayama, Meguro-ku, Tokyo 152-8550, Japan

11.1 Introduction

Melt spinning is the most commonly used method for manufacturing commercial synthetic fibers. A trend in polymer melt spinning is variation of fiber morphology by bicomponent (conjugated) spinning [1, 2], one of the most interesting developments in the field of synthetic fibers [3–6]. A bicomponent fiber is made from two or more polymers of different chemical (e.g. composition, additives) and/or physical (e.g. average molecular weight, crystallinity) nature, extruded from one spinneret to form a single fiber [7]. The polymer flows are kept separate up to the spin pack, where they meet and exit together through the spinneret. When the filament leaves the spinneret, it consists of non-mixed components that touch at the interface. It is worth noting that the dynamics at the interface play a significant role in affecting the behavior and performance of the process.

Bicomponent fibers pass through common melt-drawing processes similar to conventional synthetic fibers [4]. In consequence, both thermoplastic polymers have to withstand extrusion temperature and shear stress without degradation (thermal stability), the average molecular weight has to be high (melt strength), and the molecular weight distribution narrow (constant flow) for the fiber not to break under draw-off strain, and the mobility of the molecular chains has to be high enough to disentangle and unfold under stress and to orient in fiber direction under strain [7, 8]. In addition, coextrusion of different polymers into bicomponent fibers with special cross sections requires that design of spin pack and spinneret is well understood, that respective processing temperatures and viscosities are well balanced, and that adhesion/bonding between the two components is well controlled.

Handbook of Fibrous Materials, First Edition. Edited by Jinlian Hu, Bipin Kumar, and Jing Lu.
© 2020 Wiley-VCH Verlag GmbH & Co. KGaA. Published 2020 by Wiley-VCH Verlag GmbH & Co. KGaA.

11 Bicomponent Fibers

The main objective of bicomponent melt spinning is to exploit capabilities not existing in either polymer alone, as advantageous mechanical, physical, or chemical properties of two materials can be combined in one fiber, expanding the range of possible applications [9–11]. Depending on the characteristics of the different polymers, bicomponent fibers are predominately commercialized as bonding elements in thermobonded nonwoven fabrics, as self-crimping fibers to achieve textured yarn, or as fibers with the surface functionality of special polymers and additives at reduced cost [5, 6, 12]. In addition, the two components can mutually influence their thinning and solidification behavior along the spinline during melt spinning, which can be utilized to influence the molecular structure formation of both components in a desired way [13–15].

The first commercial bicomponent fiber was a PA66 – random copolymer side-by-side hosiery yarn produced by DuPont in the mid-1960s, which on retraction formed a highly coiled elastic fiber [16]. The first commercial core–sheath application was a binding fiber manufactured by ICI for carpets and upholstery, consisting of a higher melting temperature core (PA66) and a lower melting temperature sheath (PA6) [11]. Since then bicomponent technology progressed significantly, especially by combining commodity and engineering polymers and diverse additives to achieve tailored characteristics.

11.2 Bicomponent Fiber Spinning Technologies

11.2.1 Spin Pack Design

In bicomponent fiber spinning, one can differentiate between single die and multiple die spinneret, depending on the position where the molten polymers meet (Figure 11.1). In the first, by far most common type, two pressurized polymer

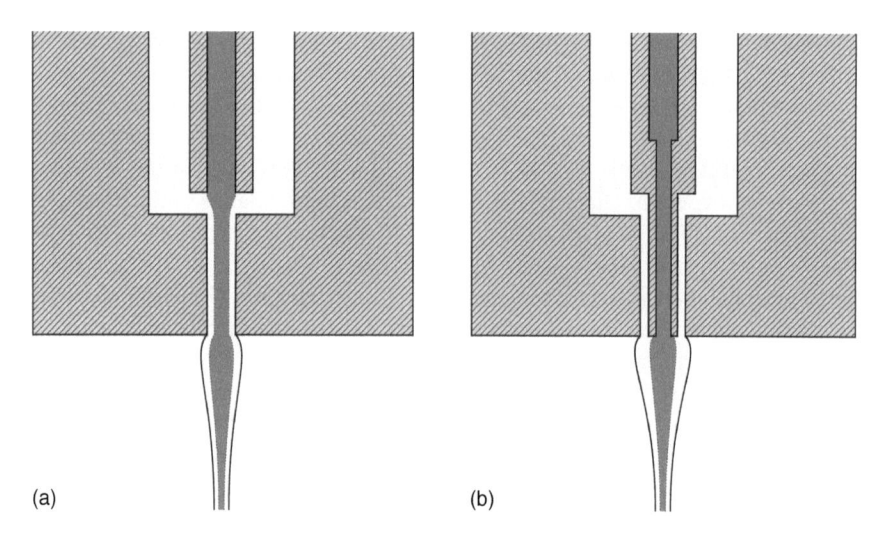

Figure 11.1 Two schematic examples of bicomponent spinnerets: (a) single die and (b) multiple die.

melts meet within the spinneret. Ensuring a laminar flow prevents mixing of the two phases. The development of etched stacked plates to create polymer channels in the spin pack led to smaller packs with higher hole densities and shorter residence times, as well as larger number of segments or islands in the fibers [17]. In the second, less prominent type, each polymer is extruder separately, and the melts meet just at the capillary exit. This geometry is complex to realize but can lessen rheological disparities between the components and enable an exact position of the core within the sheath polymer.

Central elements of bicomponent spinning equipment are the melt distribution plates. The polymer melts conveyed by the two spin pumps are fed to the respective filter packs. In the case of multifilament spinning, the two polymer melts have to be evenly distributed to each die in the required bicomponent configuration. Configurations like core–sheath and side by side can be served from simple geometries like concentric or alternating parallel melt channels from where boreholes guide the melt to the spinning capillaries. The higher the number of sectors in the cross section of each single filament, the more complicated the distribution design is. For drilled holes 16 segments are already a reasonable limit. Figure 11.2 shows a die for producing single islands-in-the-sea filaments. Even for this type of die, a spinneret for multiple filaments can be designed, but the number of filaments will be limited.

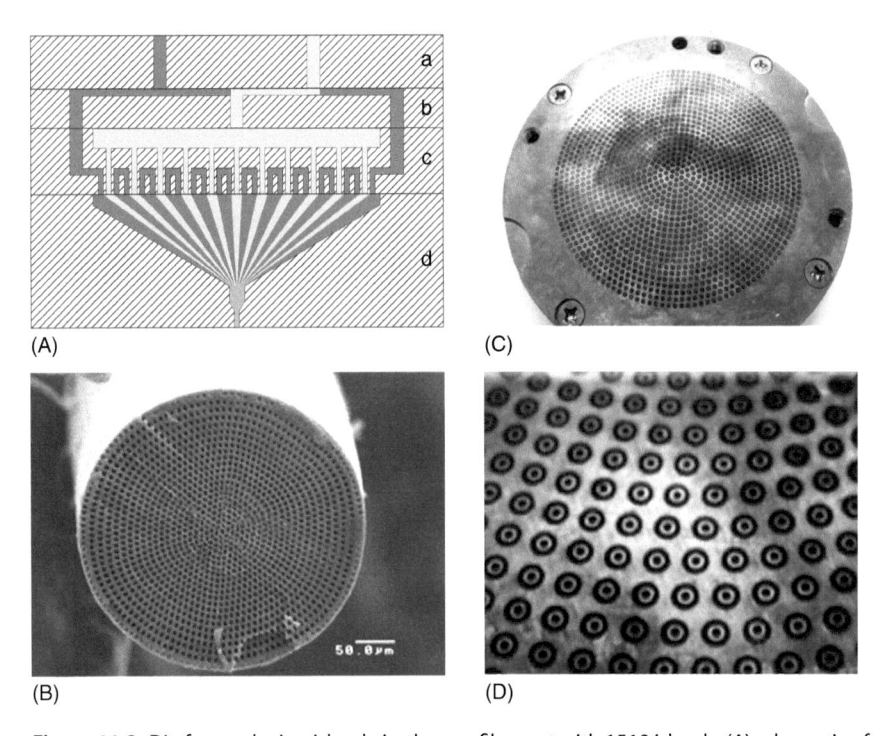

Figure 11.2 Die for producing islands-in-the-sea filament with 1519 islands: (A) schematic of die design, where a, b and c are melt distribution plates, and d is the spinneret plate, (B) cross section of fiber produced with respective die, (C) bottom side of plate c, and (D) magnified image of bottom side of plate c.

Islands-in-the-sea filaments with a large number (up to 10 000) islands need other solutions, like the Hills technology. The melt flow channels in thin distribution plates (0.1–1.5 mm) are produced by a proprietary etching process [18, 19]. This not only allows a comparably cheap production, almost optimal flow ducts, but also a high density of melt flow channels, which are required for high number of islands, resulting in sub-microstructures in the drawn filament.

11.2.2 Cross-Sectional Geometries

The three main geometries of multicomponent fibers are side-by-side, core–sheath, and multiple core configurations (Figure 11.3).

The core–sheath approach enables a variety of surfaces while maintaining major fiber and textile properties, because most thermoplastic polymers can be applied as a sheath over a core that provides the requested tensile strength [20]. Core–sheath types are predominately used as binder fibers for nonwovens, with a standard polymer as core and a low softening-point polymer as sheath [11, 21–24]. When applied in nonwoven production, the core–sheath fibers are heated to a temperature high enough to cause the sheath to soften, and consequently they will adhere to one another and stabilize the fabric [25, 26].

Side-by-side and eccentric core–sheath bicomponent fibers are most commonly used to produce self-crimping yarns applied in voluminous products. By combining two polymers that undergo differential shrinkage, the yarn curls up after thermal treatment or relaxation and develops crimp contraction [27, 28].

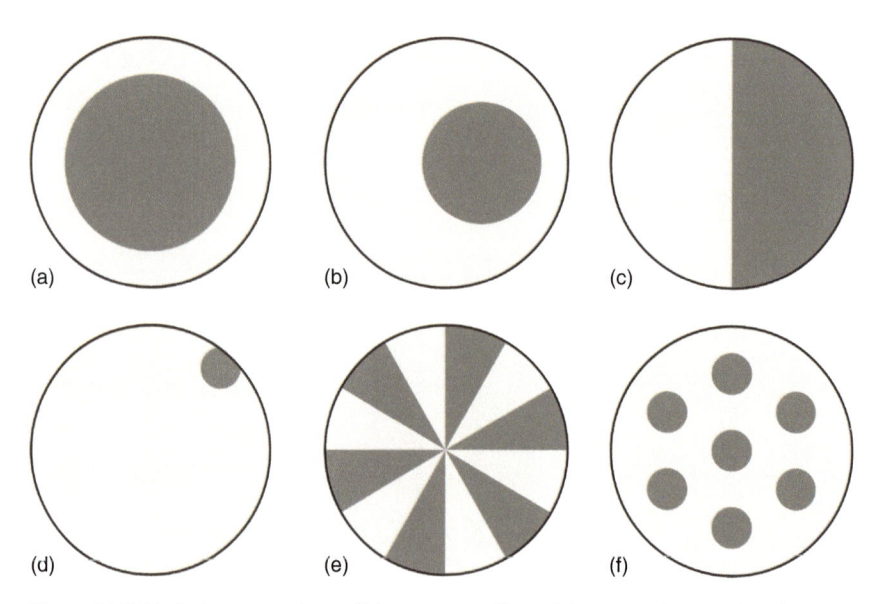

Figure 11.3 Typical cross sections of bicomponent fibers: (a) concentric core–sheath, (b) eccentric core–sheath, (c) 50/50 side by side, (d) unequal side by side, (e) segmented pie, and (f) islands in the sea.

Multiple core configurations, like islands-in-the-sea and segmented pie, are mainly applied to produce microfibers with diameters, smaller than those obtained via conventional melt spinning [29]. Kamiyama et al. [30] reported the production of nanofibers through the bicomponent spinning of islands-in-the-sea fibers with about 1000 islands. Islands-in-the-sea fibers comprise fibrils dispersed in a dissolvable matrix polymer that will be removed in a follow-up process [20]. In the segmented-pie technique, a bicomponent fiber is spun from two incompatible polymers that adhere poorly and split into microfibers when subjected to mechanical stress [12]. Prahsarn et al. [31] describe the splitting behavior of segmented-pie bicomponent fibers of polypropylene (PP)/polyethylene (PE). Fabrics made from microfibers are very flexible, and the high density of fibers makes them inherently windproof, while water vapor from perspiration can evaporate easily.

Hollow fibers can be considered as a type of core–sheath fibers in which the core is composed of a gaseous fluid (e.g. air). They have specialized uses in medical devices [32], but the largest volume market of melt-spun hollow fibers is in thermal insulation products [20]. Hollow fibers can be melt-spun using two different types of spinneret geometries. Most commonly used are "segmented arc" spinnerets, where the polymer melt flows through three or more separated arc slits, and subsequently fuses together below the spinneret to form a tubular shape [33–35]. Here, the fusing mechanism is driven by die swell and controlled by deflection of the melt streams, and it depends on the interplay of inertial and viscous forces [36]. In a "tube in orifice" spinneret, the polymer melt is spun through an annular opening, while inside the annulus a "lumen fluid," usually nitrogen or CO_2, is injected through a coaxial conduit [37, 38]. The second, less common case, indeed involves a bicomponent spinning process.

11.2.3 Melt Spinning Equipment

A typical bicomponent melt spinning plant comprises two screw extruders, a spin pack with discrete polymer conduits, and a set of spinnerets allowing for elaborate fiber cross sections. A schematic drawing of such a melt spinning plant is shown in Figure 11.4.

The polymers are melted using two single screw extruders (1, 2); for instance, one extruder is used for the core component and the other for the sheath component of a bicomponent fiber. Two separate spin pumps (3) and a special spin pack (4) combining the polymers in the spinneret (5) enable a well-balanced polymer flow. Leaving the spinneret the further run of the filaments is equal to conventional melt spinning. For safety reasons, evaporating monomers and oligomers are sucked in by an exhaust (6). The extrudate is spun into the quenching chamber (7) with adjustable air flow and quenching temperature (8). After cooling and application of a spin finish (9), the filaments (10) are drawn (online or offline) by several heated godets (11) with variable revolving speeds. The draw ratio, i.e. the ratio of speeds of draw and feed godets, can be chosen accordingly. Finally a winder (12) is used to spool the filaments on a bobbin.

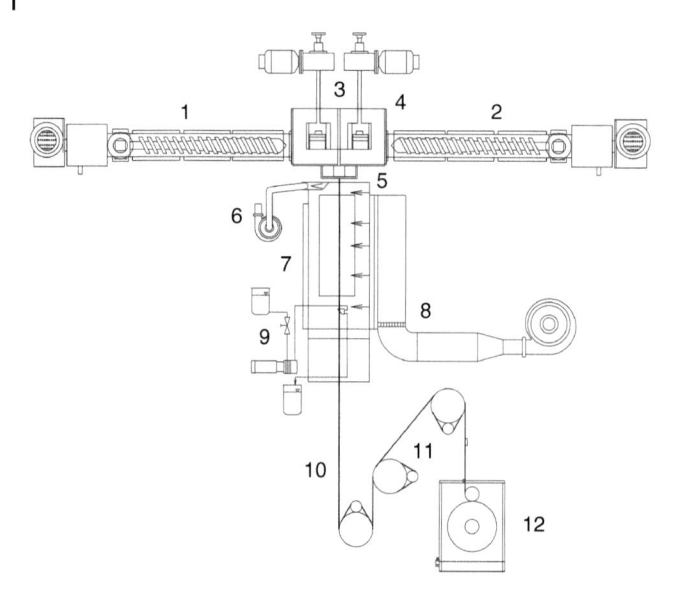

Figure 11.4 Schematic assembly of a bicomponent melt spinning plant (see text for an explanation of the numbered parts).

11.2.4 Special Spin Pack Designs

As nonlinear and elastic flow properties of polymer melts would lead to flow instabilities in coextrusion of a molten polymer and a liquid core [39], a special spin pack was designed that enabled the stable melt spinning of a bicomponent fiber with a liquid core. Figure 11.5 shows a 3D cross section of the customized spin pack [40]. The polymer is supplied from the gear pump through the inlet on the right-hand side; it flows into a conical chamber before entering an annular co-flow channel. The well-sealed inner injection capillary conveys the core liquid, injected by an attached high-pressure pump that prevents polymer backflow into the liquid-filled capillary. A suitable liquid has to meet several elementary requirements: It must exhibit a relatively low vapor pressure (e.g. $<10^5$ Pa) at the processing temperature to prevent exit spraying, the thermal stability of the liquid should be high enough, the liquid has to be noncorrosive to the extrusion equipment, and the liquid must not act as solvent for the polymer component.

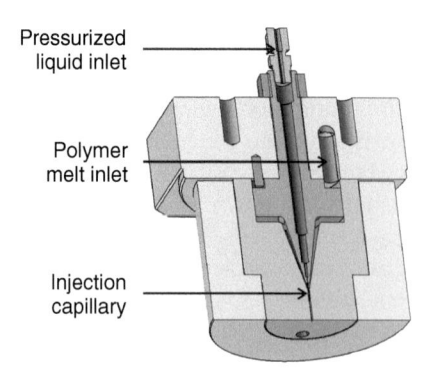

Figure 11.5 Cross section of the liquid-polymer coextrusion spin pack with concentric co-flow region. Source: Hufenus et al. 2016 [40]. Reproduced with permission of Elsevier.

Figure 11.6 Schematic of wire coating die used for overjacketing high tenacity filaments with a thermoplastic polymer. Source: Leal et al. 2016 [41]. Reproduced with permission of Elsevier.

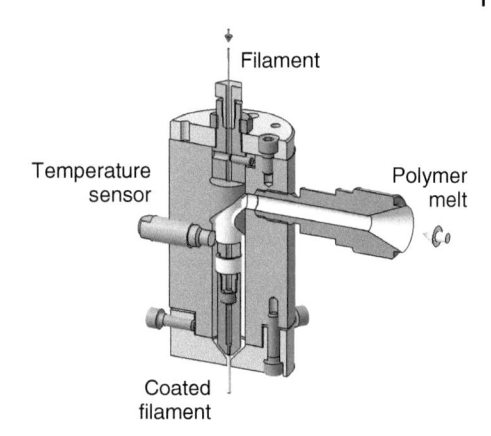

In order to produce bicomponent thermoplastic core–sheath fibers with a high tenacity core, a wire-coating spin pack (Figure 11.6) was modified to enable coating a high tenacity thermoplastic filament with a polymeric sheath [41]. The two materials are allowed to come in contact inside the die, mimicking a pressure-type die in which the pressure exerted by the polymer mass is used as an aid to enhance the adhesion between the two materials. Given that a filament lacks the rigidity of a metallic wire used in the conventional wire coating process, a steel capillary tube threaded vertically into the wire coating die serves as a guide.

The other important part in bicomponent spinning is to maintain and control the polymers as long as possible at their optimal temperatures during extrusion. One of the well-known technologies for bicomponent fiber extrusion with the concept of temperature separation is patented by Hills Inc. [42]. With the same goal, a highly integrated spin pack allowing the separate heating or cooling of up to three polymer channels was developed [43]. It comprises a separate oil cooling and heating system for each channel to keep the polymer melt flows at different temperatures down to the spinneret plate. Each of the three channel sections is thermally isolated toward the other two sections by an internal narrow grid of convection channels [44–46] (Figure 11.7). The spin pack was produced by

Figure 11.7 Schematic of a highly integrated spin pack for three different polymers with double helix heating/cooling channels and thermal insulation grid. Source: Hufenus et al. 2012 [43]. Reproduced with permission of John Wiley & Sons.

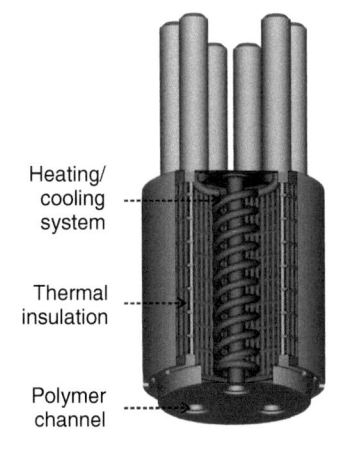

selective laser melting (SLM), an additive manufacturing process used for the fabrication of three dimensional metal parts [47].

11.3 Principles and Theories of Bicomponent Spinning

11.3.1 Structure Formation During Spinning and Drawing

In general, flexible-chain polymers can only be partially oriented in the fluid state during the melt spinning step. Their orientation has to be completed in the solid state, i.e. in the post-spinning drawing step, and for such fibers, drawing is the principal mean of building up their tensile properties. The orientation is a function of the actual strain, and a high degree of orientation can be produced without application of high stress if the strain rate is low. During the drawing process, the as-spun filament is irreversibly stretched up to several times its original length. The elongation is accompanied by extension and parallelization of macromolecules and crystallites along the fiber axis [48]. Increasing both the orientation and the degree of crystallinity through the drawing process finally leads to a fiber with greatly improved tenacity and modulus [49–51].

Significant fiber structure development also can be achieved in the melt spinning process if spinning speed is extremely high, where molecular orientation can develop in the molten state. In this case, tensile stress that originates from the strain rate and not from strain itself is the principal reason for the imposed orientation. Since viscosity increases with the decrease of temperature in cooling, the development of orientation in the spinline completes at the moment of solidification. High tensile stress can be achieved simply by applying high spinning speed, because the stresses originating from inertia and air-friction forces are the key factors. It also needs to be noted that crystallization can be accelerated significantly under the effect of molecular orientation.

In order to develop bicomponent fibers with good mechanical properties, it is essential to have information about crystallinity and orientation of both polymers. Wide-angle X-ray diffraction (WAXD) using a two-dimensional detector can provide such data [52]. Raman spectroscopy provides alternative yet supplemental information about fiber structure and morphology [53]. By comparison of the spectroscopic pattern of the orientated fiber with that of the bulk polymer, one can perform quantitative analysis of the fiber structure, e.g. to calculate the Herman's orientation factor. Coaxial distribution of molecular orientation and crystallinity in the cross section of circular fibers can be analyzed using an interference microscope [34, 54]. This technique can also be applied for the analysis of the structure of core and sheath polymers in bicomponent fibers [15].

11.3.2 Mutual Influence of Components on Orientation and Crystallinity

In addition to the high controllability of cross-sectional configuration of fibers, utilization of bicomponent melt spinning for the control of the structure development of individual polymers also attracts much attention [15]. In the melt spinning process, the polymers mutually modify their thinning behaviors along the spinline and therefore alter the tensile stress development, leading to different structure (orientation and crystallization) and properties when compared with the respective single-component fibers [55].

This mutual influence can be very well demonstrated in the case of core–sheath bicomponent fibers composed of polymers that do not crystallize in the melt spinning process. In poly(ethylene terephthalate) (PET)/polystyrene (PS) bicomponent spinning, orientation development of the polymer with higher glass transition temperature (T_g), PS, is enhanced, while that of the polymer with lower T_g, PET, is suppressed. When the spinline cools down to the T_g of PS, where the molecular chain movement of PS stops, there can be an enhancement of deformation, because the coexistence of PET in the fiber cross section leads to stress concentration on PS. When the spinline temperature further cools down to the T_g of PET, the orientation of PET can be suppressed because thinning of the spinline is already completed at T_g of PS. This mechanism of mutual influence can be found irrespective of where the two components are situated within the fiber. The same principle is applicable to the high-speed spinning of poly(methyl methacrylate) (PMMA)/PET, where enhancement and suppression of orientation of PMMA and PET, respectively, were observed [56].

Even when one of the polymers in the bicomponent spinning possesses good crystallizability, in many cases, the same mechanism can be applied. Examples are the combinations of poly(butylene terephthalate) (PBT)/poly(butylene adipate-co-terephthalate) (PBAT) [1], PET/PE [57, 58], PET/PP [15], and thermotropic liquid crystalline polymer (TLCP)/PP [59]. In these cases, when PBT, PET, or TLCP loose fluidity in the spinline, the counterparts are considered to be still in a molten state. Therefore, structure formation of the former was enhanced, while that of the latter was suppressed.

For PBT/PBAT core–sheath fibers, the mutual interaction in the bicomponent melt spinning process also improved the high-speed processability (in terms of highest take-up speed) of both polymers, when compared with the corresponding monocomponent spinning [1]. For PET/PE bicomponent fibers, structural formation of the PET component was promoted both when acting as the core [57] and when positioned in the sheath [58], whereas the structure development of PE was suppressed in both cases. Enhanced structure formation of the PET core can be noticed with PP sheath [15]. For TLCP/PP bicomponent fibers, enhancement and suppression of structure development of TLCP and PP, respectively, as well as improvement of spinnability, were also reported [59].

When solidification of the spinline starts to occur at high speed because of the orientation-induced crystallization of one component, structure development becomes much more complicated. Even for the combination of polymers with similar T_g, when one component starts to crystallize earlier than the other component, orientation development of one part can be significantly suppressed by its counterpart. Suppression of orientation becomes more prominent when crystallization of the counterpart proceeds at higher temperatures, which generally occurs at higher take-up speed. In the case of the combination of the amorphous atactic polystyrene (a-PS), and crystalline syndiotactic polystyrenes (s-PSs), orientation of a-PS was significantly suppressed at high take-up speed because of the starting of crystallization of s-PS. The spinnability of both s-PS and a-PS was improved by core–sheath bicomponent spinning [60]. On the other hand, in case of the combination of high and low molecular weight PETs, orientation and crystallization of low molecular weight PET was suppressed when orientation-induced crystallization of high molecular weight PET started to occur at high speed [13].

When both polymers have the capability of crystallization in the bicomponent melt spinning process, the structure formation behavior can reverse at a certain take-up speed. In the case of PP/PE, orientation development of PE was promoted while that of PP was suppressed at low speed [31]. This is because of the higher crystallization rate of PE compared with PP. When the take-up speed exceeds a certain level, however, orientation-induced crystallization of PP starts to occur at a temperature above the crystallization temperature of PE. In this case, orientation of PE can be significantly suppressed. The reversed structure development behavior occurs because the melting temperature of PP is higher than that of PE, while the crystallization rate of PE in the quiescent state is faster than that of PP.

In bicomponent spinning of PET with biodegradable aliphatic polyesters, i.e. PET/poly(butylene succinate/L-lactate) (PBSL) and PET/poly(L-lactic acid) (PLLA), enhanced structure formation of PET was observed even after applying high draw ratios of 5 or 7 [61]. When take-up speed in the melt spinning process is relatively high, such effect can be more intense.

When drawing is applied after solidification of the bicomponent spinline, structure development of both polymers can essentially be controlled by the drawing conditions. However, mechanisms for the mutual interaction of components in the melt spinning process stated above may still influence the structure of the final product. Accordingly, the structures of individual polymers in a bicomponent fiber are different from those of melt-spun and drawn monocomponent fibers.

In PET/poly(phenylene sulfide) (PPS) bicomponent fibers [14], structure formation of the polymer that solidifies first, i.e. PPS, is enhanced in the as-spun fibers due to higher tensile stress applied in the melt spinning process, whereas stress relaxation of PET after the solidification of PPS, which corresponds to the cessation of thinning of the spinline, suppressed the orientation development in PET. Such structural difference may lead to highly developed fiber structure for PPS and a more uniform crystal growth in the PET component after applying the drawing process. Similar tendency was found for the PET/polyamide 6 (PA6) bicomponent fibers with spinning speed of 3200 m/min and draw ratio of 1.53. Side-by-side, segmented-pie, and islands-in-the-sea configuration showed that the crystallinity of the PET component was enhanced, whereas the crystallinity of the PA6 component was suppressed. The total (weight percent) crystallinity of the fibers was a function of the interfacial area: the side-by-side type with the smallest interfacial area between the two polymers showed the highest crystallinity [62].

11.3.3 Interfacial Adhesion

For the combination of different polymers in bicomponent fibers, the melting temperatures and viscosities of the materials have to be considered, as well as the bonding strength of boundary surfaces [5, 6]. In the melt state, interdiffusion

Figure 11.8 Example of interaction of nonmiscible polymers with local segmental diffusion (interdiffusion) of macromolecules across the interface of two polymers [63, 64].

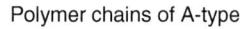

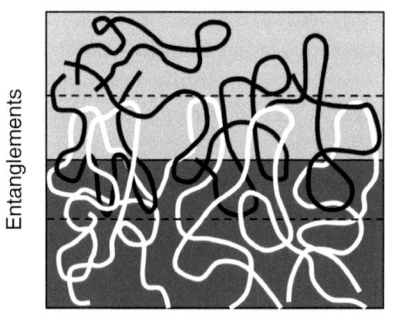

occurs at the interface between the two polymers. For two nonmiscible polymers, the diffusion coefficient is fairly small, and it is unlikely for a chain to migrate completely from one side to the other side of the interface. However, due to local segmental diffusion [63], an interfacial adhesion layer between two incompatible polymers can develop in bicomponent fibers (Figure 11.8). Interdiffusion at polymer interfaces depends on a number of parameters including, but not limited to, temperature, composition of the interface, molecular weight, polydispersity, chain orientation, and molecular structure of the polymers [65]. The interfacial adhesion strength (adhesive bond strength) increases with compatibility, molecular weight, and chain flexibility of the polymers, as well as contact time, temperature, and pressure of the co-flowing melts [64, 66].

To further increase the adhesion between the two polymers of a bicomponent fiber, and thus prevent drawing and service wear, compatibilizers can be applied. In the case of bonding a polyolefin with a polyamide (PA), a maleic anhydride-grafted polyolefin is commonly added to induce compatibilization in situ during the spinning process [9, 67, 68]. Other methods have also been reported for improvement of interfacial adhesion in bicomponent spinning. For example, interfacial reaction such as transesterification between different polyester components, e.g. PET and poly(trimethylene terephthalate) (PTT), can be used to boost the interfacial adhesion [69]. Additionally, mechanical interlocking at the interface, e.g. by employment of a contoured interface, can be used to improve bonding between the two components [70, 71].

Segmented-pie fibers are intended to split. Thus the interfacial adhesion should be limited to allow easy splitting yet being high enough to ensure safe spinning and consecutive drawing, crimping, and textile processing.

11.3.4 Coextrusion Instabilities

In melt spinning, flow instabilities can hinder the production of filaments with uniform cross section [72]. The typical melt instability that affects the flow of a polymer melt in a capillary is melt fracture: entangled polymer melts extruded through a spin pack at high shear rates can exhibit a wavy or spiral distortion of the extrudate [73–75], which can become worse with an inappropriate spinneret design. The occurrence of this instability depends on many factors including polymer chemistry, molecular weight and distribution, and surface interaction between polymer and die [76, 77].

Another unwanted spinning instability is the so-called draw resonance [78, 79], which is a finite-amplitude periodic fluctuation of the cross-sectional area [73] that starts at a critical drawdown ratio (ratio between take-up speed and extrusion velocity at the exit of the spin pack) [74, 80]. Draw resonance occurs when the mass flow rate is not constant between spinneret and take-up godet. To accommodate the variation, small fluctuations in the fiber cross-sectional area occur and produce oscillations in the fiber tension that amplify the fluctuations [81]. Once started, this resonance may not correct itself, requiring a complete shutdown of the extrusion line. For example, PP is particularly susceptible to draw resonance when quenched in a water bath [80]; usually draw resonance does not occur with polymers like PP, PET, or PA when quenched by air [82]. The critical drawdown ratio is known to be affected by polymer properties, extrusion temperature, filament nominal diameter, and distance between spinneret and water bath [83].

In the coextrusion process, two or more fluids meet in a die and flow with a common interface. This type of flow can quite easily generate all sorts of interface instabilities [84], which occur because of differences between the fluids in some property such as density, viscosity, or elasticity [81]. The viscosity difference is considered more important in driving interface deformation than elasticity effects [85, 86].

The melt viscosities of the two polymers at the die must be similar in order to avoid buckling in the fiber. Otherwise the high-viscosity polymer will lose more momentum than the low-viscosity polymer, and the latter will bend toward the former [87]. Extrudate bending occurs when the bicomponent stratified system is in a side-by-side configuration when it emerges from the die, the exit angle decreases as encapsulation progresses, and no extrudate bending occurs when the flow is in a core–sheath configuration [86]. Suggested mechanisms for instabilities in coextrusion include a jump in viscosity and/or first normal stress difference across a flat interface and a coupling of normal stresses with streamline curvature in the region where the two streams merge [88]. When the fluid with higher viscoelasticity is the minority component, i.e. occupies less than half of the channel, a jump in the normal stress difference at the interface is destabilizing; otherwise it is stabilizing (elastic stratification); this stabilizing effect of elasticity can be used to stabilize flow that would otherwise be unstable because of density or viscosity stratification [81].

11.3.5 Encapsulation

Coextrusion experiments [89–97] show that the less viscous melt tends to wrap around the higher viscosity polymer (Figure 11.9). The degree of this interfacial distortion depends primarily on the extent of the viscosity difference and the residence time, with an increasing length-to-diameter (L/D) ratio of the extrusion capillary amplifying the effect [98]. When the less viscous component migrates toward the die wall, the flow minimizes its viscous dissipation (promoted by thermodynamics) [99]. This can be significant for extruding highly viscous

Figure 11.9 Two polymer melts introduced side by side into a capillary experience interfacial distortion during flow of the melts, if the viscosities are mismatched: the lower viscosity polymer tends to encapsulate the higher viscosity polymer [85].

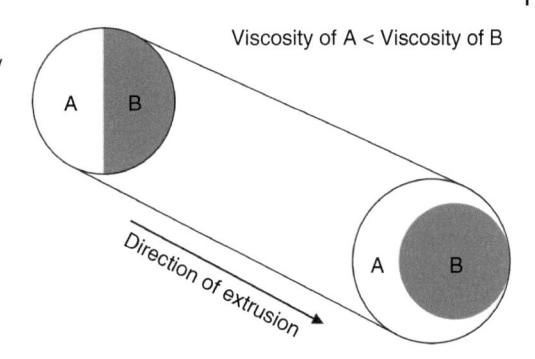

polymers, where the exceedingly high pressure drop in a die is a bottleneck to production [93]. Viscosities of polymers are dependent on extrusion temperature and shear rate, both of which may vary within the coextrusion die [85].

11.3.6 Volatiles

Polymers for man-made fibers can contain dissolved and dispersed gases, as well as volatile liquids and solids (e.g. monomers, water) [100]. When the polymer melt exits the die, respective gas-phase bubbles can evolve, which impair the quality of the fiber [101]. In the case of a bicomponent fiber with a liquid-crystalline-polymer (LCP) core and a PET sheath, a disruption of the core–sheath interface by bubble-like craters was found. The craters could be explained by expanding gas bubbles – degassing products of LCP – trapped in the interface between core and sheath (Figure 11.10). Significant mechanical deformation of the interface was observed, and in some cases the bubbles appeared to push through the sheath interface as the force of the expanding gas overcame the elastic resistance of the sheath layer (indicated by arrow in Figure 11.10b).

11.3.7 Simulation and Modeling

Simulation and modeling of co-flow behavior of polymers and fluids with different rheological properties are discussed in several reports. Mainly surface tension, viscosity, and elastic normal stresses influence the structure of a melt-spun bicomponent fiber, where the viscosity ratio of the two polymers has the most important effect on the final structure of the fiber [51, 102].

Park and coworker [103, 104] simulated bicomponent melt spinning processes and identified the Deborah number as an important parameter regarding the flow behavior of two co-spun polymers of different viscosities. In a follow-up study by Ramos [105–107], the co-flowing behavior of two polymer phases was investigated to predict velocity and radius distribution, assuming Newtonian liquids.

Kikutani et al. [13, 15] developed a one-dimensional model for the spinline dynamics of the bicomponent melt spinning process and discussed the effects of the activation energy for temperature dependence of viscosity, the effects of spinline solidification through glass transition and crystallization, and the effect of molecular weight (absolute value of viscosity) difference for the mutual

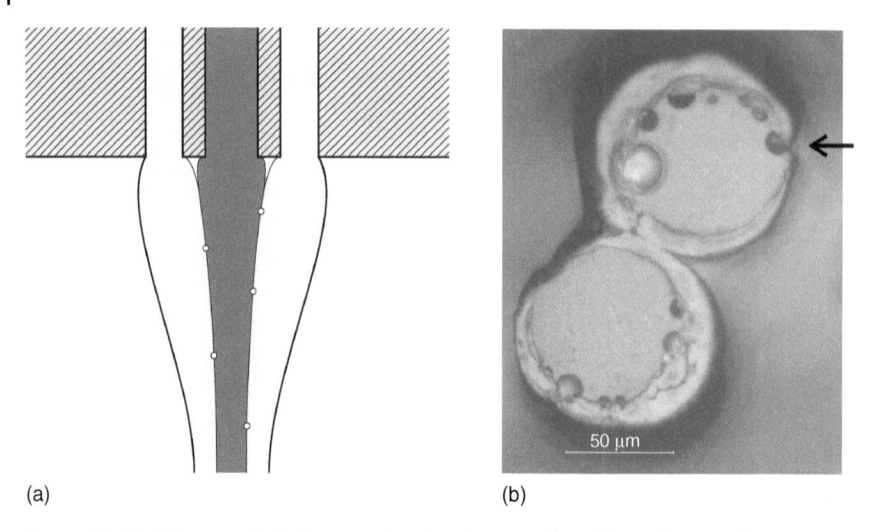

Figure 11.10 (a) In a multiple die core–sheath spinneret, the polymer flow produces a vacuum between core and sheath at the die exit. This vacuum is filled by dissolved/dispersed gases and volatile liquids/solids evaporating from the polymers. As a result, bubbles can accumulate between core and sheath. The bubbles remain when the polymers solidify before the content dissolves or condensates. (b) Optical microscopy images of the cross sections of bicomponent fibers – LCP (core) and PET (sheath) – with bubble-like craters disrupting the core–sheath interface.

interaction of two components for orientation development and relaxation behaviors in the spinning line. Ji et al. [108] investigated different parameters influencing viscosity and flowing instability and developed a one-dimensional model by ignoring surface tension and gravity forces.

Simulation and modeling of two-phase flows is of particular importance when it comes to semicrystalline polymers [109]. Fiber spinning from such polymers is often accompanied by the flow-induced crystallization, which necessitates a model capable to represent both the crystalline and amorphous phases. A pioneering study on the coexistence of these phases was carried out by Kulkarni and Beris [110] who followed the onset and development of crystals by mixing a viscoelastic fluid (representing the amorphous phase) and an anelastic solid (representing the crystalline phase) in a model. A macroscopic continuum model was developed by Doufas et al. [111] who also utilized the Avrami model. In their work, a Giesekus model was incorporated for the amorphous phase, while the crystalline phase was represented by a collection of beads forming rigid rods that could grow and orient in the flow field. The stress and momentum balance as well as the relation of the crystallinity with the relaxation time of the system were later used to couple the descriptions of the two phases in a general mathematical model [112, 113]. The resulting model was shown to be capable to correctly predict the experimental spinline data [114–117]. The latest updates on such approaches can be found in the works of Francisco et al. [118, 119] who developed a two-dimensional model to estimate the temperature distribution, crystallinity, and orientation. Moreover, Zhu and coworkers [120] have studied the dynamics of spinline formation during bicomponent fiber spinning using a log-additive model for elongational viscosity.

The flow inside a bicomponent die is typically a high Weissenberg number and high Deborah number flow. Therefore, nonlinear viscoelastic effects can significantly affect the evolution of the interface. Existing work in modeling and simulation of bicomponent spinning, however, mostly focused on viscous constitutive models. It is prudent to investigate more realistic viscoelastic constitutive models in prediction of the spinline development. In addition, the elastic membrane stress of the interface can significantly alter the flow development. Such elastic stress should also be incorporated into flow modeling and simulation.

11.4 Post-treatment of Bicomponent Fibers

Usually fibers undergo diverse processing steps to increase the strength, to texturize yarns, or to crimp and cut fibers for a staple fiber or wet-laid process. Heat setting is often applied to crystallize the fibers in order to avoid shrinkage. There are standard processes available, which in most cases can be applied for bicomponent fibers as well. Limitations may occur, for example, when the core polymer needs a certain temperature for heat setting (e.g. PET: 180 °C) and the polymer in the sheath is a PE (melting temperature about 135 °C) for easy bonding. It is obvious that the heat setting must be achieved without a hot contacting surface, but by hot air flow at reduced temperature and low speed.

In the very most cases, splittable fibers are produced by segmented-pie technology (Figure 11.3e). Easy splitting is desired, which may be achieved by hollow segmented-pie fibers [121]. Yet during fiber melt spinning, drawing, crimping, and carding, the fibers must not split, as this would significantly impair these processes. To avoid splitting during processing, but enabling it in posttreatment, the polymer melts should be compatible, but show almost no interdiffusion of macromolecules across the interface, and the polymers should have similar drawing behavior and extensibility to allow deformation during the crimping process at low force. Steam supports the crimping process but can lead to shrinkage and thereby to splitting.

Splittable fibers are processed into knits, woven fabrics, or nonwovens. For splitting, woven and knitted materials are brushed, needled, or treated by water jet, where the mechanical force separates the segments. The filaments are bound in the fabric, which allows a harsh procedure yet limits a spreading of the mechanically induced splitting. In the case of nonwovens, the fibers are being split and entangled by needling or water jets (Figure 11.11). At first the fibers are loose and able to move; they will draw aside the force, thus limiting the effect of splitting. The more the fibers are entangled, the more power each stroke applies to the fibers. Apart from the desired fiber splitting, also fiber break may happen in the needling process. Yet the split fiber bundles are immobilized due to the entanglement.

Other mechanisms to split are heat treatment through air or infrared heating if both of the fiber components have different shrinkage behavior. Chemical

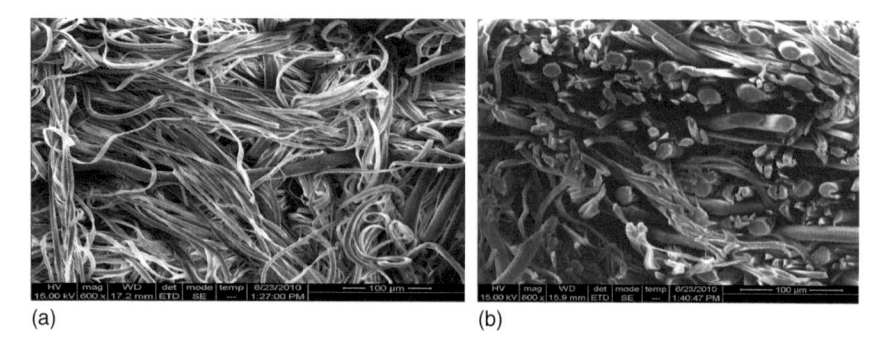

Figure 11.11 Segmented-pie (PET/PA6) spun-laid, moderately needle punched, and finally hydroentangled nonwovens. (a) Nonwoven surface with completely split fibers. (b) Cross section of the inner part of the nonwoven (200 g/m^2) with only partially split fibers.

splitting is realized by hot water, sodium hydroxide, caustic soda, and benzyl alcohol solutions, eventually supported by ultrasonic force [122].

Hot drawing and heat setting for bicomponent fibers are more complex than single-component fiber. In addition to the increased number of process parameters due to the introduction of a second polymer, the interaction between the two components during drawing and heat setting can greatly limit the process window. Unfortunately, there has relatively been little work about drawing and heat setting of bicomponent fibers. However, some simple rules can be derived from existing work. In the case of core–sheath fibers, one in general can use the core as the stronger component and therefore can select drawing conditions to promote strengthening of the core. When an amorphous component is combined with a semicrystalline component, one can draw the fiber above the glass temperature of the amorphous component and optimize the process condition to promote orientation and crystallization of the crystalline component.

11.5 Applications of Bicomponent Fibers

11.5.1 Fibers as Bonding Elements in Nonwovens

In the through-air thermal bonding of bicomponent fibers for the production of nonwoven fabrics [123, 124], a fiber web mixed with thermo-bondable core–sheath bicomponent fibers is treated by blowing hot air through the web. The temperature of the air needs to be higher than the melting temperature of the sheath component but lower than that of the core component, so that the shape of the fiber can be maintained. With this technique, nonwovens of excellent flexibility, high bulkiness, and good absorbency can be produced. In addition, the product safety is ensured because the use of adhesives can be avoided. Further advancement of thermo-bondable fibers is accomplished through the development of a PP/PE bicomponent fiber with the potential of spontaneous elongation. This fiber is utilized for the production of nonwoven fabrics with soft touch, which are applicable for diapers and hygiene products [125].

11.5.2 Microfibers

One of the first and best-known microfiber products is the artificial leather called Alcantara®. Here, islands-in-the-sea fibers with polyester islands and polystyrene sea are processed to a felt. The sea (matrix) can be dissolved by trichloroethylene [126]. But from an ecological point of view, this process is not acceptable, but could be replaced by water-soluble matrix polymers (e.g. polyvinyl alcohol (PVA), EastOne). The most successful technical application was the combination of PET as islands and EastOne as matrix [127]. Starting from a 1.3 dtex filament, the resulting island fiber diameter after removal of the matrix was 1.8 µm (Figure 11.12). Meanwhile EastOne is replaced by Cyphrex™, shortcut PET fibers (1.5 mm) with diameters of 2.5–4.5 µm, or ribbon shapes of 2.5 × 18 µm. The best-known fabric produced from these fibers is the wet-laid filter Captimax™ from Ahlstrom.

The islands-in-the-sea and segmented-pie approaches have been applied to produce canals for water transportation parallel to the fiber axis. For islands in the sea, the common approach is to use an insoluble polymer as islands and a soluble polymer as matrix and then extract the islands (in the case of islands in the sea through the matrix). Aiming at an environmentally friendly approach, a water-soluble polymer as islands with a water-insoluble polymer as the sea (matrix) polymer can be considered [128].

Under the brand Nanofront™, Teijin produces islands-in-the-sea filaments and staple fibers with PET as islands and modified polyester as matrix. The matrix is dissolved by an alkali solution, which is less economic and ecologic. Teijin even offers such fibers with diameters down to 0.7 µm, with possible applications in functional sportswear, inner wear, skin care products, filters, and precision grinding cloths [129].

A solvent-free alternative to produce microfibers is the application of mechanical stress to separate the different parts of segmented-pie fibers (see Section 11.4).

11.5.3 Fibers with Special Cross Sections

A special feature of the islands-in-the-sea technology is the logotype fiber, where the islands polymer has a different color (Figure 11.13) or has different dyeability

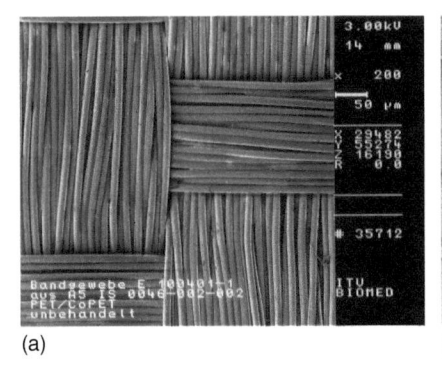

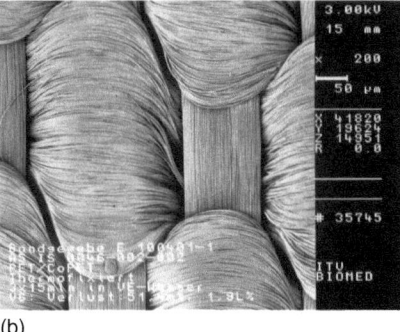

(a)

(b)

Figure 11.12 Woven textile of PET/EastOne islands-in-the-sea fibers; (a) before and (b) after removal of the water-soluble matrix.

Figure 11.13 Logotype of ITV; islands-in-the-sea fiber by Hills.

compared with the matrix (e.g. PA vs. PET). The picture quality depends on the number of islands, i.e. the pixel count. On the one hand, a fiber with respective cross section can be used as marketing tool, and on the other hand, it can provide an effective copy protection.

One of the most fascinating fibers is Morphotex® by Teijin. The fiber is a biomimetic example copied from a morpho butterfly, which get the colorful shine of their wings not as a result of pigmentation, but of light diffraction by submicrometer structures inside the fiber [130]. A PET sheath is filled by interpenetrated lamellae from PET and PA, or concentric rings of alternating polymers or winged lamellae close to the natural archetype (Figure 11.14).

11.5.4 Fibers with High-Performance Core

The reinforcement of concrete with fibers can be an economical alternative to conventional steel bar reinforcement. Polyolefin-based bicomponent fibers, with high tensile strength and elastic modulus in the core, nanoparticles and other additives in the sheath, and a structured fiber surface were successfully applied to enhance the mechanical properties of concrete [131]. The superior performance of such fiber-reinforced concrete was shown regarding pullout characteristics of the fiber and bending behavior of the concrete, making these fibers interesting for applications in precast elements, industrial floors, and earthquake protecting systems [131].

Artificial turf is a lower maintenance all-weather alternative to natural turf [132]. While PA carpets have excellent resilience but provoke abrasion injuries (friction burn), PE monofilaments are skin friendly but prone to permanent

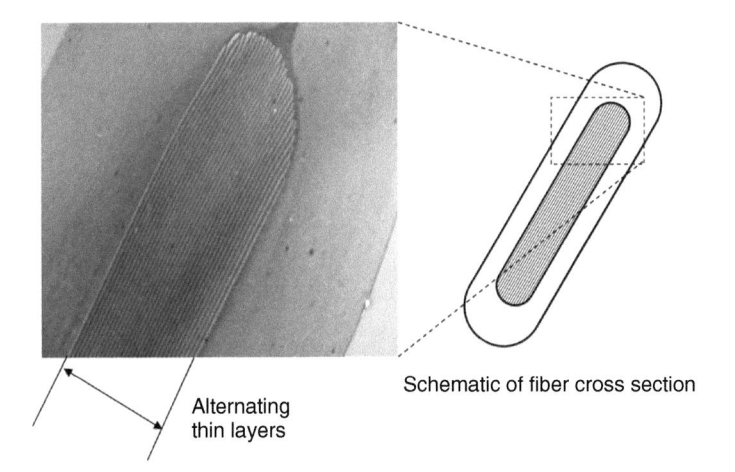

Figure 11.14 Cross section of Morphotex® fiber by Teijin, consisting of 61 alternating thin layers of polyamide and polyester. The interference color of the fiber can be defined by choosing the thickness of each layer between 70 and 100 nm [130].

Figure 11.15 Cross section of a PA/PE bicomponent monofilament applied in skin-friendly artificial turf with good resilience [133].

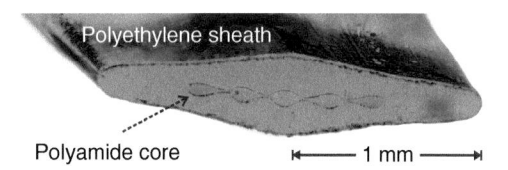

deformation. With the development of PA/PE core–sheath fibers with optimized cross section (Figure 11.15), an artificial turf that shows a better resilience than up-to-date synthetic grass, without cutback in skin friendliness, could be produced [9].

11.5.5 Fibers with Functional Surface

Core–sheath fibers offer the chance to modify the surface while leaving the bulk unchanged. The majority of commercialized bicomponent fibers are binder fibers with a low melting temperature sheath. Typically such fibers are treated in a through-air oven where the binder melts, concentrates at fiber crossings, and glues the fibers together, which finally form a stable network. Another common application is antimicrobial fibers, where the effect of the expensive active substance is required on the fiber surface, i.e. in the sheath only.

Usually easy-to-clean surfaces are achieved by coating of textiles. The so-called lotus effect requires a structured hydrophobic surface. The conventional coating is prone to wear and limits breathability. An alternative approach is to structure the fiber surface, which can be realized by a high filling of the sheath polymer by particles [134]. Again, the function is required on the surface only (Figure 11.16a); the unfilled bulk provides strength and required textile properties. For the required easy-to-clean function, the fibers require an additional hydrophobic treatment. It was shown that a breathable woven textile produced

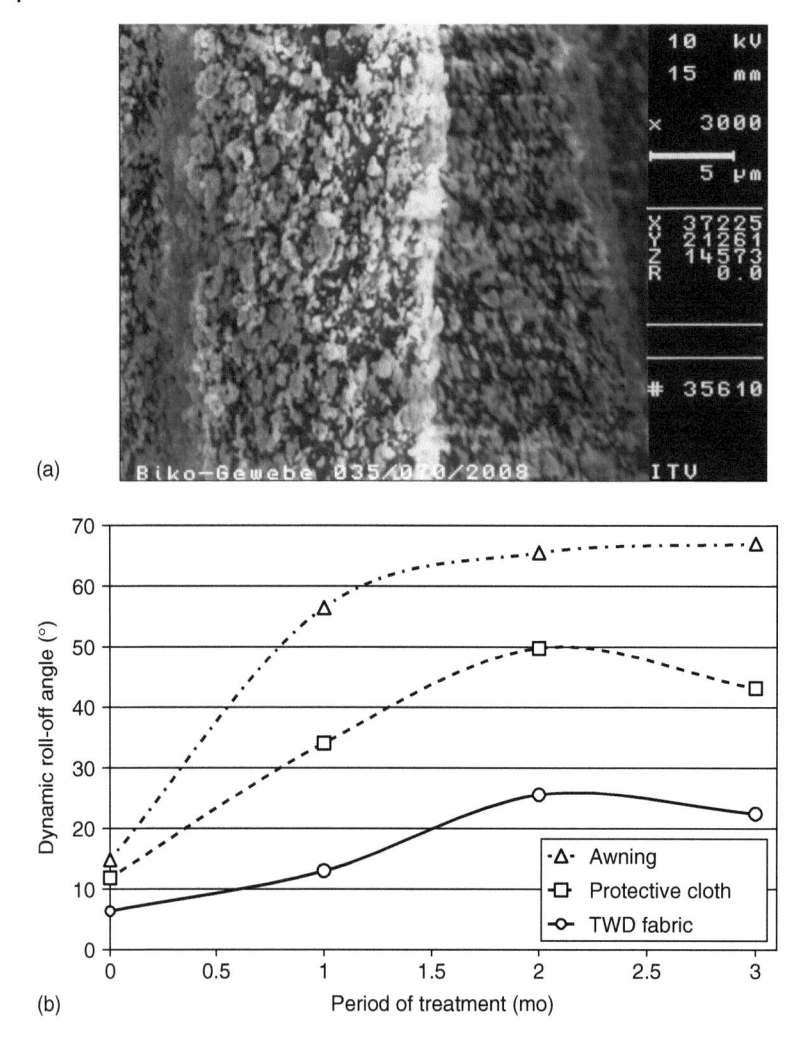

Figure 11.16 Easy-to-clean fiber. (a) Fiber structure and (b) diagram of environmental aging, contact angle vs. aging time [135].

from this novel fiber was superior to a selection of easy-to-clean coated textiles suitable for awnings and protective textiles, respectively (Figure 11.16b).

11.5.6 Biodegradable Fibers

For temporary textile implants, fibers from biocompatible and biodegradable polymers are preferable [52]. The commercially available aliphatic polyesters polylactide (PLA) and polyhydroxyalkanoate (PHA) combine these aspects. PHA/PLA core–sheath fibers were produced, where the PLA component alone was responsible for the tensile strength [43]. In vitro biocompatibility studies with human dermal fibroblasts showed that cells adhered on the fibers, making them good candidates for medical therapeutic approaches. By adjusting

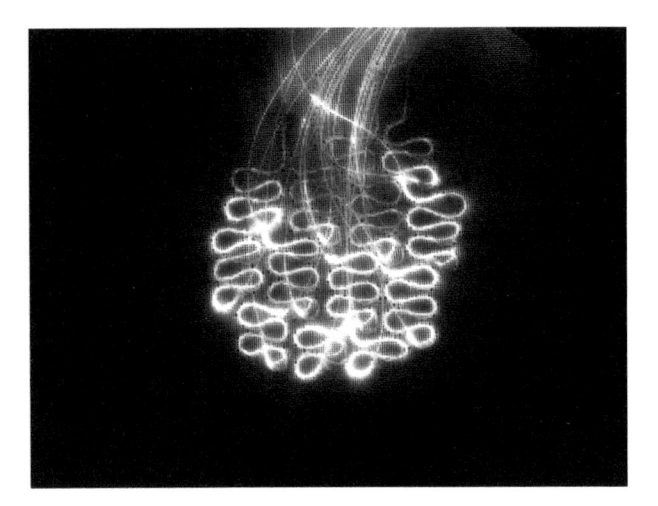

Figure 11.17 Embroidery produced from flexible bicomponent polymer optical fibers.

core–sheath ratio of the bicomponent fiber, a better controlled degradation process could be achieved.

11.5.7 Polymer Optical Fibers

To obtain thin and flexible polymer optical fibers (POFs) for textile applications, bicomponent melt-spun fibers with a cyclic olefin polymer (COP) as the core and a tetrafluoroethylene-hexafluoropropylene-vinylidene fluoride (THV) terpolymer as the sheath have been coextruded on the pilot scale [136, 137]. The bicomponent arrangement could promote the formation of a regular core surface in POFs, because thermal shielding by the sheath component could prevent unevenness at the core–sheath interface. Light propagation and tensile properties of the fibers turned out to be adequate to enable their use in industrially produced luminous textiles (Figure 11.17).

There are some papers and patents on the fabrication of graded-index (GI) POFs utilizing the diffusion of a low molecular weight polymer with high refractive index in the extrusion process of bicomponent fibers. In this process, PMMA blended with diphenyl sulfide (DPS) and virgin PMMA are used as core and sheath components, respectively. With the help of a long capillary die, diffusion of DPS is promoted to produce fibers with a refractive index variation over their cross section [138].

11.5.8 Electrically Conductive Fibers

In order to make fibers antistatic or conductive, electrically conducting materials like carbon black, carbon nanotubes (CNTs), graphene, or metal powders are used as additives. Depending on the type of additive, a polymer compound with 10–30 wt% of conductive additive is generally required to achieve conductivity [139]. As a fiber melt-spun from a respective compound would have a very low

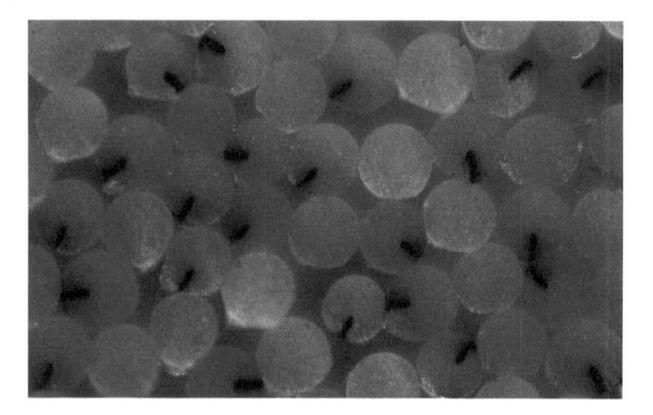

Figure 11.18 Yarn with 50% antistatic side-by-side filaments melt-spun on the pilot-scale (filament diameter approx. 50 µm). The main component is pure polyamide; the side component is a polyamide mixed with carbon black.

tenacity, bicomponent fibers with an antistatic compound as minor element have been introduced, e.g. "Belltron" (side by side) by Kanebo or "Antistat" (core–sheath) by Perlon Nextrusion. Side-by-side bicomponent fibers with the conductive element buried partially inside the filament can attenuate the unrequested dark color coming from carbon black (Figure 11.18).

Lund and Hagström, Nilsson et al., and Glauß et al. [140–143] reported about bicomponent fibers developed for piezoelectric applications, with carbon-black or CNT-filled polyolefin core and a polyvinylidene fluoride (PVDF) sheath. Straat et al. [144] developed an electrically conducting fiber with carbon-black and MWNT-filled PE core and PA sheath. They developed a solution for a common problem: as-spun fibers with reasonable conductive properties loose the conductivity at drawing. So one has to choose between acceptable strength or conductivity, respectively. The reason for the loss of conductivity is the orientation of the fillers and especially of CNT at the drawing. Straat et al. [144] produced the bicomponent fibers and drew them. After a heat treatment at 190 °C, far above its melting point, the molecules of the PE core relax and coil back resolving the orientation of the carbonaceous fillers. This gains the conductivity back by up to factor 2. Conductivity after heat treatment reached values $>200 \times 10^{-3}$ S/cm. Strength up to 36 cN/tex was reported only before heat treatment. From other data one can conclude that heat treatment leads to a decrease in tensile strength of about 20%. Convincing data of fibers were reported for both conductivity (24×10^{-3} S/cm) and strength (31.1 ± 4.2 cN/tex) after heat treatment.

11.5.9 Liquid-Core Fibers

Hufenus et al. [40] demonstrated the continuous production of liquid-filled polymeric fibers in a stable melt spinning process at the pilot plant scale (Figure 11.19). The ability to produce a continuous liquid-core fiber (LCF) is attractive since post-filling of a hollow fiber with similar dimensions is not practical [145]. As a

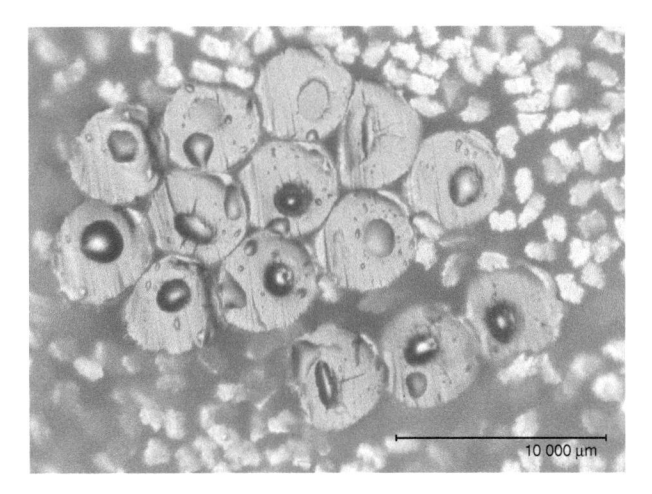

Figure 11.19 Cross sections of liquid-core fibers (LCFs) with 18% liquid content, recorded by optical microscopy (the bright viscose fibers are fillers for supporting sample preparation).

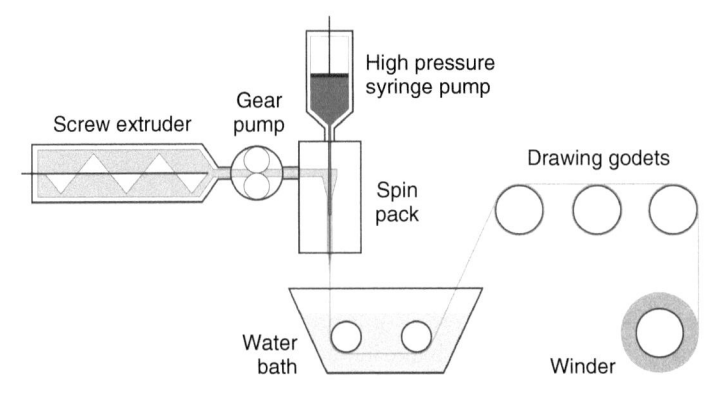

Figure 11.20 Scheme of the LCF coextrusion equipment. Design of spin pack (see Figure 11.5).

result of model experiments [39], a special extrusion line including a quenching, drawing, and winding unit was built (Figure 11.20). A LCF can exhibit significantly enhanced damping properties compared with plain polymeric filaments of same dimensions. This new type of fiber could find future applications in the enhancement of the damping factor of fiber-reinforced lightweight structures, as a fiber can precisely and seamlessly be blended into a composite material, without compromising its structural integrity. Such bicomponent fibers also have potential applications in self-healing composites [146, 147].

11.5.10 Fibers for Fully Thermoplastic Fiber-Reinforced Composites

Core–sheath and islands-in-the-sea bicomponent fibers can be utilized for the production of fully thermoplastic fiber-reinforced composites. In this case, the core or island component, which is designed to become the reinforcing fiber, must have a higher melting temperature than the sheath or sea component, which is the designated matrix material. Such a fiber-reinforced composite can be fabricated by compression molding of aligned bicomponent fibers at a temperature in between the melting temperatures of both components. The fabrication of respective composites with a structural gradient mimicking the structure of bamboo was also proposed [59].

More recently, the fabrication of fiber-reinforced composites through the compression molding of fabrics from core sheath bicomponent fibers of high and low molecular weight PET was proposed. In this case, a high-speed bicomponent spinning process was utilized to fabricate fibers with highly oriented and crystallized core and low oriented amorphous sheath (see Section 11.3.1). These fibers can be compression-molded at a temperature higher than the glass transition temperature but lower than the melting temperature of PET [148].

Leal et al. [41] introduced a novel approach for the development of ultralight, fully thermoplastic fiber-reinforced composites, based on hybrid yarn produced with a modified wire coating (overjacketing) process (see Figure 11.6). The composite material consisted of a polyolefin plastomer (POP) matrix reinforced with ultrahigh molecular weight polyethylene (UHMWPE) fibers. The interfacial affinity of both polyolefins was enhanced by the deposition of a nanometer-scale polar functional plasma polymer film on the surface of the filaments within a reel-to-reel continuous process. The activated UHMWPE yarn was subsequently coated (overjacketed) with a layer of the matrix material. Alternate layers of woven hybrid yarn and woven pure UHMWPE yarn were then stacked, and the layup was consolidated by hot compaction, resulting in a composite laminate with a fiber volume fraction of 0.54 and a density of 0.93 g/cm^3.

11.5.11 Shape Memory Fibers

Two polymers with different phase transition temperature can be combined to form a composite material with a shape memory character [149]. Bicomponent spinning provides a useful platform for engineering shape memory composites or blends. Spontak and coworkers recently demonstrated a bicomponent core–sheath fiber with a shape memory character [150]. The core is made of a thermoplastic elastomer (TPE), poly[styrene-*block*-(ethylene-*co*-butylene)-*block*-styrene], and the sheath PE. A new shape can be programmed by deforming the fiber above the soften temperature of PE and quenching to preserve the shape. Upon reheating above PE's softening temperature, the original shape can be recovered.

References

1 Shi, X.Q., Ito, H., and Kikutani, T. (2006). Structure development and properties of high-speed melt spun poly(butylene terephthalate)/poly(butylene adipate-*co*-terephthalate) bicomponent fibers. *Polymer* 47 (2): 611–616.

2 Huang, J., Baird, D.G., Loos, A.C. et al. (2001). Filament winding of bicomponent fibers consisting of polypropylene and a liquid crystalline polymer. *Composites Part A: Applied Science and Manufacturing* 32 (8): 1013–1020.

3 Kathiervelu, S.S. (2002). Bicomponent fibers. *Synthetic Fibres* 31 (3): 11–16.
4 Koslowski, H.J. (2009). Bicomponent fibers: processes, products, markets. *Technical Textiles* 52 (5): E202–E204.
5 Houis, S., Schreiber, F., and Gries, T. (2008). Fiber table: bicomponent fibers (Part 1). *Chemical Fibers International* 58 (1): 38–45.
6 Houis, S., Schreiber, F., and Gries, T. (2008). Fibers-fiber table: bicomponent fibers (Part 2). *Chemical Fibers International* 58 (3): 158–165.
7 Fourné, F. (1999). Synthetic Fibers. Munich: Hanser Publishers.
8 Walczak, Z.K. (2002). *Processes of Fiber Formation*, 1e. Amsterdam: Elsevier Science Ltd.
9 Hufenus, R., Affolter, C., Camenzind, M., and Reifler, F.A. (2013). Design and characterization of a bicomponent melt-spun fiber optimized for artificial turf applications. *Macromolecular Materials and Engineering* 298 (6): 653–663.
10 Mukhopadhyay, S. (2014). Bi-component and bi-constituent spinning of synthetic polymer fibres. In: *Advances in Filament Yarn Spinning of Textiles and Polymers* (ed. Dong Zhanged), p. 113. Cambridge, UK: Woodland Publishing.
11 Mochizuki, M. and Matsunaga, N. (2016). Bicomponent polyester fibers for nonwovens. In: *High-Performance and Specialty Fibers* (The Society of Fiber Science and Technology), pp. 395–408. Tokyo: Springer Japan.
12 Yang, H.H. (2007). Polyamide fibers. In: *Handbook of Fiber Chemistry* (ed. M. Lewin), 31–137. CRC Press.
13 Radhakrishnan, J., Kikutani, T., and Okui, N. (1997). High-speed melt spinning of sheath-core bicomponent polyester fibers: high and low molecular weight poly(ethylene terephthalate) systems. *Textile Research Journal* 67 (9): 684–694.
14 Perret, E., Reifler, F.A., Hufenus, R. et al. (2013). Modified crystallization in PET/PPS bicomponent fibers revealed by small-angle and wide-angle X-ray scattering. *Macromolecules* 46 (2): 440–448.
15 Kikutani, T., Radhakrishnan, J., Arikawa, S. et al. (1996). High-speed melt spinning of bicomponent fibers: mechanism of fiber structure development in poly(ethylene terephthalate)/polypropylene system. *Journal of Applied Polymer Science* 62 (11): 1913–1924.
16 Tippetts, E.A. (1967). Fiber engineering to meet end use requirements. *Textile Research Journal* 37 (6): 524–533.
17 Hagewood, J. (2014). 3 – Technologies for the manufacture of synthetic polymer fibers A2. In: *Advances in Filament Yarn Spinning of Textiles and Polymers* (ed. Z. Dong), 48–71. Woodhead Publishing.
18 Hills, W.H. (1996). Distribution plate for spin pack assembly. USPTO, US 5562930 A, USA.
19 Robson, T. (2013). *Multicomponent Fiber Extrusion Technology Applied to Precursors*. Buffalo, NY: Carbon Fiber R&D Workshop.
20 Jaffe, M. and East, A.J. (2007). Polyester fibers. In: *Handbook of Fiber Chemistry* (ed. M. Lewin), 1–29. CRC Press.
21 Purane, S.V. and Panigrahi, N.R. (2007). Microfibers microfilaments and their applications. *Autex Research Journal* 7 (3): 148–158.
22 Jeffries, R. (1971). *Bicomponent Fibres*. Merrow Publication.

23 Zhao, R.R., Wadsworth, L.C., Sun, C., and Zhang, D. (2003). Properties of PP/PET bicomponent melt blown microfiber nonwovens after heat-treatment. *Polymer International* 52 (1): 133–137.
24 Tomasino, C. (1992). *Chemistry & Technology of Fabric Preparation & Finishing*. North Carolina State University NC.
25 Baker, B. (1998). Bicomponent fibers: a personal perspective. *International Fiber Journal* 13: 26–42.
26 Yaida, O. (2016). Current status and future outlook for nonwovens in Japan. In: *High-Performance and Specialty Fibers* (The Society of Fiber Science and Technology), pp. 375–393. Tokyo: Springer Japan.
27 Oh, T.H. (2006). Melt spinning and drawing process of PET side-by-side bicomponent fibers. *Journal of Applied Polymer Science* 101 (3): 1362–1367.
28 Prahsarn, C., Klinsukhon, W., Roungpaisan, N., and Srisawat, N. (2013). Self-crimped bicomponent fibers containing polypropylene/ethylene octene copolymer. *Materials Letters* 91: 232–234.
29 Robeson, L.M., Axelrod, R.J., Vratsanos, M.S., and Kittek, M.R. (1994). Microfiber formation: immiscible polymer blends involving thermoplastic poly(vinyl alcohol) as an extractable matrix. *Journal of Applied Polymer Science* 52 (13): 1837–1846.
30 Kamiyama, M., Soeda, T., Nagajima, S., and Tanaka, K. (2012). Development and application of high-strength polyester nanofibers. *Polymer Journal* 44 (10): 987–994.
31 Prahsarn, C., Matsubara, A., Motomura, S., and Kikutani, T. (2008). Development of bicomponent spunbond nonwoven webs consisting of ultra-fine splitted fibers. *International Polymer Processing* 23 (2): 178–182.
32 Sonnenschein, M.F. (1999). Hollow fiber microfiltration membranes from poly(ether ether ketone) (PEEK). *Journal of Applied Polymer Science* 72 (2): 175–181.
33 Oh, T.H., Lee, M.S., Kim, S.Y., and Shim, H.J. (1998). Studies on melt-spinning process of hollow fibers. *Journal of Applied Polymer Science* 68 (8): 1209–1217.
34 Takarada, W., Ito, H., Kikutani, T., and Okui, N. (2001). Studies on high-speed melt spinning of noncircular cross-section fibers. I. Structural analysis of As-spun fibers. *Journal of Applied Polymer Science* 80 (9): 1575–1581.
35 Rwei, S.P. (2001). Formation of hollow fibers in the melt-spinning process. *Journal of Applied Polymer Science* 82 (12): 2896–2902.
36 Su, Y.Y., Rwei, S.P., Wu, L.Y. et al. (2011). Shaping conjugated hollow fibers using a four-segmented arc spinneret. *Polymer Engineering and Science* 51 (4): 704–711.
37 Huang, Q., Seibig, B., and Paul, D. (1999). Polycarbonate hollow fiber membranes by melt extrusion. *Journal of Membrane Science* 161 (1–2): 287–291.
38 De Rovère, A. and Shambaugh, R.L. (2001). Melt-spun hollow fibers: modeling and experiments. *Polymer Engineering and Science* 41 (7): 1206–1219.
39 Heuberger, M., Gottardo, L., Dressler, M., and Hufenus, R. (2015). Biphasic fluid oscillator with coaxial injection and upstream mass and momentum transfer. *Microfluidics and Nanofluidics* 19 (3): 653–663.

40 Hufenus, R., Gottardo, L., Leal, A.A. et al. (2016). Melt-spun polymer fibers with liquid core exhibit enhanced mechanical damping. *Materials and Design* 110: 685–692.

41 Leal, A.A., Veeramachaneni, J.C., Reifler, F.A. et al. (2016). Novel approach for the development of ultra-light, fully-thermoplastic composites. *Materials and Design* 93: 334–342.

42 Johnston, B., Haggard, J., Wilkie, A. et al. (2007). Temperature control system to independently maintain separate molten polymer streams at selected temperatures during fiber extrusion. USPTO, US 7252493 B1. USA: Google Patents.

43 Hufenus, R., Reifler, F.A., Maniura-Weber, K. et al. (2012). Biodegradable bicomponent fibers from renewable sources: melt-spinning of poly(lactic acid) and poly[(3-hydroxybutyrate)-co-(3-hydroxyvalerate)]. *Macromolecular Materials and Engineering* 297 (1): 75–84.

44 Dimla, D.E., Camilotto, M., and Miani, F. (2005). Design and optimisation of conformal cooling channels in injection moulding tools. *Journal of Materials Processing Technology* 164: 1294–1300.

45 Dormal, T. (2008). Rapid manufacturing of mould inserts for tooling improvements. *International Conference on Additive Technologies*, Ptuj, Slovenia.

46 Villalon, A.V. (2005). Electron beam fabrication of injection mold tooling with conformal cooling channels. MSc. Raleigh: North Carolina State University, p. 68.

47 Levy, G.N., Schindel, R., and Kruth, J.P. (2003). Rapid manufacturing and rapid tooling with layer manufacturing (LM) technologies, state of the art and future perspectives. *CIRP Annals – Manufacturing Technology* 52 (2): 589–609.

48 Kikutani, T. (2009). Structure development in synthetic fiber production. In: *Handbook of Textile Fibre Structure – Fundamentals and Manufactured Polymer Fibres* (eds. S. Eichhorn, J.W.S. Hearle, M. Jaffe and T. Kikutani). Elsevier.

49 Smith, K.J. Jr., (1990). The breaking strength of perfect polymer fibers. *Polymer Engineering and Science* 30 (8): 437–443.

50 Spruiell, J.E. (Munich, 2001). Structure formation during melt spinning. In: *Structure Formation in Polymeric Fibers* (ed. D.R. Salem), 5–93. Carl Hanser Verlag.

51 Ziabicki, A. (1976). *Fundamentals of Fibre Formation*. London: Wiley.

52 Hearle, J.W.S. (2008). Fibers, Structure. In: *Ullmann's Fibers: Fiber Classes, Production and Characterization*, vol. 1, 39–79. Weinheim: Wiley-VCH.

53 Lewis, I.R. and Edwards, H.G.M. (2001). *Handbook of Raman Spectroscopy – From the Research Laboratory to the Process Line*. New York: Marcel Dekker.

54 Shimizu, J., Okui, N., and Kikutani, T. (1981). High-speed melt spinning of poly(ethylene terephthalate): radial variation across fibers. *Sen'i Gakkaishi* 47 (4): T135–T142.

55 Kikutani, T., Arikawa, S., Takaku, A., and Okui, N. (1995). Fiber structure formation in high-speed melt spinning of sheath-core type bicomponent fibers. *Sen'i Gakkaishi* 51 (9): 408–415.

56 Yoshimura, M., Iohara, K., Nagai, H. et al. (2003). Structure formation of blend and sheath/core conjugated fibers in high-speed spinning of PET, including a small amount of PMMA. *Journal of Macromolecular Science, Part B Physics* 42 (2): 325–339.

57 Cho, H.H., Kim, K.H., Kang, Y.A. et al. (2000). Fine structure and physical properties of polyethylene/poly(ethylene terephthalate) bicomponent fibers in high-speed spinning. I. Polyethylene sheath/poly(ethylene terephthalate) core fibers. *Journal of Applied Polymer Science* 77 (10): 2254–2266.

58 Cho, H.H., Kim, K.H., Kang, Y.A. et al. (2000). Fine structure and physical properties of poly(ethylene terephthalate)/polyethylene bicomponent fibers in high-speed spinning. II. Poly(ethylene terephthalate) sheath/polyethylene core fibers. *Journal of Applied Polymer Science* 77 (10): 2267–2277.

59 Furuta, T., Radhakrishnan, J., Ito, H. et al. (1999). Fabrication of continuous fiber-reinforced thermoplastic composites with structural gradient from sheath-core type bicomponent fibers. *Composite Interfaces* 6 (5): 451–466.

60 Hada, Y., Shikuma, H., Ito, H., and Kikutani, T. (2005). High-speed melt spinning of syndiotactic-polystyrene; improvement of spinnability and fiber structure development via bicomponent spinning with atactic-polystyrene. *Journal of Macromolecular Science, Part B Physics* 44 (4): 549–571.

61 El-Salmawy, A. and Kimura, Y. (2001). Structure and properties of bicomponent core-sheath fibers from poly(ethylene terephthalate) and biodegradable aliphatic polyesters. *Textile Research Journal* 71 (2): 145–152.

62 Choi, Y.B. and Kim, S.Y. (1999). Effects of interface on the dynamic mechanical properties of PET/nylon 6 bicomponent fibers. *Journal of Applied Polymer Science* 74 (8): 2083–2093.

63 Paul, D.R. (1978). Fibers from Polymer Blends. In: *Polymer Blends* (eds. D.R. Paul and S. Newman), 167–217. New York: Academic Press.

64 Sun, C., Zhang, D., Liu, Y., and Xiao, R. (2004). Preliminary study on fiber splitting of bicomponent meltblown fibers. *Journal of Applied Polymer Science* 93 (5): 2090–2094.

65 Jablonski, E.L. (2002). Interdiffusion phenomena at partially miscible polymer interfaces. PhD thesis. Retrospective Theses and Dissertations: Iowa State University.

66 Cassidy, P.E., Johnson, J.M., and Locke, C.E. (1972). The relationship of glass transition temperature to adhesive strength. *The Journal of Adhesion* 4 (3): 183–191.

67 Ide, F. and Hasegawa, A. (1974). Studies on polymer blend of nylon 6 and polypropylene or nylon 6 and polystyrene using the reaction of polymer. *Journal of Applied Polymer Science* 18 (4): 963–974.

68 Godshall, D., White, C., and Wilkes, G.L. (2001). Effect of compatibilizer molecular weight and maleic anhydride content on interfacial adhesion of polypropylene-PA6 bicomponent fibers. *Journal of Applied Polymer Science* 80 (2): 130–141.

69 Yang, Z., Wang, F., and Xu, B. (2016). Key factors affecting binding tightness between two components of PTT/PET side-by-side filaments. *Industria Textila* 67 (4): 226–232.
70 Ando, S., Tanaka, Y., and Ogata, F. (1969). Specific conjugate composite filament. USPTO, US 3, 458, 390.
71 Levine, M., O'Connor, J., DiTaranto, F. et al. (2005). Durable highly conductive synthetic fabric construction. USPTO, US 20050095935 A1.
72 Gupta, V.B. and Bhuvanesh, Y.C. (1997). Basic principles of flow during fibre spinning. In: *Manufactured Fibre Technology* (eds. V.B. Gupta and V.K. Kothari), 31–66. London: Chapman & Hall.
73 Hatzikiriakos, S.G. and Migler, K.B. (2005). Overview of processing instabilities. In: *Polymer Processing Instabilities* (eds. S.G. Hatzikiriakos and K.B. Migler), 1–12. New York: Marcel Dekker.
74 Denn, M.M. (2004). Fifty years of non-Newtonian fluid dynamics. *AIChE Journal* 50 (10): 2335–2345.
75 Baird, D.G. and Collias, D.I. (1998). *Polymer Processing: Principles and Design*. Wiley.
76 Denn, M.M. (2001). Extrusion instabilities and wall slip. *Annual Review of Fluid Mechanics* 33: 265–287.
77 Migler, K.B., Son, Y., Qiao, F., and Flynn, K. (2002). Extensional deformation, cohesive failure, and boundary conditions during sharkskin melt fracture. *Journal of Rheology* 46 (2): 383–400.
78 Petrie, C.J.S. and Denn, M.M. (1976). Instabilities in polymer processing. *AIChE Journal* 22 (2): 209–236.
79 Jung, H.W. and Hyun, J.C. (2004). Fiber spinning and film blowing instabilities. In: *Polymer Processing Instabilities: Control and Understanding*, Marcel Dekker Series, vol. 102 (eds. S.G. Hatzikiriakos and K.B. Migler), 321–382.
80 Yoo, H.J. (1987). Draw resonance in polypropylene melt spinning. *Polymer Engineering and Science* 27 (3): 192–201.
81 Larson, R.G. (1992). Instabilities in viscoelastic flows. *Rheologica Acta* 31 (3): 213–263.
82 Kase, S. (1974). Studies on melt spinning. IV. On the stability of melt spinning. *Journal of Applied Polymer Science* 18 (11): 3279–3304.
83 Mei-Fang, Z. and Yang, H.H. (2006). *Polypropylene Fibers. Handbook of Fiber Chemistry*, 3e, vol. 8. CRC Press.
84 Mitsoulis, E. (2005). Secondary flow instabilities. In: *Polymer Processing Instabilities* (eds. S.G. Hatzikiriakos and K.B. Migler), 43–71. New York: Marcel Dekker.
85 Dooley, J. (2005). Coextrusion instabilities. In: *Polymer Processing Instabilities* (eds. S.G. Hatzikiriakos and K.B. Migler), 383–426. New York: Marcel Dekker.
86 Karagiannis, A., Hrymak, A.N., Vlachopoulos, J., and Vlcek, J. (1995). Coextrusion of polymer melts. In: *Rheological Fundamentals of Polymer Processing* (eds. J.A. Covas, J.F. Agassant, A.C. Diogo, et al.), 113–137. Kluwer Academic Publishers.
87 Yu, J.-P. (1998). Producing bicomponent Fibers: spinneret design is crucial. *International Fiber Journal* 13: 49–53.

88 Reis, T., Sahin, M., and Wilson, H. (2008). Co-extrusion instabilities modeled with a single fluid. In: *XV International Congress on Rheology: Society of Rheology* (eds. A. Co, L.G. Leal, R.H. Colby and A.J. Giacomm), 150–152.

89 Karami, A. and Balke, S.T. (2000). Polymer blend de-mixing and morphology development of immiscible polymer blends during tube flow. *Polymer Engineering and Science* 40 (11): 2342–2355.

90 Lee, B.L. and White, J.L. (1975). Notes: experimental studies of disperse two-phase flow of molten polymers through dies. *Journal of Rheology* 19 (3): 481–492.

91 Utracki, L.A., Dumoulin, M.M., and Toma, P. (1986). Melt rheology of high density polyethylene/polyamide-6 blends. *Polymer Engineering and Science* 26: 34–44.

92 Lohfink, G.W. and Kamal, M.R. (1993). Morphology and permeability in extruded polypropylene/ethylene vinyl-alcohol copolymer blends. *Polymer Engineering and Science* 33 (21): 1404–1420.

93 Han, C.D. (1975). A study of coextrusion in a circular die. *Journal of Applied Polymer Science* 19: 1875–1883.

94 Stone, H.A., Stroock, A.D., and Ajdari, A. (2004). Engineering flows in small devices: microfluidics toward a lab-on-a-chip. *Annual Review of Fluid Mechanics* 36: 381–411.

95 Minagawa, N. and White, J.L. (1975). Co-extrusion of unfilled and TiO_2-filled polyethylene: influence of viscosity and die cross-section on interface shape. *Polymer Engineering and Science* 15 (12): 825–830.

96 Kerswell, R. (2011). Exchange flow of two immiscible fluids and the principle of maximum flux. *Journal of Fluid Mechanics* 682: 132–159.

97 Ayad, E., Cayla, A., Rault, F. et al. (2016). Influence of rheological and thermal properties of polymers during melt spinning on bicomponent fiber morphology. *Journal of Materials Engineering and Performance* 25 (8): 3296–3302.

98 Southern, J.H. and Ballman, R.L. (1975). Additional observations on stratified bicomponent flow of polymer melts in a tube. *Journal of Polymer Science* 13: 863–869.

99 Borzacchiello, D., Leriche, E., Blottière, B., and Guillet, J. (2014). On the mechanism of viscoelastic encapsulation of fluid layers in polymer coextrusion. *Journal of Rheology* 58 (2): 493–512.

100 Rosato, D.V., Schott, N.R., Rosato, D.V., and Rosato, M.G. (2001). *Plastics Engineering, Manufacturing and Data Handbook*. Kluwer Academic Publishers.

101 Perepelkin, K.E. (1972). The effect of the liquid and gas phase transitions on the extrusion stability in man-made fibre and film processes. *Fibre Chemistry* 3 (2): 115–123.

102 Ward, I.M. and Sweeney, J. (2005). *An Introduction to the Mechanical Properties of Solid Polymers*. Wiley Publication.

103 Lee, W.S. and Park, C.W. (1995). Stability of a bicomponent fiber spinning flow. *ASME Journal of Applied Mechanics* 62: 511–516.

104 Park, C.W. (1990). Extensional flow of a two-phase fiber. *AIChE Journal* 36 (2): 197–206.

105 Ramos, J. (1999). Asymptotic analysis of compound liquid jets at low Reynolds numbers. *Applied Mathematics and Computation* 100 (2): 223–240.
106 Ramos, J. (2001). Drawing of annular liquid jets at low Reynolds numbers. *Computational and Theoretical Polymer Science* 11 (6): 429–443.
107 Ramos, J. (2002). Compound liquid jets at low Reynolds numbers. *Polymer* 43 (9): 2889–2896.
108 Ji, C.C., Yang, J.C., and Lee, W.S. (1996). Stability of Newtonian-PTT coextrusion fiber spinning. *Polymer Engineering and Science* 36 (22): 2685–2693.
109 Ramos, J.I. (2016). Models of liquid crystalline polymer fibers. In: *Liquid Crystalline Polymers* (eds. V.K. Thakur and M.R. Kessler), 411–451. Springer International Publishing.
110 Kulkarni, J.A. and Beris, A.N. (1998). A model for the necking phenomenon in high-speed fiber spinning based on flow-induced crystallization. *Journal of Rheology* 42 (4): 971–994.
111 Doufas, A.K., Dairanieh, I.S., and McHugh, A.J. (1999). A continuum model for flow-induced crystallization of polymer melts. *Journal of Rheology* 43 (1): 85–109.
112 Doufas, A.K., McHugh, A.J., and Miller, C. (2000). Simulation of melt spinning including flow-induced crystallization: Part I. Model development and predictions. *Journal of Non-Newtonian Fluid Mechanics* 92 (1): 27–66.
113 Doufas, A.K. and McHugh, A.J. (2001). Two-dimensional simulation of melt spinning with a microstructural model for flow-induced crystallization. *Journal of Rheology* 45 (4): 855–879.
114 Doufas, A.K., McHugh, A.J., Miller, C., and Immaneni, A. (2000). Simulation of melt spinning including flow-induced crystallization: Part II. Quantitative comparisons with industrial spinline data. *Journal of Non-Newtonian Fluid Mechanics* 92 (1): 81–103.
115 Doufas, A.K. and McHugh, A.J. (2001). Simulation of melt spinning including flow-induced crystallization. Part III. Quantitative comparisons with PET spinline data. *Journal of Rheology* 45 (2): 403–420.
116 Kohler, W.H. and McHugh, A.J. (2007). Sensitivity analysis of low-speed melt spinning of isotactic polypropylene. *Chemical Engineering Science* 62 (10): 2690–2697.
117 Kohler, W.H. and McHugh, A.J. (2008). Prediction of the influence of flow-enhanced crystallization on the dynamics of fiber spinning. *Polymer Engineering and Science* 48 (1): 88–96.
118 Francisco, J., Rodríguez, B., and Ramos, J. (2011). Melt spinning of semi-crystalline compound fibers. *Polymer* 52 (24): 5573–5586.
119 Ramos, J. (2005). Modelling of liquid crystalline compound fibres. *Polymer* 46 (26): 12612–12625.
120 Chen, L., He, H.K., Zhang, Y. et al. (2015). Studies on melt spinning of sea-island fibers. *Fibers and Polymers* 16 (2): 449–462.
121 Prahsarn, C., Klinsukhon, W., Padee, S. et al. (2016). Hollow segmented-pie PLA/PBS and PLA/PP bicomponent fibers: an investigation on fiber properties and splittability. *Journal of Materials Science* 51 (24): 10910–10916.

122 Wang, X., Yao, J., and Pan, X. (2009). Fiber splitting of bicomponent melt-blown nonwovens by ultrasonic wave. *International Journal of Chemistry* 1 (2): 26–33.

123 Kim, H.S., Ito, H., Kikutani, T., and Okui, N. (1997). The thermal-bonding behaviour of polyethylene/poly(ethylene terephthalate) bicomponent fibres. *Journal of the Textile Institute* 88 (1): 37–51.

124 Kim, H.S., Ito, H., Kikutani, T., and Okui, N. (1999). Computational analysis on the thermal bonding behaviour of bicomponent fibres. *Journal of the Textile Institute* 90 (1 Part IV): 508–525.

125 Nakano, Y. (2011). Development of nonwovens for sanitary products. *Sen'i Gakkaishi* 67 (10): P288–P292.

126 DOW. Trichloroethylene: Industrial Use as Process Chemical in Alcantara Material Production. Socio-Economic Analysis. https://echa.europa.eu/documents/10162/18584504/afa_tce-0024-02-sea_en.pdf.

127 Funk, A., Dauner, M., Hoss, M., Rieger, C., Planck, H. (2011). Entwicklung von 1 Mikrometer-Supermikrofilamenten mit der Bikomponenten-Spinntechnik, International Techtextil-Symposium, Frankfurt am Main: 1-28. https://www.tib.eu/de/suchen/id/tema%3ATEMA20110600625/Entwicklung-von-1-Mikrometer-Supermikrofilamenten.

128 Rieger, C., Funk, A., Hoss, M. et al. (2010). Special fiber structures through bicomponent technology. *Chemical Fibers International* 3: 162–163.

129 Kamiyama, M. (2012). Industrial application of polyester nanofiber. *51st Man-Made Fibers Congress*. Dornbirn.

130 Asano, M., Kuroda, T., Shimizu, S. et al. (2001). Fiber structure and textile using same. USPTO, US 6326094 B1.

131 Kaufmann, J., Lübben, J.F., and Schwitter, E. (2007). Mechanical reinforcement of concrete with bi-component fibers. *Composites Part A: Applied Science and Manufacturing* 38: 1975–1984.

132 McLeod, A. (2008). The management and maintenance of second generation sand-filled synthetic sports pitches. National Resources Department, Doctor of Engineering. Cranfield: Cranfield University, p. 351.

133 Hufenus, R., Halbeisen, M., Camenzind, M. et al. (2014). Kunststofffaser für einen Kunstrasenbelag. Europäische Patentschrift, EP 2520696 B1, p. 9.

134 Dauner, M., Hundt, W., Stegmaier, T. et al. (2011). Textile product and production thereof. EP 2061926 B1, TWD Fibres Gmbh.

135 Dauner, M., Hundt, W., Funk, A. et al. (2012). Fibers with easy to clean surfaces. *51th Man-made Fibers Congress*, Dornbirn.

136 Reifler, F.A., Hufenus, R., Krehel, M. et al. (2014). Polymer optical fibers for textile applications: Bicomponent melt spinning from cyclic olefin polymer and structural characteristics revealed by wide angle X-ray diffraction. *Polymer* 55 (22): 5509–5846.

137 Quandt, B.M., Ferrario, D., Rossi, R.M. et al. (2017). Body-monitoring with photonic textiles: a reflective heartbeat sensor based on polymer optical fibers. *Journal of the Royal Society Interface* 14 (128): 20170060.

138 Sohn, I.-S. and Park, C.-W. (2001). Diffusion-assisted coextrusion process for the fabrication of graded-index plastic optical fibers. *Industrial and Engineering Chemistry Research* 40 (17): 3740–3748.

139 Manabe, T. (1997). Speciality polyamide and polyester yarns: an industrial approach to their production and rheology. In: *Manufactured Fibre Technology* (eds. V.B. Gupta and V.K. Kothari), 360–405. Chapman & Hall.

140 Lund, A. and Hagström, B. (2011). Melt spinning of β-phase poly(vinylidene fluoride) yarns with and without a conductive core. *Journal of Applied Polymer Science* 120 (2): 1080–1089.

141 Lund, A., Jonasson, C., Johansson, C. et al. (2012). Piezoelectric polymeric bicomponent fibers produced by melt spinning. *Journal of Applied Polymer Science* 126 (2): 490–500.

142 Glauß, B., Steinmann, W., Walter, S. et al. (2013). Spinnability and characteristics of polyvinylidene fluoride (PVDF)-based bicomponent fibers with a carbon nanotube (CNT) modified polypropylene core for piezoelectric applications. *Materials* 6 (7): 2642–2661.

143 Nilsson, E., Lund, A., Jonasson, C. et al. (2013). Poling and characterization of piezoelectric polymer fibers for use in textile sensors. *Sensors and Actuators A: Physical* 201: 477–486.

144 Strååt, M., Rigdahl, M., and Hagström, B. (2012). Conducting bicomponent fibers obtained by melt spinning of PA6 and polyolefins containing high amounts of carbonaceous fillers. *Journal of Applied Polymer Science* 123 (2): 936–943.

145 Leal, A.A., Naeimirad, M., Gottardo, L. et al. (2016). Microfluidic behavior in melt-spun hollow and liquid core fibers. *International Journal of Polymeric Materials* 65 (9): 451–456.

146 Blaiszik, B., Kramer, S., Olugebefola, S. et al. (2010). Self-healing polymers and composites. *Annual Review of Materials Research* 40: 179–211.

147 Thakur, V.K. and Kessler, M.R. (2015). Self-healing polymer nanocomposite materials: a review. *Polymer* 69: 369–383.

148 Zhang, Y.J., Takarada, W., and Kikutani, T. (2015). Fabrication of fiber-reinforced single-polymer composites through compression molding of bicomponent fibers prepared by high-speed melt spinning process. *Sen'i Gakkaishi* 71 (5): 172–179.

149 Meng, Q. and Hu, J. (2009). A review of shape memory polymer composites and blends. *Composites Part A: Applied Science and Manufacturing* 40 (1): 1661–1672.

150 Tallury, S.S., Pourdeyhimi, B., Pasquinelli, M.A., and Spontak, R.J. (2016). Physical microfabrication of shape-memory polymer systems via bicomponent fiber spinning. *Macromolecular Rapid Communications* 37: 1837–1184.

12

Superabsorbent Fibers

Nuray Ucar and Burçak K. Kayaoğlu

Istanbul Technical University, Faculty of Textile Technologies and Design, Department of Textile Engineering, Inonu Caddesi, No: 65, Gumussuyu, Beyoglu, Istanbul, 34437, Turkey

12.1 Introduction

Superabsorbent polymer (SAP) is a commodity product used in a variety of industries over the last 30 years, from hygiene to filtration and from agriculture to sportswear and food packaging [1]. These materials have added great value to their product applications and improved the lives of millions of people every day. They serve for holding and retaining extremely large volumes of water and aqueous solutions inside, relative to its own mass [1–3]. Highly absorbing polyelectrolyte polymer-based materials are able to absorb up to 50 g of fluid per gram of dry mass; on the other hand, superabsorbent fibers can absorb 100 times its own weight in water [4].

As the population grows, the global demand for absorbent products increases that triggers the growth in global SAPs market. According to Future Market Insights' report [5], by 2020, the value of global SAP market is projected to be slightly over US$9 billion and reach 2,892,400 tons by volume. Another report addresses the expected growth of global SAPs market to reach US$11.03 billion by 2022. BASF SE, Nippon Shokubai, and Evonik Industries are the key manufacturers of SAPs employing 51% of the production [6].

There are many ways to use SAPs such as SAP powder, SAP granule, and SAP fiber. Among these, superabsorbent fiber can be handled easily during processing and may have higher absorption speed than SAP powder/granule, due to their fibrous shape. Thus, superabsorbent fiber can be transformed into yarn to be knitted or woven in single or blended form with other fiber types in order to meet end use expectations. Superabsorbent fiber can also be produced in micro- or nanoscale that provides higher surface area, leading to improvement in absorption and vapor transmission properties.

12.2 Overview of Superabsorbent Fibers

12.2.1 History of Superabsorbent Fibers

The first water-absorbent polymer was synthesized in 1938 by the thermal polymerization of acrylic acid and divinylbenzene. In the 1950s, the hydroxyalkyl methacrylate-based hydrogels were manufactured, finding application in contact lenses. In the 1970s, hydrolyzed starch-*graft*-polyacrylonitrile product was produced as the first commercial SAP. Commercial production and utilization of SAP in feminine napkins and baby diapers started in 1978 in Japan and in 1980 in Germany and France, which was followed by Asian countries, United States, and Europe. The worldwide production of SAP reached more than 1 million tons in the late 1990s [7].

Besides the granular SAP, in the early of 1990s, ARCO Chemical developed a superabsorbent fiber [8, 9]. In the 1993, Technical Absorbents in the United Kingdom that was a joint venture between Allied Colloids and Courtaulds has also been one of the manufacturers of Super Absorbent Fibre (OASIS SAF®) [8, 10–13]. Today, Technical Absorbents (TAL) is a subsidiary company of China National BlueStar (Group) Company Limited, who in turn are wholly owned by ChemChina [14].

12.2.2 Main Principle of Superabsorbency

SAPs are natural or synthetic cross-linked polyelectrolytes, comprising an insoluble polymer matrix with about 96% water content, starting to swell upon contact with water or aqueous solutions, resulting in a rubbery hydrogel formation [15–17]. They can also be tailored to be soluble in water and other solvents [14].

SAPs are able to absorb up to several hundred times water per gram of SAP [16, 18]. They are able to swell or shrink in aqueous solutions due to the association, dissociation, and binding of various ions to polymer chains. SAPs may swell in water until an equilibrium state is reached and retain their original shape [15, 16]. The cross-linking degree of SAP directly affects its mechanical strength, total absorbency, and swelling degree. Low density cross-linked SAP has a soft and cohesive gel formation with high absorbent capacity, whereas high density cross-linked polymer shows a firmer gel formation with lower absorbent capacity [1, 8].

Figure 12.1 shows a commercial cross-linked polyacrylic acid (PAA)-based SAP, where negative carboxylate groups are partially neutralized with hydroxides of alkali metals, usually sodium [3, 16, 17]. The main driving force responsible for swelling of SAPs includes hydration due to the presence of hydrophilic chemical groups such as $-OH$, $-COOH$, $-CONH_2$, $-CONH-$, and $-SO_3H$ and capillary areas and differences in osmotic pressure inside and outside the gel [15–17].

When a superabsorbent material comes in contact with water, hydration of sodium ions reduce their attraction to the carboxylate ions and allow them to move freely within the network, which increases the osmotic pressure within the

Figure 12.1 SAP based on polyacrylic acid.

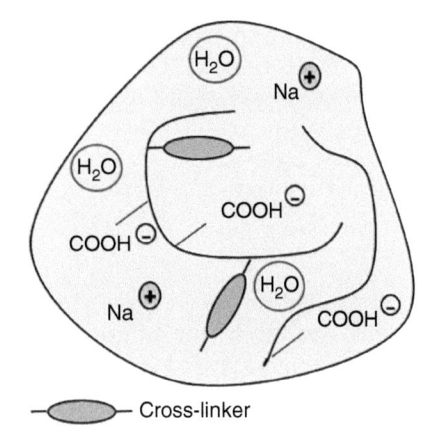

gel [1, 17]. The osmotic pressure is proportional to the concentration of ions in SAPs [16].

The forces that make hydrogel dissolution impossible and help to maintain superabsorbent material's integrity or control the extent of swelling when wet are the presence of cross-links between polymer chains forming a three-dimensional network and hydrophobic and electrostatic interactions [15, 17, 19].

Hydrophilic sodium polyacrylate powder is able to absorb up to 800 times its weight in distilled water and takes the form of a coiled chain (Figure 12.2). When the polymer interacts with water, the sodium (Na) on its polymer chains detaches from the carbonyl (COOH) group creating water-attracting carboxyl (COO^-) and sodium (Na^+) ions. In the presence of water, due to repulsion between the negatively charged carboxyl groups, the sodium polyacrylate chain uncoils and generates a gel substance. The dry coiled SAP molecules form hydrogen bonding during interaction with water, which leads to unfolding and extension of their chains (Figure 12.2) [20].

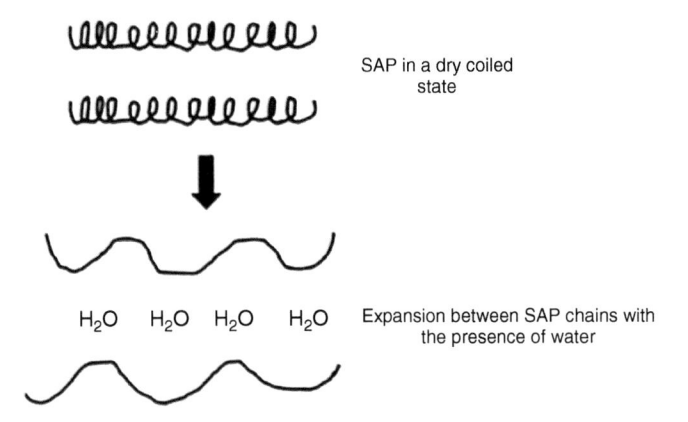

Figure 12.2 Schematic representation of the interaction of a dry superabsorbent polymer with water.

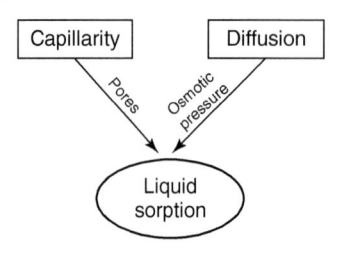

Figure 12.3 Liquid sorption mechanism of a superabsorbent textile-based product.

Capillarity (Pores), Diffusion (Osmotic pressure) → Liquid sorption

The liquid sorption capacity of superabsorbent textile-based products is controlled not only by the diffusion of liquid into the superabsorbent fibers but also transport to the pores between the fibers by capillary action [1, 21] (Figure 12.3).

When the solvent is water, hydrophilic interactions between water molecules and hydrophilic groups of the superabsorbent material and electrostatic interactions between ionic groups of an aqueous liquid and superabsorbent material take place [1]. Nevertheless oil superabsorbent materials, based on polyolefins, are oleophilic (oil attracting) and hydrophobic (water repellent). A high oil absorption (up to 45 times of its weight) was reported for a polyolefin-based oil SAP (oil-SAP) [22]. Since the superabsorbent material has affinity to oil, during absorption, hydrophobic interactions occur between oil and fatty liquid and the material [1].

Superabsorbent fibers are superior to conventional wood pulp and cotton linter since they can absorb up to 50 times their mass in water; in turn, wood pulp and cotton linter absorbents absorb approximately six times their mass in water. Their small diameter, around 30 μm, and high surface areas allow these superabsorbent fibers to typically absorb 95% of the ultimate capacity within 15 seconds [23, 24].

The textile-based superabsorbent products are commonly designed to absorb and retain bodily fluids and exudates [1, 25, 26]. Under 0.5–7 kPa load, representing real-use conditions, swollen SAP particles in absorbent products were reported to hold 20–50 g of body liquid per gram of SAP dry mass [19]. When the absorbent products are worn next to the skin, superabsorbent materials help to keep the skin dry, which is a necessity for comfort [3].

Das et al. [21] reported that useful combination of normal plus superabsorbent materials is necessary to achieve enhanced liquid sorption property. Because as the liquid diffuses into the superabsorbent fibers, since pore space is reduced due to gel blocking [19] and extensive swelling of the fibers, the rate of liquid absorption is reduced, preventing further transport of liquid. Therefore Das et al. [21] developed liquid-absorbent composite nonwoven media by combining superabsorbent fibers made from acrylic acid, methylacrylate, and a small quantity of special acrylate/methylacrylate monomer, with normal polypropylene (PP) and polyester fibers. It was reported that nonwovens produced by random mixing of equal proportion of superabsorbent fibers and fine denier circular PP fibers yielded highest sorption capacity and highest sorption rate [21]. Combination of equal proportion of superabsorbent fibers and fine denier (2.5 den) PP fibers compared to the combination of superabsorbent fibers and higher denier (6 den) PP fibers resulted in higher liquid sorption capacity and rate [21]. As a result, better sorption properties can be achieved by the use of finer fibers in the absorbent structure. It was also reported that the noncircular (trilobal and deep-grooved) fibers displayed poorer sorption characteristics than the circular fibers [21].

12.2.3 Polymer Materials

SAPs are produced by free radical polymerization of partially neutralized acrylic acid and/or other comonomers such as methacrylic acid, blended with sodium hydroxide, in the presence of a suitable cross-linking agent such as N,N'-methylene bisacrylamide and 1,1,1-trimethylol-propane triacrylate or ethylene glycol diacrylate [27]. Polyacrylamide copolymer, ethylene maleic anhydride copolymer, and polyvinyl alcohol (PVA) copolymers are among different examples of these polymers [20].

There are different polymerization techniques such as bulk polymerization, solution polymerization, and inverse suspension polymerization. Among the mentioned techniques, a high molecular weight polymer with high purity may be obtained by bulk polymerization [27].

For the production of SAP, solution-based polymerization is most commonly used, which is an efficient and low-cost method. In this method, a mass of reactant polymerized gel is obtained from a water-based monomer solution. Then the final granule size is obtained after chopping and drying processes. The suspension process provides a good production control during polymerization step where a water-based reactant is suspended in a hydrocarbon-based solvent [20].

Research groups have currently focused on producing SAPs from natural polymers such as cellulose and starch by converting them into carboxymethyl derivatives followed by structural cross-linking [4].

Among natural polymers, cotton, rayon, wood pulp are superabsorbent by nature [1, 28]. These fibers may be chemically modified to render them superabsorbent. Cotton fibers can be chemically modified, i.e. grafting with acrylonitrite, to produce superabsorbent grafted cellulosic fibers. Absorbency of rayon fiber was enhanced by introducing additives such as polyacrylic salt and cellulose sulfates to spinning solution [4]. A superabsorbent material was produced from pretreated flax yarn waste by grafting acrylic acid and acrylamide (AM) by free radical graft copolymerization [29].

12.2.4 Production Methods

Besides in the form of powder, foil, or foam, superabsorbent materials are produced in the form of fibers. Superabsorbent fibers are commonly cross-linked acrylic copolymers neutralized by a sodium salt [28].

Because of its fibrous shape, superabsorbent fiber can be mixed with other fibers and can be easily processed through spinning, weaving, and nonwoven technology, leading to better and easier processability compared with conventional SAP powder and granules [30]. In some structures, it is also possible to get 8–10 times higher absorption speed than the SAP powder/granules [31]. OASIS superabsorbent yarn (polyacrylate (copolymer superabsorbent)) which is a type of superabsorbent fiber yarn has a specific gravity of $1.4\,g/cm^3$, a tenacity

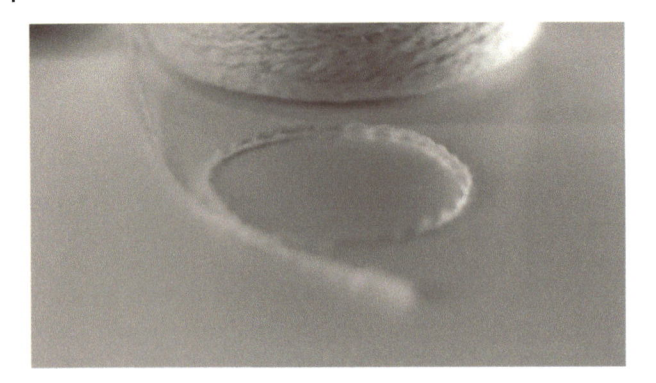

Figure 12.4 Superabsorbent yarn (containing SAF) [14] (having a permission from Technical Absorbents dated May 5, 2017). Source: Reproduced with permission from http://techabsorbents.com/.

Figure 12.5 Swollen superabsorbent yarn (containing SAF) [14] (having a permission from Technical Absorbents dated May 5, 2017). Source: Reproduced with permission from http://techabsorbents.com/.

of 0.6–0.7 g/denier, and an elongation at break of 15–30% [32]. In Figures 12.4 and 12.5, superabsorbent yarn can be seen [14].

Superabsorbent fiber is mainly produced by two different ways, i.e. mixing the superabsorbent material with hydrophobic/hydrophilic material or directly using a SAP [33].

12.2.4.1 Mixing the Superabsorbent Material with Hydrophobic/Hydrophilic Material

Some superabsorbent fibers could be obtained by mixing an SAP with any other material. For example, Lanseal® F has an outer layer with SAP and an inner layer of acrylic fiber itself. Whet it contacts with water, outer surface (SAP) absorbs water and swells (Figure 12.6) [34].

Beskisiz et al. [35] blended polyester staple fiber (PES) with superabsorbent fiber to produce a knitted fabric. They have concluded that water absorption performance of knitted fabric produced by blended yarn (superabsorbent fiber and PES) changes slightly during repeated water absorption test. Although dry cleaning treatment applied on the knitted fabric results in slight change on water absorption performance, washing and drying treatments applied to knitted fabric result in very dramatic change and decrease of water absorbance performance.

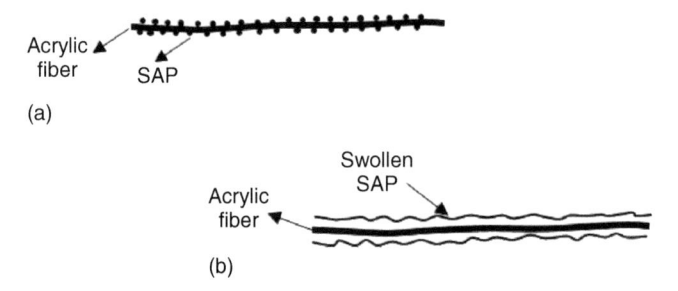

12.2 Overview of Superabsorbent Fibers

Figure 12.6 (a) Schematic representation of SAP-coated acrylic fiber. (b) After contact with liquid.

Figure 12.7 "C" cross-sectional shape in which the outer surface is polypropylene (PP) and inner cavity is filled by superabsorbent powder.

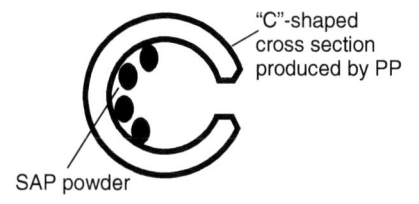

In one study, it has also been pointed out that as the percentage of SAP fiber in the blend of PES and SAP fiber increases, the absorbent capacity (cm^3/g) increases while the absorbency rate (cm^3/g-sec) decreases because of the transverse diffusion of fluid in the fibers that leads to a reduction in forward velocity and the swelling of fibers, resulting in a decrease of pore size in the structure. However, an increase of saline concentration from 0% to 2% has an opposite effect, i.e. the presence of saline causes a decrease of capacity and increase of rate. The decrease of capacity has been explained by the reduction in the coulombic repulsion in the polymer network, which limits the swelling and penetration of fluid into the fiber [36].

Ucar et al. [37] produced a filament with "C" cross-sectional shape in which the outer surface is PP and inner cavity is filled by superabsorbent powder (Figure 12.7). It has been pointed out that the filament with SAP could absorb water vapor due to the SAP filled into the cavity of C-shaped filament, while PP at the outer surface provides dry feeling.

Eskin et al. [38] also studied on the "C"-shaped filament yarn whose outer layer is consisted of a nonpolar polymer such as PP and maleic anhydride PP (MAPP) and inner cavity is filled by a hydrophilic polymer such as superabsorbent powder. Authors reported that the filament can absorb water (approximately 23%) because of SAP powder without creating any feeling of wetness due to the hydrophobic polymer at the outer layer and completely dries in 20 minutes. The water absorbance performance after washing and drying treatment could be improved because of MAPP.

In these studies [37, 38], the theory was to absorb moisture and liquid into SAP powder placed into cavity of C-cross-sectional-shaped filament that outer surface of filament is hydrophobic polymer (Figure 12.8). Thus, while filament absorbs liquid such as sweat, the skin that contacts with filament will not feel any wetness.

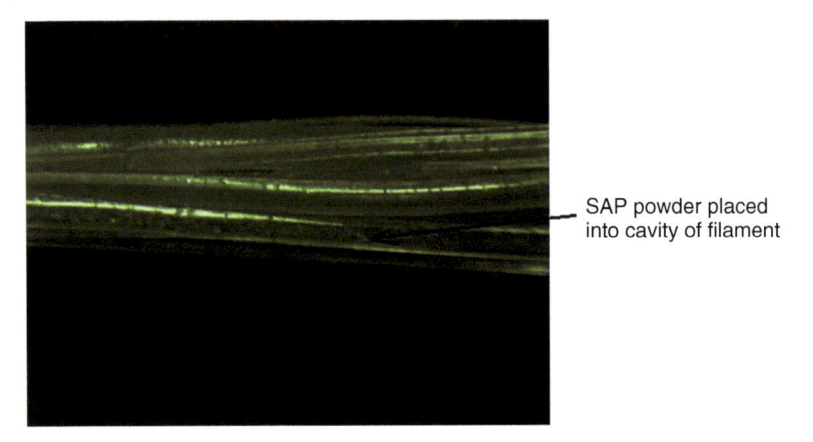

Figure 12.8 C-cross-sectional-shaped filament bundle where cavity of filament was filled by SAP powder.

12.2.4.2 Directly Using a Superabsorbent Polymer for Superabsorbent Fiber Production

In textile industry, there are many ways to produce synthetic fiber such as melt spinning, wet spinning, dry spinning, dry–wet spinning, etc. For micro-superabsorbent fiber, dry spinning and wet spinning methods have been widely used, while it is also possible to produce nano-superabsorbent fiber by the use of an electrospinning technique.

In dry spinning that can have faster production rate than wet or melt spinning, the polymer is dissolved in a volatile solvent, and after many times of filtering, spinning solution is injected through spinneret holes into hot area/drying tower where hot air or other hot gases can be used (Figure 12.9). Hot air evaporates the solvent, leading to the solidification of filament (fiber). Before winding on a take-up cylinder (drawing roller), filaments are drawn for improvement of orientation and crystallization. Evaporated solvent is absorbed or condensed and then recycled that can be costly. Solvent selection is very important with respect to boiling point, toxicity, inertness, vaporization, and thermal stability. Nonpolar

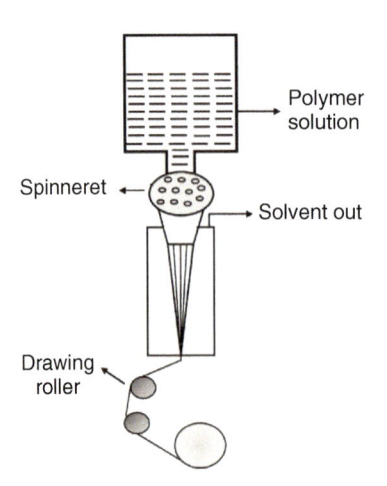

Figure 12.9 Schematic representation of dry spinning.

Polymer solution

Drawing roller

Take-up roller

Coagulation bath

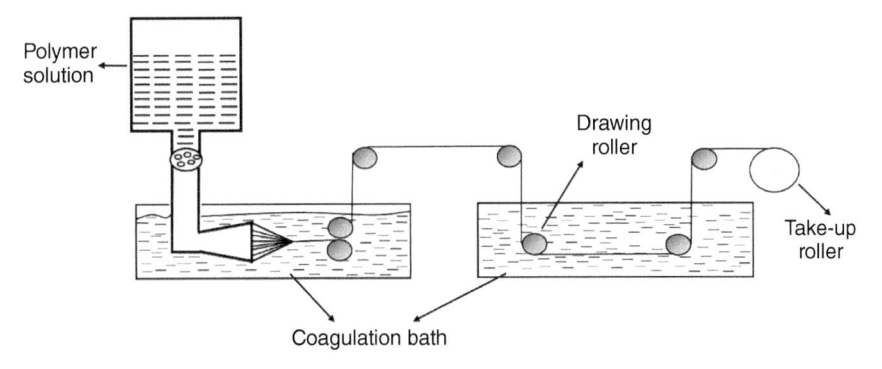

Figure 12.10 Schematic representation of wet spinning.

solvent can be preferred for their low boiling point; however its use can be risky due to building up of electrostatic charges [39, 40].

In wet spinning, the polymer that is dissolved in a nonvolatile solvent is injected into a coagulation bath for solidification of filament/fiber (Figure 12.10). During solidification steps, filament/fiber is also drawn in order to improve the orientation and crystallization. The solvent can then be recovered from wastewater in coagulation bath; however this process can be costly [40].

One of the commercial Super Absorbent Fibre that belongs to Technical Absorbents, UK, is called SAF™ and is produced by solution spinning technique (Figure 12.11). It has been announced that SAF, which is a cross-linked

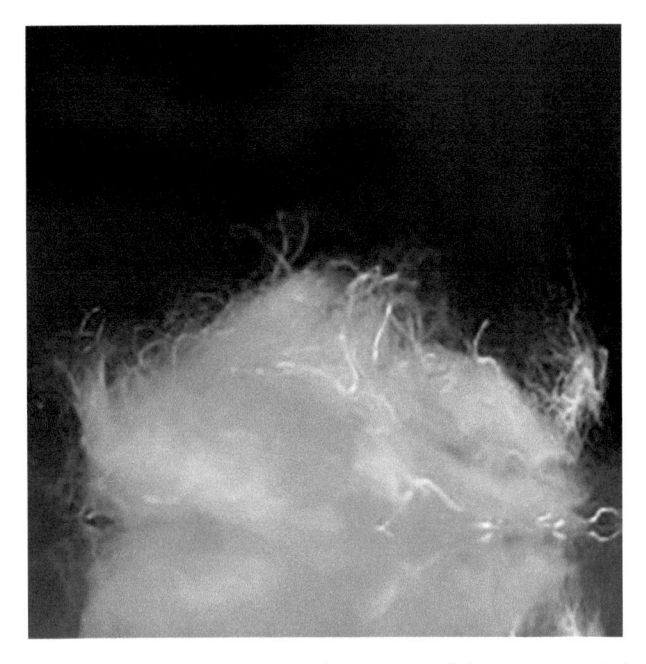

Figure 12.11 SAF by Technical Absorbents [14] (having a permission from Technical Absorbents dated May 5, 2017). Source: Reproduced with permission from http://techabsorbents.com/.

polyacrylate polymer, is produced by solution spinning through the water evaporation. It has an ability to absorb up to 200 times its own weight in water. SAF with approximately 1.4 g/ml density and 10 cN/tex tenacity with 5% extensibility has a length range of 6–80 mm, a linear density range of 2–20 dtex, 15 minutes free swelling capacity of 60 g·g^{-1} in saline, and 200 g·g^{-1} in distilled water and does not melt and decompose at a temperature higher than 200 °C [10, 14, 41].

There are several studies on superabsorbent fiber produced by wet spinning or dry spinning technique. For example, Kim et al. [42] prepared superabsorbent fiber based on sodium alginate using glutaraldehyde as a cross-linking agent. Alginate that is dissolved in water and aged for 24 hours was extruded into hydrochloric acid (HCl, 35%) coagulation bath in order to produce alginic acid gel filament by wet spinning technique. The gel fiber was then placed in dioxane solution containing glutaraldehyde and 0.1% HCl (35%) in order to cross-link, decreasing the solubility. Then, it has been neutralized to increase an absorbency. It has been pointed out that decrease of glutaraldehyde concentration up to a certain concentration results to an increase of absorbency of alginate fibers in synthetic urine solution and in saline solution.

Liu et al. [30] studied on acrylic-based superabsorbent fiber manufactured by solvent spinning. It was observed that water absorbency of superabsorbent fiber is affected by ionic valence, ion concentration, and pH values of saline solution. Maximum absorbency of superabsorbent fiber is about 117 g·g^{-1}. An increase of concentration of saline solution results to a decrease of absorbency, and maximum absorbency could be obtained in neutral solution.

Fei-Ming et al. [43] studied the absorption ability of PVA/PAA-AM-based superabsorbent fibers. It has been reported that AM content affects the absorption ability and a moderate amount of AM can improve the absorption ability.

In addition to conventional micro-superabsorbent fiber production, it is possible to produce nano-superabsorbent fiber by electrospinning. Because of nanostructure with higher surface area leading to improvement in absorption and vapor transmission properties, they can be used in various applications such as food preservation, filtration, personal hygiene products, etc. [44].

At the electrospinning system (Figure 12.12), a polymer solution is subjected to an electric field so that a charged jet of solution is ejected from the nozzle of electrospinning system. The solvent evaporates during traveling of jets through the air, resulting in solidified nanofibers (Figure 12.13) collected onto a collector [44].

There are several studies on nano-superabsorbent fiber produced by electrospinning system [44, 45]. For example, Islam et al. [45] produced nano-superabsorbent fibrous web by using the natural polysaccharide pullulan (PULL), PVA, and montmorillonite (MMT) clay and double distilled water as a solvent. They have reported that these materials have greater environmental and commercial values and low production cost together with better biodegradability. During electrospinning, applied voltage was set to 15 kV, and the distance between nozzle and aluminum foil collector was set to 15 cm. Then, nanofiber web was heated for four minutes at 150 °C without direct contact with

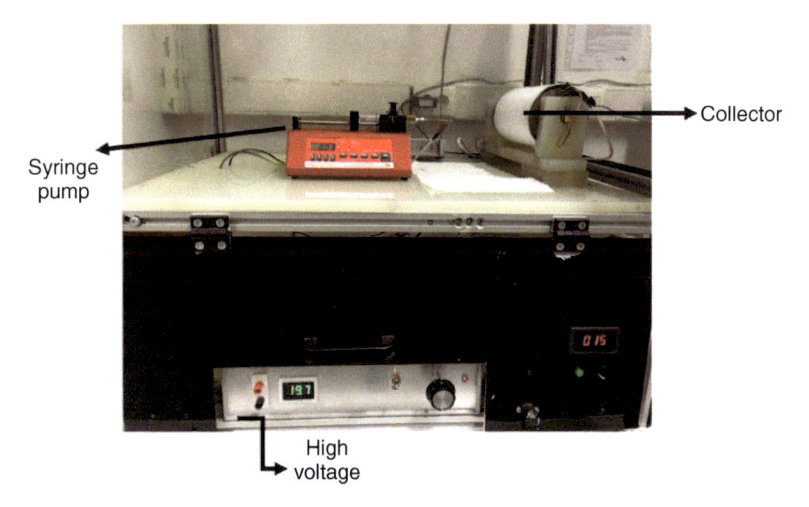

Figure 12.12 Photographic image of an electrospinning system.

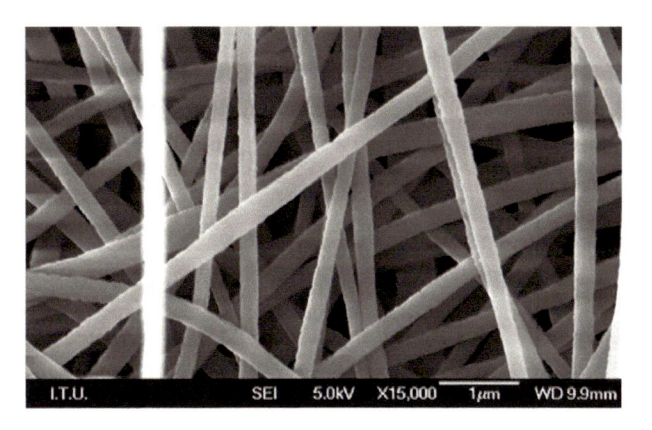

Figure 12.13 Scanning electron micrograph of electrospun nanofibers.

hot plate. Researchers have pointed out that the heat-treated nanocomposite superabsorbent fiber showed water absorbency of 143.42 g·g^{-1} and 39.75 g·g^{-1} in distilled water and a 0.9 wt% NaCl solution. They have pointed out that the heat treatment improved the stability by improvement of the crystallinity in addition to an improvement in water absorbency due to MMT. It has been concluded that an increase of MMT results in an increase of water absorbency.

12.2.5 Test Methods

The performance of superabsorbent textiles is determined by liquid absorption rate and retention measurements. The water/liquid absorption capacity can be measured by gravimetric, dynamic wetting, and topographic measurements. Liquid absorption capacity is expressed in grams of liquid retained per gram of dry superabsorbent sample (g·g^{-1}). Another parameter related to superabsorbent material's ability to absorb liquid at a certain temperature is the absorbency under load (AUL), where there is predetermined load acting on the superabsorbent material. This parameter is important to prevent the leakage of liquid from the absorbent core of a hygiene product under body loads and provide improved surface dryness [1, 4, 46].

Liu et al. [30] used the following equation to calculate water absorbency (Q) of acrylic-based superabsorbent fibers, where U_d and U_s are the weight of the dry and swollen superabsorbent fibers, respectively:

$$Q\,(g/g) = \frac{U_s - U_d}{U_d} \qquad (12.1)$$

The liquid retention measurement involves entire wetting and swelling of the superabsorbent material, which is followed by centrifugation at a fixed rotation speed for a given period of time. This parameter is expressed with the following equation [1, 29], where U_c and D are the weights of the sample after centrifuging and when dry:

$$R\,(g/g) = \frac{U_c - D}{D} \qquad (12.2)$$

Higher wicking values imply higher liquid transport capability of a superabsorbent textile. Wicking parameter may be determined gravimetrically by measuring the amount of liquid drawn into capillaries of a superabsorbent product (in percentage) after a certain period of time or the vertical height (in cm) that the liquid reaches above the level of the reservoir during a certain period of time [47].

High absorbency, protection against leakage, and low rewet are required for absorbent hygiene products. Therefore retention, rewet and absorption time under load are measured for these products. In absorption time under load test, the absorbent product, i.e. a diaper, is placed on a polyurethane foam base onto which a cover plate and weights are placed. The foam base helps to divide the pressure equally across the diaper. Synthetic urine is pumped into an application tube through a hole on the cover plate. An electrode on the application tube connected to a computer measures the time required for the absorption of all the liquid of the tube by the diaper. Absorption rates are reported as ml/s [3].

Rewet property is related to keeping the wearer's skin dry, which shows the amount of absorbed liquid forced out again through the facing under pressure. A weight is applied on a liquid loaded diaper onto which an absorbent material such as filter paper is placed. After the removal of the weight, the amount of liquid absorbed by the absorbent material in grams or milligrams is calculated from the change in material's weight. Besides the abovementioned tests, the time in seconds for liquid to penetrate the top sheet of the diaper is measured that is known as strikethrough [3].

For ionic sensitivity of SAP materials, a dimensionless swelling factor, f, is measured with the following formula [7]:

$$f = 1 - (\text{Absorption in a given fluid}/\text{absorption in distilled water}) \qquad (12.3)$$

SAP materials with lower swelling factor are desired since it implies lower absorbency loss of the sample swollen in salt solutions.

Free swell capacity of SAP material implies absorption capacity of SAP without pressure. For this test, pure SAP is placed in a tea bag and submerged in 0.9% saline. Then excess solution is removed by letting the bag drip off for 10 minutes. Then swelling capacity is calculated from the difference between the weights of tea bag and initial weight of SAP sample. In centrifuge retention capacity test, the solution between particles is removed by centrifugation of the tea bag. The swelling capacity is calculated by the following formula:

$$S = (W_2 - W_0 - W_1)/W_1 \tag{12.4}$$

where W_1 is the weight of SAP material, W_2 is the weight of the tea bag, and W_0 is the weight of an empty bag [7, 48].

12.3 Application

Superabsorbents may be placed in the product as layers or enclosed in between textile layers in the form of compressed particulate powder [1]. Absorbent structures comprise an absorbent core in which SAPs and/or mixtures of fibrous absorbents from cellulose or fluffed wood pulp and SAPs have been incorporated [2, 3]. Superabsorbent textiles including superabsorbent fibers and their blends with other natural and synthetic fibers can be produced by spinning, weaving, or nonwoven manufacturing techniques [30]. It is also possible to bond or laminate superabsorbent fiber-based fabrics to spunbond fabrics or films for special product applications such as food packaging and filtration. In Figure 12.14, several application areas done by SAF can be seen [14].

Superabsorbent fibers have been utilized mainly in liquid absorbent products such as disposable infant diaper, feminine sanitary napkin, underpad, adult incontinence pad, wound dressing, and absorbent wipes for hygiene and healthcare industries [3, 21, 26, 49]. These products can be very thin yet absorbent due to the existence of SAPs [46]. For instance, Mextra® superabsorbent dressing, produced as a wound care product, has a four-layered structure comprising polyacrylate superabsorbent particles [50]. Xtrasorb® is a superabsorbent

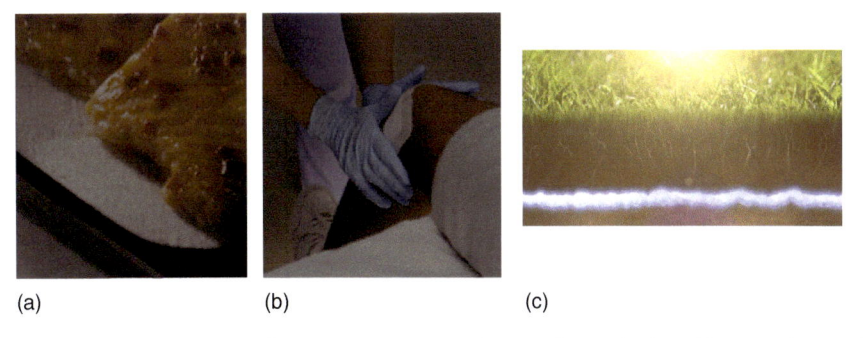

(a)   (b)   (c)

Figure 12.14 Application of SAF nonwoven. (a) Packaging, (b) medical, and (c) agriculture [14] (having a permission from Technical Absorbents dated May 5, 2017). Source: Reproduced with permission from http://techabsorbents.com/.

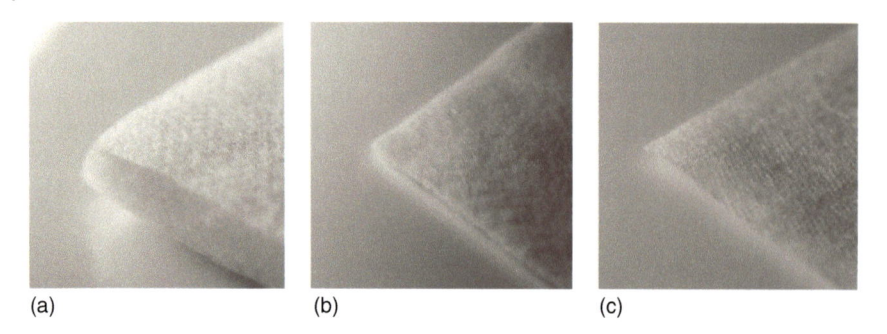

Figure 12.15 Nonwoven structures. (a) Airlaid fabric, (b) needlefelt fabric, and (c) carded fabric [14] (having a permission from Technical Absorbents dated May 5, 2017). Source: Reproduced with permission from http://techabsorbents.com/.

wound dressing that comes in the form of pad and foam and helps to take the wound fluid away from the wound and turn it into gel [51].

Superabsorbent textiles also find application in sportswear and footwear for heat and moisture management [1]. SAF [52] has been converted into 2.0 m wide, 150–1000 gsm fabrics through needle punching technology (Figure 12.15). Airlaid nonwoven fabrics, 50–1000 gsm, have also been produced by blending short staple SAF with wood pulp and low melt fibers [52].

SAF is also blended with man-made fibers such as polyester and nylon to produce spun yarns using traditional yarn spinning technologies, i.e. open end and ring. These yarns may also be used to produce woven, knitted, and net structures.

Swellcoat™ from Fiberline [53] is a water-blocking finish comprising a SAP. It may be applied on fibers, yarns, and textile substrates and generates a stable gel and can absorb up to 10 times its own weight. Universal Carbon Fibres Ltd. [54] produced staple yarns from OASIS Super Absorbent Fibre (SAF), which can absorb high volumes of moisture rapidly and have been used by cable manufacturers.

Baby diapers were reported to have 85% share among hygiene products that depend on superabsorbent materials, while adult incontinence products followed with 10% share [28]. In Figure 12.16, swollen structures of superabsorbent fibers can be seen [14].

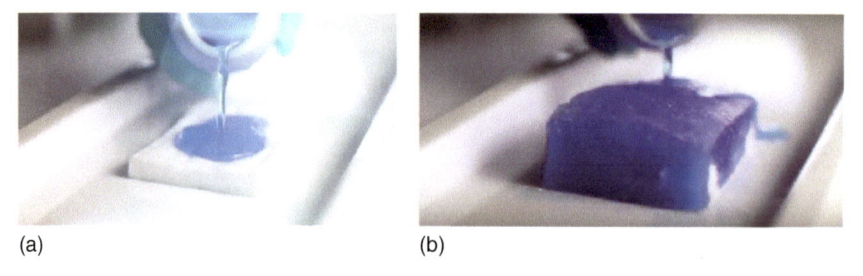

Figure 12.16 Nonwoven. (a) Before swollen and (b) after swollen [14] (having a permission from Technical Absorbents dated May 5, 2017). Source: Reproduced with permission from http://techabsorbents.com/.

According to Future Market Insights' report, the highest global demand for SAP was in disposable diaper industry with more than 74% share in 2014, and the adult incontinence products accounted for almost 12%, while it was over 9% for feminine hygiene products [5]. Commercial brand names like Pampers® and Huggies® use tiny SAP crystals scattered inside the layers of the absorbent core of a baby diaper in order to absorb and trap fluids such as urine and wet poopy [55].

In biomedical applications such as wound dressings, hydrogels present water to the wound site and maintain a moist environment, absorb a degree of wound exudate, and help in the wound healing process [15]. Aquaform, Intrasite, Granugel, Nu-Gel, Purilon, and Sterigel are some of the examples of hydrogels used in biomedical applications [15].

Superabsorbent materials have also been used in technical applications such as oil recovery and air and water purification [56, 57] to absorb other types of fluids such as organic liquids and gases, heavy metal ions, and dyes [1, 4]. Arkema and SNF Floerger are the manufacturers of AM and acrylic acid-based technical SAPs used for cable isolation [58], landscaping, and food packaging [24, 59].

In food pads, superabsorbent materials are used to keep the meat or vegetables fresher for a longer period of time by absorbing their liquids. Dry SAP particles have been reported to be used in concrete mixture as well, serving for absorbing water and finally forming empty pores in the cement paste [16].

Different applications for SAF include filter fabrics, used to remove contaminants and separate water from fuel and oil and also in active sportswear garments, which cools down the user with the evaporation of water [52]. Another field of application for superabsorbent materials is protective clothing, for instance, for fire protection [60]. Bartkowiak [23] developed needled nonwoven materials with high liquid sorption capacity, containing superabsorbent fibers, designed for sweat-absorbing inlays inside protective clothing. They combined "Oasis 102" superabsorbent fibers [61] with cotton, polyester, and PP fibers. These fibers were reported to have high salt water absorption under pressure, which is important for absorption of sweat and human liquids [28].

Superabsorbent textiles were used to enhance the release of agrochemicals and water in agriculture application while they serve for drainage or soil reinforcement in geotextiles application [1]. Aquasorb [62] is a SAP, based on an anionic polyacrylamide, used to increase the water holding capacity of the soil, which is beneficial in growing trees, shrubs, and plants. Another commercial SAP containing potassium, phosphorus, and nitrogen was developed by Ma's Group for agriculture application [63]. Table 12.1 summarizes the common application fields of SAPs.

12.4 Future Scope and Challenges Ahead

There is a growing global demand for improved superabsorbent materials as they find applications in medical, food, and personal care articles representing a multibillion-dollar market [19].

Table 12.1 Main applications of superabsorbent materials.

Superabsorbent powders
Disposable infant diaper
Underpad
Adult incontinence pad
Oil recovery
Air and water purification
Cable isolation
Landscaping
Food packaging

Superabsorbent fibers
Sportswear
Footwear
Protective clothing
Agrotextiles
Geotextiles
Disposable infant diaper
Wound and surgical dressing
Feminine sanitary napkin
Absorbent wipes for hygiene and healthcare industries
Food packaging
Cable isolation
Automotive filter cartridges
Filtration

The introduction of inorganic fillers and nanocomposites such as antiseptic and healing agents to the superabsorbent systems to have added value and functionalities and the use of natural polymers with lower environmental impact and higher biocompatibility are currently the subjects of interest [1]. Research and development efforts on efficient absorbent materials for medical and personal care purposes have shifted toward the development of new absorbent materials with biodegradability and an ability to decompose in landfills [4, 19, 29]. With respect to superabsorbent fibers, research efforts have been carried out to reduce the fiber size to submicron and nanoscales in order to enhance the absorption properties.

12.5 Summary

High volumes of superabsorbent materials have been manufactured worldwide as they have a variety of consumer and industrial product applications such as wound dressings, hydrogels, biomedical applications, hygienic products, oil recovery, air and water purification, cable isolation, landscaping, food packaging, protective clothing, agriculture, etc.

Superabsorbent materials that are commonly cross-linked acrylic copolymers are produced in the form of fibers in addition to powder form, foil, or foam. Superabsorbent fibers can be mixed with other natural and synthetic fibers or polymers. They can be easily processed in the fiber form through spinning, weaving, and nonwoven technology.

Superabsorbent materials, either natural or synthetic based, will continue to expand their application areas by the modification of their morphological, surface, and chemical characteristics and the introduction of new additives to build composite structures and develop new functionalities.

References

1 Glampedaki, P. and Dutschk, V. (2015). Superabsorbent finishes for textiles. In: *Functional Finishes for Textiles: Improving Comfort, Performance and Protection*, 1e (ed. R. Paul), 283–302. Woodhead Publishing.
2 Gupta, B.S. and Smith, D.K. (2002). Nonwovens in absorbent materials. In: *Absorbent Technology*, 1e (eds. P.K. Chatterjee and B.S. Gupta), 372–384. Elsevier Science B.V.
3 Wiesemann, F. and Adam, R. (2011). Absorbent products for personal health care and hygiene. In: *Handbook of Medical Textiles*, 1e (ed. V.T. Bartels), 316–328. Woodhead Publishing.
4 Chinta, S.K., Mhetre, S.B., and Daberao, A.M. (2014). Superabsorbent fibers and antimicrobial activity: a textile review. *Man-Made Textiles in India* 42 (1): 13–17.
5 http://www.futuremarketinsights.com/reports/super-absorbent-polymer-market (accessed 28 April 2017).
6 Global SAP Market Expected to Grow (2015). Nonwovens Industry.
7 Zohuriaan-Mehr, M.J. and Kabiri, K. (2008). Superabsorbent polymer materials: a review. *Iranian Polymer Journal* 17 (6): 451–477.
8 Gooch, J.W. (2010). *Biocompatible Polymeric Materials and Tourniquets for Wounds*, Topics in Applied Chemistry Series, 1e, 34. New York: Springer-Verlag.
9 http://www.m2polymer.com/html/history_of_superabsorbents.html (accessed 28 April 2017).
10 Technical Absorbents Ltd (2015). Super absorbent fibres provide greater flexibility. *Filtration & Separation* 52 (6): 32–34.
11 http://pdfs.findtheneedle.co.uk/9098-21018827-MSDS-International.pdf (accessed 28 April 2017).
12 http://www.innovationintextiles.com/super-absorbent-fibre-technology-well-received-at-cinte-techtextil/ (accessed 28 April 2017).
13 http://www.technicaltextile.net/featureproduct/super-absorbent-fibre-saf (accessed 28 April 2017).
14 http://techabsorbents.com/ (accessed 28 April 2017).
15 Gupta, B., Agarwal, R., and Alam, M.S. (2011). Hydrogels for wound healing applications. In: *Biomedical Hydrogels: Biochemistry, Manufacture and Medical Applications*, 1e (ed. S. Rimmer), 184–227. Woodhead Publishing.

16 Friedrich, S. (2012). Superabsorbent polymers (SAP). In: *Application of Superabsorbent Polymers (SAP) in Concrete Construction*, 1e (eds. V. Mechtcherine and H.-W. Reinhardt), 13–20. Springer.

17 Elliott, M. (2004). *Superabsorbent Polymers*, 1–13. BASF.

18 http://www.venturecenter.co.in/techrx/pdfs/TechShowcase_SAP.pdf (accessed 28 April 2017).

19 Dutkiewicz, J.K. (2002). Superabsorbent materials from shellfish waste - a review. *Journal of Biomedical Materials Research* 63 (3): 373–381.

20 Gooch, J.W. (2011). Super absorbent fibers. In: *Encyclopedic Dictionary of Polymers* (ed. J. W. Gooch), 2e, pp. 712–714. Springer.

21 Das, D., Rengasamy, R.S., and Kumar, M. (2013). Liquid sorption behavior of superabsorbent fiber based nonwoven media. *Fibers and Polymers* 14 (7): 1165–1171.

22 Chung, M. (2015). Polyolefin from commodity to specialty. *Journal of Material Science & Engineering* 4 (2): 1–2.

23 Bartkowiak, G. (2006). Liquid sorption by nonwovens containing superabsorbent fibres. *Fibres & Textiles in Eastern Europe* 14 (1): 57–61.

24 Wiesemann, F. and Adam, R. (2016). An overview of medical textile products. In: *Medical Textile Materials*, 1e (ed. Y. Quin), 14–15. Woodhead Publishing.

25 Tijing, L.D., Ruelo, M.T.G., Amarjargal, A. et al. (2012). Antibacterial and superhydrophilic electrospun polyurethane nanocomposite fibers containing tourmaline nanoparticles. *Chemical Engineering Journal* 197: 41–48.

26 Watson, D.E. (2012). Compositions comprising honey and a super-absorbent material. US Patent 9180219 B2, filed 12 September 2013.

27 http://nptel.ac.in/courses/116102006/17 (accessed 28 April 2017).

28 Bartkowiak, G. and Frydrych, I. (2011). Superabsorbents and their medical applications. In: *Handbook of Medical Textiles*, 1e (ed. V.T. Bartels), 505–546. Woodhead Publishing.

29 Liu, H., Zhang, Y., and Yao, J. (2014). Preparation and properties of an eco-friendly superabsorbent based on flax yarn waste for sanitary napkin applications. *Fibers and Polymers* 15 (1): 145–152.

30 Liu, Q., Ding, Z., and Dong, Z. (2012). Swelling behaviors of acrylic-based superabsorbent fibers. *Advances in Materials Research* 476–478: 1331–1335.

31 Nantong Jiangchao Fiber Products Co., Ltd., China. http://www.jcxw.cn/en.asp (accessed 28 April 2017).

32 http://www.patrickyarns.com/files/687b1d39-8af3-4331-95e7-29cc47823af0--6f55a7f1-820b-400d-bfee-e8f10915a42a/fiber-fact-sheet-2013-lettersize.pdf (accessed 28 April 2017).

33 Kavitha, A., Arjun, D., and Farheen, M.N. (2013). Super absorbent fibre-overview. *International Journal of Advanced Research in Engineering and Technology* 4 (7): 85–91.

34 http://www.toyobo-global.com/seihin/ap/lanseal_f/lanseal_f.html (accessed 28 April 2017).

35 Beskisiz, E., Ucar, N., and Demir, A. (2009). The effects of super absorbent fibers on the washing, dry cleaning and drying behavior of knitted fabrics. *Textile Research Journal* 79 (16): 1459–1466.

36 Chatterjee, P.K. and Gupta, B.S. (2002). *Textile Science and Technology, Absorbent Technology*, 115–117. Elsevier Science B.V.
37 Ucar, N., Beskisiz, E., and Demir, A. (2009). Design of a novel filament with vapor absorption capacity without creating any feeling of wetness. *Textile Research Journal* 79 (17): 1539–1546.
38 Eskin, B., Ucar, N., and Demir, A. (2011). Water vapor absorption properties of a novel filament composed of maleic anhydride polypropylene, polypropylene and super absorbent polymer. *Textile Research Journal* 81 (14): 1503–1509.
39 http://www.polymerprocessing.com/operations/dspin/ (accessed 28 April 2017).
40 http://encyclopedia.che.engin.umich.edu/Pages/PolymerProcessing/FiberSpinning/FiberSpinning.html (accessed 28 April 2017).
41 http://www.packplus.in/images/Technical-Absorbents.pdf (accessed 28 April 2017).
42 Kim, Y.-J., Yoon, K.-J., and Ko, S.-W. (2000). Preparation and properties of alginate superabsorbent filament fibers crosslinked with glutaraldehyde. *Journal of Applied Polymer Science* 78 (10): 1797–1804.
43 Fei-Ming, C., Xian-Zhong, Z., and Hui-Xia, Y. (2013). Effect of the pore fractal dimensions on absorption ability in superabsorbent fibers. *Applied Mechanics and Materials* 345: 205–208.
44 Ali, A.A. (2008). New generation of super absorber nano-fibroses hybrid fabric by electro-spinning. *Journal of Materials Processing Technology* 199: 193–198.
45 Islam, M.S., Rahaman, M.S., and Yeum, J.H. (2015). Electrospun novel super-absorbent based on polysaccharide–polyvinyl alcohol–montmorillonite clay nanocomposites. *Carbohydrate Polymers* 115: 69–77.
46 Buchholz, F.L., Pesce, S.R., and Powell, C.L. (2005). Deswelling stresses and reduced swelling of superabsorbent polymer in composites of fiber and super-absorbent polymers. *Journal of Applied Polymer Science* 98: 2493–2507.
47 Fangueiro, R., Filgueiras, A., Soutinho, F., and Meidi, X. (2010). Wicking behavior and drying capability of functional knitted fabrics. *Textile Research Journal* 80: 1522–1530.
48 Brown, P. (2012). Superabsorbent principles & properties, *VISION Consumer Products Conference*, New Orleans, Louisiana (23–26 January 2012), pp. 1–25.
49 Yu, C., Yu, L., Yiguli, A. et al. (2012). Structure and drug release of superabsorbent sponge prepared by polyelectrolyte complexation and freezing-induced phase separation. *Journal of Applied Polymer Science* 126: 1307–1315.
50 http://www.molnlycke.com.au/wound-care-dressings/absorbent-dressings/mextra-superabsorbent/#confirm (accessed 28 April 2017).
51 http://www.dermasciences.com/xtrasorb (accessed 28 April 2017).
52 https://www.environmental-expert.com/companies/technical-absorbents-42860/products (accessed 28 April 2017).
53 http://www.fiber-line.com/en/products/water-blocking-yarn (accessed 28 April 2017).
54 http://ucfltd.co.uk/water-absorption/ (accessed 28 April 2017).

55 http://www.babygearlab.com/a/11113/What-Is-Inside-Those-Disposable-Diapers (accessed 28 April 2017).
56 Kalayci, V.E., Doyle, J., Jones, D.O. et al. (2011). Superabsorbent-containing web that can act as a filter, absorbent, reactive layer or fuel fuse. US Patent 7, 988, 860 B2, filed 13 March 2008.
57 Yuan, X. and Chung, T.C.M. (2012). Novel solution to oil spill recovery: using thermodegradable 19. Polyolefin oil superabsorbent polymer (Oil–SAP). *Energy Fuels* 26: 4896–4902.
58 Flautt, M.C., Priest, J.R., Stotler, D.V., and Hager, T.P. (2012). Superabsorbent water-resistant coatings. US patent 8, 313, 833 B2, filed 13 February 2009.
59 Mango, P. (2011). The future of superabsorbents in food packaging: nonwovens powders or fibers? Nonwovens Industry, October.
60 Rogers, M.E., Phillips, J.P., and Koene, B. (2008). Flame retardant synthetic textile articles and methods of making the same. US Patent 7, 423, 079 B2, filed 17 May 2005.
61 http://www.m2polymer.com/pdf/Tech%20Sheet_Oasis%20SAF%2052mm_M2PT.pdf (accessed 28 April 2017).
62 http://www.kntp.ru/en/produkty/reagenty/vodouderzhivajushhie-polimery-aquasorb.html (accessed 28 April 2017).
63 http://www.socochem.com/super-absorbent-polymer-for-agriculture.html (accessed 28 April 2017).

13

Elastic Fibers

Lu Jing

The Hong Kong Polytechnic University, Institute of Textiles and Clothing, QT807a, Hung Hom, Kowloon, Hong Kong, China

13.1 Introduction

13.1.1 Definition

Elastic fiber can deform to a relative large elongation in proportion to the load applied and recover its original length after removal of the applied stress. Elastic fiber is an important class of polymer fiber with elastic and recovery performance, which can be achieved by wet spinning, dry spinning, or melt spinning of elastomers or other polymers with specific molecular structures, which can provide high elongation and recovery after post-treatment. The stretch and recovery property of elastic fiber can be controlled by adjusting the polymer structures and post-treatment like heat setting, drawing, or cross-linking. Besides the stretch and recovery performance such as elongation, recovery percentage, and recovery force, other physical properties have also been studied including modulus, strength, heat resistance and anti-chloride, dye ability and UV resistance, etc.

As the idea of most thinking, larger elasticity and recovery is better for elastic fibers. However, diversity and controllable combination of the elasticity and recovery performance is just what the elastic fiber is.

13.1.2 Classification

Naturally, the elastic fiber is commonly classified according to the elastic elongation, that is, high elastic fiber with elongation of 400–800%, medium elastic fiber with elongation of 150–390%, low elastic fiber with elongation of 20–150%, and micro-elastic fibers with elastic elongation below 20% [1].

Based on the polymer structures, elastic fibers consist of polyurethane (PU) elastic fiber, polyester-ether elastic fiber, polyester elastic fibers, olefin-based elastic fiber like XLA, and others such as hard elastic fibers and bicomponent elastic fiber. Shape memory fiber also exhibits elasticity property to some extent, but it will not be included in this chapter.

The most well-known and most widely used elastic fiber is PU elastic fiber, and spandex is the representative, which has been commercialized for many years.

Handbook of Fibrous Materials, First Edition. Edited by Jinlian Hu, Bipin Kumar, and Jing Lu.
© 2020 Wiley-VCH Verlag GmbH & Co. KGaA. Published 2020 by Wiley-VCH Verlag GmbH & Co. KGaA.

13 Elastic Fibers

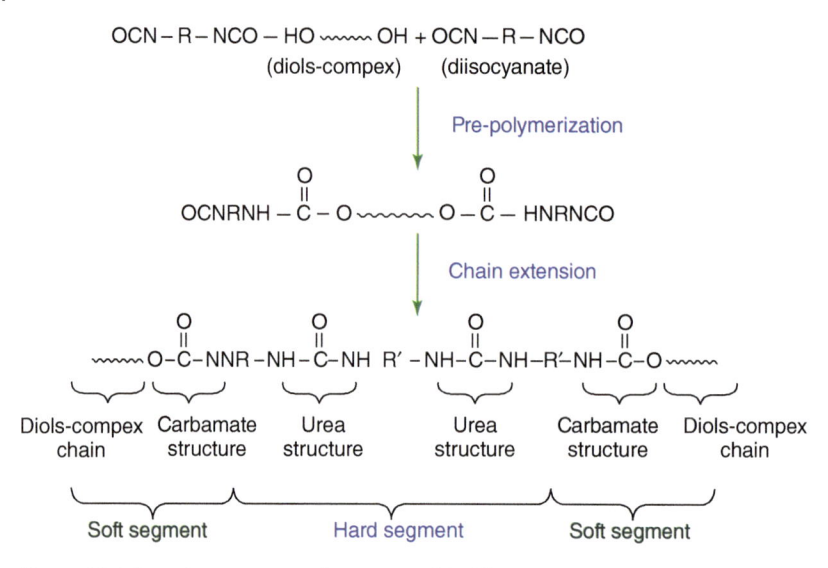

$$\text{OCN-R-NCO} - \text{HO}\mathord{\sim\!\sim\!\sim}\text{OH} + \text{OCN-R-NCO}$$
$$\text{(diols-compex)} \quad \text{(diisocyanate)}$$

↓ Pre-polymerization

$$\text{OCNRNH} - \overset{\text{O}}{\underset{\|}{\text{C}}} - \text{O}\mathord{\sim\!\sim\!\sim}\text{O} - \overset{\text{O}}{\underset{\|}{\text{C}}} - \text{HNRNCO}$$

↓ Chain extension

$$\mathord{\sim\!\sim}\text{O}-\overset{\text{O}}{\underset{\|}{\text{C}}}-\text{NNR}-\text{NH}-\overset{\text{O}}{\underset{\|}{\text{C}}}-\text{NH}\;\;\text{R}'-\text{NH}-\overset{\text{O}}{\underset{\|}{\text{C}}}-\text{NH}-\text{R}'-\text{NH}-\overset{\text{O}}{\underset{\|}{\text{C}}}-\text{O}\mathord{\sim\!\sim}$$

| Diols-compex chain | Carbamate structure | Urea structure | Urea structure | Carbamate structure | Diols-compex chain |

Soft segment   Hard segment   Soft segment

Figure 13.1 Reaction process and structure of PU fibers.

By definition, these fibers have an elongation to break more than 200%, usually 400–800%, and, on release of the deforming stress, return quickly and almost completely to their original length [1, 2]. A major advantage of PU fibers over rubber yarns is that they are easily spinnable into thin fibers, making them suitable for textile applications. PU elastic fiber can be developed by wet spinning, dry spinning, reactive spinning, and melt spinning process from the PU solution or PU polymer with the structure shown in Figure 13.1. The basis for PU elastic fibers in commercial production was the diisocyanate polyaddition technology, which has been discovered in 1937 by Fabricius and Wulfhorst [2].

H. Rinke firstly succeeded in spinning fibers from the high molecular weight PU resulting from the reaction of butylene glycol and hexamethylene diisocyanate. In 1939, P. Schlack obtained high molecular PU by diisocyanate polyaddition process, which has been used since 1941 to synthesize high value elastomers having higher tenacities and better end-use properties than the previously known diene polymers. Reaction spinning was first investigated by H. A. Pohl in 1942 [3]. Windemuth developed a chemical spinning process in which the chemical synthesis of high molecular weight PU occurred simultaneously with extrusion and fiber formation. W. Brenschede succeeded in solution spinning PU elastomers in 1951. Based on earlier work by M. D. Snyder, J. C. Shivers achieved the first large-scale technical production of PU fibers by the dry spinning. The final development of the fibers was worked out by scientists at DuPont and the US Rubber Company. DuPont used the brand name Lycra and began full-scale manufacture in 1962. All large producers attempted to find a melt spinning route for elastane yarns. In 1967 Nisshinbo Industries introduced a melt-spun elastane into the Japanese market. This was followed by Kanebo Ltd. in 1977 and Kuraray Co. Ltd. in 1991. A single-stage synthesis was next sought, in which the PU raw materials are obtained in the form of granulate and are subsequently melt-spun as

Table 13.1 Comparison of four spinning processes of polyurethane fiber.

Item	Dry spinning	Wet spinning	Reactive spinning	Melt spinning
Spinning speed (m/min)	200–900	50–150	50–150	400–1000
Spinning temperature (°C)	200–230	<90	—	160–220
Liner density (tex)	2.22–124.4	4.4–44.0	4.4–38.0	2.2–110
Fiber quality	Excellent	Common	Common	Good
Capacity factor (%)	~70	~1	<5	~20
Pollution	High	High	High	Low

reacted PU. A two-stage reaction process is more modern and provides filaments of improved properties. Despite these developments, dry spinning remains the most widely used production process with high production and better fiber quality [4]. Table 13.1 compares the four spinning methods of PU fibers. Melt spinning of PU fiber has been developed rapidly recently because of the low cost and less pollution.

The polyester elastic fibers of polytrimethylene terephthalate (PTT) fiber and polybutylene terephthalate (PBT) are discovered in the 1940s [5]. PBT fiber production for textiles first started between the end of the 1970s and the mid-1980s in the company of Toray, Teijin, Kuraray, and Unitika in Japan and the company of Celanese in the United States. PTT fiber [5] was first patented in 1941, but until the 1990s, Shell Chemicals developed a low-cost production method. The polyester elastic fibers also show a very good recovery rate under low loads and create a wearing comfort. The fibers can recover 100% from 120% strain. Melt spinning is the main production method for this kind of elastic fiber [6].

Polyether-ester (PEE) fiber was developed early as a thermoplastic elastomer in the 1970s and industrialized in plastic trade. Recently some research center and factories have done some research about the PEE, but do not produce volume. PEE elastic fiber was obtained by melt spinning technology from the polyester and polyether block copolymer [6]. It shows good elastic recovery rate under the elongation of 100%. It has good dye ability and anti-chemical properties. The polyester-ether elastic fibers are promising fibers with wide range uses and lower cost than PU elastic fibers. By controlling the molecular structure, different functional polyester-ether fibers will be developed. The advantages of PEE fiber are the cheap raw materials and nontoxicity and spinning on conventional melt spinning machines for polyester fiber. So the PEE fiber may replace PU elastic fiber in some applications.

Olefin-based stretch fiber is naturally resistant to harsh chemicals, high heat, and UV light. It is a good replacement to PU fiber in some end-use conditions of chloride or high temperature processing. The most well-known olefin-based elastic fiber is XLA fiber developed by Dow Company. Dow XLA elastic fiber can withstand temperature of more than 220 °C. Because of the inherent properties of olefins, it can be bleached, mercerized, stone-washed, or thermosol-dyed using standard processes without special care.

Bicomponent elastic fibers will produce twisting curl effects like natural wool with a textured appearance. The curl effects are formed from the fiber with two parallel components of differing shrinkage or expansion properties. This kind of elastic fiber shows more durable elasticity and recovery than the fibers made of elastomers.

The hard elastic fiber is a small part of elastic fiber family from the production or application view. It shows a low elasticity around less than 50–100%. But the recovery is very good under the lower elongation [6] but deteriorates when the elongation exceeds 200–300%. It was discovered in the 1960s that polypropylene (PP), as well as other highly crystalline polymers, can be produced as elastic fibers. Because of the higher modulus, this kind of fibers was termed as "hard" elastic fiber to distinguish from common elastomer fibers. Hard elastic PP fiber is made by melt spinning and processed by annealing under tension to form lamellar crystallization structure. Between the lamellae, microfibrils are oriented parallel to the draw direction. Under stress, the lamellae tend to separate and voids bridged by fibrils appear [7].

13.2 Structure, Principles, and Characteristics

13.2.1 Structure and Principle in Elasticity

The unique stretch and recovery characteristics of elastic fibers are based on the polymer structure and fiber morphology.

For most kinds of elastic fibers such as PU fiber, PEE elastic fiber, and olefin-based stretch fibers, the elasticity performance comes from the combination of soft segments and the hard segments of the polymer structure (Table 13.2). As shown in Figure 13.2, the soft segments are commonly the long, soft amorphous regions with random molecular structures in natural state, which can be stretched easily under low force and recover to the original coil morphology state due to the elastic entropy. The hard segments are the short, rigid crystalline regions or cross-linking points, which keep the fiber structure stable when stretching, heating, and so forth.

The soft segments of polyether type PU include the polyether as shown in Figure 13.3, such as polyoxyethylene, PP oxide, or polytetrahydrofuran. The interaction between them is mainly the van der Waals force, which is too weak to form hydrogen bonds. The soft segments of polyester-type PU (Figure 13.4) comprise

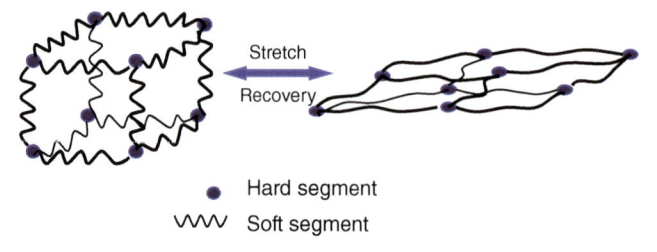

Figure 13.2 Scheme of principles of elasticity with hard and soft segments.

Table 13.2 Elasticity principle of the fibers with hard and soft segments.

Type	Crystalline (%)	Hard segments	Soft segments	Elongation (%)	Recovery at 100% elongation (%)	Fiber examples
Polyurethane elastic fiber	~40	Hydrogen bonds/crystals	Amorphous polyol chains	Maximum ~800	>95	Spandex; Lycra
Polyolefin elastic fiber	~14	Crosskicking and crystals	Amorphous polyolefin chains	Maximum ~500	>90	XLA
Polyether-ester elastic fiber	25–35	Polyester segments	Polyether segments	Maximum ~600	>85	Polybutylene terephthalate–polytetramethylene glycol (PBT–PTMG) fiber; polytrimethylene terephthalate (PTT) fiber; polybutylene terephthalate (PBT) fiber

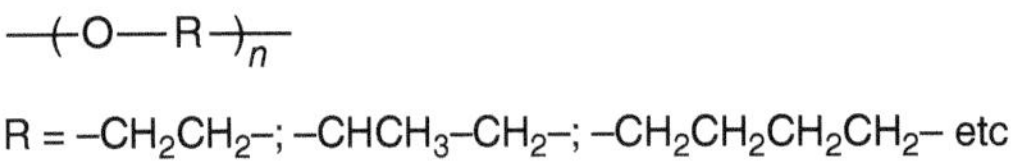

$$\text{---}(\text{O}\text{---}\text{R})_n\text{---}$$

R = –CH$_2$CH$_2$–; –CHCH$_3$–CH$_2$–; –CH$_2$CH$_2$CH$_2$CH$_2$– etc

Figure 13.3 Molecular structure of polyether type PU.

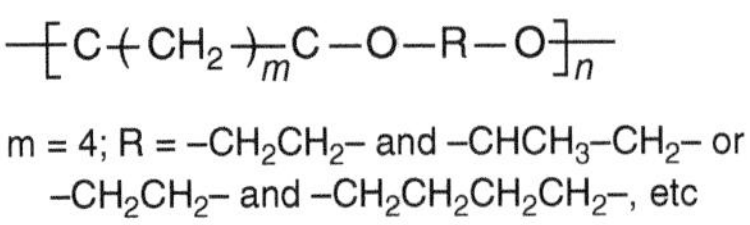

$$\text{---}[\text{C}(\text{CH}_2)_m\text{C}\text{---}\text{O}\text{---}\text{R}\text{---}\text{O}]_n\text{---}$$

m = 4; R = –CH$_2$CH$_2$– and –CHCH$_3$–CH$_2$– or –CH$_2$CH$_2$– and –CH$_2$CH$_2$CH$_2$CH$_2$–, etc

Figure 13.4 Molecular structure of polyester type PU fiber.

polyadipate of mixed dihydric alcohol, which forms the polyester chains. For PU fibers, when the external force is applied to stretch the fibers, the amorphous segments straighten out. This makes the amorphous segments longer, thereby increasing the length of the fiber. When the fiber is stretched to its maximum length, the rigid segments still keep the crystalline states and resist the slippage of amorphous segment chains. After the force is removed, the amorphous segments recoil, and the fiber returns to its original relaxed state [2, 6, 8]. The formation of soft segments and hard segments in PU fibers is due to the thermodynamic incompatibility between the two segments, and PU s undergo micro-phase separation, resulting in the phase-separated heterogeneous structure consisting of hard and soft domains and interphases between them. Better phase separation leads to higher stretch and recovery performance [9, 10]. The hard segments of PU fibers include the main chains of carbamin acid ester and diisocyanate, which have the high cohesive energy and form large amount of hydrogen bonds with each other and give the fiber structure.

Fiber of polyether ester was obtained by melt spinning from the polyester and polyether block copolymer [4, 11]. As shown in Figure 13.5, the PEE fiber is composed by crystalline aliphatic polyester as hard segment A and amorphous aliphatic polyether as soft segment B. Among them, Ar is aryl group of aromatic bi-carboxylic acid, D is the alkyl group of dihydric alcohol, n is the degree of polymerization of polyether, and x is the average polymerization degree of hard segment. The hard segment A and soft segment B are alternated and linked by chemical bonds to form a copolymer. At present, industrialized PEE polymer mainly use polytetramethylene oxide (PTMO) as soft segment and PBT as hard segment. Besides PTMO, polyethylene oxide (PEO), nylon, and polydimethylsiloxane (PDMS) are also used as soft segments. The hard segment can adopt polyethylene (PE) glycol terephthalate or PBT.

The polyester part forms crystalline regions as hard segments. Larger crystal size and polymerization degree of hard segments are helpful to the recovery. To achieve the elasticity property, the contents of polyglycol will be at least 60% or 70–80%, and the polyester contents are only 20–30%. It was a high elastic fiber with breaking elongation of 500%, and the elastic recovery is above 85% at the

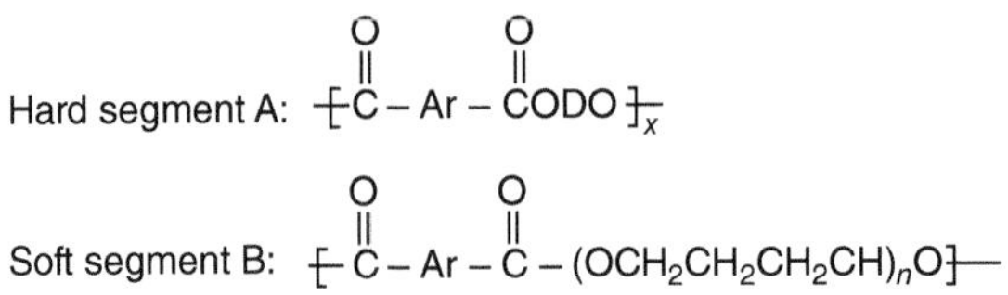

Hard segment A: $\text{---}[\overset{\text{O}}{\overset{\|}{\text{C}}}\text{---}\text{Ar}\text{---}\overset{\text{O}}{\overset{\|}{\text{C}}}\text{ODO}]_x\text{---}$

Soft segment B: $\text{---}[\overset{\text{O}}{\overset{\|}{\text{C}}}\text{---}\text{Ar}\text{---}\overset{\text{O}}{\overset{\|}{\text{C}}}\text{---}(\text{OCH}_2\text{CH}_2\text{CH}_2\text{CH})_n\text{O}]\text{---}$

Figure 13.5 Structure of hard segment and soft segment of polyether-ester fiber.

13.2 Structure, Principles, and Characteristics

elongation of 100%. It has good dye ability and anti-chemical properties. With the increase of hard segments, under the same spinning parameters, the crystallinity will increase, but the crystal size becomes smaller, which leads to the higher tenacity but less elasticity and recovery. For a kind of PEE fiber with PBT as hard segments and polytetramethylene glycol (PTMG) 2000 as soft segments, when the PBT weight fraction increasing from 40% to 60%, the elastic recovery decreases from 80% to 50%. Heat drawing and heat setting will improve the crystal size and molecular orientation and accordingly increase the elasticity and elastic recovery. After heat setting, the elastic recovery will increase from 80% to 90% [12].

Polyolefin elastic fiber consists of at least 95% polyethylene by mass and a little other olefins. The polyethylene macromolecules partially cross-link and partially crystallize, which make the hard segments. The soft, long amorphous polyethylene macromolecules form the soft segments, which is easily stretched. XLA is a representative polyolefin fiber. The special microstructure of XLA combines long, flexible chains with crystallites, and covalent bonds or cross-links, forming an intricate network. The length of the chains and number of crystallites are specifically controlled to give XLA fiber a unique elastic profile. High stretch is achieved with low levels of force, allowing garments to stretch and flex effortlessly and still return to their original shape.

The polyester elastic fiber mainly includes PTT fiber and PBT fiber. Comparing with PU fibers, the polyester elastic fiber displays lower elongation. The elasticity of this fiber is due to the configuration changing of the molecular chains. As shown in Figures 13.6 and 13.7, molecule chains of polyester fiber are fully extended with two carboxyl groups of each terephthaloyl group in opposite directions, and all open-chained bonds are trans with successive phenylene groups at the same inclination along the chain. PTT has a conformation with bonds of the $-OO(CH_2)_3OO-$ unit having the sequence of trans–gauche–gauche–trans, leading to a concentration of the repeating unit. The opposite inclinations of successive phenylene groups along the chain force the molecular chains to take on an extended zigzag shape, which can be extended and recover easily [6].

Hard elastic fibers exhibit 50–95% recovery from 100% extension. They are generally produced by spinning at relatively high spinline stress levels, and the elastic effect is enhanced significantly by annealing. Traditionally, the recovery of elastic fibers has been explained by the entropic mechanisms of elasticity network theory. The elasticity of the hard elastic fibers, on the other hand, is primarily contributed by energetic rather than entropic mechanisms. Hard elastic fibers may be

Figure 13.6 Molecular structure comparison of PET, PTT, and PBT fiber.

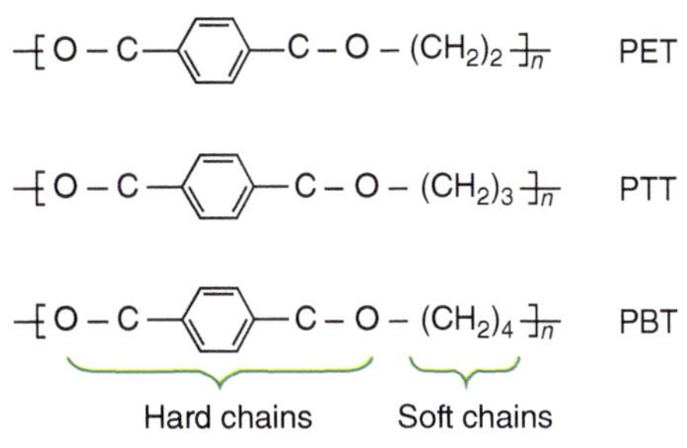

$\{O-C-\langle\bigcirc\rangle-C-O-(CH_2)_2\}_n$ PET

$\{O-C-\langle\bigcirc\rangle-C-O-(CH_2)_3\}_n$ PTT

$\{O-C-\langle\bigcirc\rangle-C-O-(CH_2)_4\}_n$ PBT

Hard chains   Soft chains

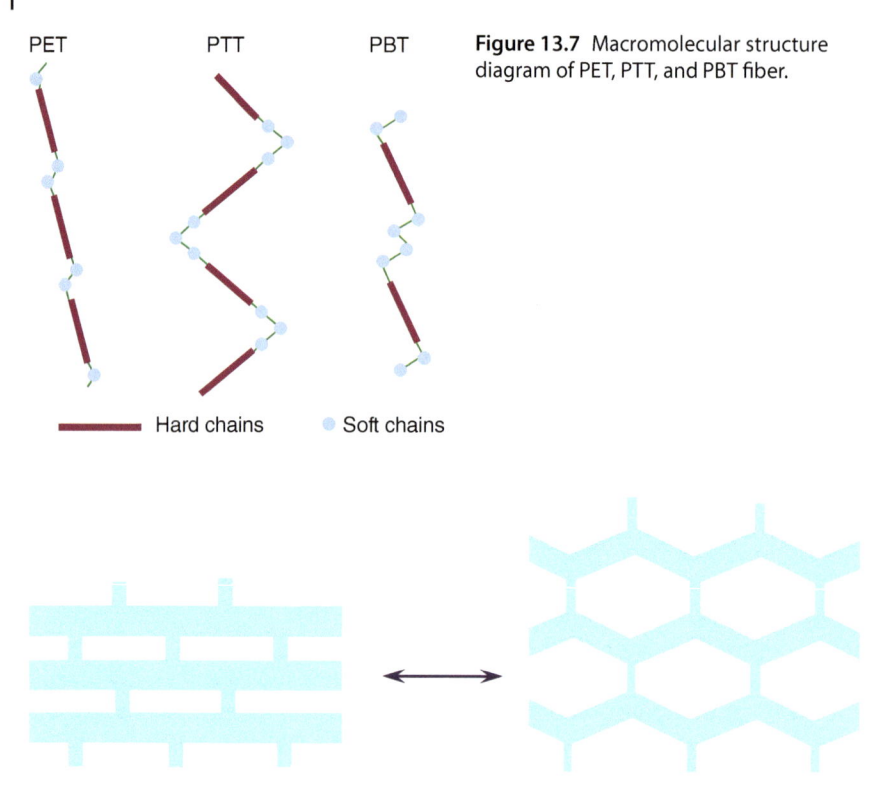

Figure 13.7 Macromolecular structure diagram of PET, PTT, and PBT fiber.

Figure 13.8 Model of lamella structure and microvoids created of hard elastic fiber.

produced from many polymers, including polypivalotactone, acetal copolymer, polybutene-1, poly-4-methyl pentene, polyethylene, and PP [13].

Some structural models have been proposed to explain the elastic properties of these fibers. Clark [13] has given a reasonable model for the mechanism of energy-driven elasticity. The feature of this model is the existence of short tie molecules connecting adjacent lamellae normal to the fiber direction. The short tie molecules comprise the load-bearing elements, transferring stress from one lamella to the next. Figure 13.8 proposed a weblike system that is created by large numbers of lamellae to form a row structure. The interconnecting fibrils will distribute an applied strain in a cooperative action to deform each lamella slightly. Small deformations of the folded-chain lamellae are reversible. If an assembly of lamellae and fibril links is highly oriented, as in a row structure, this effect is additive and produces elasticity in the macroscopic fiber. Figure 13.8 also predicts that a large number of microvoids will develop during fiber extension. The diameter of a fiber with this type of structure should remain constant for a considerable elongation. A plot of relative specific volume versus extension for elastic PP and a classical elastomer confirm the same. The PP volume increases, and the diameter remains constant up to an extension of nearly 100%.

The fibers annealed above 135–140 °C showed complete reversibility for extension as large as 100%. The annealing results in additional ordering of the initial row-nucleated materials. This ordering also gives higher recoveries after repeated cycles [12].

Because of a higher modulus, the time effects, and the limited extensibility and recovery in comparison to classical elastomers, energy-driven elastic fibers have not been applied widely in textiles.

Although the elastic fiber with soft segment and hard segment shows extraordinary elongation property and good recovery performance, with the cycling use or stretch, the recovery rate will decrease gradually. This is inevitable because of the polymer macromolecule fatigue. The elasticity of bicomponent fibers is with the self-crimp-textured appearance, which is formed from a composition of two parallel but attached fibers with differing shrinkage or expansion properties made by melt spinning method. Since the crimp is naturally formed by heating and not mechanically induced, the fiber gives greater, more durable stretch and recovery than textured yarns that are used for stretch.

13.2.2 Structure and Principle in Other Performances

The thermal behavior of PU fibers at use and processing temperatures are governed by the soft segments. PU with polyester-type soft segments has worse heat resistance property than that of polyester PU fiber. For the polyester-type PU fiber, it will become yellow at 150 °C, and strength will decrease at 170 °C. For polyester-type PU fiber, the strength decreases at about 190 °C. The anti-chemical property of PU fiber is good. PU fibers are soluble in highly polar solvents, such as dimethylacetamide. While PU fibers containing polyether soft segment are less subject to hydrolysis, polyester elastanes are more resistant to oxidation. But the polyester-type PU fiber is poor in alkali resistance. All the traditional PU fibers are with worse anti-chloride performance. Elastane fibers have good resistance to oxygen and ozone. Nitrogen oxides cause a color change to yellow or yellow brown, the intensity depending on the concentration, ambient temperature, and relative humidity. PU fibers are more resistant to aging and abrasion than rubber yarns. Long exposure, particularly to UV radiation, leads to a change in color of the fiber and to photochemical degradation. Polyester urethane has higher resistance to photooxidation than polyether urethane [10].

The polyester elastic fibers have the similar physical properties to the common polyester fibers. However, they also shows some specific performances besides the large elongation and recovery. PTT fiber is a typical polyester elastic fiber. The diacid group affects the mechanical property, and its process ability depends on the type of the diol group. In particular, PTT is a highly crystalline polymer. Its melting temperature is lower than that of polyethylene terephthalate (PET) by 20–30 °C. Therefore, the process ability of PTT is superior to that of PET. The highly flexible PTT fibers are obtained as a result of its low initial modulus. The elasticity and dye ability of PTT are better than those of PET or PBT. It is well-known that the number of methylene unit influences the physical properties of many polycondensation polymers such as polyamides (PAs) and polyesters, which is called the odd–even effect. PTT have the best soft hand of all, as PBT will be close to PET. PTT can be dyeable at low temperature (100 °C) with the same or slightly deeper shades than PET at 130 °C. Disperse-dyed PTT shows excellent colorfastness to laundering, crocking test, and UV light and ozone. It also has the better abrasion resistance and dimensional stability.

There are many similar properties of PEE fiber and PU fiber. In chemical structure, both fibers have long-chain polyether and the elasticity root from the entropy change. It has good dye ability and anti-chemical properties. But there are many ether bonds, which lead to the poor thermal stability of the PEE fiber. Therefore the temperature in the dry and melt spinning process should be controlled reasonably. The water and oil resistance properties can be enhanced by the larger hard segment concentration in the copolymer. The hot water shrinkage of PEE fiber is large. Heat setting is a useful method to decrease the hot water shrinkage. Under 60 °C relax heat setting, the shrinkage in boiling water decreases from 12% to 2.5% and after heat setting at 100 °C, the shrinkage will decrease to 1%. From the Small-angle X-ray (SAX) data, after relax heat setting, the crystal size of 010 lattice increases from 5.5 to 8.5 nm, and the size of 100 lattice increases from 4.7 to 6.0 nm [9, 10].

The olefin-based elastic fiber is with superior heat resistance because of the cross-links formed in the fiber's molecular structure. The olefin elastic fiber can maintain its integrity at 220 °C for three minutes; however PU fiber comes apart. The fibers are inherently resistant to strong acid or alkali or chloride. But olefins have an affinity to hydrocarbon solvents and mineral oils; the fiber should not be exposed to these classes of chemicals for extended periods of time. UV/xenon resistance makes the fibers to deliver stretch in outdoor applications.

Polymers that can be used for producing hard elastic fibers include semicrystalline polymers as polyoxymethylene (POM), PP, poly-(4-methyl-1-pentene) (PMP) (TPX™), PE, etc. [14, 15]. All the above hard elastic fibers have the higher modulus than PU fibers. The physical properties are based on the chemical structures. For PP hard elastic fibers, the physical properties depend on the chemical structure. The PP hard elastic fiber has high strength and initial modulus and worse moisture absorption and dye ability. It can resist acid and alkali. But it is with worse light resistance and heat resistance. The poly(vinylidene fluoride) (PVDF) hard elastic fiber shows higher initial modulus, and it decreases with the increasing stretching rates when spinning.

According to the above description, Table 13.3 summarizes the physical properties of these elastic fibers.

13.3 New Development of Elastic Fibers

13.3.1 Polyurethane Elastic Fiber

PU elastic fibers are superior in elasticity and elastic recovery. Due to these advantages, PU elastic fibers are widely used as materials for stockings, women's underwear, and flexible fabrics, and their applications continue to be extended to aerobic clothing and swimming suits. However, PU fibers have worse heat and chemical resistance, which are the enduring conditions during the textiles process. Some techniques have been recently developed for increasing the performance of PU fibers.

Table 13.3 Physical properties of elastic fibers.

Fiber type	Tenacity (cN/dtex)	Modulus (cN/dtex)	Melting point (°C)	Heat shrinkage (%) (100 °C water)	Chemical resistance	Atmosphere resistance
PU, polyester type	0.4–0.6	~0.45	~250	~10	Cool and dilute acid resistance; hot alkali and chloride labile	Tenacity loss color change
PU, polyether type	0.6–1.0	~0.11	200–290	~10	Acid resistance; yellowing in H_2SO_4 or HCl; chloride labile	Tenacity loss under light
Polyester ether	0.4–4.0	0.2–0.5	200–230	~20	Good anti-oils; hydrolysis under hot water	Degradation under UV light
PTT	~3.0	~1.0	220–230	~12	Good chemical resistance	Good anti-light aging
XLA	~1.0	<0.1	Decomposition at 300	~20	Chloride and acid and alkali resistance	UV resistance
Hard elastic fiber (PP)	0.4–0.8	~4.5	160–180	~2.5	Good acid and alkali resistance	Tenacity loss under light
PTT/PET bicomponent elastic fiber	~2.0	~0.5	~225	N/A	Good anti-chemical	Good anti-light

PU fibers have alkali resistance property to some extent but are likely to easily degrade in caustic soda solution at a high concentration of about 25% to about 30% and applied at a temperature as high as 150 °C. For some fabric treatments such as burning-out and printing process should be conducted under high temperature and alkali solutions. Accordingly, it should separate these processes for the fabric containing PU fibers, and there is still no technique established that simultaneously proceed to both. Thus, there is a need for PU elastic fibers having superior resistance to heat and alkali and without any problem associated with nonuniform physical properties.

A patent [16] has found that when adding a cellulose acetate (diacetate or triacetate) in PU polymer solution for spinning, the final PU fiber will have a high modulus and superior resistance to heat without changing polymerization viscosity or physical properties of the fibrous product. This kind of PU fiber can resist 25% aqueous NaOH solution and be heated to 150 °C. This PU fiber increases the heat resistance of about 25% and extends the alkali resistance time to 20 minutes than the common PU elastic fiber.

PU elastic fiber is known to display heavy tackiness as compared with conventional, nonelastic fibers. Because of the tackiness, PU fiber may cohere to each other or alternatively adhere to various surfaces. High tackiness becomes a problem in unwinding during weaving. The pressure caused by elasticity leads to the adjacent fibers to cohere to each other, particularly near the core of the cone. It is difficult to remove the fiber from the cone package without breaking. Unusable fiber commonly occurs at the core and is referred to as "core waste." After winding and packaging, fiber tackiness may become even worse during storage depending on time and temperature. Longer storage time and higher temperatures lead to higher tackiness and more core waste than freshly spun and packaged PU fibers. Accordingly, a reduction in PU fiber tackiness would reduce core waste and increase cost-effectiveness.

To increase the anti-tackiness property of PU fiber, an embodiment including cellulose ester (CE), calcium stearate, magnesium stearate, organic stearates, silicon oil, mineral oil, and mixtures thereof can be added into the PU polymer before spinning or during spinning. The soluble anti-tack CE may be a specific cellulose acetate butyrate (CAB) or cellulose acetate propionate (CAP). The PU fiber with 1% of CE embodiment by weight shows a significant reductions in yarn tackiness by about 20% or more and improvements in spinning performance such as fewer breaks [17].

PU elastic fiber is usually used together with other natural or synthetic fibers for weaving or knitting. However PU fibers are with worse color fastness, especially some dark color like black. After several washing with hot water or prolonged use, the fading of PU fibers may occur due to the "grin" phenomenon. This is mainly because of the dyeing process; most of the dye is only attached to the surface of the PU fibers and cannot penetrate the interior of the fibers. Using colored PU fiber itself can effectively solve this problem. Black spandex is most commonly used as a colored spandex, which is mainly used in black underwear, stockings, and swimwear.

The black PU fiber is typically made of adding black fillers such as carbon black into polymers or spinning solutions directly. The key issues include how to achieve uniform dispersion of black fillers, thus ensuring stable spinning process and uniform product quality, and how to strengthen the interaction between black filler and PU urea, thereby enhancing the coloring effect and color fastness. A China patent [18] mentioned the use of surfactants to improve the carbon black dispersion, but it did not establish an effective chemical bond between the carbon black and the polymer to increase the force between the two phases. Therefore, the carbon black particles and the polymer molecule chains rely only on intermolecular physical force or a mechanical combination lock, which also cause discoloration of PU fabric after long uses. Another China patent [19] mentioned the use of carbon black particles with active surface hydroxyl groups with the isocyanate groups to form chemical bonds to increase the color fastness of PU fiber and mechanical strength. But carbon black itself is with limited reactive sites on surface; therefore it is still difficult to build enough chemical bonds between the carbon black and the polymer chains. Both of these patents are not good solutions to how to improve the binding between carbon black and PU urea.

The invention [20] adopts single-walled or multiwalled carbon nanotubes with diameter of 10–50 nm and a length of 1–10 μm as fillers to increase the bindings between two phases and solves the dispersion problems. The present invention deals with the surface modification of carbon nanotubes by carboxylic acid and then dispersion of carbon nanotubes in black PU urea for improved adhesion. Only 0.01% by mass of carbon nanotubes can effectively increase the chemical binding.

Besides the black PU fiber made by this method has a good color fastness. They also have high mechanical strength, elongation, and low volume resistivity and at same time excellent antistatic effect and excellent heat temperature resistance, which can withstand prolonged high temperature boiling and still maintain good performance.

Besides the improvement of physical property and chemical resistance property of PU fibers, some studies also have developed other high performance PU fibers, for example, anion PU fibers. Anion is beneficial to health and can facilitate body growth and prevent disease. Anion can also improve sleep quality and ease pain and tension. Anion can improve reflex system and endothelial system and enhances body's anti-disease ability. CN 205501464 U [21] reported a hollow PU fibers with anion core structures, which has high elasticity and a good anion effect.

PU fibers are with poor moisture absorption property due to lack of hydrophilic groups, which maybe limit the textile applications requiring high hygroscopicity. The patent CN 104153037 [22] published a PU fiber with higher moisture absorption through adding natural silk fibroin in the PU polymers. Natural silk fibroin is made from silk, and it is a natural polymer and nontoxic and non-irritating and has good biocompatibility and biodegradability. Silk fibroin has the forms of solution, fiber, powder, or film. A patent of CN 1560136A [23] published a method of making silk fibroin from silk, which can produce nanoscaled silk fibroin with average particle size of 30–60 nm. In terms of chemical structure, silk fibroin as a natural polymer contains carboxyl, amino, hydroxyl, and other polar groups, which have good water absorption property.

At present, the commonly used hydrophilic modification methods include surface modification, graft, copolymer, and blends. Among these methods, blending is relatively easy to implement in the industry. In this patent, the silk fibroin powder is blended with PU urea polymer and then for dry spinning. When the silk fibroin content is 20% by weight, the moisture absorption rate of the fibers increases from 0.8% to 4.6%, and breaking elongation just decreases to 7% compared with the fiber without silk fibroin.

As we know textiles are excellent substrate for bacterial growth. The fabric provides a good base for attachment, and sweat provides nutrients necessary for bacterial growth. The bacteria do not lead to some diseases but break down sweat into fatty and acids and accordingly cause smelly odor. Some mildew will damage the fabric and the body health. Particularly, women's underwear, such as girdles and braziers, are commonly made of PU elastic fibers; therefore some attempts to impart antimicrobial properties to PU elastic fibers have been proposed.

For example, A US patent [24] discloses a process for imparting antimicrobial properties to PU fiber by using poly(pentane-1,5-carbonate)diol or poly(hexane-1,6-carbonate)diol, which is a polycarbonate diol selected among aliphatic diols, or a copolymer thereof, as a soft segment. This method has the advantage that the antimicrobial property is achieved by inherent physical properties of the raw materials and without adding additional antimicrobial agent, but it has the poor antimicrobial effect.

Further, Korean Patent Publication No. 93-5099 [24] teaches a method with a porous inorganic antimicrobial agent of porous crystalline aluminosilicate zeolite to impart antimicrobial properties to a PU elastic fiber. However, the zeolite has strong water absorption property, which will cause too much cross-linking during the preparation of PU. This cross-linking increases the viscosity of the polymer and causes the formation of a gel, which results in high spinning pressure and frequent occurrence of fiber breakage during the spinning.

Another Korean Patent No. 103406 describes the use of a nonporous inorganic ceramic containing silver or zirconium as antimicrobial components in order to impart antimicrobial properties to PU fiber. However, the silver causes undesirable yellowing of the elastic fiber during spinning at a high temperature of 200 °C or more.

Further, Korean Patent No. 445313 describes the use of a glass metal compound as an antimicrobial agent, containing ZnO, SiO_2, and an alkali metal oxide in order to impart antimicrobial properties to PU elastic fiber. Although no yellowing of the fiber occurs during spinning at a high temperature of 200 °C or more, the antimicrobial agent tends to agglomerate in dimethylacetamide or dimethylformamide, which is a polar solvent for PU. As a consequence of the agglomeration, the pressure of spinning head increases, and fiber breakage frequently occurs during spinning, which make it difficult to maintain stable spinning of the antimicrobial fiber for a long period of time. So the particle size of antimicrobial agent must be suitable for addition to the PU. To this end, extending the milling time of the antimicrobial agent is required. At this step, the antimicrobial agent will be discolored to gray, which also renders the final antimicrobial elastic fiber gray [24].

Therefore, there is a need to develop a process for preparing an antimicrobial elastic fiber without affecting the spinnability or the color of the fiber while maintaining superior antimicrobial properties. US 20070292684 A1 [24] describes a method for antimicrobial PU fiber producing with a glass metal compound containing ZnO, SiO_2, and an alkali metal oxide and having an average particle size of 0.1–5 μm.

It is preferred that the antimicrobial agent is nonporous and preferably contains 50–78 mol% of ZnO. When the ZnO content exceeds 78 mol%, it is difficult to form the antimicrobial agent into a glass compound. On the other hand, when the ZnO content is below 50 mol%, the antimicrobial properties of the antimicrobial agent are insufficient. The antimicrobial agent preferably contains 21–49 mol% of SiO_2, which is a component for glass formation. When the SiO_2 content exceeds 49 mol%, the water solubility of the antimicrobial agent is high, thus causing poor antimicrobial properties. Meanwhile, when the SiO_2 content is less than 21 mol%, it is difficult to obtain a stable glass compound. SiO_2 is used as an essential component for glass formation in the antimicrobial agent, but a portion of SiO_2 may be replaced with other components for glass formation, such as P_2O_5, Al_2O_3, TiO_2, and ZrO_2. The preferred amounts of the components range from 0.1 to 19 mol%.

The antimicrobial agent preferably contains 1–10 mol% of an alkali metal oxide. When the content of the alkali metal oxide is less than 1 mol%, the antimicrobial properties of the antimicrobial agent are degraded. On the other hand, when the content of the alkali metal oxide exceeds 10 mol%, the water solubility of the antimicrobial agent is high, and thus poor antimicrobial properties are caused with impairing the discoloration resistance of the antimicrobial agent. In the process of the present invention, the antimicrobial agent is preferably added in an amount of 0.2–5% by weight, relative to the weight of yarn. If the antimicrobial agent is added in an amount of less than 0.2% by weight, the antimicrobial effects cannot be ensured. If the antimicrobial agent is added in an amount exceeding 5% by weight, the physical properties of the antimicrobial elastic fiber may be deteriorated. According to the process of the present invention, a dispersant, such as a fatty acid, a fatty acid salt, a fatty acid ester, or an aliphatic alcohol, is added to improve the dispersibility of the antimicrobial agent in the polar solvent, thereby shortening the milling time and preventing the discoloration of the antimicrobial agent. The antimicrobial PU fibers prepared by the process of this invention exhibit excellent spinnability while maintaining superior antimicrobial properties and remaining color whiteness.

Electronic, stretchable, and flexible PU fiber also has been studied for wearable electronics or smart textile applications. A novel and environment-friendly method was presented for fabricating silver plating PU fiber, which exhibits high conductivity and excellent elasticity. An adherent polydopamine film was coated on the surface of PU filaments by in situ polymerization reaction of dopamine. The silver particles and silver plating were reduced on the surface of PU fibers by in situ reduction of polydopamine and glucose in turn. The silver particles could be combined on the surface of fiber by the catechol groups of polydopamine. The testing results show the electrical resistivity of silver plating PU fiber can reach the minimum value of 4.5 ± 0.1 Ω/cm, when the concentration of silver nitrate and dopamine are, respectively, 55 and 3 g/l. The loss of the breaking strength and breaking elongation are, respectively, 5% and 11.9%. The nonlinearity error and the hysteresis of this PU fiber strain sensor are, respectively, less than 29.3% and 34.3% [25].

13.3.2 Bicomponent Elastic Fiber

Elasticity, comfort, and durability make polyester bicomponent elastic fibers have broader applications. But the study of this kind of fiber still focuses on a single function. The diversity development is still in a blank state. The present inventions address the deficiencies of the prior words and develop some multifunctional polyester bicomponent elastic fibers.

The bicomponent parallel fiber technology began in the late 1980s, but the progress has been slow. The parallel bicomponent fiber has a permanent self-crimping performance, which provides bulkiness, flexibility, and resilience and feel good of the textile products. The two components for this kind of elastic fibers should have different heat shrinkage, or different relative viscosity, which will lead to the three-dimensional crimps of fiber after heat treatment.

The stretch and recovery property of bicomponent elastic fibers comes from the three-dimensional spiral crimps due to the different shrinkage rate of two polymers. Patent CN 1962968A [26] discloses a preparation method and application of a PBT/PET three-dimensional crimped fibers, using PBT chips with an intrinsic viscosity of 1.0 dl/g and PET chips with intrinsic viscosity of 0.64 dl/g as raw materials. Because of the small intrinsic viscosity of two components, the final curl fibers show small elongation and elastic recovery. In addition, patent CN101126180A [27] discloses a bicomponent elastic fibers with peanut cross section but not the common circular cross section. Because the distance between the centers of two components is larger than that of the round fiber, this leads to less number of crimps per unit and large crimp radius and accordingly the lower elastic and recovery.

The objective of the CN104342802A [28] is to provide a flexible and elastic recovery bicomponent elastic fiber with high rate of production. The invention is as follows: a melt-spun two-component composite elastic fibers, which are made of PET and PBT under the ratio of 70 : 30 to 30 : 70. The intrinsic viscosity of two components is controlled between 0.5 and 1.0 dl/g. The cross section is oval, and the boundary between two components is along the long axis of the oval. The center distance of two components is smaller than the common round cross section or other shapes, which is beneficial to the crimp formation and elasticity. In the final bicomponent the fibers have a crimp of 55–75/25 mm, the elastic elongation is about 100%, and recovery rate is above 95%.

CN104726946A [29] reported a thermal insulation bicomponent elastic fiber, which comprises two parallel components with hollow structure, and the far-infrared powders are added in at least one component. The elastic or cured spiral structure of fiber is formed through the different shrinkage rate of two different polyester ingredients, which are selected form PET, PBT, PTT, etc. This far-infrared bicomponent elastic fiber has excellent heat retention effect and is widely applied to underwear, sportswear, and warm clothing, which have the advantages of elasticity, comfort, lightness, and thermal insulation performance. With addition of 2–4% by weight, the final fiber will increase the thermal insulation of 30% than common bicomponent elastic fiber, and the elastic elongation is 35% or more, and elastic recovery rate is more than 70%.

The fabric sheerness directly affects the garment appearance, especially for some thin, light color fabrics for summer or swimwear. If the fabric has poor shading property, the unwanted see-through effect will occur, which is very ungraceful. The bicomponent elastic fiber as a widely used material in garment also needs the excellent shading property. CN 102534861 B [30] reported a high-resistance parallel shading bicomponent elastic fiber, which comprises two different parallel components selected from PET, PTT, PA, and PBT. TiO_2 powder, with particle size below 200 nm and refractivity index of 2.6–2.9, will be added into the two components 5–10% by weight. The white fabric made of these fibers shows a good shading effect than common white fabric.

13.3.3 Polyolefin Elastic Fiber

The common polyolefin elastic fiber made of PE and Polyolefin Elastomers (POE) has less elastic recovery and lack of flexibility. The high-performance polyolefin elastic fiber has good stretch and recovery property but with a high cost. CN 103276584 B [31] describes a technology, which provides a way to improve the elastic recovery property of polyolefin fibers, and the process is simple and low cost.

The method of the this invention includes the following steps: a radiation-sensitive agent such as vinyl-terminated PDMS or octamethylcyclotetrasiloxane siloxane of 0.1–5.0% by mass is added in the polyolefin polymer for melt spinning to obtain a polyolefin-spun filament; radiate the polyolefin filament under the radiation dose of 10–100 kGy to prepare elastic polyolefin fibers. After radiation, the polyolefin macromolecular can be cross-linked to form a three-dimensional network structure, which increases the elastic recovery property of the fiber, to avoid additional residual shrinkage force, resulting in better flexibility polyolefin elastic fiber and improve the weathering resistance, aging resistance, UV resistance, and toughness.

CN 105088403 A [32] publishes a core–shell melt-spun polyolefin elastic fiber, which does not need radiation cross-linking process but still has excellent elasticity and recovery. The core layer is made of a kind of polyolefin with melt temperature at most 160 °C, and the core layer is made of another polyolefin with melt temperature at most 120 °C. The elongation will be above 600%, and the recovery rate after five cycles under 300% elongation will be more than 90%. Other advantages of this core–shell polyolefin elastic fiber comprise antiaging, acid and alkali resistance, anti-oxygen, and heat setting under low temperature (100–140 °C). It has a wide application in wrinkle-free shirt, swimwear, underwears, socks, denim fabrics, etc.

A China patent [33] publishes a simple preparation process and low-cost method for preparing polyolefin composite elastic fibers with good elastic recovery. It comprises a polyolefin elastomer, a thermoplastic resin, silicone rubber, radiation-sensitive agent, and an antioxidant, which are mixed and fed to a twin screw extruder to obtain a polyolefin mixer master batch. The amount of polyolefin elastomer is 60–80% of the total mass; the amount of the thermoplastic resin makes up the total mass of 5–25%; the amount of the silicone rubber is 10–20% of the total mass; the amount of the radiation-sensitive agent shares 0.1–0.5% in the total mass. The polyolefin mixer master batch is processed by melt spinning to get as spun fiber and then placed under irradiation treatment at 50–500 kGy to obtain the polyolefin composite elastic fiber, which has a elongation about 900% and the recovery rate above 95% under 100% elongation and a breaking strength above 1.0 cN/dtex. CA 2642462 A1 [34] describes a method to produce cross-linked polyethylene elastic fibers with higher tenacity to further reduce the fiber breaking during downstream process. It has been discovered that a polyolefin composite comprises two or more polyolefin components, in which one imparts the elasticity and cross-linking to meet the stretch and recovery performance and other component with higher molecular weight improves the tenacity.

13.3.4 Polyether-Ester Elastic Fiber

PEE elastic fibers contain a high crystalline polyester such as a polyalkylene terephthalate as a hard segment and a polyalkylene glycol as a soft segment and are capable of being melt-spun. It has some advantages such as high productivity, excellent heat resistance, and excellent heat setting property. Furthermore, the future development of the PEE elastic fibers has been expected as elastic fibers suitable for the multifunctional applications.

The present [34] invention relates to a PEE elastic that has a good moisture absorbing or releasing property and reversibly expanded or contracted by the absorption or release of water. It especially gives a comfort for sportswear and intimates applications. This patent discloses the PEE elastomer, which uses PBT as a hard segment and polyoxyethylene glycol as a soft segment, characterized by having a coefficient of moisture absorption of not less than 5% at 35 °C and not less than 10% at a relative humidity of 95%. The PEE elastomer is copolymerized with a metal organic sulfonate, which consists of an aromatic hydrocarbon group or an aliphatic hydrocarbon group, an ester-forming functional group, and an alkali metal or an alkaline earth metal. In the present invention, the weight ratio of the hard segment and soft segment is preferably in a range of 70 : 30 to 30 : 70, more preferably in a range of 60 : 40 to 40 : 60. When the weight ratio of the hard segment exceeds 70%, the elongation is lowered, and it becomes difficult to use it for high stretch textiles. The moisture absorption property is liable to be lowered. When the weight ratio of the hard segment is less than 30%, the strength is liable to be lowered, because the rate of the crystal portion of the polyester is small. It is difficult to copolymerize other added polyoxyethylene glycol, and the colorfastness is to be deteriorated in post-processing such as scoring and dyeing.

PEE elastic fiber has better chemical resistance property than PU fiber, but the elastic recovery is worse than it. So improvement of the elastic recovery is also a development direction of the PEE elastic fiber. The common ways to improve the elastic recovery rate of the PEE elastomer fibers generally include changing the spinning parameters and processing conditions, such as spinneret draw ratio, after-stretching ratio and temperature, and heat setting time and temperature, increasing more hard segment crystalline regions to prevent the slipping between macromolecules of the soft segments such as blended spinning with PBT, or adding nucleating agents. These methods are good attempts to enforce physical cross-linking of the hard segment to improve elastic recovery; however, they cannot solve the problem well.

CN 101100768 A [35] discloses a method to improve the elastic recovery rate by adding nano-TiO_2 particles, which has high surface energy. When dispersed in polymer, it can absorb the macromolecular chains to reduce its surface energy as physical cross-linking points, which can form a network structures between macromolecular chains, thereby improving the elastic recovery performance of the fibers. The preferable ratio of nano-TiO_2 is 0.3–1.0% by mass in the polymer. Among this range, the TiO_2 can be dispersed evenly and stably, and the elastic recovery rate will be 97% under 100% elongation and 88% under 300% elongation.

Method for preparing glycerol micro-cross-linking PEE elastic fiber has been published by CN 101100769 A [36], which is another method to improve the micro-cross-link and elastic recovery rate of PEE elastic fiber. The final PEE elastic fiber will be with 90% elastic recovery rate at 100% elongation and 88% recovery rate at 300% elongation.

13.3.5 Polyester Elastic Fiber

Polyester elastic fibers have excellent physical and chemical properties, in particular chemical, heat stability, high melting points, and high strength. Fabrics made of polyester elastic fibers exhibit softness hand feeling, resiliency, and stretch recovery, among other desirable properties.

PTT has recently received much attention as a polymer for use in textiles, flooring, packaging, and other end uses. There are also some special demands on this fiber such as in automotive interiors textiles, outdoor uses including housing, garden and patio furniture, or personnel protective equipment, which demand the PTT fiber to have a good UV resistance property. US 7196125 B2 [37] provides the colored PTT fibers suitable for use in transportation areas, in which fibers can be subjected to high UV exposures and stringent heat conditions. The processes include the use of a dye bath consisting of 3.0% benzotriazine derivative UV absorber by weight, 0.5% disperse dye, 0.25% sequestering agent, 0.5% alcohol ethoxylate surfactant, and water. The dyeing temperature is 132–145 °C for 30 minutes. The final dyed PTT fiber shows at least grade 4 light fastness after exposure to 488 kJ of UV radiation.

13.4 Evaluation and Application

Elastic fibers have been widely used in different areas from apparel, home textiles, personal protective equipment, furniture, automobiles, outdoor textiles, medical textiles, etc. Among these, textile applications are the main market. In textiles, elastic fibers are usually applied in woven fabrics to increase the stretchable and comfortable property such as in two-way or four-way stretchable denim. Elastic fibers also are used in knitted fabrics to achieve a more large elasticity for applications of underwear, sportswear, and yoga garments. Elastic fibers with high recovery force are often adopted in medical socks for the prevention and treatment of varicose vein and sports pants for prevention of sports injury and fatigue recovery. Due to the excellent resilience of some elastic fiber, it is a preferable choice for floor rugs, packaging, and furniture covers.

As most consideration, the most import performances of elastic fibers are elongation and recovery rate index; however the elastic force at specified elongation also plays an important role in applications. The most common method for elastic properties of elastic fibers is ASTM D 2731-2015 [38], which is applicable to elastic fibers having a range of 40–3200 dtex. The elastic properties include force at specified elongations, permanent deformation, and stress decay, and the hysteresis-related properties can be calculated. The specimen is mounted in a constant rate of extension (CRE)-type machine, and a series of five loading/unloading cycles is subjected in which the specimen is extended and relaxed between 0% and 75% of the elongation at first fiber break. During the fifth cycle, the specimen is held at the maximum extension point for 30 seconds and then unloaded to allow a return to its original gauge length position. The specimen is then subjected to a sixth load/unload cycle. Forces at specified elongations are calculated form the force–elongation curve for the first and fifty loadings and for the fifth unloading. Stress decay is calculated on the fifty cycle. Extension at a specified force is determined on the sixth loading and is used to calculate the permanent deformation. In the previous, elongation rate and permanent deformation are the main concerns. With the developing of varied products and endues, the force at specified extension and stress decay attract more and more attentions.

As shown in Figure 13.9, fibers 1 and 2 have almost the same elongations and excellent recovery rate. However, considering the elastic forces, these two kinds of fibers have different application areas. Fiber 1 displays higher extension force and recovery force than that of fiber 1. The large force is beneficial to some textiles need high pressure such as compression socks or compression sportswear wears.

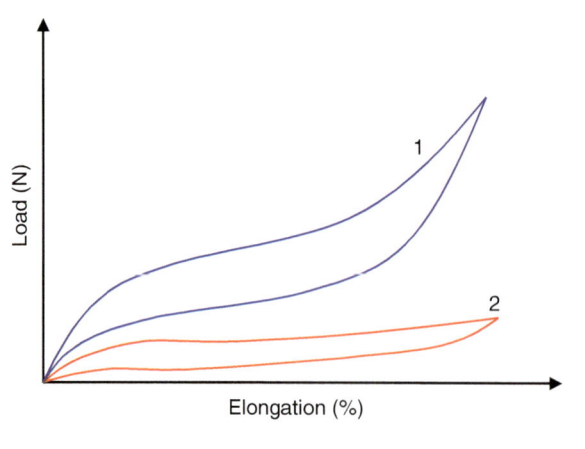

Figure 13.9 Load–elongation curves of different elastic fibers.

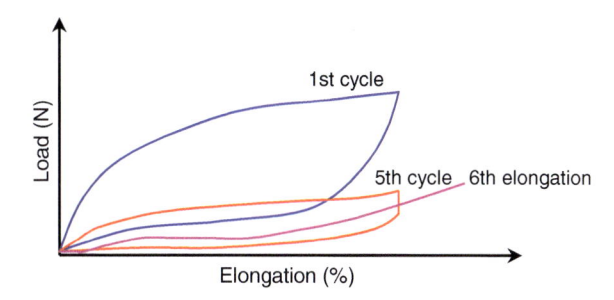

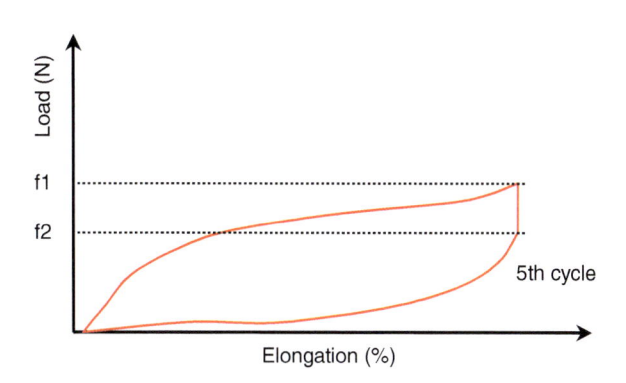

Figure 13.10 Elastic property test of elastic fiber.

Figure 13.11 Force at specified elongation of elastic fiber.

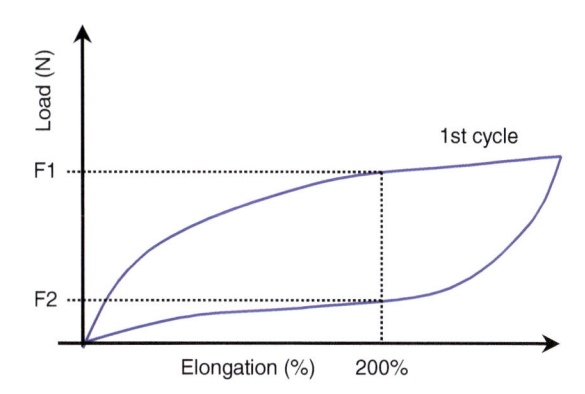

Fiber 2 shows a relative flat load and elongation/recovery curves, which is more suitable for comfortable textiles with large elongation but low pressure such as intimate, sock welts, etc.

Figure 13.10 displays the elongation and load curves of the first cycle, the fifth cycle, and the sixth elongation based on ASTM D 2731. From the curves, the first cycle shows a higher elongation and recovery force than that of the fifth cycle. From Figure 13.11 the force at specified elongation can be determined as F1 and F2. The stress decay of the elastic fibers can be calculated based on the force f1 at maximum extension and the force f2 after holding at maximum extension for 30 seconds (Figure 13.12) on the fifth loading cycle. Higher stress decay rate means the elastic fiber easily to relax, and the elastic force cannot maintain, which is not good for high elastic stress applications such as compression

Figure 13.12 Stress decay of elastic fiber.

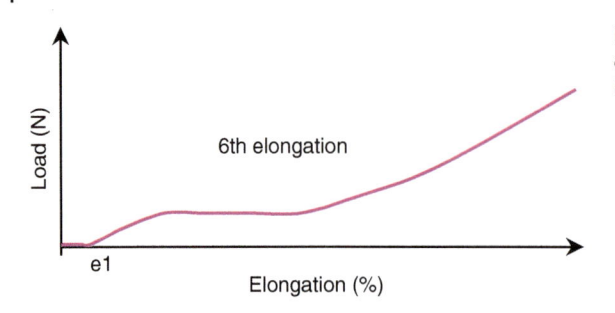

Figure 13.13 Permanent elongation of elastic fiber below e1.

socks. No matter what kind of elastic fibers, the elasticity fatigue exists. After several cycles of extension, the elastic recovery rate will decrease, and accordingly the permanent deformation occurs. Figure 13.13 shows the determination of the permanent deformation from the sixth extension curve. The force before e1 elongation is almost zero on the sixth cycle that means the fiber can be elongated before the elongation of e1. The permanent deformation is calculated by the e1 divided by the gauge length. Higher permanent deformation means worse elasticity recovery property.

The applications of elastic fibers mainly depend on the elastic property of fibers. The fiber with large elongation, high stress, and low stress decay is suitable to compression socks, pressure pants, or shape wears. The fiber with small elongation but higher stress and low stress decay is suitable to some carpets. The underwears need the elastic fiber with large elongation and low stress, which supplies comfort elasticity on human body. As we know, the stress will increase with the extension. But for comfortable textiles, the customers want it easily extended but without large compression. Therefore, this kind of elastic fiber shows a relative flat load–elongation curves compared to others.

Besides the elastic property, other physical or chemical properties also determine the application areas of elastic fibers. The chloride resistance elastic fibers such as XLA or polyester elastic fibers are good choice for swimwear. The low temperature heat setting PU elastic fiber is suitable to anti-laddering hosiery or fabrics containing wool fibers, which are sensitive to heat. The UV resistance elastic fibers are suitable for some outdoor textiles like outdoor car coverage or awning. Hard elastic fiber shows excellent stretch and recovery property under cold environment than other elastic fibers, so it can be used in some industry or engineering areas in outdoor. Because of the microvoids when stretching, the hard elastic fiber can be processed to make filtration hollow fibers for water or blood filtration and separation.

13.5 Future Trends

The principles of difference types of elastic fibers have been studied thoroughly. So the future works for elastic fibers mainly focus on the fiber modifications to achieve functional performances or overcome the present shortcomings and improve the production efficiency.

The PU fiber will continue to improve the chemical or chlorine resistance, heat setting efficiency, and easy dyeing properties and develop comfortable fiber with lower elongation force, high tenacity, and stress fibers or other functional fibers.

The bicomponent elastic fibers will aim to multifunctional development such as thermal and elasticity property, temperature controlled, and elasticity property for high-performance textile applications.

The polyolefin fibers will enhance the elastic durability and focus on the advantage of lower elongation stress than PU fibers.

The PEE elastic fibers are promising fibers with a wide range of uses and lower cost than PU elastic fiber. By controlling their molecular structure, different functional polyester-ether fibers will be developed.

The developments of hard elastic fiber will emphasize on the production of hollow fiber and filter materials in medical applications and the fibers with excellent resilience for furniture endues.

References

1 Xin-yuan, S. (2006). Structure and elasticity of elastic fiber and its dyeing and finishing (I). *Printing and Dyeing* 8: 44–46.
2 Fabricius, M. and Wulfhorst, B. (1995). Elastane fibers (spandex). *Chemical Fibres International* 45 (10): 400.
3 Hicks, E.M. (1963). Lycra spandex fiber I structure and properties. *American Dyestuff Reporter* 7: 33–35.
4 Ibrahim, S.M. (1967). Fibers, elastomeric. In: *Encyclopedia of Polymer Science and Technology* (eds. Herman Francis Mark, Norman Grant Gaylord and Norbert M. Bikales), vol. 6, pp. 573–593. Interscience Publishers.
5 Hoe, H.C. (1998). A new polyester fibre. *Textile Magazine* 2 (1): 12–14.
6 Hu, J., Lu, J., and Zhu, Y. (2008). New developments in Elastic fibers. *Polymer Reviews* 48: 275–301.
7 Chodák, I. (1999). Hard-elastic or 'springy' polypropylene. In: *Polymer Science and Technology* (ed. Karger-Kocsis, J.), vol. 2, 291–294. Dordrecht: Springer Science+Businessmedia.
8 Bonart, H. (1968). X-ray investigations concerning the physical structure of cross-linking in segmented urethane elastomers. *Journal of Macromolecular Science, Part B Physics* 2 (1): 115–138.
9 Lee, H.S., Ko, J.H., Song, K.S., and Choi, K.H. (1997). Segmental and chain orientational behavior of spandex fibers. *Journal of Polymer Science Part B: Polymer Physics*: 1882–1832.
10 Liang, Z., Wenke, W., and Dongyan, S. (2001). Development of high performance polyurethane elastic fiber. *Polyester Industry* 14 (2): 12–16.
11 Zhao, Q. and Xia, H. (1994). Polyether ester elastic fibers. *Textile Science Research* 1: 18–22.
12 Yan, T. (2016). Relationship between structure, properties and processing technology of copolyester-ether fiber. Marter dissertation. China: Donghua University.
13 Lewin, M. and Pearce, E.M. (1998). *Handbook of Fiber Chemistry*, 2e, 258.

14 Sprague, B.S. (1973). Relationship of structure and morphology to properties of "hard" elastic fibers and films. *Journal of Macromolecular Science, Part B: Physics*, 8 (1–2): 157–187.

15 Youyi, X., Changhui, X., and Baiming, X. (1996). Research development of crystalline hard elastic polymer material. *Functional Material* 27 (1): 12–31.

16 Kang, Y.S., Jin, J.S., Seo, S.W., and Kwon, I.H. (2005). Process for preparing elastic fiber having high modulus, alkali-resistance and heat-resistance. WO 2005021847 A1, published 10 March 2005.

17 Pardini, S.P., Bing-Wo, R.D., and Teerlink, T.W. (2016). Methods of making and using elastic fiber containing an anti-tack additive. US 9315924 B2, published 19 April 2016.

18 Liu, G., Zhou, Z., and Zheng, Y. (2012). Method for producing polyether black spandex fibers. CN 101984158B, published 4 July 2012.

19 Xue, S., Liu, Y., Wen, Z. et al. (2014). Method for preparing black polyurethane elastic fiber via in situ polymerization. CN 102899739 A, 20130130.

20 Xu, T., Liang, H., Feng, Y. et al. (2015). A kind of black polyurethane urea elastomer adding CNT and preparation method thereof. CN103726127B, published 12 August 2015.

21 Shen, J., Shen, D., and Shen, X. (2016). High elastic anion fibre. CN 205501464 U, published 24 August 2016.

22 Xue, S., Fei, C., Wang, X. et al. There is the preparation method of the polyurethane elastomeric fiber of humidity absorption and release performance. CN 104153037 A, published 19 November 2014.

23 Zhang, Y. (2005). Manufacture process of nano fibroin particle. CN 1560136 A, published 5 January 2005.

24 Song, B., Seo, S., Kim, J., and Kwon, I. (2007). Process for preparing antimicrobial elastic fiber. US 20070292684 A1, published 20 December 2007.

25 Liu, H., Zhu, L., He, Y., and Cheng, B. (2017). A novel method for fabricating elastic conductive polyurethane filaments by in-situ reduction of polydopamine and electroless silver plating. *Materials and Design* 113: 254–263.

26 Gong, J. (2007). A method of making PBT/PET fibers and its applications. CN 1962968A, published 16 May 2007.

27 Cui, H., Zhang, S., Wang, G. et al. (2008). Elastic polyester fibre and preparation method thereof. CN 101126180 A, published 20 February 2008.

28 Cao, H. and Fujimori, A. et al. (2015). A kind of bi-component elastic fiber. CN 104342802 A, published 11 February 2015.

29 Zhao, M., Chen, L., Qian, J. et al. (2015). A thermal insulation bi-component elastic fiber. CN 104726946 A, published 24 June 2015.

30 Shi, M., Gan, G., Qiu, Y. et al. (2014). A high sheerness bi-component elastic fiber. CN 102534861 B, published 2 April 2014.

31 Wang, X., Zhang, H., Huang, N. et al. (2015). A method of making polyolefin elastic fiber by radiation. CN 103276584 B, published 22 April 2015.

32 Xiang, H. (2015). Two-component polyolefin elastic fiber. CN 105088403 A, published 25 November 2015.

33 Wang, X., Zhang, H., Huang, N. et al. (2016). A component elastic fiber. CN 103668550 B, published 17 August 2016.

34 Mizohata, S., Makino, S., Morioka, S. et al. (2006). Polyether ester elastic fiber and fabric, clothes made by using the same. US 20060177655 A1, published 10 August 2006.

35 Song, X., Li, G., Jiang, J. et al. (2008). Method for preparing nano TiO_2 polyester ether elastic fiber. CN 101100768 A, published 9 October 2008.

36 Song, X., Li, G., Yang, S. et al. (2008). Method for preparing glycerol crosslinked polyether ester elastic fiber. CN 101100769 A, published 9 January 2008.

37 Chang, J.C. and Nolan, D.P. (2007). Poly(trimethylene terephthalate) fibers useful in high-UV exposure end uses. US 7196125 B2, published 27 March 2007.

38 ASTM D 2731-2015 (2015). *Elastic properties of elastomeric yarns (CRE type tensile testing machines)*.

14

Smart Fibers

Dong Wang[1,2], Weibing Zhong[2], Wen Wang[2], Qing Zhu[2], and Mu Fang Li[1]

[1] Hubei Key Laboratory of Advanced Textile Materials & Application, Wuhan Textile University, 1st Yangguang Road, Wuhan 430200, China
[2] Donghua University, College of Chemistry, Chemical Engineering and Biotechnology, 2999 North Renmin Road, Shanghai 201620, China

14.1 Introduction

14.1.1 Definition

Smart fibers, as well as intelligent fibers, are special designed fibers or fiber-based materials that could realize the biomimetic response to stimulus with dynamic property changes [1, 2]. For more intelligent fibers, judgment and self-control abilities to vary stimulus are also necessary [3]. Generally, sensing and responding are typical features of smart fibers [3, 4]. Sensor systems provide a perception to the stimulus, such as thermal [5, 6], mechanical [7, 8], chemical [9], optical [10], electromagnetic [11], and humidity changes [12, 13], while the response systems exhibit the appropriate responses that can be measured in mild method [14]. It is the perfect cooperation between the two systems that the whole response process completes spontaneously. On the other words, smart fibers can implement functions of stimulus-controlled switching or self-adjustment and self-healing. Figure 14.1 presents the distinguish features of stimuli-responsive fibers, intelligent fibers, and smart fibers. Smart fiber exhibits dynamic response including perception and response ability, while the stimuli-responsive fiber can only present discriminative characteristic. Additionally, intelligent fiber possesses central processor to deal with varies of stimulus.

14.1.2 Research Status

Smart fiber is a kind of developed fiber variety derived from the smart materials or intelligent materials that was proposed in the 1990s. After that, smart fiber is gradually investigated due to their potential applications in smart textiles [15–18] due to its flexibility, spinnability, and wearability. The earliest smart fiber that was considered as the shape memory silk was studied in 1979. During the latest decades, intensive investigations about smart fibers were carried out because amounts of high quality and multifunctional materials and the preparation

Handbook of Fibrous Materials, First Edition. Edited by Jinlian Hu, Bipin Kumar, and Jing Lu.
© 2020 Wiley-VCH Verlag GmbH & Co. KGaA. Published 2020 by Wiley-VCH Verlag GmbH & Co. KGaA.

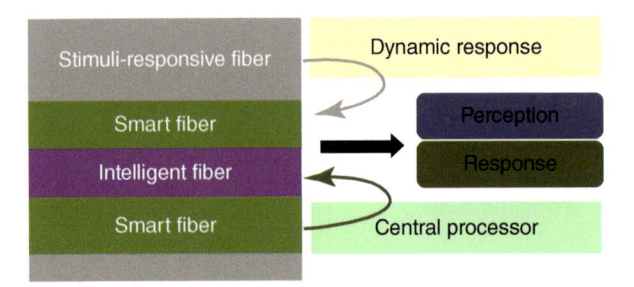

Figure 14.1 Factors of smart fibers differ from stimuli-responsive fibers with presenting dynamic response and from intelligent fibers with unnecessarily possessing of central processor.

methods were discovered [11, 19–26]. Cultivars of smart fibers were extremely expanded with the gradually progressing theories of preparation methods. For example, Cheng et al. [12] prepared a fiber-based motor by rotary processing of freshly spun GO fiber hydrogel. Chen et al. [27] reported a hierarchically arranged helical fibrous actuators fabricated by dry-spun from multi-wall carbon nanotube (MWCNT) arrays. Consequently nonnegligible effects of microstructure on the performance of smart fibers were perceived gradually, and plenty efforts were taken place on the researching of micromorphologies. For example, Jiang et al. [5] utilized electrospinning method to prepare a rapid thermo-responsive actuator with nanofibrous structure. So far, quantities of investigations are still devoting to develop high-performance smart fibers.

14.1.3 Classification

With more and more novel researches about the smart fibers have been raised up, different kinds of mechanisms and applications make it complicated and difficult to classify. Conventional classifications were mostly based on the materials composition that was summarized into metal related species [28, 29], inorganic nonmetallic species [30–32], and polymers species [33, 34]. Here we would like to give a division according to their response behaviors.

The response behaviors vary according to different stimulus source such as thermal, mechanical, chemical, optical, electromagnetical, and humidity changes. In details, smart fibers can be divided into shape memory fibers, piezoelectric fibers, fiber-shaped actuators, color-changed fibers, thermoregulated fibers, smart antibacterial fiber, etc. Figure 14.2a shows the variable responsive behaviors, and rest of Figure 14.2 exhibit several reports of representative smart fibers.

14.2 Raw Materials and Preparation

14.2.1 Raw Materials

Smart fibers set extreme requirements to raw materials and preparation procedure due to the unique response mechanism [2, 39]. Intelligent unit is necessary

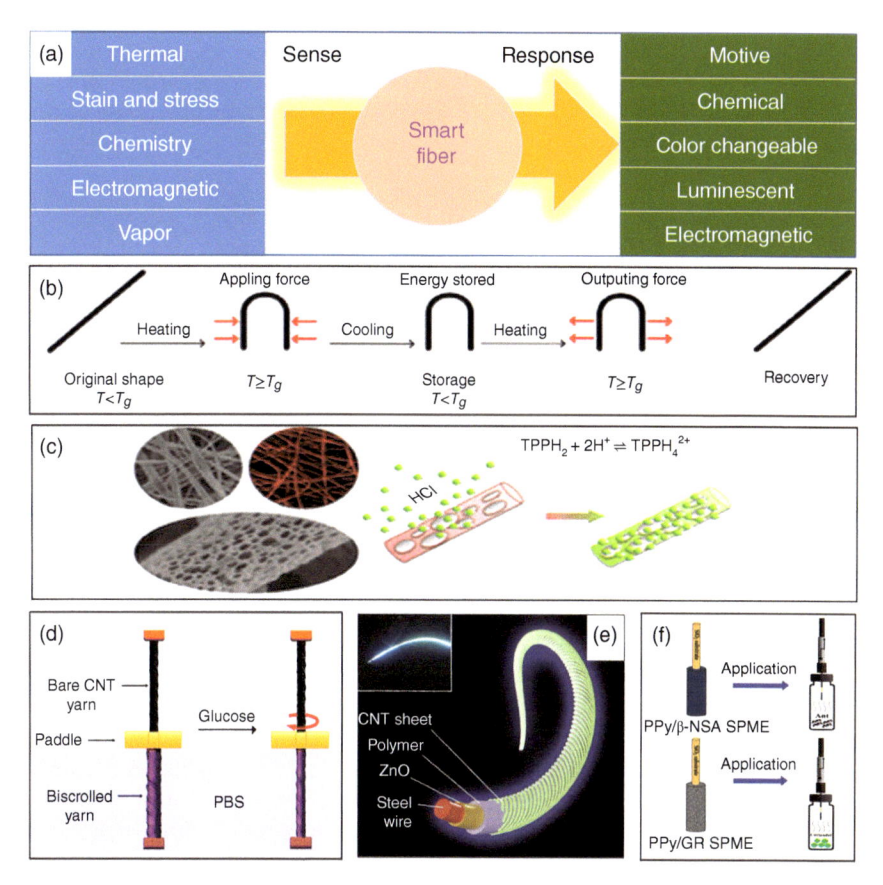

Figure 14.2 (a) Typical components of smart fiber and conventional stimulus and response. (b) Schematic diagram of thermal induced shape memory. Source: AIAA 2009 [35]. Reproduced with permission of Institute of Physics. (c) Sensitive and fast HCl sensor with porous structure. Source: Hu et al. 2016 [36]. Reproduced with permission of Springer Nature. (d) Biothermal sensitive torsional artificial muscle. Source: Lee et al. 2016 [37]. Reproduced with permission of Royal Society of Chemistry. (e) Fiber-shaped color turnable electroluminescence. Source: Zhang et al. 2015 [28]. Reproduced with permission of Springer Nature. (f) Polypyrrole composite solid-phase microextracted fiber for the trace analysis of volatile organic. Source: Zhang et al. 2013 [38]. Reproduced with permission of Royal Society of Chemistry.

for raw materials, which can quickly sense and response to the external stimulus. Otherwise, the materials should be easy to be intelligentized [4]. Generally, the raw materials can be concluded in metal related [40–42], inorganic nonmetallic materials [12, 23, 43, 44], and functionalized polymers [45–47]. They will be demonstrated, respectively, below. Figure 14.3 presents the classifications.

14.2.1.1 Metal Related

Metal-related smart fibers usually use metal, alloy, and metallic compounds as the functional elements to functionalize fibers with stimuli responsibility. Pure metal, alloy, and partial metallic compounds are expected to be ideal materials with high electrical and thermal conductivity, so they are frequently utilized

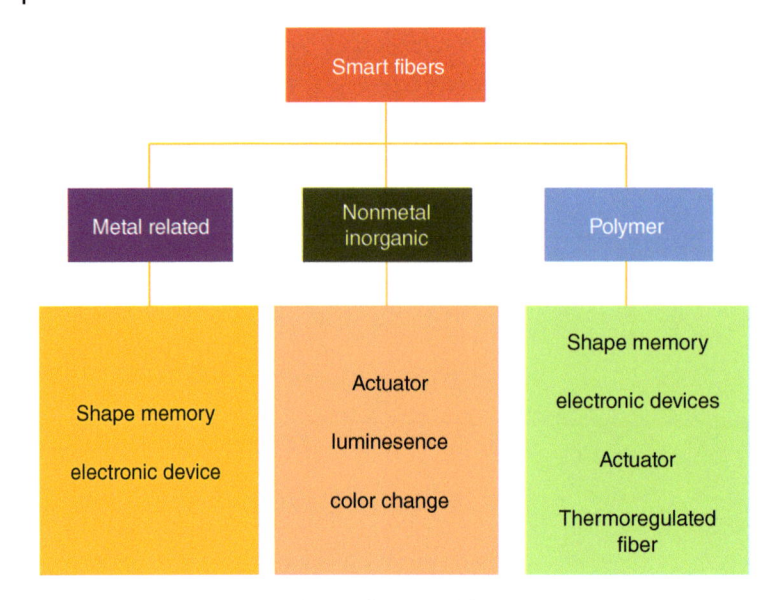

Figure 14.3 Raw materials of smart fibers and their main application.

as the main materials or smart ingredients in thermal- and electricity-induced smart fibers. For example, Seung Hwan Ko group [48] reported a transparent and highly stretchable Ag nanowire (NW)-based heater used for flexible and wearable instrument, which could produce and transfer heat sufficiently. Besides, stimulus induces complex crystal transitions that exist in alloys extensively and may lead to series of motions subsequently. So they were successfully applied in shape memory fiber and actuators. For instance, Choi et al. [49] used NiTi alloy to make a superelastic shape memory alloy (SMA) fiber with different anchorages. Li et al. [50] applied Cu/Ni/Al for main parts to get a shape memory fiber that have temperature-induced martensitic transformation.

Except that, many kinds of metallic compounds are recommended for the excellent stimulus responsibility. For example, ZnS and 8-tris-hydroxyquinoline aluminum (Alq_3) were discovered to be good candidates for high-efficiency electroluminescence. Highly designed ZnO nanowire arrays were constructed to the implementation of smart nanogenerator, which sufficiently convert mechanical energy into electricity [51].

14.2.1.2 Inorganic Nonmetallic and Polymer Related

Inorganic nonmetallic materials mainly include certain elements of oxide, nitride, carbide, halogen compounds, boride, silicate, aluminum acid salt, phosphate and boric acid salt, etc. Single-crystal silicon fibers and silicon oxide fibers are frequently used for their wondrous optical performance, which contributes to hold the crucial position in high-performance communication optical fibers. What's more, crystalline silicon (c-Si) is preferred for electronic devices because of its superior thermal and electronic properties compares with amorphous counterpart [52]. Lagonigro et al. [53] reported a kind of silicon fibers used for photonics field with low loss. Besides, carbon materials were intensively studied for their superb chemical and physical properties [54, 55]. Graphene fibers and oriented carbon nanotube (CNT) fibers with the superior electric conductive rate (10^5 S/cm), thermal conductive rate (3500 W/m·K), and mechanical strength (0.9 TPa) were adopted to fabricate high-performance smart fibers [56–59].

Polymer is one of the top drawer candidates to manufacture smart fibers for their flexibility and easy processing [60, 61]. A plenty of smart fibers can be fabricated with different kinds of polymers or modified polymers directly [14, 62]. In addition polymeric fibers can also be selected to be fiber-shaped substrates for some smart fibers. Usually, the manufacture procedure of being smart fibers is easily to achieve. Firstly, functional modification can be implemented easily due to the abundant reaction sites on the backbone of polymers. After that, responsive agents can be embedded homogeneously for the proper viscoelasticity of polymer melt and solution. For example, microcapsule and sea-island fibers were two typical formations, which were obtained for controllable release, temperature regulation, and self-heal. Lastly, polymer-based fibers can be designed and controlled, manufactured into varies of micromorphologies that enable them to combine with responsive agents in special way [5, 14, 63].

14.2.2 Preparations

Preparations of smart fibers can be summarized into two processes, which refer to spinning and intelligentization [4]. The order of the two processes can be adjusted according to the featured structure of target product. In other words, the intelligentization process can be brought into effect by the master batches or spinning solution so that smart fibers can be straight acquired after spinning. And the intelligentization can also be carried out on manufactured fiber products [4]. Obviously, spinning after intelligentization is a prior choice for the commercial production due to its simple technology and the predominance of nonpollution additional. However, simplified preparation method leads to univocal structure, which eventually results in response limitation. Naturally, many reports rise up a novel scheme to get smart fibers by intelligentizing primary fibers. With this method, the implementation of smart fibers is the combination of common fiber and responsive agent. Respectively, the spinning method and intelligentization method will be introduced in detail below.

14.2.2.1 Spinning Methods

As we all know, conventional spinning processes are melt spinning and solution spinning.

Melt spinning method is a straightforward process for fiber acquisition by cooling and drawing the melts. Most of the alloy fibers are prepared with melt blending method. Smart fibers can also be obtained by inputting responsive agents during the melt spinning process. Some kind of piezoelectric fibers were obtained by melt blending with piezoelectric ceramic ultrafine powder, which is an ideal material with excellent piezoelectric performance. However, if the responsive agents are added that could not bear the melting temperature, it will lead to degradation and result in dysfunction eventually. Moreover, extra efforts such as adding compatibilizers or subtle controlling should be taken to improve the

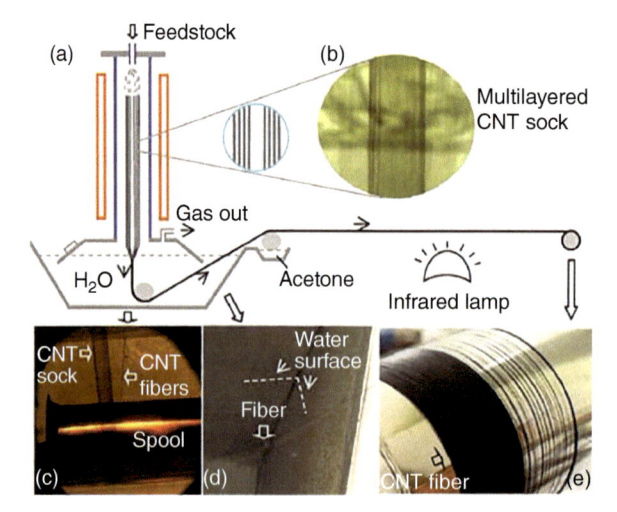

Figure 14.4 Schematic diagram of wet-spun method to prepare multilayered CNT sock. (a) Schematic diagram of the synthesis and spinning for CNT fiber, (b) photograph of CNT aerogel formed in the gas flow, (c) transformation of the CNT aerogel into a fiber, (d) followed by water densification, and (e) spinning of the resultant CNT fiber on a spool.

uniformity of melt blends, which is also critical to smart performance. Although every process was perfectly manifested, performance of responsive agents cannot work perfectly due to the wrapping and isolation effect of substrate melts. Solution spinning can largely ignore the disadvantages because the porous structure is built by solvent evaporation (dry spun) or double diffusion process (wet spun). Figure 14.4 shows schematic diagram of wet-spun method to prepare multilayered CNT sock. What's more, high temperature induced degradation of responsive agents, and it is no longer the limitation of mixture. Wet spinning, which manufactures fiber products by flocculation effect in poor solvent, can be the optimal method to acquire high quality CNTs and graphene fibers.

Electrospinning is a developed preparation method, which shares similar principle with dry spun. The fibers are drawn by extra high voltage electricity field, and the diameter of fiber decreases to micrometer or even nanometer [64, 65]. The scale effect entrusts the fibers with new properties. However, electrospinning is not suit for mass production due to the difficulties in fabrication facilities and production speed, so the melt extrusion or solution spinning method may be the best choice in the case of mass production. For example, polyolefin and cellulose acetate butyrate nanofibers, which share the same principle of sea-island fibers, are usually fabricated by melt extrusion.

14.2.2.2 Intelligentization Method

Treatments on primary fibers were selected as a mild fabrication of smart fibers. Abundant treatment methods lead to the diversity of smart fibers. Taking the chemical structure of the primary fibers into consideration, conventional treatment methods can be divided into chemical modification, coating, self-assembling, absorption, etc.

Chemical modification is a common method that uses the chemical reaction to alter the physical or chemical properties of materials. For example, Zhu and Sun [66] reported a method that organic photocatalyst anthraquinone-2-carbon acid was grafted onto the nanofibers to get the antibacterial nanofibrous membrane.

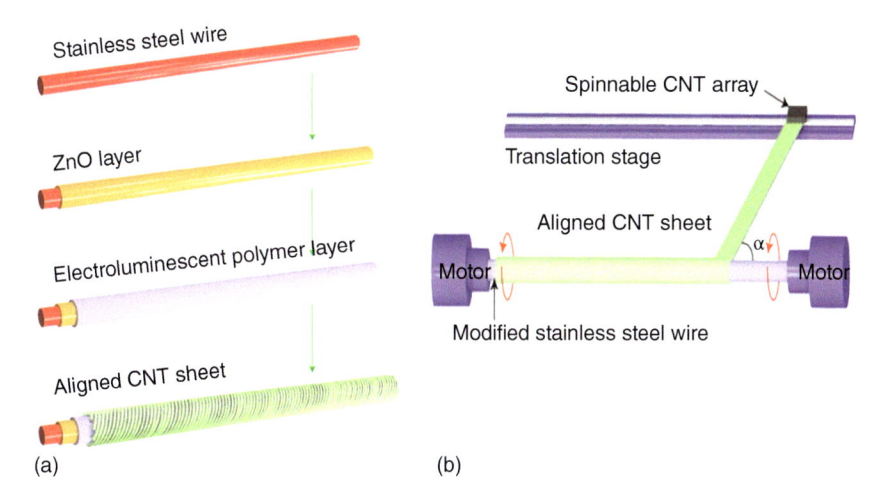

Figure 14.5 (a) Schematic illustration of the fabrication of a fiber-based light-emitting electrochemical cells (PLEC). (b) Schematic of wrapping an aligned CNT sheet around a modified stainless steel wire. Source: Zhang et al. 2015 [28]. Reproduced with permission of Springer Nature.

Coating is a frequently used method to obtain hierarchical-structured smart fibers. For example, electricity-induced responsive behaviors usually need electrodes, which were fabricated by adding conductive materials on the basic fiber substrate [23, 28, 67]. Commonly used methods include vacuum evaporation, chemical vapor deposition, and in situ polymerization [63, 68–70]. Figure 14.5a shows schematic illustration of the fabrication of fiber-based devices with coating method.

Self-assembling is an easy-operating method to acquire featured structure by utilizing unique chemical or physical properties. It would be an excellent candidate to prepare gel-based smart fibers. Typical process can be applied on particular materials, which can provide Coulomb force, hydrogen bond, coordinate bond, and even steric effect.

14.3 Structure and Properties

Structure design is crucial to smart fibers by determination of the responsive properties. Therefore, the structures of smart fibers diverse kind to kind by using the different preparation methods introduced above. Currently, diverse micromorphologies of smart fibers were designed so that they can response to different kinds of stimulus [24, 71, 72]. So the diameters of fibers are usually diminished to microscale or even nanoscale, and porous structured fibers are put forward. Majority of smart fibers are composite fibers comprising functional agents and substrates [73]. Totally, structures of smart fibers can be classified into primary fiber structure, multilayered structure, and three-dimensional fiber-based structure.

14.3.1 Primary Fiber Structure

As already introduced in paragraph of Section 14.2.2, if the melts or spinning solution has been functionalized, the smart fibers will be obtained directly after spinning. With this method, the structures can be manifested by changing the formula of raw materials and the shape of spinneret plate. As whole, primary fiber structures include conventional fibers, profiled fibers, skin-core structure, hollow fibers, and sea-island fibers. Among them, secondary structure of multi-holes is usually a method to construct smart fibers with higher performance.

14.3.1.1 Conventional Fiber Structure

Conventional fiber structure merges in fibers that are obtained from smart raw materials or chemical-intelligentized fibers. Alloys always use such a structure with secondary structure of densification for the excellent ductility of metal melts. However, solution-spun fibers are often porous so that they can be applied in actuators [12, 30] and artificial muscles [31, 74, 75]. Figure 14.6 demonstrates the schematic diagram of CNT-PDDA artificial muscles. Porous structure brings higher specific surface area, which could be helpful for reacting to external stimulus. For example, Cheng et al. [12] managed to prepare a humidity-driven graphene oxide actuators, which achieved rotary speed of 5190 revolutions/minute. Gu et al. [31] reported a twisting hybrid CNT fiber artificial muscle; also, fiber-shaped devices were usually prepared into primary monofilament structure with an easy way.

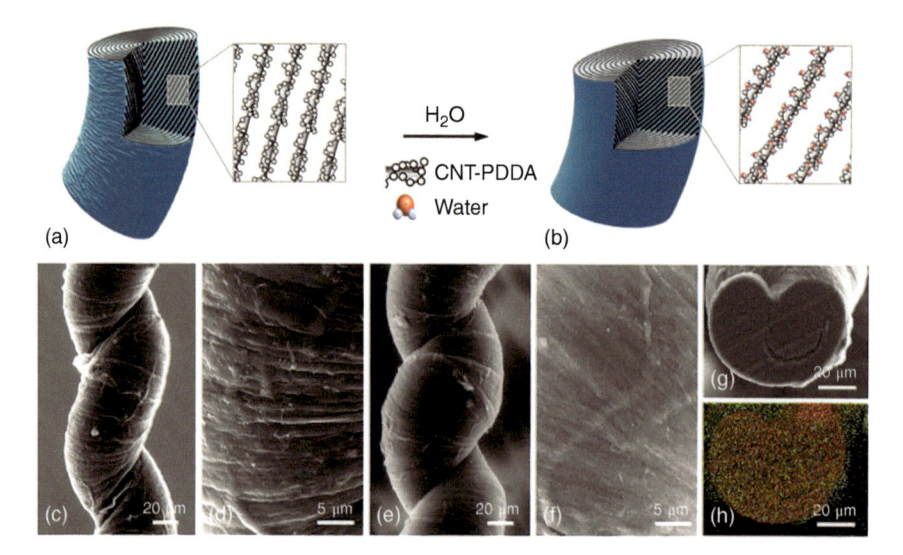

Figure 14.6 Schematic diagram of moisture-driven actuator (a) before and (b) after water absorption. (c, d) Morphology of yarns at 10% RH. (e, f) Morphology of yarns at 90% RH. (g) Cross-sectional SEM image. (h) Elemental mapping by EDAX over the image. Source: Kim et al. 2016 [74]. https://www.nature.com/articles/srep23016#rightslink. http://creativecommons.org/licenses/by/4.0/. Licensed under CCBY 4.0.

14.3.1.2 Skin-Core Structure

Skin-core fibers, which present some unique performance than conventional fibers, are a kind of differential fibers with at least two components incompatibly aligned into concentric or eccentric shape along the radial direction of the fiber. From the view of designer, skin-core fibers can be a desirable candidate for thermal regulating fibers, because the skin can prevent the phase change material from leakage. Moreover, if external stimulus activates the skin as a switch, the core would be spontaneous response against the stimulus. For instance, Shi et al. [76] prepared electrospinning skin-core nanofiber with temperature-responsive poly(N-isopropylacrylamide) (PNIPAM) as skin and polycaprolactone (PCL)-loaded nattokinase (NK) as the core; as Figure 14.7 demonstrated, the switchable skin would be turned on, and the anti-thrombus agent NK would be released and worked.

14.3.2 Multilayered Structure

Compared with primary fiber structure, multilayered structure is more adopted for the whole intelligenize process, which is accomplished only by finishing. The multilayered structure is constructed owing to the different properties between coatings and substrates such as electronic parameters and mechanical strength. Besides, for fiber-shaped electronics, several layers such as electrodes layers, carrier transmission layers, and functional layers are needed. Common methods to get multilayered structures are by means of chemical vapor deposition (CVD) [29], in situ polymerization [33], layer-by-layer self-assembling [77] and core spun [78]. For example, Huang et al. [79] reported yarn-based resistance sensors prepared with single and double wrapping method. Shim et al. [81] designed

Figure 14.7 Schematic procedure for fabrication of PCL/PNIPAAm core–sheath nanofibers containing nattokinase (NK) to capture and release red blood cells (RBCs) from the blood. Source: Shi et al. 2016 [76]. Reproduced with permission of Royal Society of Chemistry.

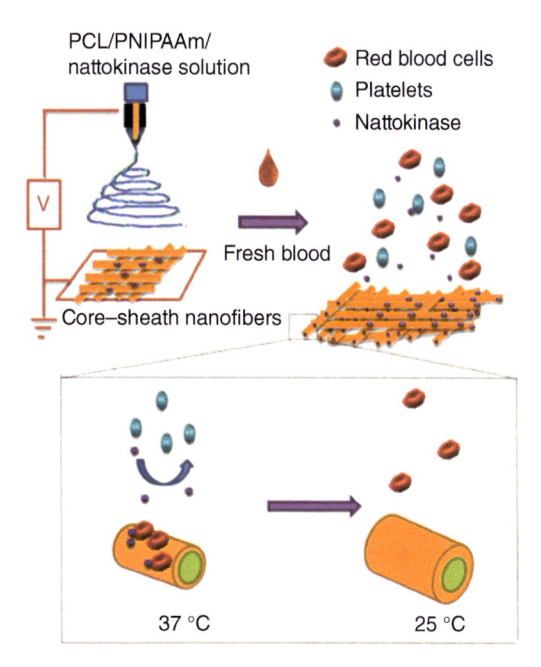

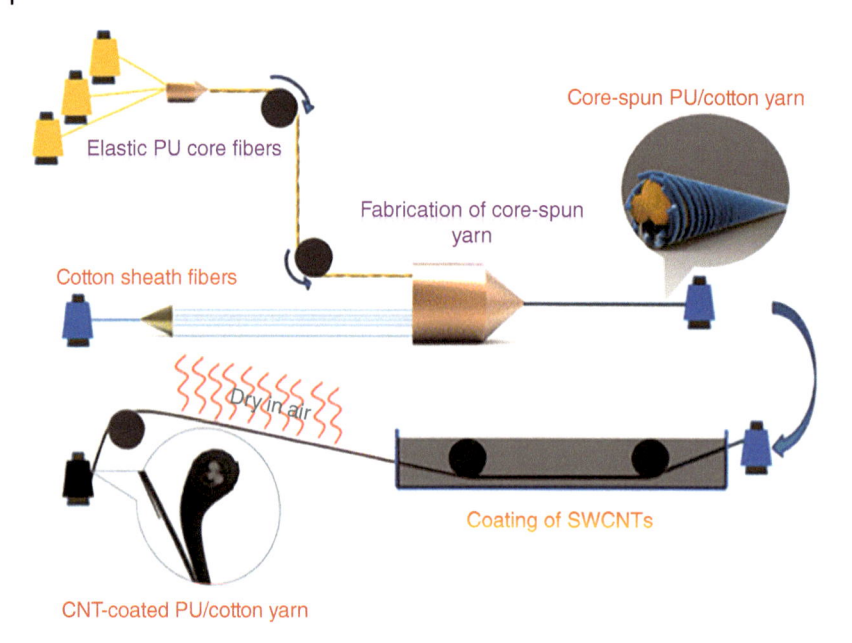

Figure 14.8 Schematic diagram of preparation of core-spun CNT-coated PU/cotton fiber. Source: Wang et al. 2016 [78]. Reproduced with permission of American Chemical Society.

a CNT/electrolyte-coated cotton yarn to fabric e-textiles. Correspondingly, the diversity of preparation leads to wide applications. Figure 14.8 gives the preparation process of single-walled carbon nanotube (SWCNT)-modified core-spun PU/cotton yarn.

14.3.2.1 Coating with Single Layer

Single coating is sufficient for most smart fibers, which is applied in sensors and primitive electronics. There are two kinds of situations. One is that the responsive agents are attached to the fiber-shaped substrates and the responsive agents accomplish the whole procedures of perception and response. For example, fiber-shaped photoluminescence [80] could be the representations. The other one is comprehensive utilization of substrates and coating. Stimulus will lead to the change of unique property of coating (substrate), which affects the structure of substrate (coating); as a result, the changed substrate (coating) delivers measurable functional changes. For instance, ultrathin conductive layer wrapped polyurethane yarns were fabricated though layer-by-layer assembling to monitoring tiny human motions [14]. Figure 14.9 shows the SWNT-coated cotton yarn.

14.3.2.2 Coating with No Less Than Double Layers

Coating with no less than double layers is a typical structure for electricity-induced devices such as electroluminescence and electrochromic fibers. Usually, layers of electrodes such as anodes and cathodes, layers of electron transmitting and carrier transmitting, and layers of electroluminescence or electrochromic

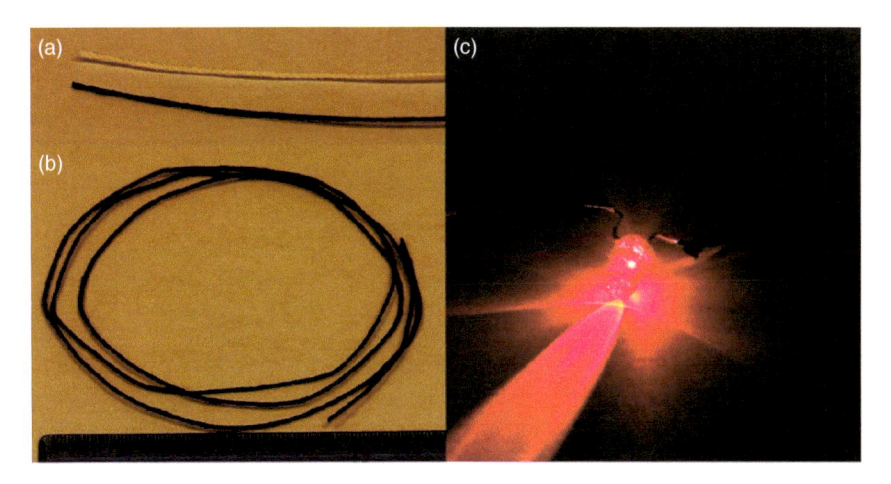

Figure 14.9 Photographs of SWNT-cotton yarn. (a) Comparison of the original and surface modified yarn. (b) 1 m long piece as made. (c) Demonstration of LED emission with the current passing through the yarn. Source: Shim et al. 2008 [81]. Reproduced with permission of American Chemical Society.

chemicals are required for archetypal smart devices, which are simplified later. The first fiber-shaped electroluminescence device was considered to be constructed in 2007 by vacuum thermal evaporation of organic active materials onto swiveling polyimide-coated silica fiber [82]. In this application, the conductivity and flatness may heavily influent the performance of devices [28]. Peng et al. [83] prepared an electrochromic fiber with multilayer structure. Zhang et al. [28] constructed a fiber-based polymer light-emitting cell of which the luminescence changed no more than 6% with different observation angles. Figure 14.10 exhibits typical structure of fiber-shaped organic light-emitting device (OLED).

Figure 14.10 (a) An illustration of a typical organic light-emitting device (OLED) structure, along with (b) an illustration of a fiber-based OLED structure analyzed in this report. (c) A photograph of a flexed fiber having a 1 mm green light-emitting "pixel" turned on. Source: O'Connor et al. 2007 [82]. Reproduced with permission of John Wiley & Sons.

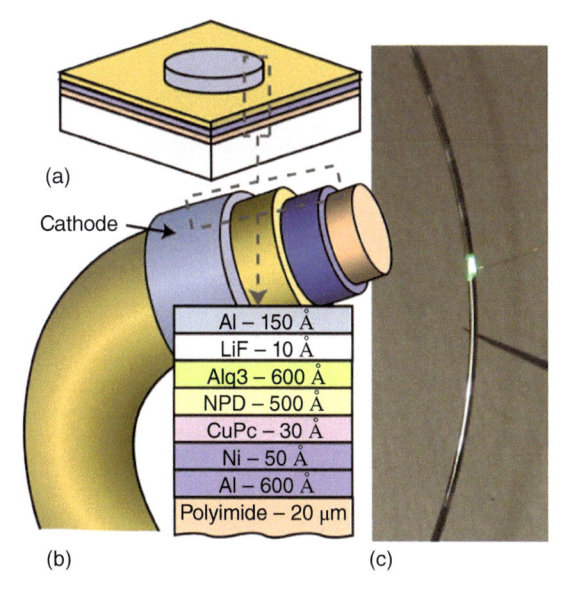

Al – 150 Å
LiF – 10 Å
Alq3 – 600 Å
NPD – 500 Å
CuPc – 30 Å
Ni – 50 Å
Al – 600 Å
Polyimide – 20 μm

14.3.3 Three-Dimensional Fiber-Based Structures

Since the high-speed development of nanotechnology, sub-microfibers, and nanofibers are intensively investigated because of the amazing high specific surface area and additional properties. Currently, these fibers with superfine diameters are designed into membranes and aerogels to construct novel fiber-based smart material.

14.3.3.1 Fabrics and Fiber-Based Membrane

Smart fabrics and fiber-based membranes are typical fiber-based structures, which are fabricated by weaving or stacking of smart fibers or direct treatment to conventional fabrics and fiber-based membranes. Generally, smart fabrics show spiffy softness and wearability, which have promising applications like color-changed textiles [84], smart antibacterial textiles [85], smart robots [86], and oil/water separation [87]. For example, Zhong et al. [63] constructed ultrasensitive pressure sensor by using patterned hybrid nanofiber membrane. Figure 14.11 shows the fiber-based functional fabrics.

Figure 14.11 (a) Photograph of a blank fabric. (b–d) SEM images of the cotton side of the blank fabric with increasing magnification. (e) Cross-sectional SEM image of the blank fabric. (f–h) Polyester side of the blank fabric with increasing magnification. (i) Photograph of a single-sided conductive fabric. (j–l) SEM images of the cotton side of the single-sided conductive fabric with increasing magnification. (m) Cross-sectional SEM image of the single-sided conductive fabric. (n–p) Polyester side of the single-sided conductive fabric with increasing magnification. Source: Huang et al. 2016 [84]. Reproduced with permission of Royal Society of Chemistry.

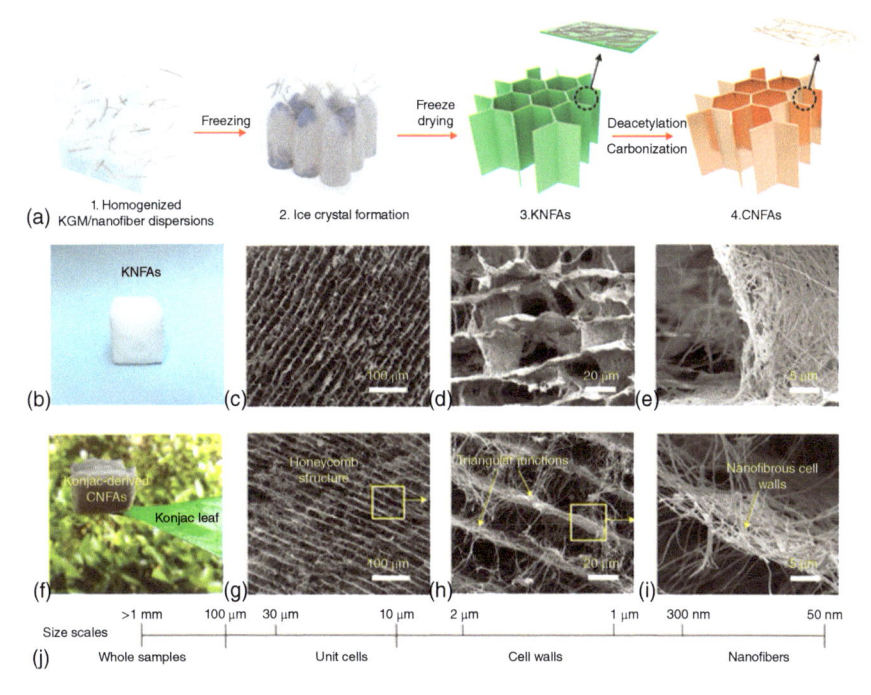

Figure 14.12 (a) Schematic showing the synthetic steps: (1) Homogenized nanofiber dispersions are prepared through high-speed homogenization. (2) Frozen dispersions are prepared by freezing in liquid nitrogen. (3) KNFAs are fabricated via freeze-drying the frozen dispersions. (4) The resultant CNFAs are fabricated by the deacetylation and carbonization of KNFAs. (b) An optical photograph of a KNFA. (c–e) Microscopic structure of KNFAs at different magnifications. (f) An optical photograph showing a 20 cm³ CNFA ($\rho = 0.14$ mg/cm³) standing on the tip of a konjac leaf. (g–i) Microscopic structure of CNFAs at various magnifications, demonstrating the biomimetic honeycomb cellular fibrous architecture. (j) Schematic showing the four levels of hierarchy at three distinct length scales. Source: Si et al. 2016 [88]. Reproduced with permission of John Wiley & Sons.

14.3.3.2 Aerogel

Aerogels are well known for their high porosity up to 95–99% and outstanding mechanical property. Si et al. [88] demonstrated a creating of ordered honeycomb-like structure aerogel with konjac glucomannan and silicon dioxide nanofiber to establish high sensitive pressure sensor of which are demonstrated in Figure 14.12.

14.4 Principles and Theories

14.4.1 Shape Memory Fiber

Shape memory fibers are smart and intelligent fibers, which can realize the cycle from other shape to memorized shape under the control of stimulus such as temperature, pH, and light [89, 90]. There are many different types of shape memory fibers. However, the most common types among them are SMAs and shape memory polymers (SMPs) [91–93]. Typically, shape memory materials include fixed phase and revisable phase. The fixed phase is the provider of recovery force that contributes to avoidance of unacceptable damage and memory of the original shape, while the shape of revisable phase can be changed and locked in another status. For alloys, the fixed phase is the crystal transformation of thermal elastic martensitic, which would balance the thermal effect and elastic behaviors [90, 94], while for polymers, the fixed phase would be chemical cross-linking or supermolecular entanglements; the undefined structure can be transformed into glass state by cooling. Beyond the softening temperature, the recovery force provided by fixed segments contributes to the shape recovery. Besides, a new kind of shape memory fibers, which is made from at least two polymers that share the same mechanism as the polymers, was reported and named with shape memory hybrids (SMHs) [91, 95].

Except that the differences of expansion coefficient can also lead to shape memory effect (SME) [96]. The expansion extent of one phase can be revisable controllable; with stretch or extrusion, the shape of the fiber will change.

14.4.2 Fiber-Based Actuator

Fiber-based actuator is a kind of materials that possess the dynamic motor behaviors in response to environmental stimulus including temperature, humidity, pH, electricity, magnetic field, and so on [13, 97]. Among these different kinds of stimulus, the responsive mechanism can be roughly summarized as the following two categories: one is discriminative expansion/contraction effect that is very common in hydrogels, and the other is deformation that includes structure reformation and crystal transformation induced by microstructure.

Discriminative expansion/contraction effect can be induced by variety of stimulus such as temperature, humidity, pH, and solvent [98, 99]. Generally, the fiber-based actuators are manufactured with more than two different materials, and the different swell and deswell performances between different materials could cause the actuate behavior. Among them, the thermo-responsive hydrogel is widely used for thermo-responsive actuator. For example, Jiang et al. [5] reported a fiber-based superfast thermo-triggered actuator with a double-layer structure including thermo-responsive hydrogel layer and hydrophobic PU layer. The hydrogel phase (PNIPAM) could transform from hydrophilicity to hydrophobicity above the critical temperature, which resulted in actuate behavior because of the unbalanced swell/deswell in water. Besides, hydrogels with weak acid or alkaline groups, such as carbonyls or amino, can also be used to fabricate pH-active actuator [100, 101]. Mahdavinia et al. [102] reported a pH-responsive hydrogel with a polyacrylic acid system.

There are also some fiber-based actuators with one phase that show up obvious expansion under the featured environment. For example, Cheng et al. [12] reported a method to prepare moisture-activated graphene fiber-shaped actuator; the pre-twisting graphene oxide fiber would expanse/contract due to the absorption/desorption of water molecular from the rich oxygen groups. Besides, Chen et al. [27] demonstrated a method to create solvent- and vapor-responsive MWCNT-wrapped primary fiber with large actuation (Figure 14.13).

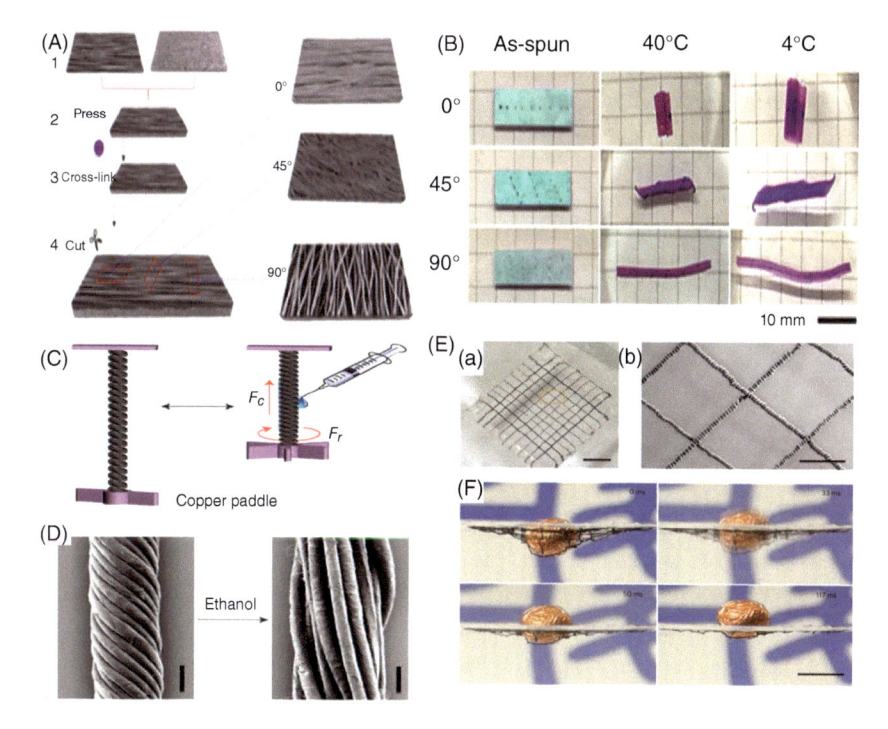

Figure 14.13 (A) Schematic of formation of parallel-arranged PNIPAM fiber actuator along different directions of PNIPAM-0°, PNIPAM-45°, and PNIPAM-90° by electrospinning. (B) The fiber-based actuator was actuated under different temperatures (0, 40, 4 °C). Source: Niu et al. 2016 [47]. Reproduced with permission of Elsevier. (C) Schematic illustration of the contractive and rotary actuations of a hierarchically arranged helical fibers. (D) SEM images of twisted primary fibers before and after ethanol infiltration. The helical structure broke down due to excessive untwisting. (E) Photograph of an actuating textile woven ((a) actuating textile woven fabricated by 18 hierarchically arranged helical fibers (HHF) and (b) the HHF was twisted from 20 primary fibers)). (F) Photographs of the smart textile lifting a copper ball (240 mg) at 0, 33, 50, and 117 ms after being sprayed with ethanol. Source: Chen et al. 2015 [27]. Reproduced with permission of Springer Nature.

The field-induced deformation of the microstructure is the typical mechanism of electric actuator and magnetic actuator. Under the influence of electric field or magnetic field, the microstructures of the materials are forced to adjust along the featured direction with mechanism of crystal transition and the rearrangement of the dipoles and spontaneous magnetization of the magnetic domain. For example, Gu et al. reported a liner actuator of polymeric nanofibrous bundle that can produce a high strain under electric stimuli [26]. Figure 14.13 gives several kinds of specially designed actuators.

14.4.3 Luminescence Fiber

Luminescence fibers are highly functionalized fiber-shaped devices that possess the ability of light emitting. Photoluminescence and electroluminescence are the most investigated branches.

Photoluminescence is a phenomenon of photo-induced light emitting after stock shift or anti-stock shift, which is determined by inner dipole moment.

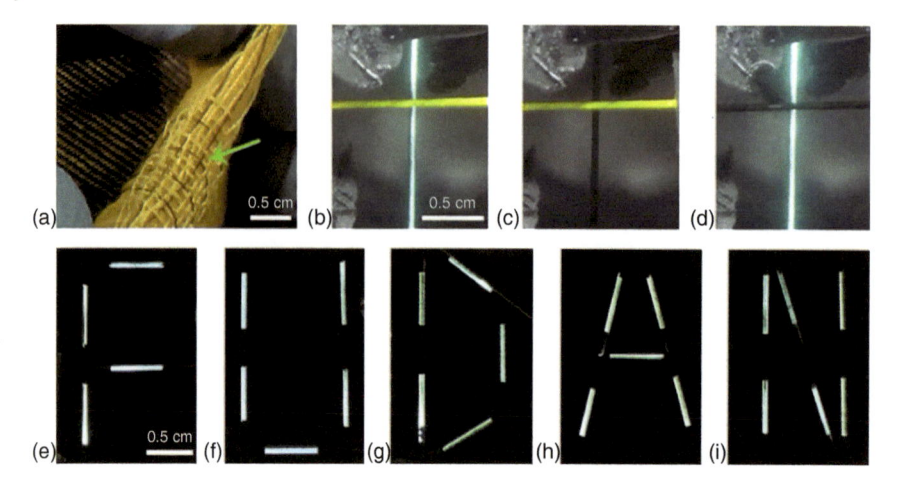

Figure 14.14 (a) Electroluminescent fiber was fabricated into textile under bending and twisting. (b–d) Two fiber-based light-emitting electrochemical cells (PLECs) with different colors (biased at 10 V). (e–i) Fiber-shaped light-emitting electrochemical cells (PLECs) were woven into letter shapes (biased at 9 V). Source: Zhang et al. 2015 [28]. Reproduced with permission of Springer Nature.

From the view of energy, excited electrons spent extra energy on overcoming the barrier and return to ground state meanwhile, and during this process, energy is released by emitting visible lights. Besides, the electrons can be trapped in unique designed chemical structure before their journey back, and only stimulated by temperature the trapped electrons can be reactivated to transited back and luminescent. This phenomenon is named lasting phosphorescence. Lasting phosphorescence phenomenon is often applied in environmental friendly lighting fixture.

Electroluminescence is one of the energy conversion forms. The electrons from cathode and carriers from anode are forced to overcome the interfacial potential barrier by the electricity field. After passing the transporting layer, electroluminescence layer was the final site of incoming electrons and carriers. They will combine in very near distance to form excitons, which can release energy by luminescence. Typically, the electrodes, transporting layer, and the light-emitting layer are critical factors to the efficiency of luminescence. To obtain the high-performance electroluminescence, the work function of the electrodes and the interfacial potential barrier should be concerned and dedicatedly designed. Figure 14.14 shows the electroluminescence fiber with woven ability.

14.4.4 Color-Changed Fiber

Color-changed fibers are unique structured fibers that can rapidly change their color or transparency under the thermal, electrical, and optical impact [103–105]. Generally, the mechanism of color-changed fibers can be roughly summarized as redox reaction, crystal transformation, and geometric configuration changes of ligand [1, 106].

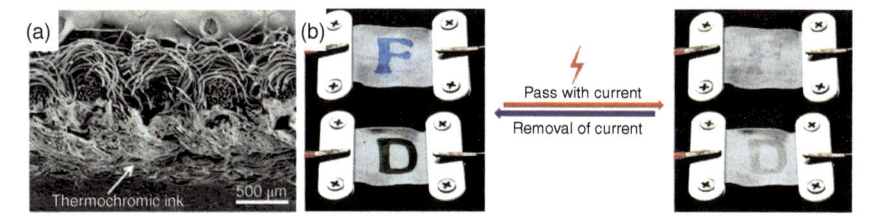

Figure 14.15 (a) Cross-sectional SEM image of color-changed smart textile. (b) The smart textile shows color-changed phenomenon under currents. Source: Huang et al. 2016 [84]. Reproduced with permission of Royal Society of Chemistry.

Redox reaction mechanism often takes place in electrochromic and photochromic [104]. It requires different colors between reduction state and oxidation state. The electrons are obtained or lost in order to alter the oxidation–reduction state results in the change of absorption spectrum with visible behavior of color change. Amounts of chemicals can expose the property; so far, the most popular candidates are metal ions, dyes, and conductive polymers.

Crystal transformation mechanism often exerts in thermal-induced color-changed phenomenon [107]. Amounts of iodine–iodides change their color due to the crystal transformation. For example, α crystal form of Cu_2HgI_4 appears a red color, while the β form gives purple color. Additionally, α form can transform into β form under a higher temperature beyond 68 °C.

The geometric configuration changes of ligand are often used in titration. The fiber can give a similar color-changed performance when the ligand compound is immobilized on. Figure 14.15 exhibits a sample of color-changed textile.

14.4.5 Thermoregulated Fiber

Thermoregulated fiber is microstructure-designed fiber with the function to store heat and adjust temperature. The thermal behaviors are controlled by phase change, which can store or release plenty of energy. In detail, the material will absorb energy and store by means of phase change to prevent the temperature from increasing; on the contrary, it inverses phase change causing heat release to prevent the temperature from decreasing. The often used phase change material were polyethylene glycol (PEG) and wax.

14.5 Applications

With the further research of the smart fibers, multifunctional, and high intelligent fibers have been developed, and its applications have been extended to all walks of life.

14.5.1 Smart Fibers Used in Textiles

14.5.1.1 Smart Clothing
Smart fibers have the properties such as flexibility, spinnability, and knittability, as well as the perception and responsiveness to the external environment, and

can be easily fabricated to different kinds of fabrics and apparels with special functions. Thermoregulated fibers with reversible temperature adjustable ability can be used in different temperature conditions. These textiles can adjust the temperature without changing the conditions or loadings to protect the body. For example, the thermoregulated fibers or fiber-based materials can be used in space suits and gloves as developed products to protect the astronauts from the hard conditions [108]. And with the dramatically development, they will gradually appears in our common garments, such as sportswear, shoes, some special kinds of bedding, and accessories [108, 109].

In recently years, photochromic and thermochromic fibers have also aroused great interests among people, for they can change their colors under different radiations or temperatures. These outstanding performances have been applied in clothing and decorative textiles, such as cap, jacket, skirt, bedspread, chimney, curtain cloth, etc. [110].

14.5.1.2 Wearable Electronics

Recently, wearable electronics have attracted lots of interests due to the potential applications on human physical signal monitoring, human robot interface, and wearable energy collector.

Artificial electronic skin consisted of fiber-based pressure sensor is the representative branch of wearable electronics. By mimicking the natural human skins, artificial e-skins can realize the real-time monitoring for the human physical signals, which can be the judgment indexes of health condition. For example, the resistance sensor prepared by patterned nanofibrous membrane could collect human pulse with high accuracy [63]. In addition, the human physiological signals, such as blood pressure, heartbeat, breathing, etc., can also be detected by wearable electronics [14, 111].

The wearable energy collector can capture and store energy from the movement behavior, which realize the effective use of energy. Wang Zhong Lin group [112] reported a self-charging power system with a fabric-hybridized system. The system could harvest the solar energy from the natural source and gather the mechanical energy from the human motion. This system realized the energy storage and conversion in wearable device. Besides, his group also reported a whole textile-based energy harvesting system that could convert low-frequency human motion energy into high-frequency current outputs [113] (see in Figure 14.16).

Besides, wearable electronics with special optical performance also show prior advantages than traditional flat devices. Peng Huisheng group has taken much effort to promote the fiber-shaped electroluminescence and electrochromic devices into practical uses.

14.5.2 Smart Fibers Used in Industrial

14.5.2.1 Medical Supplies

The smart fibers also show great potential in applications of medical care. Among them, the most common applications are controllable drug delivery. Figure 14.17 shows application for physical signal diagnose with smart textile. Smart materials used for drug delivery have been studied since the last couple of

Figure 14.16 (a) Scheme of a self-charging power textile. (SC, supercapacitor yarns; TENG, triboelectric nanogenerator cloth). Source: Pu et al. 2016 [114]. Reproduced with permission of John Wiley & Sons. A self-charging power textile woven with F-TENGs, F-DSSCs, and F-SCs under outdoor (b), indoor (c), and movement (d) conditions. Source: Wen et al. 2016 [112]. Reproduced with permission of AAAS. (e) Scheme of a self-charging power system, among them being a fiber-based dye-sensitized solar cell (F-DSSC) and a fiber-based triboelectric nanogenerator (F-TENG) act as energy harvesting devices and a fiber-shaped supercapacitor (F-SC) act as an energy-storing device.

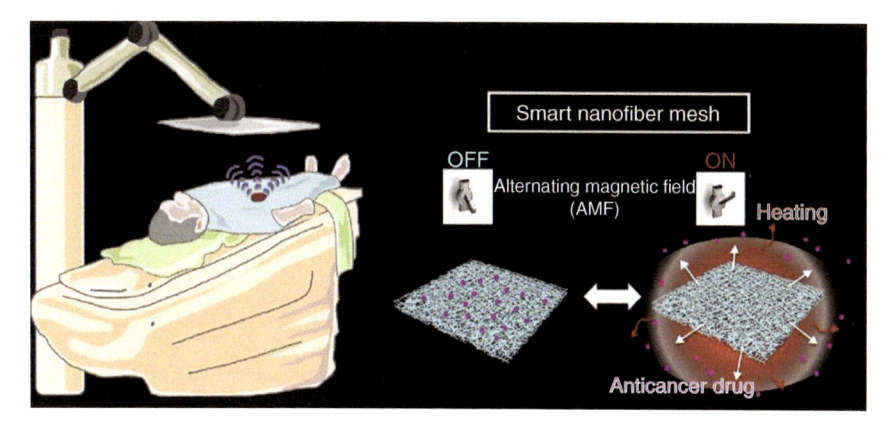

Figure 14.17 A scheme of a smart hyperthermia nanofiber system that utilizes magnetic hyperthermia and temperature-responsive polymers.

decades and substantive progress had been made in recent years especially on the automated drug delivery systems with automaticity and targeted [68, 115]. For example, the thermo-responsive hydrogel of PNIPAM and its copolymers are widely designed as automatic switch in temperature-controllable drug delivery systems. Furthermore, the biologically available phase change temperature of PNIPAM about 32 °C is the fundamental insurance of this application [5, 115]. Except this, in the field of engineering and cell engineering, smart fibers have been further developed and succeeded in using as scaffolds with required dimensions and sizes [116–118]. For example, the nonwoven biodegradable nanofiber scaffolds fabricated by electrospinning were used in bone tissue engineering. The nanofiber-based scaffolds with highly porous structure and large surface area can promote the tissue ingrowth [118, 119]. Besides artificial muscles, smart bandages also arouse great interests in medical field. Gu et al. [26] reported an artificial muscle fibers that had the function to imitate natural muscles. The smart bandage integrated with a variety of sensors could detect bacteria, humidity, and oxygen concentration that can better cure diseases [120].

14.5.2.2 Sensors

Sensors have played a significant role in promoting the automatic production that tremendously improves the industrial production efficiency. Smart fibers are good candidate to fabricate sensors especially some microsensors, since they have small size and volume that can be easily encapsulate into different kinds of devices [110]. Figure 14.18 shows the nanofiber-based pressure sensor used as large area pressure-sensitive device and human pulse detection.

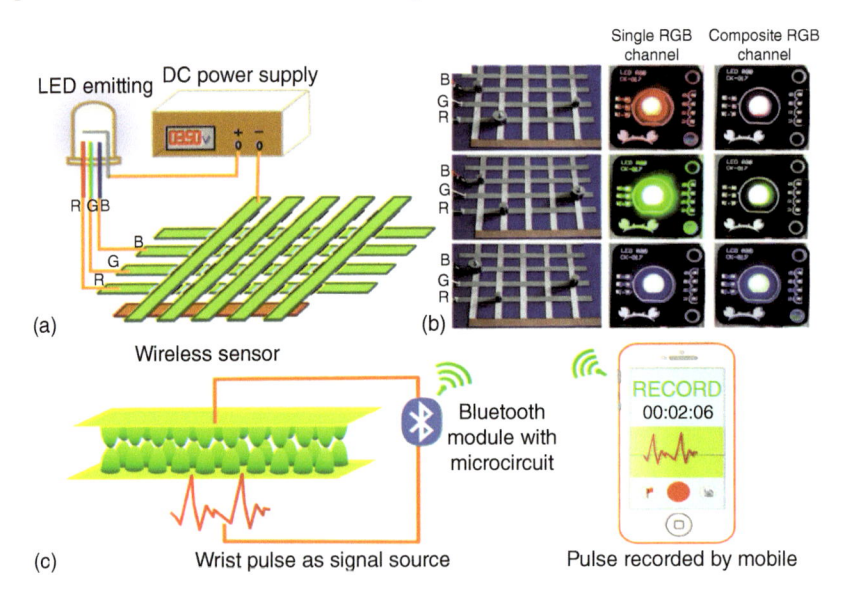

Figure 14.18 (a) A schematic illustration of the fiber-based stress sensor network with a connecting RGB-LED module. (b) When three different weights (20, 10, and 5 g) stand the distinguished stress were placed on the pixels, different color lights (red, ~2.2 V; green, ~3.0 V; and blue, ~3.1 V) will be lighted up (from individual and composite channels). (c) A schematic illustration of the wireless sensor with a Bluetooth module. The weak signals can be converted into wireless electromagnetic waves and simultaneously recorded by a smartphone.

Liu et al. prepared a nanofiber-based sensor based on conductive nickel-deposited PET membrane intertwined with PVA-*co*-PE nanofibers for alcohol concentration detection and healthcare-related industries [121]. In this sensor, the alcohol concentration detection property could be reflected by capacitance values, which had great sensitivity and efficiency.

In addition, smart fibers have also spread to the humidity sensing field. High sensitive humidity sensor based on single SnO_2 nanowire [122], oxidation of graphene (GO) [12], and NW films [123, 124] have been reported.

14.5.2.3 Energy Conversion

Energy conversion is a very important part in smart materials. Recently, with the further research of the fibrous materials, more and more fiber-based energy conversion devices are developed.

Fiber-based piezoelectric materials especially the nanogenerator is widely studied for converting low-frequency vibration and biomechanical energy into electrical energy. For example, Chen et al. [125] reported a nanogenerator using lead zirconate titanate (PZT) nanofiber for energy harvesting. Besides, Yang et al. [126] used a piezoelectric fine wire and fabricated a flexible generator with cyclic stretching–releasing properties. This generator could create an oscillating output voltage up to approximately 50 mV under a strain of 0.05–0.1%. Li and coworkers [127] reported a single-wire generator by ZnO nanowire, which can convert the biomechanical energy to electric.

CNTs and graphene oxide are popular in scientific since their emergence. They are always the good candidates to make various energy conversion devices for their excellent conductivity, moisture absorption, photothermal hearting effect, etc. [59]. For example, the twist GO hydrogel fibers act as a generator under humidity condition can reverse the kinetic energy into electrical energy [12] (Figure 14.19).

14.5.3 Other Uses

Smart fibers can also be used in the defense industry, semiconductor/electronic industry, nuclear power plant, automobile industry, sports equipment, etc. At present, the research of smart fibers has not yet reached a wide range of applications. The true sense of intelligence still needs unremitting efforts of scientific research workers.

14.6 Future Trends

Although the research of smart fiber has made great progress, there are still some challenges that restrict its further development. To realize the real intelligence, there is still a long way to go:

(1) *Sensitivity*: Sensitivity is the most important part of the smart fiber. At present, most smart fibers have the stimuli-responsive property, but the sensitivity still needs to be improved. For example, in order to improve the response speed, the temperature-triggered actuator made by hydrogel should be made into very small form, which greatly restricted its application in large-scale equipment. Besides, some wearable pressure sensing devices

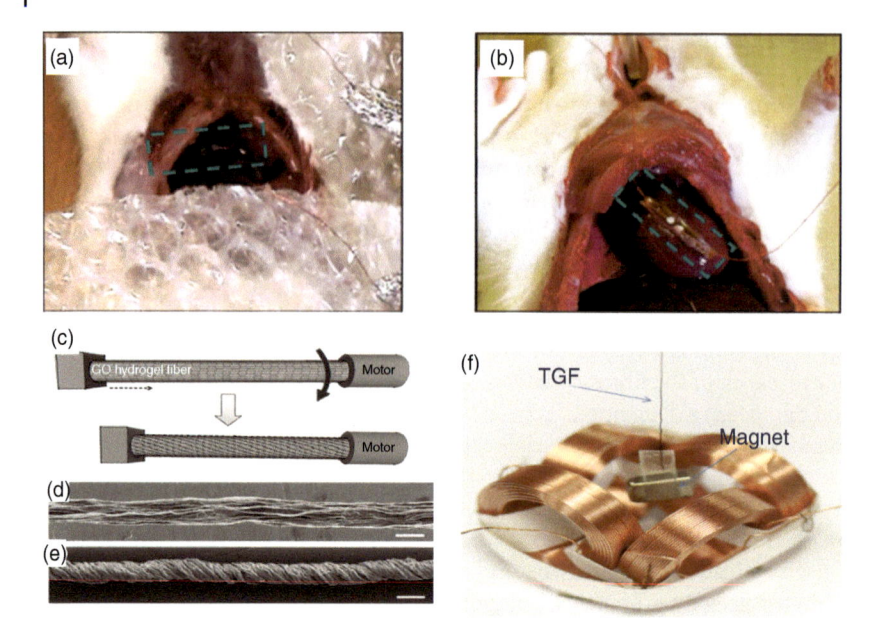

Figure 14.19 Energy harvesting from the breath (a) and heartbeat (b) of a live rat using a single-wire generator (SWG). Source: Li et al. 2010 [8]. Reproduced with permission of John Wiley & Sons. (c) Scheme of the fabrication of a twisted GO fiber; SEM images of a nontwist-dried GO fiber (d) and a twist one (e) with an applied 5000 turns/m. (f) A twisted GO fiber (TGF) used as an humidity alternating current generator. Source: Cheng et al. 2014 [12]. Reproduced with permission of John Wiley & Sons.

can detect very weak signal from human body only if they reach a high sensitivity. Therefore, continued molecular analogs and mechanics modeling are needed to optimize the molecular structure and improve the smart fiber sensitivity.

(2) *Multiple responsive*: Multiple responsive smart fibers mean to respond to multiple stimuli at the same time. At present, most smart fibers are unilateral intelligence. For fabricating multiple-stimulus-active smart fibers or fiber-based smart materials, we should combine the different stimuli-responsive structures into one single fiber. While in many cases, the incompatibility of the structure performance and processing method will make the smart process hard to continue further. For example, a sort of raw materials cannot withstand high temperature treatment or incompatible with the main solvent that causes the process hard to continue. Therefore, the material composition, size, processing method, shape design, macroscopic, or microcosmic configuration also should be optimized to obtain the optimum performance.

(3) *Stability*: Stability is one of the most important factors; it must be considered for practical application. Compared with the common fibers, smart fibers have a more complex and difficult production process. A bad stability will affect the accuracy, which finally limits the applications.

(4) *Large-scale production*: Large-scale production is the biggest challenge of fibrous intelligent materials for the future development. To achieve large-scale production, the fiber must possess the following characteristics: firstly, the technological process should be simple and easy to operate. Secondly, there are no toxic and harmful substances or gas releasing during the whole progress. Thirdly, low production cost and mass production ability should be achieved. Lastly, products should be safe and stable. The existing smart fibers have a relatively complicated manufacturing process, and the conversion cost is high. Therefore, the further exploration should be carried to optimize technique process. Besides, single fibrous energy storage devices can only transfer or store small amounts of energy. In order to improve the energy conversion rate, the fibers should be further processed into fibrous energy device. This progress still has many difficulties that need to be solved.

Last, smart fibers play an increasingly important role in our daily life. Therefore, the research and development of smart fiber and materials should be valued and make contributes to the textile industry and other fields.

References

1 Hu, J., Meng, H., Li, G. et al. (2012). A review of stimuli-responsive polymers for smart textile applications. *Smart Materials and Structures* 21 (5): 053001.
2 Pelipenko, J., Kocbek, P., and Kristl, J. (2015). Critical attributes of nanofibers: preparation, drug loading, and tissue regeneration. *International Journal of Pharmaceutics* 484 (1): 57–74.
3 Schwarz, A., Van Langenhove, L., Guermonprez, P. et al. (2010). A roadmap on smart textiles. *Textile Progress* 42 (2): 99–180.
4 Castano, L.M. and Flatau, A.B. (2014). Smart fabric sensors and e-textile technologies: a review. *Smart Materials and Structures* 23 (5): 053001.
5 Jiang, S., Liu, F., Lerch, A. et al. (2015). Unusual and superfast temperature-triggered actuators. *Advanced Materials* 27 (33): 4865–4870.
6 Yang, B., Huang, W., Li, C. et al. (2006). Effects of moisture on the thermomechanical properties of a polyurethane shape memory polymer. *Polymer* 47 (4): 1348–1356.
7 Zhang, M., Tao, J., Wang, J. et al. (2015). Single $BaTiO_3$ nanowires-polymer fiber based nanogenerator. *Nano Energy* 11: 510–517.
8 Li, Z., Zhu, G., Yang, R. et al. (2010). Muscle-driven in vivo nanogenerator. *Advanced Materials* 22 (23): 2534–2537.
9 Feil, H., Bae, Y., Feijen, J. et al. (1992). Mutual influence of pH and temperature on the swelling of ionizable and thermosensitive hydrogels. *Macromolecules* 25 (20): 5528–5530.
10 Lee, E., Kim, D., Kim, H. et al. (2015). Photothermally driven fast responding photo-actuators fabricated with comb-type hydrogels and magnetite nanoparticles. *Scientific Reports* 5: 15124.

11 Sendoh, M., Ishiyama, K., and Arai, K.I. (2003). Fabrication of magnetic actuator for use in a capsule endoscope. *IEEE Transactions on Magnetics* 39 (5): 3232–3234.
12 Cheng, H., Hu, Y., Zhao, F. et al. (2014). Moisture-activated torsional graphene-fiber motor. *Advanced Materials* 26 (18): 2909–2913.
13 Mu, J. et al. (2015). A multi-responsive water-driven actuator with instant and powerful performance for versatile applications. *Scientific Reports* 5: 9503.
14 Wu, X. et al. (2016). Highly sensitive, stretchable, and wash-durable strain sensor based on ultrathin conductive layer@polyurethane yarn for tiny motion monitoring. *ACS Applied Materials & Interfaces* 8 (15): 9936–9945.
15 Wang, Z. et al. (2007). Voltage generation from individual $BaTiO_3$ nanowires under periodic tensile mechanical load. *Nano Letters* 7 (10): 2966–2969.
16 Karlsson, J.O. and Gatenholm, P. (1999). Cellulose fibre-supported pH-sensitive hydrogels. *Polymer* 40 (2): 379–387.
17 Casper, C.L. et al. (2005). Functionalizing electrospun fibers with biologically relevant macromolecules. *Biomacromolecules* 6 (4): 1998–2007.
18 Lendlein, A. and Langer, R. (2002). Biodegradable, elastic shape-memory polymers for potential biomedical applications. *Science* 296 (5573): 1673–1676.
19 Koncherry, V., Potluri, P., and Fernando, A. (2016). Multifunctional carbon fibre tapes for automotive composites. *Applied Composite Materials* 24 (2): 477–493.
20 Descloux, A., Amitonova, L.V., and Pinkse, P.W. (2016). Aberrations of the point spread function of a multimode fiber due to partial mode excitation. *Optics Express* 24 (16): 18501–18512.
21 Wang, D., Sun, G., and Chiou, B.S. (2007). A high-throughput, controllable, and environmentally benign fabrication process of thermoplastic nanofibers. *Macromolecular Materials and Engineering* 292 (4): 407–414.
22 Li, M. et al. (2015). Highly hydrophilic and anti-fouling cellulose thin film composite membrane based on the hierarchical poly(vinyl alcohol-co-ethylene) nanofiber substrate. *Cellulose* 22 (4): 2717–2727.
23 Zhang, Z. et al. (2015). Flexible electroluminescent fiber fabricated from coaxially wound carbon nanotube sheets. *Journal of Materials Chemistry C* 3 (22): 5621–5624.
24 Di, J. et al. (2016). Carbon-nanotube fibers for wearable devices and smart textiles. *Advanced Materials* 28 (47): 10529–10538.
25 Daghash, S.M. and Ozbulut, O.E. (2016). Characterization of superelastic shape memory alloy fiber-reinforced polymer composites under tensile cyclic loading. *Materials and Design* 111: 504–512.
26 Gu, B.K. et al. (2009). A linear actuation of polymeric nanofibrous bundle for artificial muscles. *Chemistry of Materials* 21 (3): 511–515.
27 Chen, P. et al. (2015). Hierarchically arranged helical fibre actuators driven by solvents and vapours. *Nature Nanotechnology* 10 (12): 1077–1083.
28 Zhang, Z. et al. (2015). A colour-tunable, weavable fibre-shaped polymer light-emitting electrochemical cell. *Nature Photonics* 9 (4): 233–238.

29 Mandia, D.J. et al. (2015). The effect of ALD-grown Al_2O_3 on the refractive index sensitivity of CVD gold-coated optical fiber sensors. *Nanotechnology* 26 (43): 434002.

30 Kim, H. et al. (2016). Temperature-responsive tensile actuator based on multi-walled carbon nanotube yarn. *Nano-Micro Letters* 8 (3): 254–259.

31 Gu, X. et al. (2016). Hydro-actuation of hybrid carbon nanotube yarn muscles. *Nanoscale* 8 (41): 17881–17886.

32 Li, B. et al. (2008). Simultaneous growth of SiC nanowires, SiC nanotubes, and SiC/SiO_2 core–shell nanocables. *Journal of Alloys and Compounds* 462 (1–2): 446–451.

33 Wang, Y. et al. (2016). Polypyrrole/poly(vinyl alcohol-co-ethylene) nanofiber composites on polyethylene terephthalate substrate as flexible electric heating elements. *Composites Part A Applied Science and Manufacturing* 81: 234–242.

34 Guo, H. et al. (2014). Ultrafast and reversible thermochromism of a conjugated polymer material based on the assembly of peptide amphiphiles. *Chemical Science* 5 (11): 4189–4195.

35 AIAA (2009). Fiber reinforced shape-memory polymer composite and its application in deployable hinge in space. *Smart Materials and Structures* 18 (2): 1282–1294.

36 Hu, M. et al. (2016). Sensitive and fast optical HCl gas sensor using a nanoporous fiber membrane consisting of poly(lactic acid) doped with tetraphenylporphyrin. *Microchimica Acta* 183 (5): 1713–1720.

37 Lee, S.H. et al. (2016). Biothermal sensing of a torsional artificial muscle. *Nanoscale* 8 (6): 3248–3253.

38 Zhang, Z. et al. (2013). Preparation of polypyrrole composite solid-phase microextraction fiber coatings by sol-gel technique for the trace analysis of polar biological volatile organic compounds. *Analyst* 138 (4): 1156–1166.

39 Sridhar, R. et al. (2015). Electrosprayed nanoparticles and electrospun nanofibers based on natural materials: applications in tissue regeneration, drug delivery and pharmaceuticals. *Chemical Society Reviews* 44 (3): 790–814.

40 Daghia, F. et al. (2008). Shape memory alloy hybrid composite plates for shape and stiffness control. *Journal of Intelligent Material Systems and Structures* 19 (5): 609–619.

41 An, S. et al. (2016). Self-junctioned copper nanofiber transparent flexible conducting film via electrospinning and electroplating. *Advanced Materials* 28 (33): 7149–7154.

42 Sohn, J.W. et al. (2009). Vibration and position tracking control of a flexible beam using SMA wire actuators. *Journal of Vibration and Control* 15 (2): 263–281.

43 Lusiola, T. et al. (2015). Ferroelectric KNNT fibers by thermoplastic extrusion process: microstructure and electromechanical characterization. *Actuators* 4 (2): 99–113.

44 Bortolani, F. et al. (2014). High strain in $(K,Na)NbO_3$-based lead-free piezoelectric fibers. *Chemistry of Materials* 26 (12): 3838–3848.

45 Tallury, S.S. et al. (2016). Physical microfabrication of shape-memory polymer systems via bicomponent fiber spinning. *Macromolecular Rapid Communications* 37 (22): 1837–1843.
46 Nunes-Pereira, J. et al. (2013). Energy harvesting performance of piezoelectric electrospun polymer fibers and polymer/ceramic composites. *Sensors and Actuators A: Physical* 196 (7): 55–62.
47 Niu, W. et al. (2016). Photoresponse enhancement of Cu_2O solar cell with sulfur-doped ZnO buffer layer to mediate the interfacial band alignment. *Solar Energy Materials and Solar Cells* 144: 717–723.
48 Hong, S. et al. (2015). Highly stretchable and transparent metal nanowire heater for wearable electronics applications. *Advanced Materials* 27 (32): 4744–4751.
49 Choi, E. et al. (2017). Monotonic and hysteretic pullout behavior of superelastic SMA fibers with different anchorages. *Composites Part B Engineering* 108: 232–242.
50 Li, D.Y. et al. (2016). Superelasticity of Cu–Ni–Al shape-memory fibers prepared by melt extraction technique. *Journal of Minerals, Metallurgy and Materials* 23 (8): 928–933.
51 Wang, X. et al. (2007). Direct-current nanogenerator driven by ultrasonic waves. *Science* 316 (5821): 102–105.
52 Sazio, P.J. et al. (2006). Microstructured optical fibers as high-pressure microfluidic reactors. *Science* 311 (5767): 1583–1586.
53 Lagonigro, L. et al. (2010). Low loss silicon fibers for photonics applications. *Applied Physics Letters* 96 (4): 041105.
54 Zhang, W. et al. (2011). Synergistic effect of chemo-photothermal therapy using PEGylated graphene oxide. *Biomaterials* 32 (33): 8555–8561.
55 Cao, X. et al. (2013). Ambient fabrication of large-area graphene films via a synchronous reduction and assembly strategy. *Advanced Materials* 25 (21): 2957–2962.
56 Dong, Z. et al. (2012). Facile fabrication of light, flexible and multifunctional graphene fibers. *Advanced Materials* 24 (14): 1856–1861.
57 Chen, J. et al. (2010). Preparation and evaluation of graphene-coated solid-phase microextraction fiber. *Analytica Chimica Acta* 678 (1): 44–49.
58 Zhang, Q. et al. (2009). Hierarchical composites of carbon nanotubes on carbon fiber: influence of growth condition on fiber tensile properties. *Composites Science and Technology* 69 (5): 594–601.
59 Koziol, K. et al. (2007). High-performance carbon nanotube fiber. *Science* 318 (5858): 1892–1895.
60 Saheb, D.N. and Jog, J.P. (1999). Natural fiber polymer composites: a review. *Advances in Polymer Technology* 18 (4): 351–363.
61 Huang, Z.M. et al. (2003). A review on polymer nanofibers by electrospinning and their applications in nanocomposites. *Composites Science and Technology* 63 (15): 2223–2253.
62 Li, R. et al. (2008). Polyamide 11/poly(vinylidene fluoride) blends as novel flexible materials for capacitors. *Macromolecular Rapid Communications* 29 (17): 1449–1454.

63 Zhong, W. et al. (2016). A nanofiber based artificial electronic skin with high pressure sensitivity and 3D conformability. *Nanoscale* 8 (24): 12105–12112.
64 El-Hadi, A.M. et al. (2014). Enhancing the crystallization and orientation of electrospinning poly(lactic acid) (PLLA) by combining with additives. *Journal of Polymer Research* 21 (12): 1–12.
65 Matabola, K.P. and Moutloali, R.M. (2013). The influence of electrospinning parameters on the morphology and diameter of poly(vinylidene fluoride) nanofibers- effect of sodium chloride. *Journal of Materials Science* 48 (16): 5475–5482.
66 Zhu, J. and Sun, G. (2014). Fabrication and evaluation of nanofibrous membranes with photo-induced chemical and biological decontamination functions. *RSC Advances* 4 (92): 50858–50865.
67 Xin, Y. et al. (2015). Full-fiber piezoelectric sensor by straight PVDF/nanoclay nanofibers. *Materials Letters* 164: 136–139.
68 Caldoreramoore, M.E., Liechty, W.B., and Peppas, N.A. (2011). Responsive theranostic systems: integration of diagnostic imaging agents and responsive controlled release drug delivery carriers. *Accounts of Chemical Research* 44 (10): 1061–1070.
69 Reina, A. et al. (2009). Large area, few-layer graphene films on arbitrary substrates by chemical vapor deposition. *Nano Letters* 9 (8): 655–663.
70 Yang, X. et al. (2009). Anti-oxidation behavior of chemical vapor reaction SiC coatings on different carbon materials at high temperatures. *The Chinese Journal of Nonferrous Metal* 19 (5): 1044–1050.
71 Sonnenfeld, C. et al. (2011). Microstructured optical fiber sensors embedded in a laminate composite for smart material applications. *Sensors (Basel)* 11 (3): 2566–2579.
72 Bhuvana, T. et al. (2010). Contiguous petal-like carbon nanosheet outgrowths from graphite fibers by plasma CVD. *ACS Applied Materials & Interfaces* 2 (3): 644–648.
73 Muller, D. et al. (2015). Synthesis of conductive PPy/SiO_2 aerogels nanocomposites by in situ polymerization of pyrrole. *Journal of Nanomaterials* 2015: 1–6.
74 Kim, S.H. et al. (2016). Bio-inspired, moisture-powered hybrid carbon nanotube yarn muscles. *Scientific Reports* 6: 23016.
75 Lee, J. et al. (2016). Carbon nanotube yarn-based glucose sensing artificial muscle. *Small* 12 (15): 2085–2091.
76 Shi, Q. et al. (2016). A smart core-sheath nanofiber that captures and releases red blood cells from the blood. *Nanoscale* 8 (4): 2022–2029.
77 Bhattacharjee, Y. et al. (2016). Construction of a carbon fiber based layer-by-layer (LbL) assembly – a smart approach towards effective EMI shielding. *RSC Advances* 6 (113): 112614–112619.
78 Wang, Z. et al. (2016). Polyurethane/cotton/carbon nanotubes core-spun yarn as high reliability stretchable strain sensor for human motion detection. *ACS Applied Materials & Interfaces* 8 (37): 24837–24843.
79 Huang, C.-T. et al. (2008). A wearable yarn-based piezo-resistive sensor. *Sensors and Actuators A: Physical* 141 (2): 396–403.

80 Duval, Y. et al. (1992). Correlation between ultraviolet-induced refractive index change and photoluminescence in Ge-doped fiber. *Applied Physics Letters* 61 (25): 2955.

81 Shim, B.S. et al. (2008). Smart electronic yarns and wearable fabrics for human biomonitoring made by carbon nanotube coating with polyelectrolytes. *Nano Letters* 8 (12): 4151–4157.

82 O'Connor, B. et al. (2007). Fiber shaped light emitting device. *Advanced Materials* 19 (22): 3897–3900.

83 Peng, H. et al. (2009). Electrochromatic carbon nanotube|[sol]|polydiacetylene nanocomposite fibres. *Nature Nanotechnology* 4 (11): 738–741.

84 Huang, G. et al. (2016). Smart color-changing textile with high contrast based on a single-sided conductive fabric. *Journal of Materials Chemistry C* 4 (32): 7589–7594.

85 Davoudi, Z.M. et al. (2014). Hybrid antibacterial fabrics with extremely high aspect ratio Ag/AgTCNQ nanowires. *Advanced Functional Materials* 24 (8): 1047–1053.

86 Tajitsu, Y. (2016). Smart piezoelectric fabric and its application to control of humanoid robot. *Ferroelectrics* 499 (1): 36–46.

87 Li, J.J., Zhou, Y.N., and Luo, Z.H. (2015). Smart fiber membrane for pH-induced oil/water separation. *ACS Applied Materials & Interfaces* 7 (35): 19643–19650.

88 Si, Y. et al. (2016). Ultralight biomass-derived carbonaceous nanofibrous aerogels with superelasticity and high pressure-sensitivity. *Advanced Materials* 28 (43): 9512–9518.

89 Wagermaier, W. et al. (2010). Shape-memory polymers: characterization methods for shape-memory polymers. *Advances in Polymer Science* 226 (1): 97–145.

90 Sun, L. et al. (2012). Stimulus-responsive shape memory materials: a review. *Materials and Design* 33 (1): 577–640.

91 Huang, W.M. et al. (2010). Shape memory materials. *Materials Today* 13 (7): 54–61.

92 Wei, Z.G., Sandstrom, R., and Miyazaki, S. (1998). Shape memory materials and hybrid composites for smart systems: Part II Shape-memory hybrid composites. *Journal of Materials Science* 33 (15): 3763–3783.

93 Janke, L. et al. (2005). Applications of shape memory alloys in civil engineering structures—overview, limits and new ideas. *Materials and Structures* 38 (5): 578–592.

94 Fan, K. et al. (2011). Water-responsive shape memory hybrid: design concept and demonstration. *Express Polymer Letters* 5: 409–416.

95 Wei, Z., Sandstroröm, R., and Miyazaki, S. (1998). Shape-memory materials and hybrid composites for smart systems: Part I Shape-memory materials. *Journal of Materials Science* 33 (15): 3743–3762.

96 Lv, H. et al. (2008). Shape-memory polymer in response to solution. *Advanced Engineering Materials* 10 (6): 592–595.

97 Bauer, S. et al. (2014). 25th anniversary article: a soft future: from robots and sensor skin to energy harvesters. *Advanced Materials* 26 (1): 149–162.

98 Wang, C. et al. (2016). A solution-processed high-temperature, flexible, thin-film actuator. *Advanced Materials* 28 (39): 8618–8624.

99 Meng, H. and Hu, J. (2010). A brief review of stimulus-active polymers responsive to thermal, light, magnetic, electric, and water/solvent stimuli. *Journal of Intelligent Material Systems and Structures* 21 (9): 859–885.

100 Hoffman, A.S. (2000). Bioconjugates of intelligent polymers and recognition proteins for use in diagnostics and affinity separations. *Clinical Chemistry* 46 (9): 1478–1486.

101 Jocic, D. (2008). Smart textile materials by surface modification with biopolymeric systems. *Research Journal of Textile and Apparel* 12 (2): 58–65.

102 Mahdavinia, G. et al. (2004). Modified chitosan 4. Superabsorbent hydrogels from poly(acrylic acid-co-acrylamide) grafted chitosan with salt-and pH-responsiveness properties. *European Polymer Journal* 40 (7): 1399–1407.

103 Monk, P.M.S., R.J. Mortimer, and D.R. Rosseinsky (2007). *Electrochromism: Fundamentals and Applications*. John Wiley & Sons.

104 Irie, M. (2000). Photochromism: memories and switches-introduction. *Chemical Reviews* 100 (5): 1683–1684.

105 Hadjoudis, E. and Mavridis, I.M. (2005). Photochromism and thermochromism of Schiff bases in the solid state: structural aspects. *ChemInform* 33 (14): 579–588.

106 Zhao, X.Y., Wang, M.Z., and Liu, H.J. (2008). Structure dependence of photochromism of azobenzene-functionalized polythiophene derivatives. *Journal of Applied Polymer Science* 108 (2): 863–869.

107 Chance, R.R. et al. (1977). Thermochromism in a polydiacetylene crystal. *Journal of Chemical Physics* 67 (8): 3616–3618.

108 Mondal, S. (2008). Phase change materials for smart textiles – an overview. *Applied Thermal Engineering* 28 (11): 1536–1550.

109 Zhang, X. (2001). Heat-storage and thermo-regulated textiles and clothing. *Smart Fibres, Fabrics and Clothing* 2001: 34–57.

110 Shen, X. and Shen, Y. (2001). Present situation and development of intelligent fiber. *China Synthetic Fiber Industry* 24 (1): 1–5.

111 He, Z., Zhang, X., and Batchelor, W. (2016). Cellulose nanofibre aerogel filter with tuneable pore structure for oil/water separation and recovery. *RSC Advances* 6 (26): 21435–21438.

112 Wen, Z. et al. (2016). Self-powered textile for wearable electronics by hybridizing fiber-shaped nanogenerators, solar cells, and supercapacitors. *Science Advances* 2 (10): e1600097.

113 Pu, X. et al. (2016). Wearable power-textiles by integrating fabric triboelectric nanogenerators and fiber-shaped dye-sensitized solar cells. *Advanced Energy Materials* 6 (20): 1601048.

114 Pu, X. et al. (2016). Wearable self-charging power textile based on flexible yarn supercapacitors and fabric nanogenerators. *Advanced Materials* 28 (1): 98–105.

115 Stoychev, G., Puretskiy, N., and Ionov, L. (2011). Self-folding all-polymer thermoresponsive microcapsules. *Soft Matter* 7 (7): 3277–3279.

116 Xu, C. et al. (2004). In vitro study of human vascular endothelial cell function on materials with various surface roughness. *Journal of Biomedical Materials Research Part A* 71 (1): 154–161.
117 Yang, F. et al. (2005). Electrospinning of nano/micro scale poly(L-lactic acid) aligned fibers and their potential in neural tissue engineering. *Biomaterials* 26 (15): 2603–2610.
118 Yoshimoto, H. et al. (2003). A biodegradable nanofiber scaffold by electrospinning and its potential for bone tissue engineering. *Biomaterials* 24 (12): 2077–2082.
119 Hu, Y. et al. (2002). Fabrication of poly(α-hydroxy acid) foam scaffolds using multiple solvent systems. *Journal of Biomedical Materials Research* 59 (3): 563–572.
120 Hua, Z. (2007). The preparation of intelligent fiber and its application in biomedicine domanial. *New Chemical Materials* 9: 015.
121 Liu, X. et al. (2014). Surface tailoring of nanoparticles via mixed-charge monolayers and their biomedical applications. *Small* 10 (21): 4230–4242.
122 Kuang, Q. et al. (2007). High-sensitivity humidity sensor based on a single SnO_2 nanowire. *Journal of the American Chemical Society* 129 (19): 6070–6071.
123 Zhang, Y. et al. (2005). Zinc oxide nanorod and nanowire for humidity sensor. *Applied Surface Science* 242 (1–2): 212–217.
124 Wu, R.J. et al. (2006). Composite of TiO_2 nanowires and Nafion as humidity sensor material. *Sensors and Actuators B: Chemical* 115 (1): 198–204.
125 Chen, X. et al. (2010). 1.6 V nanogenerator for mechanical energy harvesting using PZT nanofibers. *Nano Letters* 10 (6): 2133–2137.
126 Wang, Z.L. (2007). The new field of nanopiezotronics. *Materials Today* 10 (5): 20–28.
127 Yang, R. et al. (2009). Converting biomechanical energy into electricity by a muscle-movement-driven nanogenerator. *Nano Letters* 9 (3): 1201–1205.

15

Optical Fibers

Hiroaki Ishizawa

Shinshu University, Institute for Fiber Engineering, Interdisciplinary Cluster for Cutting Edge Research, Ueda, 386-8567, Japan

15.1 Introduction

When we survey some database by using the keyword "optical material," we could find out a huge number of research articles. From the electrical and electronic engineering view point, over 50 thousands of them could be found nowadays [1]. Furthermore, in the case of "fibers" and "optical material," we could find thousands of them as well. Therefore, optical system based on fibers is now one of the most active research areas. For example, there have been published over 200 papers in the transactions of *IEEE*, which comprises over 170 journals, during this year. They are composed mainly by photonics, light wave technologies, electronics, and sensors. Their percentages are approximately 40%, 25%, 20%, and 10%, respectively. In turn, it is well known that optical fibers have been applied to the telecommunication technology and its related technologies. As the fiber-optic telecommunication industry has matured, the costs of its related devices have decreased. Based on these backgrounds, the new application technologies appeared for sensing. The initial application was the fiber-optic gyroscope, which are used on a various vehicles, airplanes, cars, and so on. After that, technologies have been developed for fiber Bragg grating (FBG) sensing method. It could detect multipoint strain along with the fiber length continuously. Therefore, FBG has been applied to remote sensing of the soundness of infrastructures [2]. Recently, it is assumed that technologies for FBG sensors have also matured in its fabrication and application, resulting in very wide diversity. For example, single-mode polymer optical fiber (POF) has been proposed for FBG-based materials end up to obtaining high sensitivity and high reflectivity sensors [3]. Multimode polymer waveguides are also an attractive technology for onboard optical interconnects, as they can be cost-effectively integrated onto standard printed circuit boards (PCBs) [4]. In the electronic field, the ultrasonic welding of POFs onto carbon fiber-reinforced polymers for advanced fiber-optic sensing is demonstrated [5]. Regarding with the optical fiber application to textiles, it is conceivable that sensing system on textile such as the "smart textiles" has been developed during these years [6].

Handbook of Fibrous Materials, First Edition. Edited by Jinlian Hu, Bipin Kumar, and Jing Lu.
© 2020 Wiley-VCH Verlag GmbH & Co. KGaA. Published 2020 by Wiley-VCH Verlag GmbH & Co. KGaA.

In this chapter, optical fiber sensing technologies as the textile material would be introduced. At first, this would glance at the manufacturing process of fiber optics. And secondly, fiber-optic sensing applications would be overviewed. Temperature measurement, oxygen concentration sensing, strain measurement, and biomedical measurement would be introduced mainly on these 20 years. And this chapter focuses the structures and properties of FBG sensor. After that sensing principles in order to detect human vital signs and their application would be introduced before having a view of the future based on the healthcare textile circumstances.

15.2 Fundamentals of Fiber Optics

15.2.1 Fiber Optics Classification

Basic principles of optical fibers such as guided ray, guided waves, attenuation, and dispersion could be reviewed in many references [7]. Furthermore, the engineering of fiber optics has been reviewed in outstanding books [8]. This section would glance at fiber optics from the material viewpoint. It could be categorized basically into two groups by the materials, such as silica glass and plastics.

Table 15.1 shows this classification and their typical properties. Today, glass optical fibers are well known of their types and properties, since they have been applied to the telecommunication technology and its related technologies [8]. However, they are so fragile that glass fiber devices are not easy to fabricate and end up to being quite expensive. Polymeric materials permit the mass production of low-cost high-port-count photonic circuits in parallel on a planar substrate [9]. Commercially available polymer optical fibers have been emerged now as well as the active and effective manufacturing process developments. Furthermore, plastic fiber-optic applications to telecommunication or data communication are now developed worldwide.

15.2.2 Manufacturing Process

Figure 15.1 illustrates the manufacturing process of silica fiber optics. The preform rod making is the first process. In this process mainly three kinds

Table 15.1 Classification of fiber optics.

Materials	Type/index distribution	Dimensions (µm) core/cladding	Properties (band µm)
Silica glass	Multimode/step	50/125	Hard/brittle/stable
	Multimode/graded		Mode dispersion
	Single mode	8/125	Low loss (1.55)
Plastics	Core: PMMA Cladding: fluoride acrylic	Core: 500	High loss (0.6) Flexible/stable

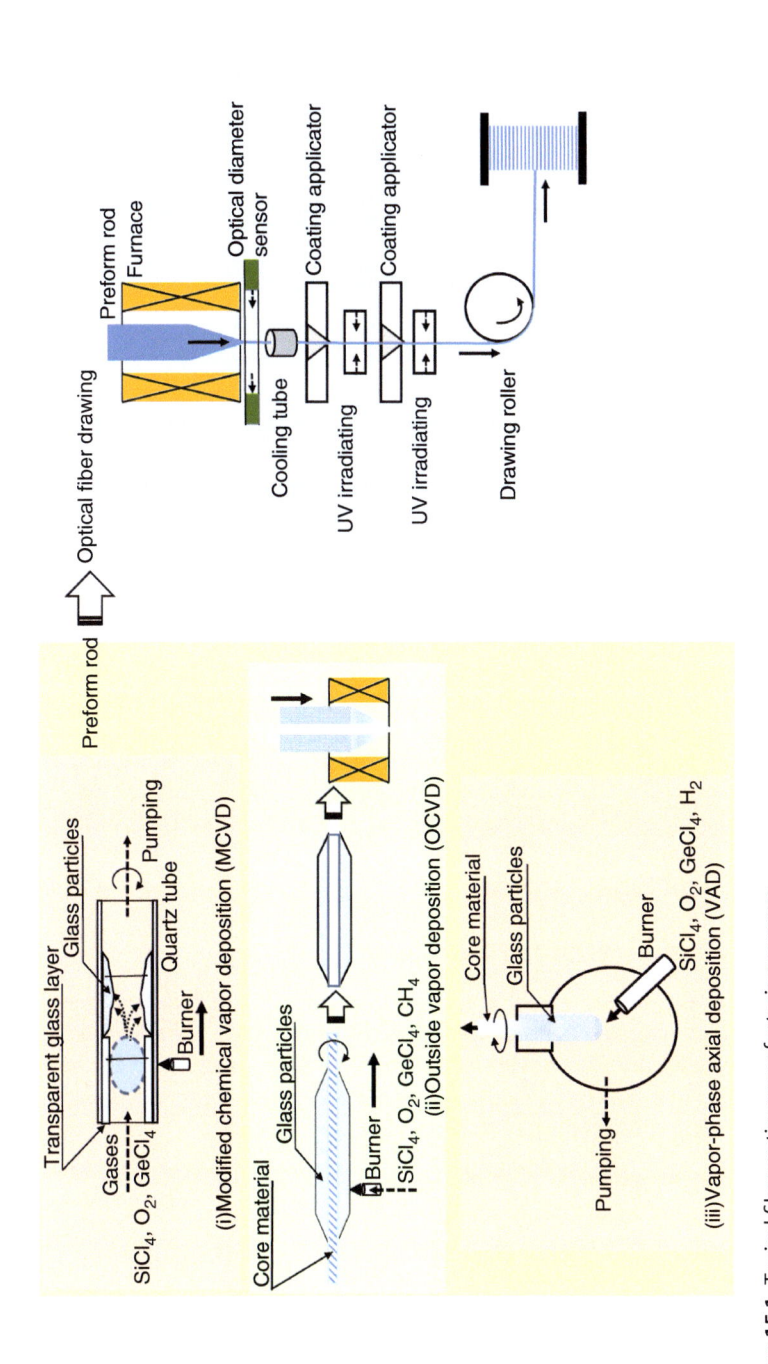

Figure 15.1 Typical fiber optics manufacturing process.

of deposition process have been developed and applied in those industries. Modified vapor deposition method (MCVD) [10], outside vapor deposition method (OVD) [11], or vapor-phase axial deposition method (VAD) [12] has been developed to make preform of glass optical fiber as in Figure 15.1. The second process is the drawing process as in right part of Figure 15.1. The preform rod is inserted into a furnace and heated over 2100 °C. This process is controlled by the diameter of fiber (125 μm), the cleanliness of the atmosphere (100 class), temperature, and rotation speed of the roller. The drawing speed depends on the conditions above mentioned, which is around 10–100 km/h.

15.2.3 General Characteristics of Fiber Optics

The compositions of silica glass remained chemically stable for a long time. Therefore, silica glass optical fiber has the same stability as the glass basically. Highly pure silica is characterized by extremely low optical loss. In most commercially available silica fibers, the optical intensity loss is about 0.2 dB/km. The intrinsic strength of glass is about 1.5×10^5 kg/cm^2, so that a typical glass fiber optics of diameter 125 μm can support a load of 18 kg. Broken off probability of a typical silica glass fiber optics is low enough that it keeps the reliability at 1% strain. It is well known that silica glass optical fiber has such robustness against electromagnetic field, since its dielectric constant is low.

POF are fibers made from polymethyl methacrylate (PMMA), polystyrene, polycarbonates, and fluorinated polymers. Optical losses in the fibers are higher than silica glass, with 100–500 dB/km. The losses are due to such factors as Rayleigh scattering and intrinsic absorption of the material itself and of the impurities [8]. POFs are generally environmental stable, and some of them are thermally stable [9].

15.3 Optical Fiber Sensor

15.3.1 Advantages of Optical Fiber Sensors

Fiber-optic sensors have shown a growth since the 1970s. Recently, they have been studied so actively, because they have various merits, which could never expected by using the electrical sensors. These have the outstanding properties of silica glass. Furthermore, their application is now on the practical phase. Generally, fiber-optic sensors have such advantages as follows;

(1) Compact lightweight
(2) No need to electrical energy supply
(3) Robust against electromagnetic noises
(4) Signal transmittable for length of kilometers
(5) Stable and high environmental durability

This section overviews the properties of fiber sensors.

15.3.2 Principles and Applications of Optical Fiber Sensors

15.3.2.1 Temperature Measurement

Fiber-optic fluorescence thermometer has been proposed by Aizawa et al. [13, 14]. Temperature dependence of emission spectra from both Tb : SiO_2 and Tb : YAG sensor head was studied for the fiber-optic thermometer application. They proposed fiber-optic thermometer equipment using Tb-doped phosphor sensor head connected with silica fiber. They found out that Tb : SiO_2 glass is suitable sensor head for an intensity-based fiber-optic thermometer at the temperature range from 300 to 1200 K. Until now, the design and development of a plastic optical fiber temperature sensor is recently proposed by Wildner and Drummer [15]. In their study, the design and development of POF temperature sensor is presented. By combining two different thermo-optic coefficients a transparent oil or polymer and glass particles, a temperature-dependent transmission can be achieved. FBG has been also applied to the temperature sensing by using metal coating FBG [16]. By the continuous chemical electroplating process of nickel layer, the temperature sensitivity of FBG sensor could be enhanced. It could be used as a temperature sensor from 0 to 300 °C.

15.3.2.2 Oxygen Concentration Measurement

Oxygen measurement could be propose by using fiber-optic fluorosensor [17]. Toba and coworkers fabricated luminous probe by using solvent green 5 as the fluorescence material. The construction of the measurement system was similar with one of temperature measurements by Aizawa et al. [13]. Fluorescent light (514 nm) could be excited by the light of which wavelength is 468 nm. Fluorescent light intensity decreases according to the concentration of oxygen. They found out that it could measure oxygen concentration from 5% to 50% in vapor and from 1 to 10 ppm in liquid. Furthermore, they also found out that a linear relation exists in vitro between oxygen concentration and optical intensity in human blood, and the fluorosensor probe could be used for in vivo measurement of oxygen partial pressure in blood.

15.3.2.3 Strain Measurement by Brillouin Optical Time Domain Reflectometry (B-OTDR) and Fiber Bragg Grating Sensors (FBG)

Table 15.2 compares Brillouin optical time domain reflectometry (B-OTDR) and FBG sensor as the strain sensor. In the case of a strain measurement by B-OTDR method, it is possible to use a commercial-based optical fibers without any additional processing to the fibers. Furthermore, it could detect the strain distribution through a long distance by detecting the Brillouin scattering in the optical fiber core. Therefore, B-OTDR has been applied to measure and monitor the strains of the large infrastructures such as railroads, bridges, and so on [18]. However, the spatial resolution of B-OTDR is several tens of centimeters long. And it is difficult to detect the spatial dynamic measurement because it is time-consuming method. As a result, it is not adequate to apply B-OTDR method to detect the human-caused strain, such as pulse wave, or chest movement.

On the other hand, FBG sensors have the grating in its core along with long axis direction as was shown in Table 15.2. This grating functions as the strain sensor.

Table 15.2 Comparison of FBG and B-OTDR as a strain sensor.

	FBG	B-OTDR
Spatial resolution	<10 mm	~1 m
Sampling period	100 ns–100 ms	~Minutes
Sensor element	Grating	Optical fiber
Advantage	Real-time measurement High wavelength precision	Low cost
Demerits	Finite measuring points	Low spatial resolution
Application	Dynamic or static analysis	Wide area measurement

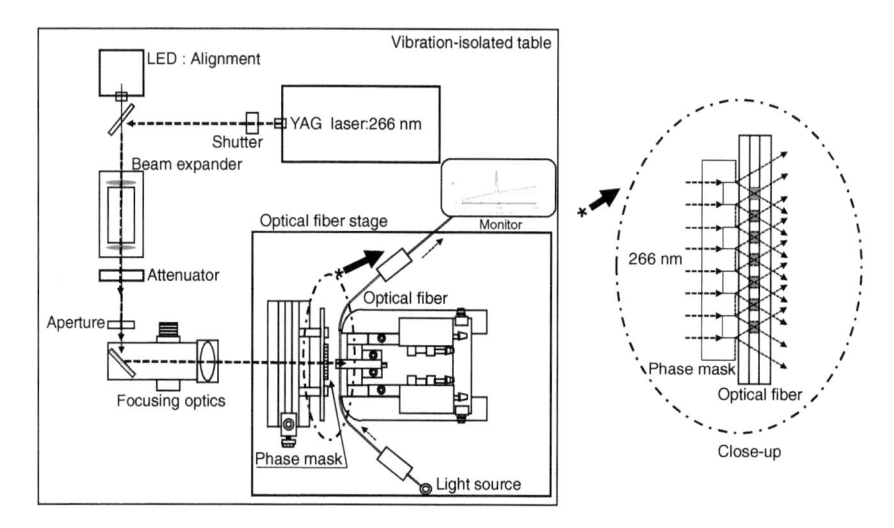

Figure 15.2 FBG making apparatus by using YAG 4th harmonic (266 nm).

As a result, FBG has gotten a very small spatial resolution as small as 10 mm or less [19].

Figure 15.2 introduces the outlook of FBG making apparatus based on phase-mask method equipped at author's laboratory in Institute for Fiber Engineering, Interdisciplinary Cluster for Cutting Edge Research, Shinshu University. Laser light (Spectra-Physics, Model HIPPO prime 266-2, 2W) is used as the drawing light source for grating formation in the fiber core. The wavelength is 266 nm that is the fourth harmonic of YAG laser. Optical fiber stage is made of low thermal expansion cast iron. It also has a set of monitoring optics to confirm the constant distance between optical fiber and phase mask. It also checks the power of laser beam by a laser power meter and monitors the grating formation process by a spectrum analyzer. This figure shows the close-up of phase mask and optical fiber. Diffraction pattern from the phase mask makes the periodic variation of optical index in the core of the optical fiber. This part works as the grating of the FBG sensor. The Bragg wavelength can be selected by

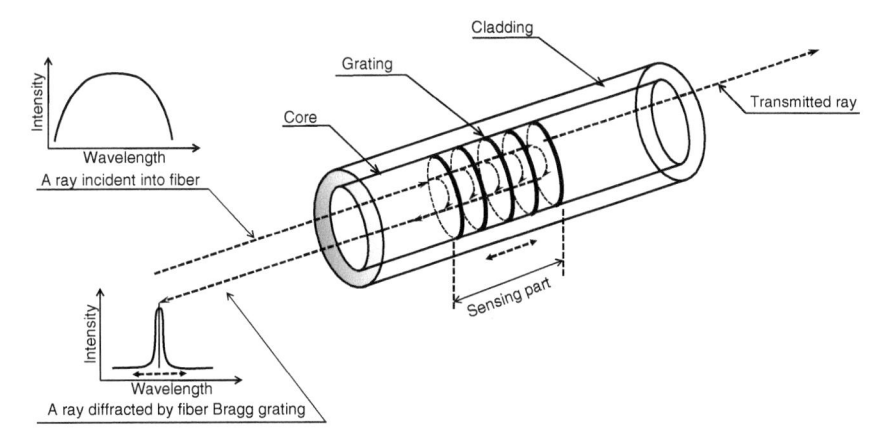

Figure 15.3 Geometry and Bragg wave diffraction of FBG sensor.

setting the optimal phase mask. It takes several minutes to make one grating in a single-mode silica optical fiber.

Figure 15.3 shows the construction and the strain detection principle of a typical FBG sensor. FBG is a sort of optical fiber that reflects the specified light wave. It is called Bragg wavelength, and it shifts according to the grating period of FBG. If some pressure changes the grating period, or as the result some strain changes, the reflected Bragg wavelength of the FBG sensor on the sample surface changes. In other words, the Bragg wavelength shift is linear to the strain caused by the pressure. It is the strain measurement principle of the FBG sensor. In order to detect the wavelength shift, the rapid optical wavelength interrogator has been used [20, 21].

15.3.2.4 Biomedical Measurement

The characteristics of optical fibers abovementioned in this section are very adequate for the biomedical sensing applications. The size of typical optical fibers makes it possible to insert it into the hypodermic needles. They could offer the minimum invasive operations, or the position-sensitive monitoring.

Physical measurement such as temperature, oxygen concentration, pressure, and so on could be possible even in vivo by applying technologies in this section. Regarding with the vital sign detection, Toba et al. proposed a new vital sign by using fiber optics [22]. They had shown that pulse waves could be detected base on the reflectometry through optical fiber bundle as in Figure 15.4. And the heart rates as well as the respiratory rates could be measured simultaneously at high accuracy. It would be one of the earliest works on this issue.

Optical fiber itself has been applied to detect the respiratory monitoring by Krehel et al. [6]. Highly flexible POF reacts to applied pressure. Throughput light intensity of POFs changes in accordance with the pressure between the human torso and the wearable sensor-embedded textiles. Continuous respiratory monitoring could be possible within ±3/min. Its ranges from 5 to 50 min^{-1}, which covers human respiratory rate in daily life [6].

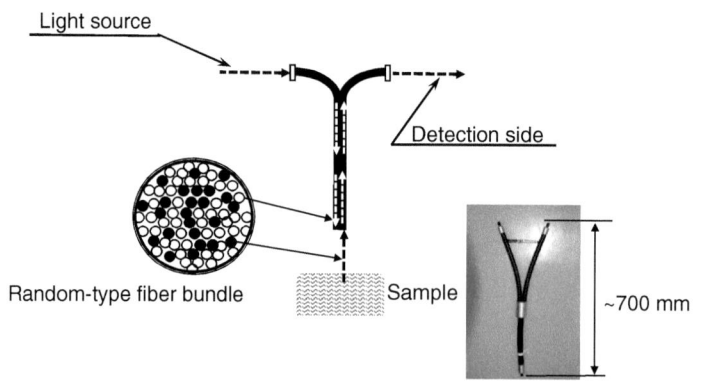

Figure 15.4 An arrangement for vital sign measurement by using fiber-optics.

15.3.3 Principles and Application of the Vital Sign Measurement by FBG Sensor

This section describes the principles of vital sign measurement by FBG sensor. FBG sensor is a fiber whose diameter is submillimeter and is flexible. These characteristics are suitable ones in order to make "a sensor on textiles" for realizing a healthcare circumstances.

15.3.3.1 FBG Sensor Interrogator to Measure Human Vital Signs

In the case of application of FBG sensor to detect human vital signs, high sensitive interrogator has to be used, because the target human pulse wave is very weak. The resultant strain is estimated as small as submicrometer. It is known that typical strain of pulse wave at wrist is sometimes less than 0.1 µm. Therefore, Mach–Zehnder interferometry is suitable [15, 23].

Table 15.3 lists the specifications of the typical interrogator (PF25-S01: Nagano Keiki, Inc.). Wideband light of which wavelength range is from 1525 to 1575 nm

Table 15.3 Specification of FBG sensor system of Mach–Zehnder interrogator.

	Type	Amplified spontaneous emission
Light source	Power	30 mW
	Wavelength range	1525–1575 nm
	Grating length	10 mm
FBG sensor	Bragg wavelength	1550 ± 0.5 nm
	Wavelength resolution	0.1 pm
	Strain resolution	0.08 µm
	Material	Quartz glass
Optical fiber	Cladding diameter	145 µm
	Core diameter	10.5 µm
Detector	Type	InGaAs PIN
	Wavelength range	900–1650 nm

from amplified spontaneous emission (ASE) is connected with the optical fiber. Through the optical circulator, light reaches to the FBG sensor and to reflecting the light of Bragg wavelength. The reflection light is interfered by a Mach–Zehnder interferometer. The optical path difference of the interferometer is set at 3.3 mm. Beam splitter makes three phase of which phase shifts were every $2\pi/3$ radian. Three phases were detected by the wavelength division multiplexing. The wavelength shift is calculated by Eq. (15.1):

$$\Delta\phi = (2\pi n d/\lambda^2)\Delta\lambda \tag{15.1}$$

where $\Delta\phi$ is the phase shift obtained, n is the optical constant of the fiber, and d is the grating period that is 500 nm in this setup. The phase resolution depends on the sampling frequency that is 10 kHz. As a result, the wavelength resolution of this measuring system is 0.1 pm as in Table 15.3, which corresponds to equivalent the resolution of strain, which is 0.08 μm [20].

15.3.3.2 Pulse Wave Measurement

In order to measure the pulse waves, FBG sensor was attached on subject's skin by a medical adhesive tape.

Figure 15.5 shows the look of typical pulse wave measurement. Subject is in supine position, and his wrist was kept as high as his heart. The radial artery exists beneath the measuring point. Sampling rate of the system setting was 10 kHz. Measurement was carried out for 20 seconds that is as long as the one cycle time of the automatic sphygmomanometer to measure the subject's blood pressure as in Figure 15.5.

Each raw pulse wave was filtered by a band-pass filter, of which passband is 0.5–5 Hz. This bandwidth covers the ordinal human heart rate. Figure 15.6 shows the typical raw signals and corresponding pulse waves filtered by band-pass filter. It is clear that FBG-detected waveforms seem to show the heartbeats.

By the comparison of typical acceleration of pulse waveform obtained by photo-plethysmograph and the waveforms measured by FBG sensor, FBG-sensed waveform resembles the acceleration of pulse waveform closely. Therefore, it is conceivable that filtered FBG-detected waves (Figure 15.6 lower) are ones of pulses of heartbeats.

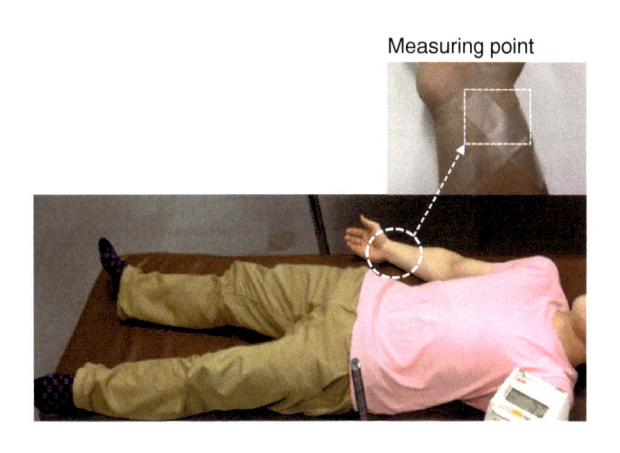

Figure 15.5 Pulse wave measurement at subject's wrist on the radial artery.

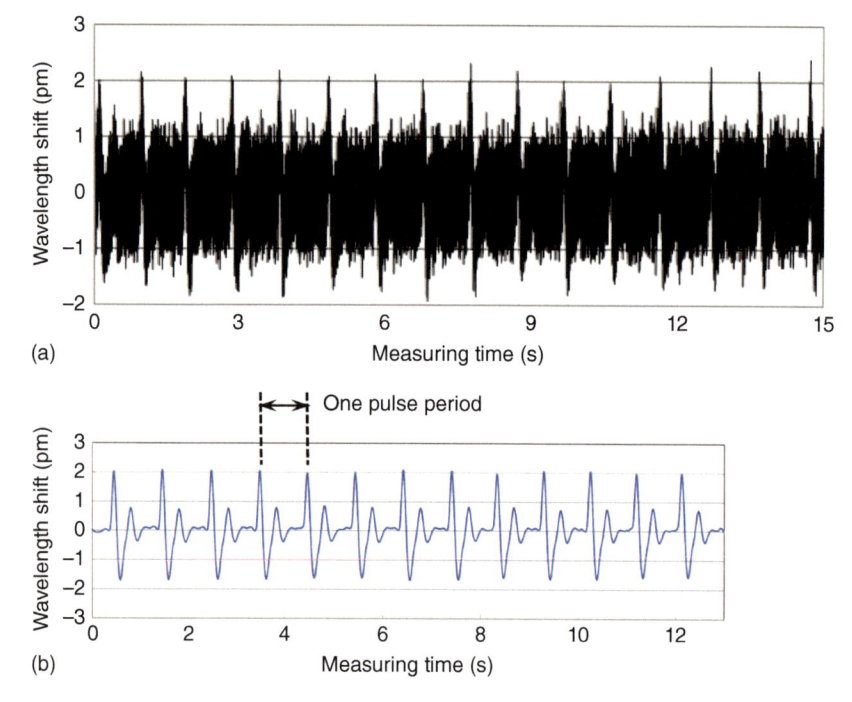

Figure 15.6 Detected raw signals of FBG (upper) and filtered waves. (a) Typical raw signal measured by FBG. (b) Corresponding waves processed by the band pass filter.

15.3.3.3 Pulse Rate Measurement

Base on the consideration above mentioned in Section 15.3.3.2, after obtaining the filtered pulse waves, mean pulse period during 20 seconds could be estimated [23]. Pulse rate per minute (beat per minute, bpm) at this measurement timing (20 seconds) could be easily obtained. Miyauchi et al. showed the typical correlation of reference values and predicted heart rates by using three subjects [23]. The correlation coefficients for subject A, B, and C are 0.83, 0.67, and 0.60, respectively. The corresponding standard errors are 3, 2, and 2 bpm, respectively [23]. Additionally, total correlation coefficient and standard error over these three subjects are 0.89 and 2 bpm. Based on these results, this method is proved to be significant ($p < 0.01$). This method takes about eight seconds to detect [23].

15.3.3.4 Respiratory Rate Measurement

Respiratory pulse (respiratory arrhythmia) is observed normally for human [24]. In breathing in, the pulse rate increases, and conversely when we breathe out, it decreases. Therefore, by monitoring the pulse waves and detecting the pulse periods simultaneously, the real-time respiratory could be easily measured.

Miyauchi et al. showed the relationship between the pulse period fluctuations and corresponding temperature fluctuations of subject's expiration breathing measured by a face mask equipped with a temperature sensor [23]. The pulse period fluctuation is synchronous with the respiration temperature fluctuation. Therefore, respiratory rate could be measured by detecting the frequency of pulse period fluctuation [23]. They reported a typical correlation of reference values and predicted respiratory rates [23]. The correlation coefficients for subject A, B, and C are reported to be 0.99, 0.99, and 0.99, respectively. The corresponding standard errors are 0.6, 0.4, and 0.4 bpm, respectively. Additionally, overall correlation coefficient and standard error of three subjects are 0.99 and 0.5 bpm. Based on these results, this method is proved to be significant ($p < 0.01$) as well.

Miyauchi et al. also reported that the signal to noise ratio in these measurement can be estimated to be 26 dB based on the wavelength shift and its resolution. The conventional pulse measurement systems have relative error of ±5%, which corresponds ±3 bpm at 60 bpm [23].

Therefore, by attaching FBG sensor on the human skin, it was found out that pulse rate can measured accurately. In the case of respiratory measurement, it is proved to be very accurate without any individual differences. As a consequence, FBG sensor could be expected to monitor heart rates and respiratory rates simultaneously.

15.3.3.5 Blood Pressure Measurement by Pulse Transit Time Detection

The arrival time lag of pulse wave between the different places of skin is called pulse transit time (PTT). PTT could be obtained by calculating the measuring points distance to pulse wave velocity (PWV) ratio. PWV is the velocity of the pulse waves from a point to the other point through the artery. It has been expressed by Eq. (15.2) by Moens and Kortberg since 1878, assuming that the artery wall is isotropic and experiences isovolumetric change with pulse pressure [24]:

$$\text{PWV} = \sqrt{\frac{E \cdot h}{2r \cdot \rho}} \tag{15.2}$$

E, h, r, and ρ are Young's modulus of the blood vessel, blood vessel wall thickness, radius of the blood vessel, and viscosity of blood, respectively. PWV is proportional to the square root of the Young's modulus. Therefore, it correlates with the blood pressure [25]. In this method, two gratings are set in one fiber as is seen in Figure 15.7 [26], so that this could measure the pulse wave at two points.

The Bragg wavelength of FBG1 and FBG2 is 1550 ± 0.5 nm and 1560 ± 0.5 nm. The length (L) between FBG1 and FBG2 is 250 mm (constant). As a result, time difference of the pulse wave (PTT) could be detected by means of this system as in Figure 15.7. As a result, if some relationship between PTT and blood pressure could be obtained, this method could be applicable to measure blood pressure without any machetes of the conventional methods.

Takagi et al. reported that by using PTT, blood pressure could be measured.

Figure 15.8 shows a typical relationship between PTT and systolic blood pressure in the experimental set of them [26]. This shows that PTT decreases according to the increase of systolic blood pressure. Takagi et al. reported that the correlation coefficients of the calibration curves obtained were around 0.70 and had high significances ($p < 0.01$) [26]. However, in this method, multipoint's measurement is inevitable. Subjects can never neglect the restraint feeling.

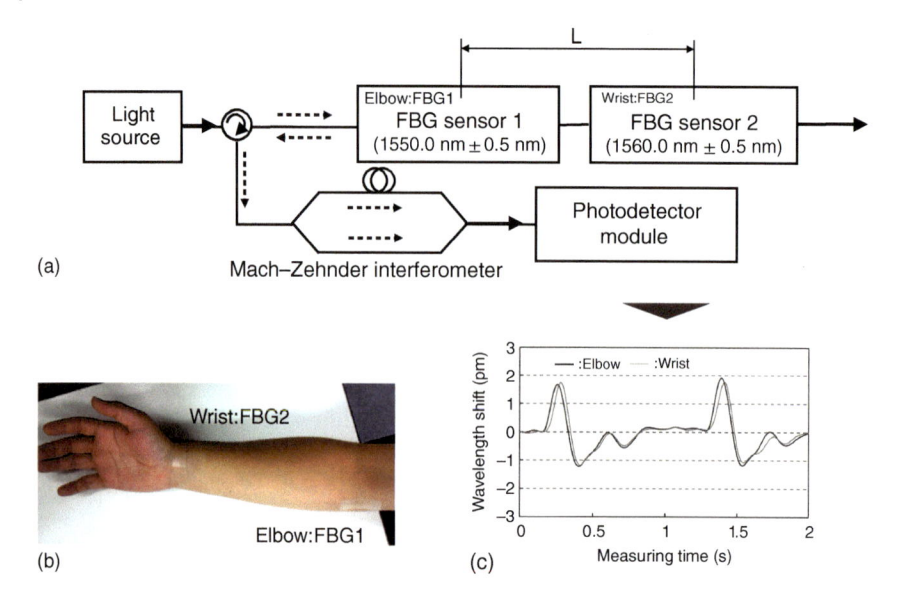

Figure 15.7 Pulse transit time measurement by FBG sensors. (a) Block diagram of PTT measurement. (b) FBG method: radial artery (L-250 mm). (c) Typical pulse waves at two points with time lag.

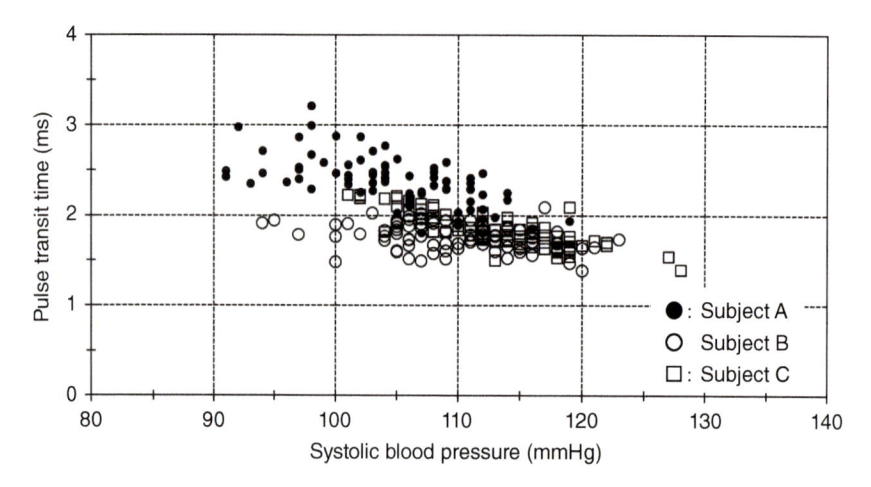

Figure 15.8 Relationship between pulse transit time and systolic blood pressure of three subject.

15.3.3.6 Partial Least Squares Regression Calibration for Blood Pressure Measurement

In order to calibrate the blood pressure level by using the pulse waves measured by FBG sensor, partial least squares regression (PLSR) could be applied to obtain the calibration curve. PLSR is a multivariate analysis that combines feature extraction by principal component analysis and multiple regression [27]. In PLSR, the elements are obtained by successive extraction of the feature values both from the objective variables and the explanatory variables. As a result, it gives feature vectors from the explanatory variables, and variances minimize the explanatory variables for explaining the objective variables. It finds for a set of latent PLS vectors. In this case, one-point measurement by FBG sensor is enough. Therefore, if PLSR is successful, FBG sensor system would be more practical than multipoint method in Section 15.3.3.5.

Pulse waves at wrist, for example, as in Figure 15.4, could be used as the explanatory variables, and the blood pressure levels measured by the automatic sphygmomanometer were used as the objective variables. In this case pulse waves were normalized, setting the first peak value one and the first valley value zero. Principle component analysis is carried out for the pulse waves, which is a vector having 10^4 elements in the case of 10 kHz sampling frequency. And the feature vector called PLS factor was extracted successively, such as factor 1, factor 2, and so on. In this process, the objective variables (blood pressure levels) were explained by a linear combination of the latent PLS factor. The residuals of this linear combination were the new set for next extraction step until the predicted residues of the objective values reach minimum [27]. In this calibration, the cross-validation of the model is carried out by the leave-one-out method. The optimal numbers of the PLS factors were tested statistically with an adequate significant level, such as 5%. In the validation of the PLSR calibration model, pulse waves that had not used in the calibration process were used to have the predicted values of blood pressure levels [28, 29].

Figure 15.9 shows the typical PLSR calibrated results of systolic blood pressure obtained by authors. They could be easily detected by a FBG sensor on human wrist. Table 15.4 shows the typical result of PLSR calibration and validation for systolic blood pressure measurement by FBG sensor at wrist. In this case, systolic blood pressure can be measured by a PLSR calibration model that has four factors. In the calibration, the mean absolute difference is within 3 mmHg. Furthermore, in the validation, it is 4 mmHg. This model has significant number of four PLS factors as in Figure 15.9b.

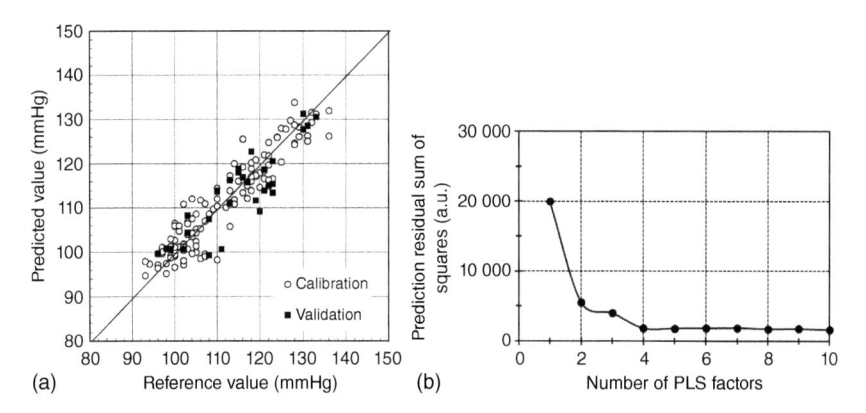

Figure 15.9 Typical PLSR model for systolic blood pressure calibration and validation. (a) PLS models for systolic blood pressure. (b) Relation between the predicted residual sum of squares and extracted PLS factors.

Table 15.4 Calibration and validation of PLS models for systolic blood pressure measurement.

Measuring point	PLS factor	Correlation coefficient	Calibration MAD (mmHg)[a]	Validation
Wrist	4	0.93	3	4

a) MAD: mean absolute difference (IEEE 1708).

Once the PLSR calibration model is obtained, it is possible to measure systolic blood pressure only by detecting pulse wave. Katsuragawa and Ishizawa presented that it can measure not only systolic blood pressure but also diastolic blood pressure [28]. They also presented that by using FBG sensor, blood pressure could be measured continuously and noninvasively. S. Koyama et al. have shown that the influence of the individual differences on the calculated blood pressure model by using FBG sensor method was very low. Therefore, this measurement method has been found to be suitable for use by many people [29].

15.4 Healthcare Monitoring by Using FBG Sensor

Multi-vital sign measurement is possible by using FBG sensor. This section introduces experimental results of vital sign measurement by using subjects. This section also views the multi-vital sign smart textiles for healthcare monitoring system.

15.4.1 FBG Sensor Application to Multi-vital Sign Measurement

Pulse rate, respiratory rate, and systolic blood pressure could be measured simultaneously by using FBG sensor. Until now, it is well known that a single function measuring system has been proposed and used for monitoring each vital sign such as pulse rate, respiratory rate, and blood pressure. Recently, POFs have applied to respiratory rate detection based on the optical intensity measurement on human torso [6]. However, FBG sensing system could measure multiple vital signs at the same time. FBG is highly flexible and reacts to the pressure waves in the blood vessel caused by the heartbeat.

On the other hand, it has been suggested by the authors that pulse wave contains the information of the blood contents such as blood glucose concentration [30, 31]. The pulse wave depends on the viscoelasticity of blood vessel as in Eq. (15.2). So, it seems to be possible to detect arteriosclerosis base on the pulse wave pattern [29].

The pulse wave intervals synchronize with the R-R intervals of electrocardiogram. Based on this discovery, pulse waves measured by FBG sensor could show the state variables of human psychological stress [32].

As mentioned in Section 15.3.2.1, FBG could be used as a temperature sensor from 0 to 300 °C [16]. It means that it also could measure body temperatures.

Therefore, FBG sensor could offer most practical healthcare monitoring smart textiles that inform users their vital signs such as body temperature, heart rate, respiratory rate, blood pressure, blood glucose levels, and/or level of psychological stress.

15.4.2 Fabrication of Smart Textiles

One of the next tasks should be introducing the FBG sensor on textiles. From the practical point of view, wearable textiles with embedded sensors are the main problem. One of the most important projects is EU FP6 project OFSETH [33]. In this project, medical textiles with embedded FBG sensors have been proposed for monitoring respiratory movement [34]. By this development, non-intrusive system for continuous measurement of abdominal and thoracic respiratory movement was proposed. In this system FBG was applied. They wrote the gratings in a silica optical fiber and stitched it on an elastic fabrics for measurement of thoracic respiratory movement. As a result, the system on textiles enables the continuous measurement of respiratory movement.

In order to embed FBG sensor on textiles, covering FBG by silk yarns has been developed by Sakaguchi et al. [35, 36] Silica optical fiber is not suitable to embed directly. They covered optical fiber by silk yarn, of which diameter was approximately 200 μm. The covering was carried out by a braiding machine. This method and covered optical fiber are illustrated in Figure 15.10. Based on this technology, FBG sensor could be easily embedded on knitted fabrics. Figure 15.11 shows the knitted fabrics embedded covered FBG sensor [35]. The covered FBG was implemented halfway round of tubular knitted fabrics. It is easily implemented by a connection of covered FBG every nine wales. Base on this breakthrough, there is some possibility of realizing wearable healthcare smart textiles [37].

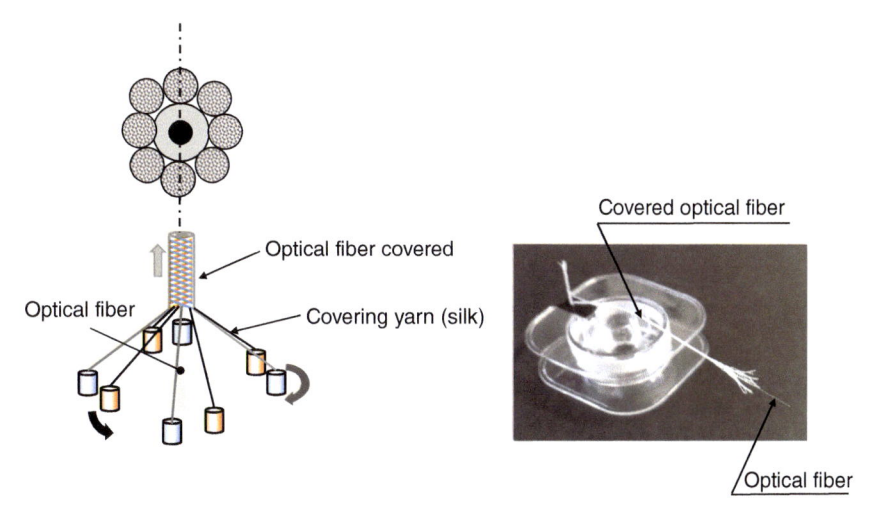

Figure 15.10 Covering method (braiding) and look of covered optical fiber(right).

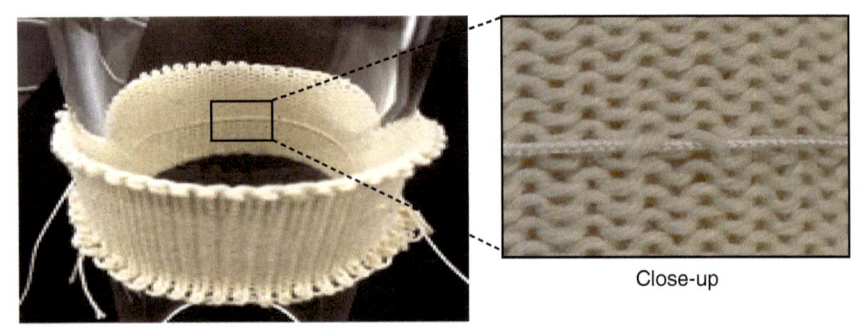

Wristband shape knitted fabric

Close-up

Figure 15.11 Covered FBG embedded knitted fabric.

15.5 Future Trends: As the Summary

Javier Andreu-Perez et al. categorized the evolution of health monitoring application of wearable system [38]. According to them, in the first generation, "a single sensing modality with wireless connectivity" had been realized. After that, "continuous monitoring with multiple sensors" has been proposed in the second generation. In the final generation until now, "combining continuous health monitoring with other sources of medical knowledge" has been studied. Healthcare monitoring devices are tending to smart implants from wearable sensors on these 10 years. FBG sensor applications to health monitoring are now in the second generation. As aforementioned, FBG can measure almost all the vital signs continuously. Therefore, FBG and its embedded smart textiles aim to combine with other sources of medical knowledge. It also should integrate the information technologies. Furthermore, FBG sensor could work as the nervous system for controlling the human states. FBG sensor could be implanted, because it does not change in quality in vivo. Also, it could be some practical wearable devices such as fabrics, caps, glasses, pillows, and beds as for monitoring and communicating the health indicators.

Figure 15.12 proposes the applications of FBG sensor-embedded smart textiles to health monitoring and the development of FBG sensor system on textiles in order to summarize this chapter. In the personal healthcare information space, smart sensors on textiles would play the most important role and meet the human needs all over the world. FBG sensor would be one of the successful solutions, because it is a flexible, easy to embed, stable, and robust material.

In this chapter, the author has focused on the newly developed applications of FBG sensor to human vital sign sensing. It is needless to say that optical fibers and/or optical fiber sensors have a great potential. Many outstanding reviews have been published as the author mentioned above [7, 8]. The author does not have enough experience or academic standing to describe the whole potential of fiber optics. Fiber-optic communication system would get the highest speed and largest capacity according to optical fiber progresses [39]. In relation to the instrumentation and measurement, POFs and/or silica glass optical fiber would play an

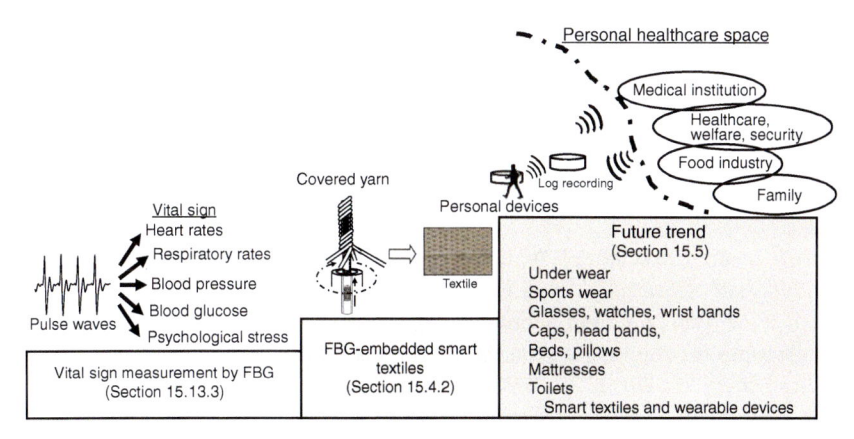

Figure 15.12 Development of smart textiles embedded FBG sensor.

important role in the process control, health monitoring of the infrastructures, and other industrial sensing systems. These should be ceded to the author of qualified somewhere. Finally, this chapter would wrap up to describing that FBG sensor and its system could curve out future of humankinds, since it could not only sense our health condition but also contribute to medical care environment.

Fiber Bragg Grating sensing (FGB) application developments to human vital sign sensing in this chapter are supported by JSPS KAKENHI Grant 16H01805 that is ongoing from 2016 to 2020.

These research results in this chapter are partially supported by the creation of a development platform for implantable/wearable medical devices by a novel physiological data integration system of the Program on Open Innovation Platform with Enterprises, Research Institute and Academia (OPERA) from the Japan Science, and Technology Agency (JST).

References

1 For example, IEEE Xplore®, Digital Library. http://ieeexplore.ieee.org/Xplore/cookiedetectresponse.jsp (accessed 16 October 2016).
2 Yau, M.H., Chan, T.H.T., Thambiratnam, D.P., and Tam, H.Y. (2013). Static vertical displacement measurement of bridges using fiber Bragg grating (FBG) sensors. *Special Issues of Advances in Structural Engineering* 16 (1): 165–176.
3 Bhowmik, K., Peng, G.D., Luo, Y. et al. (2016). Etching process related changes and effects on solid–core single-mode polymer optical fiber grating. *IEEE Photonics Journal* 8 (1): 1–9.
4 Liu, W., Romeira, B., Li, M. et al. (2016). A wavelength tunable optical buffer based on self-pulsation in an active microring resonator. *Journal of Lightwave Technology* 34 (14): 3466–3472.
5 Shimada, S., Tanaka, H., Hasebe, K. et al. (2016). Ultrasonic welding of polymer optical fibres onto composite materials. *Electronics Letters* 52 (17): 1474–1474.

6 Krehel, M., Schmid, M., Rossi, R.M. et al. (2014). An optical fibre-based sensor for respiratory monitoring. *Sensors* 2014 (14): 13088–13101.

7 For example, Kao, C. and John Russel, P.S. (2009). *Fiber Optics, Fundamentals of Photonics*, Chapter 9, 2e (eds. B.E.A. Saleh and M.C. Teich), 325–364. Wiley.

8 For example, Thyagarajan, K. and Ghatak, A. (2007). *Fiber Optic Essentials*. Wiley.

9 Eldada, L. and Shacklette, L.W. (2000). Advances in polymer integrated optics. *IEEE Journal of Selected Topics in Quantum Electronics* 6 (1): 54–67.

10 Reed, W.A., Yan, M.F., and Schnitzer, M.J. (2002). Gradient-index fiber-optic microprobes for minimally invasive in vivo low-coherence interferometry. *Optics Letters* 27 (20): 1794–1796.

11 For example, Blankenship, M. and Deneka, C. (1982). The outside vapor deposition method of fabricating optical waveguide fibers. *IEEE Journal of Quantum Electronics* 18 (10): 1418–1423.

12 For example, Takahashi, H. and Sugimoto, I. (1983). Preparation of germanate glass by vapor-phase axial deposition.

13 Aizawa, H., Takei, K., Katsumata, T. et al. (2005). Development of erbium-doped silica sensor probe for fiber-optic fluorescence thermometer. *Review of Scientific Instruments* 76: 094902.

14 Aizawa, H., Takei, K., Katsumata, T. et al. (2006). Fluorescence thermometer based on the photoluminescence intensity ratio in Tb doped phosphor materials. *Sensors and Actuators A: Physical* 126: 78–82.

15 Wildner, W. and Drummer, D. (2016). A fiber optic temperature sensor based on the combination of two materials with different thermo-optic coefficients. *IEEE Sensors Journal* 16 (3): 688–692.

16 Li, Y., Hua, Z., Yan, F., and Gang, P. (2009). Metal coating fiber Bragg grating and the temperature sensing after metallization. *Optical Fiber Technology* 15: 391–397.

17 Toba, E., Kazama, J., Tanaka, H. et al. (2000). Fiber optical fluorosensor for oxygen measurement. *IEICE Transactions on Electronics* E83-C (3): 366–370.

18 Izumita, H., Sato, T., Tateda, M., and Koyamada, Y. (1996). Brillouin OTDR employing optical frequency shifter using side-band generation technique with high-speed LN phase-modulator. *IEEE Photonics Technology Letters* 8 (12): 1674–1676.

19 Li, K., Yau, M.H., Chan, T. et al. (2013). Fiber Bragg grating strain modulation based on nonlinear string transverse force amplifier. *Optics Letters* 38: 311–313.

20 Sano, Y. and Yoshino, T. (2003). Fast optical wavelength interrogator employing arrayed waveguide grating for distributed fiber Bragg grating sensors. *Journal of Lightwave Technology* 21 (1): 132–139.

21 Todd, M.D., Johnson, G.A., and Chang, C.C. (1999). Passive, light intensity-independent interferometric method for fiber Bragg grating interrogation. *Electron Letters* 35 (22): 1970–1971.

22 Toba, E., Shimada, T., Kamoto, T. et al. (2000). Development of vital sign sensor by using fiber-optics. *Journal of Robotics and Mechatronics* 12 (3): 286–291.

23 Miyauchi, Y., Ishizawa, H., and Niimura, M. (2013). Measurement of pulse rate and respiration rate using Fiber Bragg grating sensor. *Transactions of SICE* 49 (12): 1101–1105.

24 de Boer, R.W., Karmaker, J.M., and Strackee, J. (1985). Relationship between short-term blood-pressure fluctuations and heart-rate variability in resting subjects II. *Medical and Biological Engineering and Computing* 23 (4): 359–364.

25 Scisence Pressure Technical Note. Pulse wave velocity (PWV) basics. http://www.transonic.com/resources/research/pulse-wave-velocity/. (Accessed 11 March 2017).

26 Takagi, T., Ishizawa, H., Koyama, S., and Niimura, M. (2015). Fundamental study on blood pressure measurement by FBG sensor. *Transaction of SICE* 51 (4): 274–279.

27 Martens, H. and Neas, T. (1993). *Partial Least Squares Regression (PLSR), in Multivariate Calibration*, 116–125. Wiley.

28 Katsuragawa, Y. and Ishizawa, H. (2015). Non-invasive blood pressure measurement by pulse wave analysis using FBG sensor. In: *Proceedings IEEE IIMTC 2015* (11–14 May 2015), 511–515.

29 Koyama, S., Ishizawa, H., Fujimoto, K. et al. (2017). Influence of individual differences on the calculation method for FBG-type blood pressure sensors. *Sensors* 2017 (17): 48. https://doi.org/10.3390/s17010048.

30 Ishizawa, H. and Koyama, S. (2016). Non-invasive blood glucose level measurement method and non-invasive blood glucose level measurement device. Japanese patent pending No. N15067, USA Patent Pending 15/544,677, EU Patent Pending 16 764 625.6. Chinese Patent Pending 201680006291.2, Korean Patent Pending 10-2017-7023639.

31 Kurasawa, S., Koyama, S., Ishizawa, H. et al. (2017). Verification of non-invasive blood glucose measurement method based on pulse wave signal detected by FBG sensor system. *MDPI-Sensors* 17: 2702. https://doi.org/10.3390/s17122702.

32 Koyama, S., Ishizawa, H., Hosoya, S. et al. (2017). Stress loading detection method using the FBG sensor for smart textile. *Journal of Fiber Science and Technology* 73 (11): 276–283.

33 European Commission Six framework programe 2002–2006. https://ec.europa.eu/research/fp6/index_en.cfm. (Accessed 17 March 2017).

34 Witt, J., Narbonneau, F., Schukar, M. et al. (2012). Medical textile with embedded fiber optic sensors for monitoring of respiratory movement. *IEEE Sensors Journal* 12 (1): 246–254.

35 Sakaguchi, A. and Kato, M. Japanese Patent Pending No. P151033.

36 Sakaguchi, A., Kato, M., Ishizawa, H. et al. (2016). Optical fiber embedded knitted fabrics for smart textiles. *Journal of Textile Engineering* 62 (6): 129–134.

37 Koyama, S., Sakaguchi, A., Ishizawa, H. et al. (2017). Vital sign measurement using covered FBG sensor embedded into knitted fabric for smart textile. *Journal of Fiber Science and Technology* 73 (11): 300–308.

38 Andreu-Perez, J. et al. (2015). From wearable sensors to smart implants–toward pervasive and personalized healthcare. *IEEE Transactions on Biomedical Engineering* 62 (12): 2750–2762.

39 Thyagarajan, K. and Ghatak, A. (2007). *Fiber Optic Essentials*, Chapter 8, 100–124. Wiley.

16

Memory Fibers

Harishkumar Narayana[1], Jinlian Hu[1], and Bipin Kumar[2]

[1] *The Hong Kong Polytechnic University, Institute of Textiles and Clothing, R403 Hung Hom, Kowloon, Hong Kong, China*
[2] *Indian Institute of Technology Delhi, Department of Textile and Fibre Engineering, TX135, Hauz Khas, New Delhi 110016, India*

16.1 Introduction

Fibers are fine substances with a high ratio of length to its thickness. The heritage of fibers is quite long, and they have traditionally been used in vivid cultures on Earth to meet the fundamental requirements of clothing and utilitarian products such as fishing nets and ropes. In general, fibers are obtained from natural resources, regenerated or man-made. Shape memory polymers (SMPs) are exciting class of smart materials and have been considered as an important milestone in the revolutionary path toward impactful transformation from active to smart era. SMPs have been gained a significant research interest from both academic and industries due to their fascinating behaviors leading profound applications in multidisciplinary arenas to mankind over the past few decades [1–4]. The word "shape memory" was first proposed by Vernon in the year 1941 [5]. SMPs comes under the shape memory materials (SMMs) including shape memory alloys (SMAs) and shape memory ceramics (SMCs). SMPs can undergo a significant macroscopic deformation or a temporary shape and recover spontaneously to its original conformation upon an external heat stimulus such as heat, light, solvent, water, and magnetic field [2, 6–8]. SMPs are not only able to store the shapes, but also other physical parameters could be stored and retrieved reversibly such as stress (stress memory) [9], temperature (temperature memory) [10], chrome (chromic memory) [11], and electricity (electric memory) [12]. Hence this polymeric system could also be termed as memory polymers (MPs). MPs are composed of thermodynamically immiscible fixed (hard) and reversible (soft) segments. The transition temperature of these polymers is very critical and can be tailor-made to deploy them into suitable applications. Transition is the temperature range in which a significant change of modulus and shape could occur caused by temperature change. There are two types of transition states in MPs, and they are glassy (T_g) and melting (T_m) transition type. This critical temperature range can be fine-tuned to custom needs by controlling the two phases in

Handbook of Fibrous Materials, First Edition. Edited by Jinlian Hu, Bipin Kumar, and Jing Lu.
© 2020 Wiley-VCH Verlag GmbH & Co. KGaA. Published 2020 by Wiley-VCH Verlag GmbH & Co. KGaA.

a heterogeneous structure. MPs have been developed for numerous applications such as aerospace, construction, transportation, textiles, and biomedical [2, 3].

Memory polyurethanes can be synthesized practically via bulk or solution polymerization technique. In general, MPs can be processed and utilized into several applications via different forms such as fibers, films, coating solution, gels, and foams. MPs could be spun into fibrous structure via almost all spinning methods such as melt spinning, dry spinning, wet spinning, reaction spinning, and electrospinning [13]. MPs can be easily spun into macro-, micro-, and nanofibers via aforementioned spinning methods. The mechanical and thermal properties of MP fibers can be controlled or tailor-made via controlled synthesis, chemical composition, and ratio between soft and hard segments. In addition, shape memory behavior, mechanical, thermomechanical, and thermal properties, can be further enhanced by post-spinning operations such as heat setting and drawing process. MP fibers could possess other unique behaviors apart from shape memory characteristics such as stress memory phenomenon, which is recently discovered by Hu et al. in a semicrystalline polyurethane material [9]. In this chapter, fundamentals of MPs, processing of fibers, effect of chemical composition and post-spinning operations, stress memory behavior, and their implications in medical compression and textiles are broadly discussed. The organization of contents is to let the reader comprehend the background and impactful potentials of MP fibrous materials. Fibers could be scientifically applied into multidisciplinary areas such as artificial muscles, tissue engineering, pressure or massage garments, drug-controlled release, and energy storage devices.

16.2 Morphology and Molecular Mechanism of Memory Polymers

MPs are stimulus-responsive materials that can be easily deformed above the T_{trans} (transition temperature) and fixed to a temporary shape upon cooling below the T_{trans}. This temporary shape can be spontaneously recovered back to the original conformation by reheating above the T_{trans}. This effect is generally denoted as shape memory effect (SME). The thermomechanical properties such as shape fixity and shape recovery can be easily tailored by blending MP using different types of fillers. The notable advantage of MPs is their switching, or transition temperature can be fine-tuned to a wide range from −20 to +150 °C, and they are responsive to multiple stimulus. They possess very good chemical stability, biodegradability, and biocompatibility. MPs are inherently sensitive to ambient temperature and do respond within a narrow range of temperature as shown in Figure 16.1a.

An overall 3D architecture of MPs was proposed by Hu and coworkers based on the molecular mechanisms, and it is shown in Figure 16.1b. The SME in MPs is basically induced by the conformational changes that occur in polymeric molecular chains and segmental bonds [14, 15]. Herein, Figure 16.1b shows switch and netpoint that basically represent the soft (reversible phase) and hard (fixed phase) segments. The entropic elasticity of molecular chains enables the shape recovery in the soft segment switch, whereas netpoint decides the overall structural integrity and permanent shape of MPs during the thermomechanical

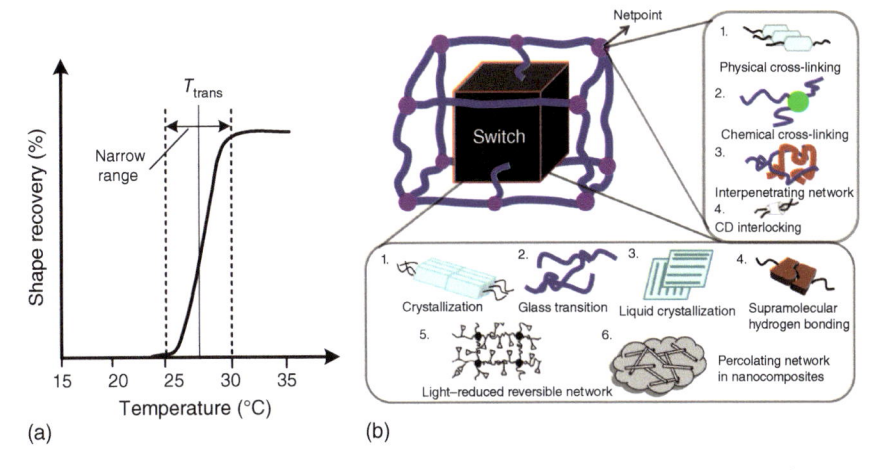

Figure 16.1 Molecular structure and mechanism of MPs. (a) Responsiveness trend of MPs in an ambient temperature; (b) the overall 3D architecture of MPs.

deformation. Netpoint or hard segments connect with molecular chains via cross-linkage such as physical, chemical, interpenetrating, or interlocked supramolecular network. The type of transition in MPs and controlling SME is decided by the soft segment switch, and it could be of amorphous, crystalline, liquid crystalline, supramolecular hydrogen bonding, light-reversible network, and percolating nanocomposite whiskers [2].

16.2.1 Nature of Transitions in MPs

As mentioned in the previous section, polyurethane-based MPs are segmental type, and the nature of the soft or reversible phase decides the transition type. A temperature range where a significant change occurs in the modulus and shape is known as transition, and this could be either a glass (T_g) or melting type (T_m) for the reversible phase. If molecular chains in the soft segment is semicrystalline type, the T_{trans} is T_m, and if it is amorphous, the T_{trans} is T_g type [15–17]. If the transition is T_m type, crystallites melt during heating with deformation and further lead to strain-induced crystallization [18] upon cooling below the T_m range. The presence of crystallites in the molecular network below the $T_{trans} = T_m$ prevents the shape recovery until it is reheated above T_m [16]. Tobushi et al. [19] and Takahashi et al. [20] mentioned that the micro-brown motions of the molecular chains will be set into glassy region and frozen when the MP is cooled below $T_{trans} = T_g$. The network will be in nonequilibrium state until it is reheated above $T_{trans} = T_g$ to activate the micro-brown motions. The transition type of MPs decides by the compositions of hard and soft segments [1, 21–30].

16.3 Evaluation of Shape Memory Properties

Before the discussion of memory fibers, the reader should know the evaluation of shape memory properties by thermomechanical cyclic testing [15, 16, 19, 31].

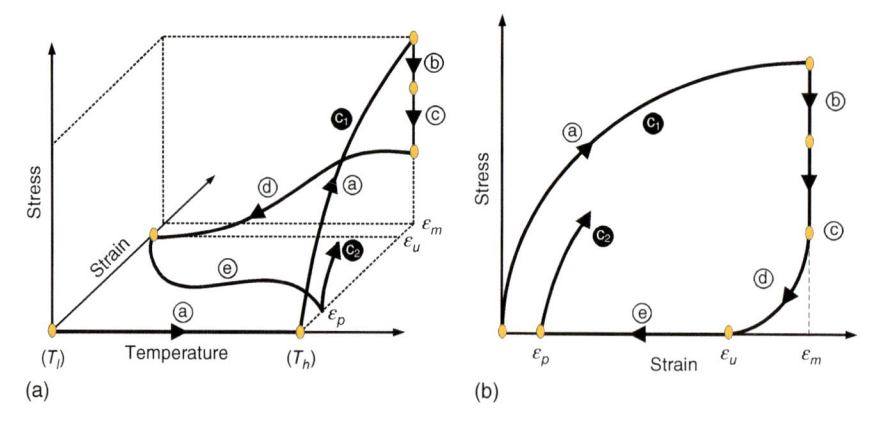

Figure 16.2 Scheme of thermomechanical cyclic test. (a) Plot of stress with strain and temperature. (b) Typical curve of stress vs strain. (T_l), low temperature ($T < T_{trans}$); (T_h), high temperature; (ε_p), plastic strain; (ε_u), fixed strain; (ε_m), programming strain; ($\varepsilon_m - \varepsilon_u$), elastic strain; $\varepsilon_r = (\varepsilon_m - \varepsilon_p)$.

So that this would assist in comprehending the further sections easily, Figure 16.2 shows the detailed steps involved in a typical thermomechanical cyclic test, and this helps to determine the shape fixity and shape recovery properties. Description for the steps are as follows:

Step a: Heating the specimen above the T_{trans} under zero strain condition and programming to a desired strain level (ε_m) under constant strain rate (e.g. 10 mm/min) at elevated temperature.
Step b: Holding the specimen under same temperature and constant strain for about 10 minutes.
Steps c and d: Cooling ($T < T_{trans}$) and unloading the specimen under a constant rate to a strain level ε_u (ε_u can be achieved due to presence of elastic strain and hinders the complete shape fixation).
Step e: Heating ($T > T_{trans}$) to achieve the shape recovery and to release the stored strain to ε_p.
Step C_2: Repeating another cycle from steps a to e.

Shape fixity and recovery ratios are calculated from Eqs. (16.1) and (16.2).

$$\% \text{ Shape fixity} = \frac{\varepsilon_u}{\varepsilon_m} \times 100 \tag{16.1}$$

$$\% \text{ Shape recovery} = \frac{\varepsilon_r}{\varepsilon_m} \times 100 \tag{16.2}$$

16.4 Memory Polymers As Fibers (MPFs)

Polyurethane-based MPs are segmented structure in which soft and hard segments are present in the same polymeric chain. Usually soft segment contains long-chain macroglycol, and hard segment possesses low molecular weight chain extenders. Segmented polyurethanes are generally synthesized by pre-polymerization techniques, in which macroglycol reacts with diisocyanate to form soft segment and chain extender reacts with diisocyanate to provide hard segment. MPs possess versatility in processing them into fibers via different spinning methods; melt spinning, dry spinning, wet spinning, reaction spinning, and electrospinning. There are several polymeric systems available for making the fibers; however polyurethane-based memory fibers are discussed in this chapter.

MPs have been investigated for few decades, and spinning them into fibers and utilizing in the textile applications were first achieved by Hu et al. [32]. Their research team have developed MPU filaments (MPFs) via different spinning methods; dry spinning, wet spinning, melt spinning, reaction spinning, and electrospinning, using polyol as soft segment and small-size polydiols/MDI as hard segment [13, 33]. For the first time, Hu and coworkers have developed memory fibers via wet spinning in 2006 and investigated their mechanical, thermal, and shape memory properties. In addition, they have compared the difference between memory fibers and other man-made fibers (Figure 16.3a). Varying the ratio of hard segment content to soft segment could change the behavior of fibers. Zhu et al. achieved complete shape recoverability in wet-spun fibers (Figure 16.3b) [25]. Investigation of dynamic analysis revealed the difference between memory fibers and other man-made fibers in terms of decrease in the elastic modulus (E') in the normal apparel using temperature range. It can be seen from Figure 16.3a that the decreased elastic modulus in memory fibers PU56-120 and PU66-120 is very much significant in the normal using temperature range. Increasing the temperature above T_g transforms glassy state into rubbery region, and this enabled to decrease the modulus. Lycra showed the decrease at −40 °C, whereas other fibers showed between 100 and 105 °C. Hence, memory fibers show the switch temperature in the normal using temperature to tune their memory behaviors, and it is advantageous to implicate for several applications.

The plastic deformation or irreversible strain is most common in the film substrates, whereas thermal shrinkage occurs in the case of fibers. Zhu et al. [25] claim that counteraction effect of two opposite factors irreversible strain and thermal shrinkage induces MP fibers to achieve complete shape recoverability as shown in Figure 16.3b.

16.4.1 Which Fibers Do Have Better Performance, Wet or Melt Spun?

Method of fiber spinning is very much crucial, which thus decides the properties of fibers for adopting them into several applications. In line with this, Meng et al. [34] have investigated a comprehensive and extensive research to reveal the relationship between phase separation, thermal, and mechanical properties and fibers spun with different spinning methods to unveil the underlying physics and property differences. Table 16.1 shows the thermal and mechanical properties of MP fibers spun by melt and wet spinning methods as a comparison. Melt spinning induces fibers to have higher crystallinity and melting transition of soft segment compared with wet spinning method (Table 16.1). It can also be seen that melt-spun fibers have significant transition peaks for both soft

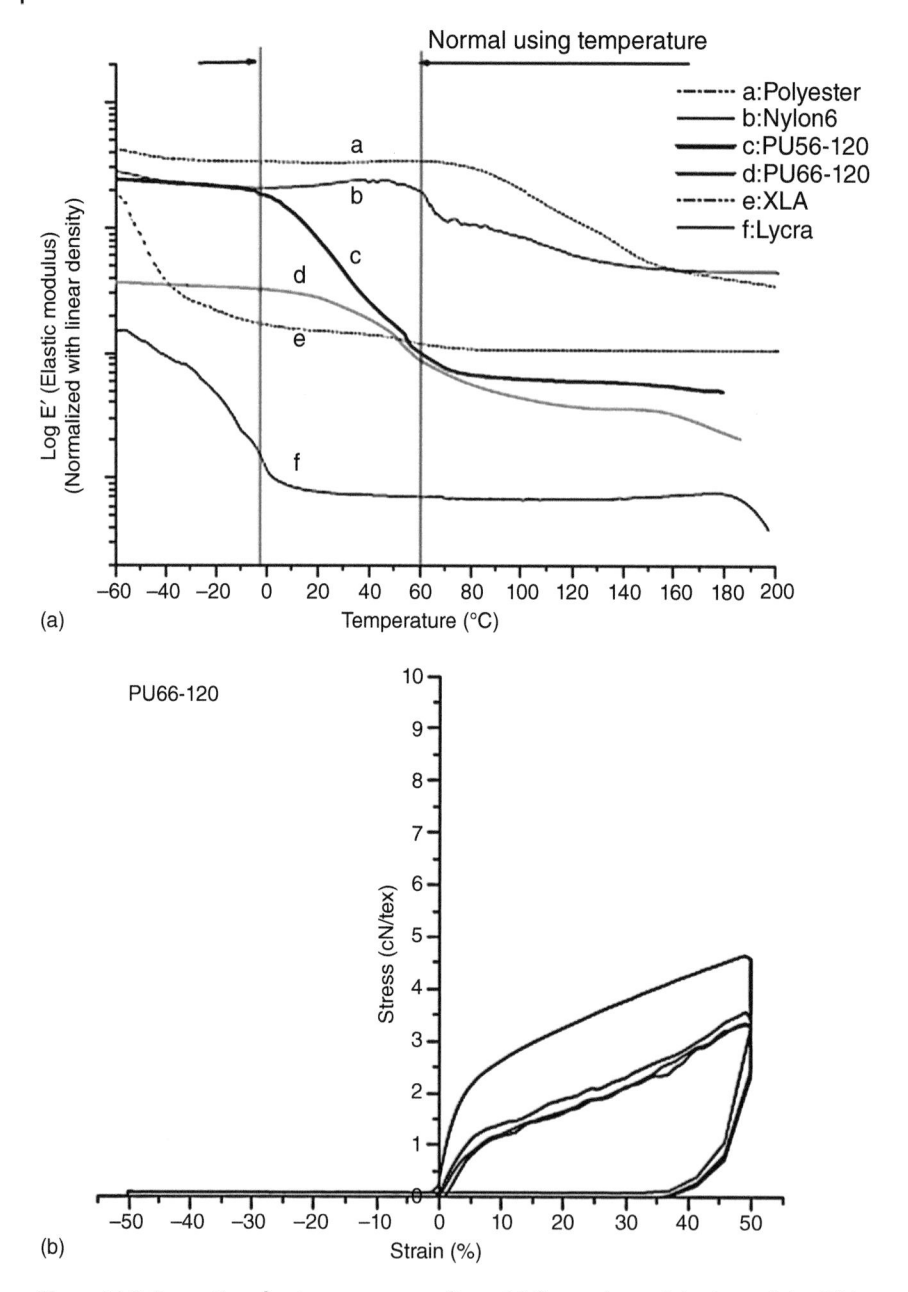

Figure 16.3 Properties of wet-spun memory fibers. (a) Comparison of elastic modulus (E') in MP fibers and other man-made fibers. (b) MP fibers showing complete shape recoverability.

segment (36–47 °C) and hard segment (214 °C) compared with no peaks in hard segment for wet-spun fibers (Figure 16.4a). Higher crystallinity and larger melting enthalpy of melt-spun fibers suggest that melt spinning induces perfect crystallinity and more ordered polymer packaging, a significant rich hard

Table 16.1 Thermal properties of melt- and wet-spun MP fibers.

	Soft segment			Hard segment	
	T_m (°C)	ΔH (J/g)	Crystallinity (%)	T_m (°C)	ΔH (J/g)
1 Melt spun	48.73	27.89	19.93	214.33	1.294
2 Melt spun	47.67	28.46	20.17	214.94	1.305
3 Melt spun	46.82	27.94	19.96	214.64	1.286
1 Wet spun	36.23	22.8	16.29		
2 Wet spun	36.62	22.82	16.3		
3 Wet spun	36.98	22.78	16.27		

Source: Meng et al. 2007 [34]. Reproduced with permission of Springer Nature.

segment phase in the fibers. The results of X-ray diffraction (XRD) profiles for fibers also show that the crystalline peaks are significant and higher for melt spun fibers and this is in accordance with the differential scanning calorimetry (DSC) traces (Figure 16.4b).

In addition, thermomechanical cyclic test results also have proven that melt-spun fibers (shape fixity, 86%; shape recovery, 98%) could provide higher shape fixity and recovery ratios compared with wet-spun (shape fixity, 82%; shape recovery, 95%) fibers (Figure 16.4c,d). The physical properties of MP fibers are shown and tabulated in Table 16.2. Preparation of fibers using melt spinning does not include any type of solvent, and it is also advantageous in terms of environmental and economy concerns. Henceforth, it is imperative to note that melt spinning could be chosen to produce MP fibers to have better performance to implicate them into vivid applications.

16.4.2 Effect of Post-spinning Operations on MP Fiber Properties

Now, we know that MP fibers can be spun using different spinning methods. Spinning may induce some internal stress and structure deficiency in the fiber along its axis, and this has to be removed or nullified to enhance their performance properties. Post-spinning operations such as heat treatment and drawing could help fiber to rearrange their molecular orientation and polymer packaging at temperature above the transition of soft segment. The influence of post-spinning operations on thermal, mechanical, and thermomechanical cyclic properties are discussed in this section.

16.4.2.1 Effect of Thermal Setting or Heat Treatment

Thermal setting or heat setting process is carried out at high temperature above the transition of soft segment in the fibers. A schematic of filament passage for roller heat setting is shown in Figure 16.5. Meng et al. have experimentally investigated and stated that thermal setting has a significant influence on the soft segment and thermal and mechanical properties [35]. In general, the transition temperature of soft segment ranges from 40 to 50 °C and hard segment with 200–240 °C. Heating the fibers above the soft segment transition helps to

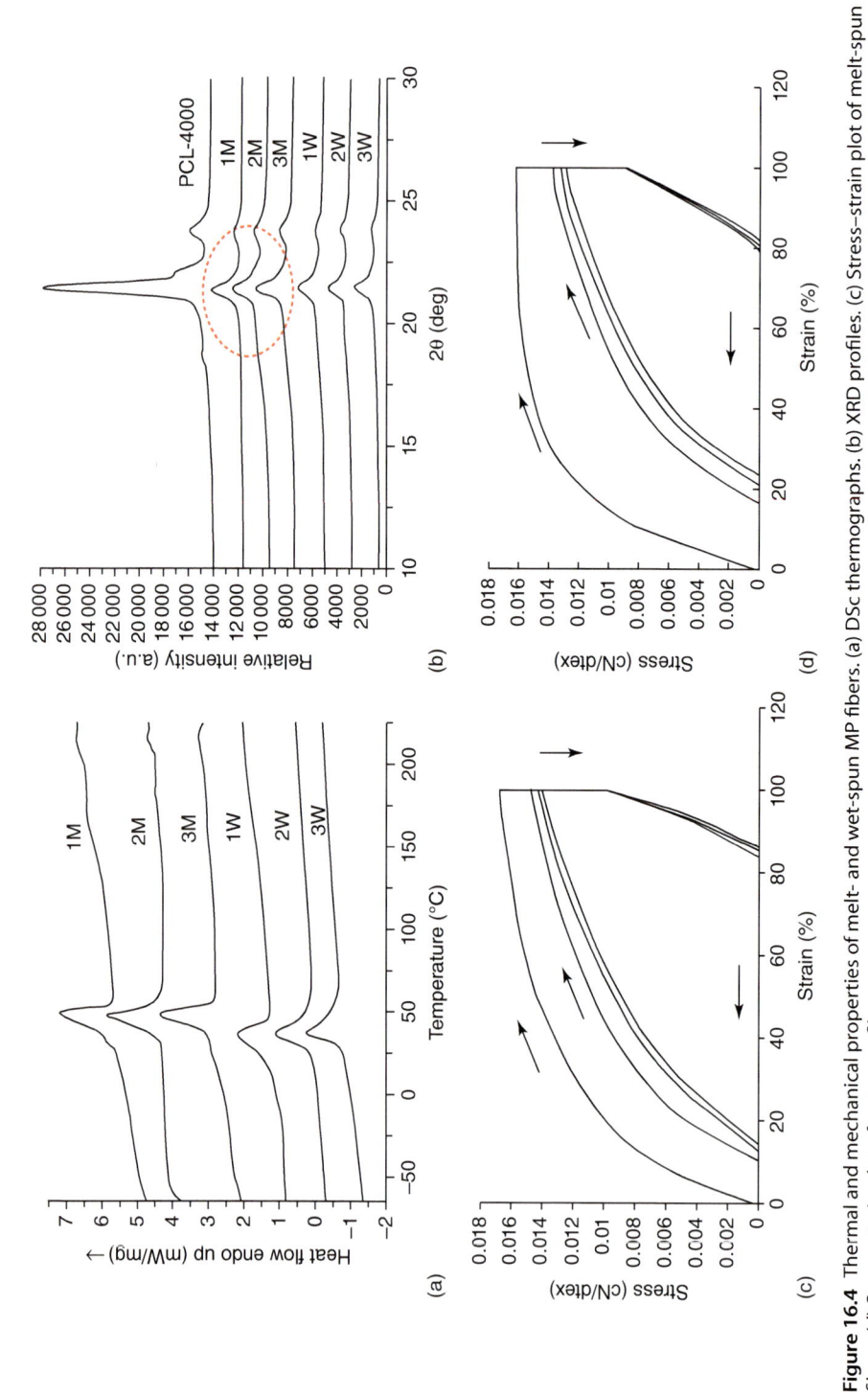

Figure 16.4 Thermal and mechanical properties of melt- and wet-spun MP fibers. (a) DSc thermographs. (b) XRD profiles. (c) Stress–strain plot of melt-spun fibers. (d) Stress–strain plot of wet spun fibers.

16.4 Memory Polymers As Fibers (MPFs)

Table 16.2 Physical properties of MP fibers.

Si no.	Particulars	
1	Tensile strength	> 0.9 cN/dtex
2	Elongation at break	350–500%
3	Initial modulus	0.08–0.3 cN/dtex
4	Shape fixity ratio	80–100%
5	Shape recovery ratio	90–100%
6	Tunable switch temperature range	0–100 °C
7	Diameter of electrospun fibers	50–700 nm

dtex: weight in grams of 10 km of fiber.

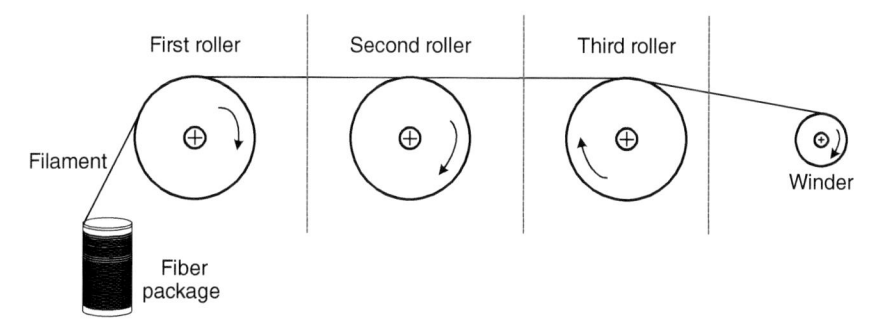

Figure 16.5 Schematic passage of MP filament through thermal setting process.

increase the crystallinity and melting enthalpy by molecular reorientation with reduced entanglement in the molecular chains. It can be seen from Figure 16.6a that transition peak of soft segment is increasing with heat setting temperature and there is a significant peak for hard segment in the fiber treated at 125 °C. This suggests that the crystallinity of soft segment increases, which is enabled by rotation of polymeric chains and ordered packaging. Heat treatment may repair the destroyed crystals during the spinning process to increase the crystallinity. Figure 16.6b shows the effect of heat treatment on the fiber tenacity; it can be noticed that at low temperature above the transition, molecular disorientation occurs, thus yielding lower tenacity. Heat treatment at higher temperature improves the phase separation and hard segment stability that is visible in Figure 16.6b. Partial molecular disorientation at low temperature can also influence the linear density by decreasing it; increasing the temperature can make the stable network and thus avoid further decrease in the linear density (Figure 16.6c).

Apart from molecular disorientation, releasing the internal stress at high temperature can also increase the breaking elongation of the fiber (Figure 16.6d). If a fiber needs to be employed in an application that needs a long-time response, stress relaxation is the most important criteria to be considered. Increasing the heat treatment would cause the fiber to improve the phase separation and strong hydrogen bond with well-ordered hard segment network. High

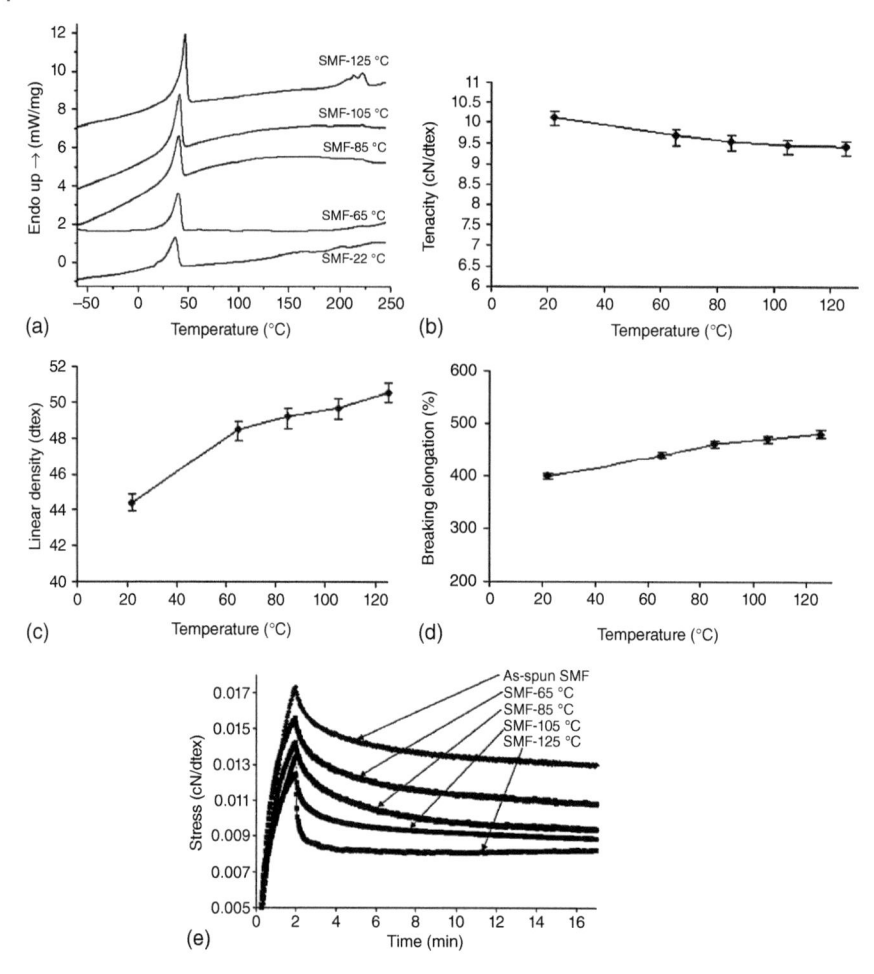

Figure 16.6 Influence of heat treatment on fiber properties. (a) DSC thermograms. (b) Effect on tenacity. (c) Effect on linear density. (d) Effect on breaking elongation. (e) Effect on stress relaxation.

temperature-treated fibers could show less decay in the stress and great stability over time (Figure 16.6e). Increasing the crystallinity, phase separation, and hydrogen bonding of hard segment would improve the shape fixity and shape recovery ratios as well. As we know, soft segment is responsible for shape fixation and hard segment stability for shape recovery during thermomechanical cyclic process [35].

16.4.2.2 Influence of Drawing Process

Drawing here refers to stretching or tensile deformation during the thermomechanical cyclic testing. Meng et al. have studied the effect of both cold and thermal drawing on the SMEs for the wet-spun fibers in comparison with commercial polyurethane fibers. Their experimental investigation reported that thermal drawn fibers could improve the shape fixation and shape recovery ratios as the soft segment crystallinity increases with hard segment stability and phase separation [36].

16.4.3 Other Type of MPU Fibers

16.4.3.1 Smart Hollow Fibers

Memory fibers can be spun into different shapes of cross section depending on the type of spinnerets being used. Hu and coworkers have developed a thermal-responsive hollow polyurethane fiber via melt spinning, which can change and recover their diameter upon thermal stimulation. The cross-sectional image is shown in Figure 16.7a and thermally adjustable internal holes in Figure 16.7b–d. The hollow fiber showed tenacity of 1.14 cN/dtex, breaking elongation of 682%, and shape fixity of 87% with 89% recovery ratio. These could be used for the applications such as smart filtration and drug-controlled release and in other suitable fields [37].

16.4.3.2 Electro-responsive Fibers

Electro-responsive MPU fibers were developed by Hu and coworkers and could recover their shapes under electrical stimulation. The shape memory

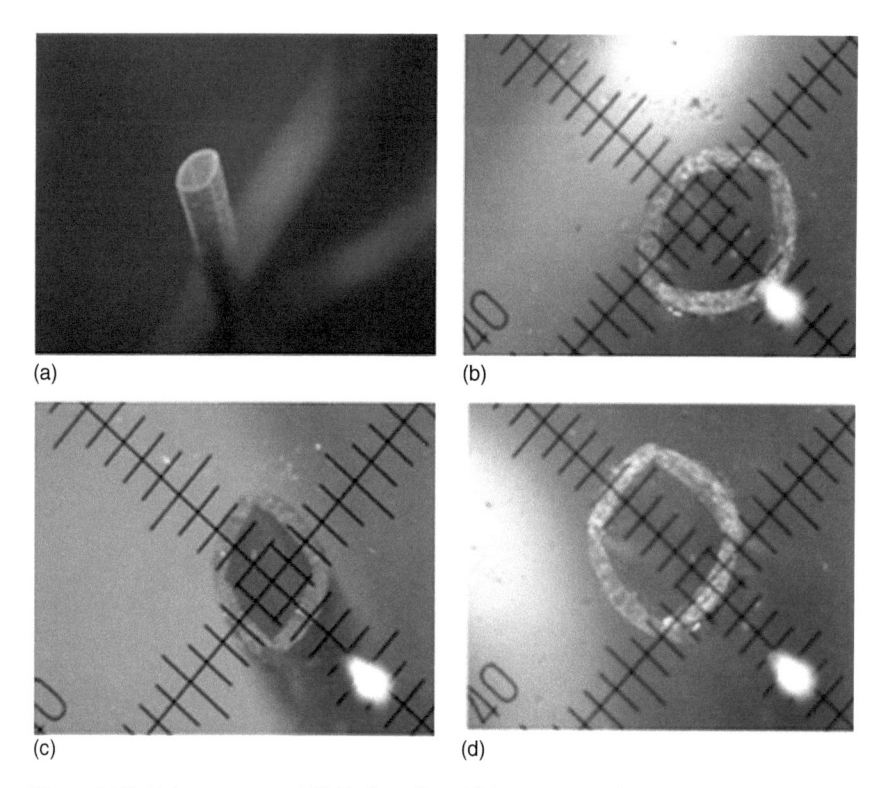

Figure 16.7 Melt-spun smart MPU hollow fiber. (a) Cross-sectional image. (b–d) Thermally adjustable internal holes.

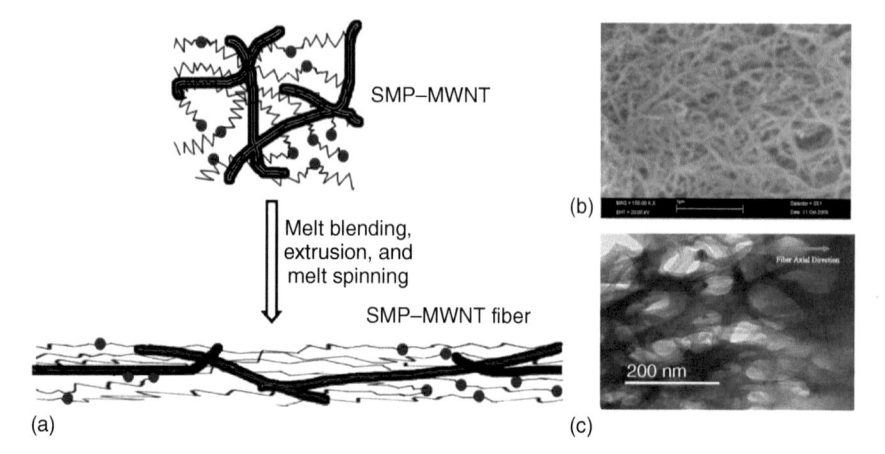

Figure 16.8 Electro-active memory fiber reinforced with MWNTs. (a) Schematic of composite fiber and MWNTs alignment. (b) SEM image. (c) TEM image.

polyurethane (SMP–MWNT) composite was prepared by in situ polymerization, and the MPU–MWNT fiber was prepared by melt spinning. Carbon nanotubes were incorporated into the MPU fibers and are charged at both ends (Figure 16.8). Voltage required to generate the heat to trigger the shape recovery was still high. Scanning electron microscopy and transmission electron microscopy (TEM) observations of the morphology revealed that the MWNTs are axially aligned and homogenously distributed in the MPU matrix, which is helpful for the fiber's electrical conductivity improvement and for the electro-active SME [38].

16.4.3.3 Electrospun Fibers

Electrospinning is used to produce ultrafine nanofibers using MP solution or melt with an external electric field. The main advantage of these fibers is they provide a large surface area to volume ratio, smaller fiber diameter, potential to carry active chemicals, filtration abilities, low weight, and high permeability. As the memory fibers have the ability to change and recover the shapes, it would allow nanofibers to some unique functions such as controlled water vapor permeability and hydro-absorbency upon an external heat stimulus [39]. Hu and coworkers have achieved shape fixity and recovery ratios of up to 80% and 98% respectively [40]. Core–shell structured nanofibers containing pyridine moieties could offer antibacterial properties [41].

16.5 Novel Stress Memory Behavior in MPs

16.5.1 What Is Stress Memory?

Stress memory is a novel phenomenon that was recently discovered by Hu et al. in a thermal-sensitive memory polyurethane material [9]. Like shape

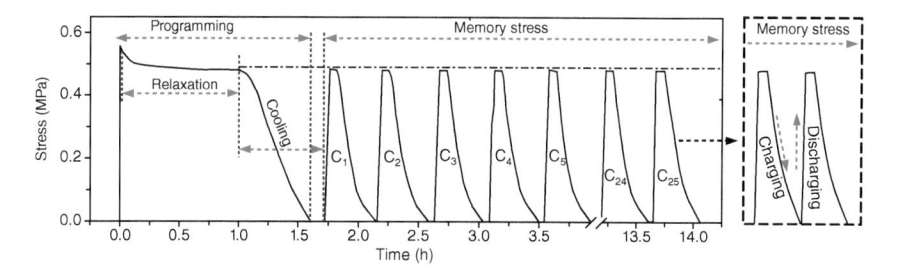

Figure 16.9 Stress memory profile of a thermal-sensitive memory polymer.

memory, stress in MPs can also be stored and retrieved reversibly upon an external heat stimulus. Figure 16.9 shows the profile of stress memory cycles in a thermal-induced memory polyurethane film substrate. When the material is stretched with a known deformation strain level, corresponding stress will be generated. In Figure 16.9, the MP is heated above the transition of soft segment and given a particular level of strain, and further it is continued to hold under the same constraint strain and temperature level to carry out the stress relaxation process. After obtaining a plateau of stress level, it is further cooled down to room temperature below the transition point of soft segment to fix the strain, and this process is called charging. During this cooling process, the generated amount of stress is being stored completely and reached zero stress level. This entire step is denoted as "stress memory programming" process, where the required amount of stress is programmed and stored. In the next step, the MP film is heated above the transition level under the same constraint strain level to trigger the stored stress, and this element is called as "memory stress," and the process is denoted as discharging. These charging and discharging cycles can be continued for several cycles without any significant loss of stress, and this is known as "stress memory" cycle.

16.5.2 Mechanism of Stress Memory

The molecular mechanism involved in stress memory process can be explained using the switch-spring-frame model as shown in Figure 16.10. Thermal-sensitive memory polyurethane used in this study comprises physically cross-linkable hard segment (fixed phase) and semicrystalline soft segment. Crystalline phase represents the switch, and amorphous region represents the spring in the model. When the MP is heated above the transition temperature, crystals in the soft segment melt and enable the modulus to decrease to assist easy deformation. Switches are open, assisting to increase the entropy elasticity of springs above the transition. Detailed steps are depicted in Figure 16.10, and step 2 shows the stretched molecular chains with amorphous network having no more crystals. After deforming above the transition, the next step is cooling the network in stretched constraint strain (step 3). During cooling process, the entropy elasticity of springs is stored in the network by closing the switches (OFF state). Crystallization or vitrification

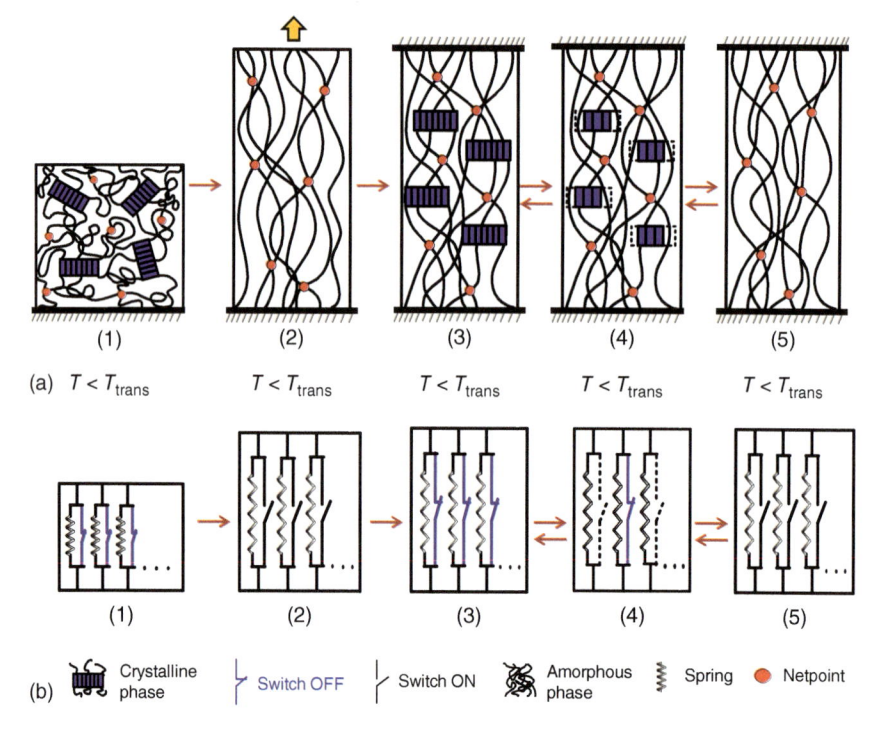

Figure 16.10 Mechanism of stress memory process. (a) Crystal-coil-cross-link model. (b) Switch-spring-frame model.

process occurs during cooling to form new crystals in the network and to close the switch to store the stress.

In step 5, MP is heated above the transition to melt the newly formed crystals and open the switch (ON state) to release the stored entropy elasticity. This results in breaking of strong internal bonds and enables conformational motion of polymeric chains in soft segment. During this step, the spring tries to recoil back, but it is kept under the constant strain, thus enabling to increase the stress upon thermal stimulus. Step 4 is shown as an intermediate step between step 3 and step 5. The more the number of switch opens, the more the recovery of memory stress will be.

16.5.3 Components of Stress Memory

Stress memory programming mentioned in this chapter is based on a tensile stretch programming method. When MPs are deformed under constant rate of strain and temperature, it induces five major components in a polymer matrix [42]. These components are plasticity (ε_p), viscoelasticity (ε_v), memory (ε_m), elasticity (ε_e), and thermal strain (ε_e) as shown in the Eq. (16.3), whereas total stress components are viscous stress (σ_v), memory stress (σ_m), elastic stress (σ_e), and thermal stress (σ_t). To obtain a pure memory stress, it is imperative to eliminate all other impeditive components in the molecular network. Plasticity is also referred as irreversible strain occurs due to permanent slippage of molecular chains in the network. As mentioned in Figure 16.2, thermomechanical process should be carried out at higher strain than the stress memory programming strain to eliminate both plasticity and elasticity. Viscoelasticity and viscous stress are nullified during the stress relaxation process at above the transition with constant strain over time. Thermal strain and thermal stress are caused by thermal expansion coefficient, which is unavoidable and in a negligible amount. The thermal stress and thermal strain components are negative or impeditive in tensile stretch programming and positive in the compression-type programming method. Narayana et al. [42] have analytically investigated to predict pure memory stress and total stress components using constitutive model via phase transition approach [43, 44].

$$\varepsilon_{Total} = \varepsilon_p + \varepsilon_v + \varepsilon_m + \varepsilon_e \pm \varepsilon_t \tag{16.3}$$

$$\sigma_{Total} = \sigma_v + \sigma_m + \sigma_e \pm \sigma_t \tag{16.4}$$

16.6 Stress Memory Behavior in Memory Fibers

Evaluation of mechanical, thermal, and shape memory properties in MPs is extensively studied by several researchers and industrial technocrats. However, revelation of other novel behavior such as stress memory is not reported anywhere till date. Stress memory behavior may differ from material to material such as film, fibers, and foam. For the first time, Narayana et al. [45] have unveiled the novel stress memory behavior in a thermal-sensitive memory fiber made of semicrystalline polyurethane. The soft segment is composed of poly(1,6-hexanediol adipate), 4,4'-diphenylmethane diisocyanate, and chain extender 1,4-butanediol as hard segment. The as-resulted semicrystalline MP is used and prepared fibers via melt spinning method. The response of memory stress in MP film and filament is slightly different.

Figure 16.11 shows the two plots of stress memory response in both film and filament. It can be noticed that the amount of memory stress generated or recovered in the filament/fiber is greater than the film. MP film is able to freeze the complete stress in the soft segment molecular chains upon cooling. Whereas filament shows some residual stress upon cooling, this could be of two reasons: First, the crystallization peak is below the room temperature; second is the incapability of filament to store huge amount of memory stress upon cooling at room temperature. This residual stress is also very much useful in the potential application such as compression stockings, which will be discussed in detail in the next section.

Figure 16.12 shows the evolution of memory stress and their relationships with thermal and X-ray properties. The response of memory stress is higher in the filaments compared with film, and it increases with temperature and strain level (Figure 16.12b). The reason for stress escalation could be well explained by the switch-spring-frame model (Figure 16.10) as discussed earlier. As the temperature increases, more number of switches actuates and

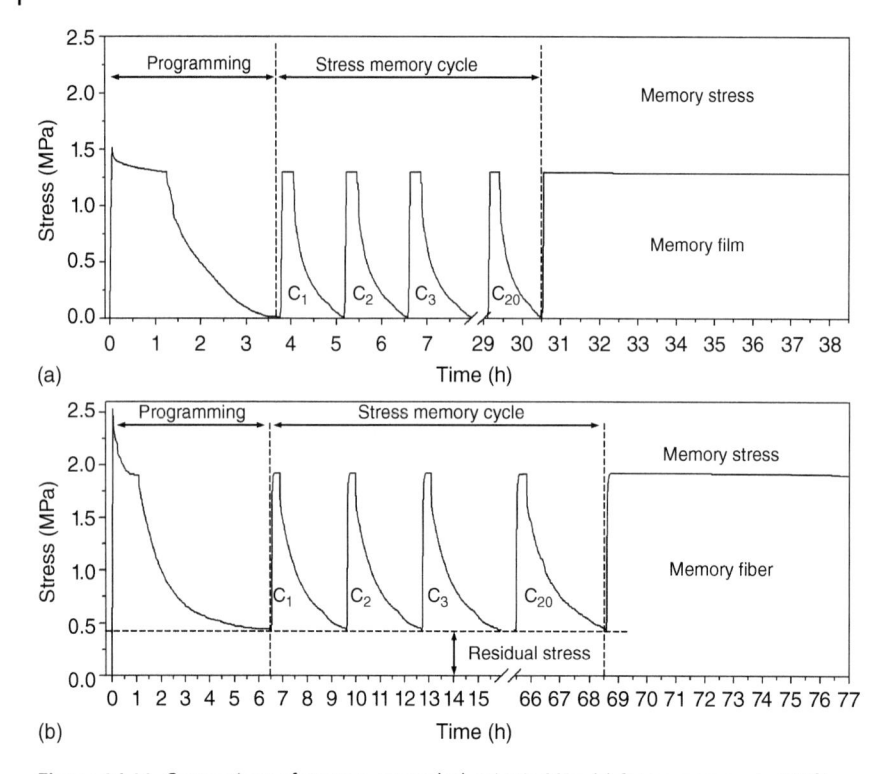

Figure 16.11 Comparison of stress memory behavior in MPs. (a) Stress memory in MP film. (b) Stress memory in MP fiber/filament.

releases the stored memory stress from the strained amorphous springs with gradual melting of crystals. The elevation of stress with strain at constant temperature level is in linear trend with correlation coefficient almost nearly 1 (Figure 16.12b). Crystallization in the soft segment is the main criteria to store memory stress upon cooling. Filament shows higher memory stress than the film, and the reason could be ascribed to higher melting enthalpy and increased crystallization. As discussed in previous section somewhere, melt spinning induces perfect crystallinity, orientation of molecular chains, and compact packaging of polymer. Figure 16.12c shows that the melting enthalpy of MP filament is higher than the film with more prominent crystallization peak obtained from the DSC analysis of stress memory programmed specimens. In addition, evidence for increased crystallinity in MP filament/fiber can be seen from wide-angle X-ray diffraction peaks (WAXD) of stress memory programmed specimens in Figure 16.12d. The difference in the evolution of memory stress in film and filaments are known, which is caused by melt spinning, and the structural changes are schematically depicted in Figure 16.13.

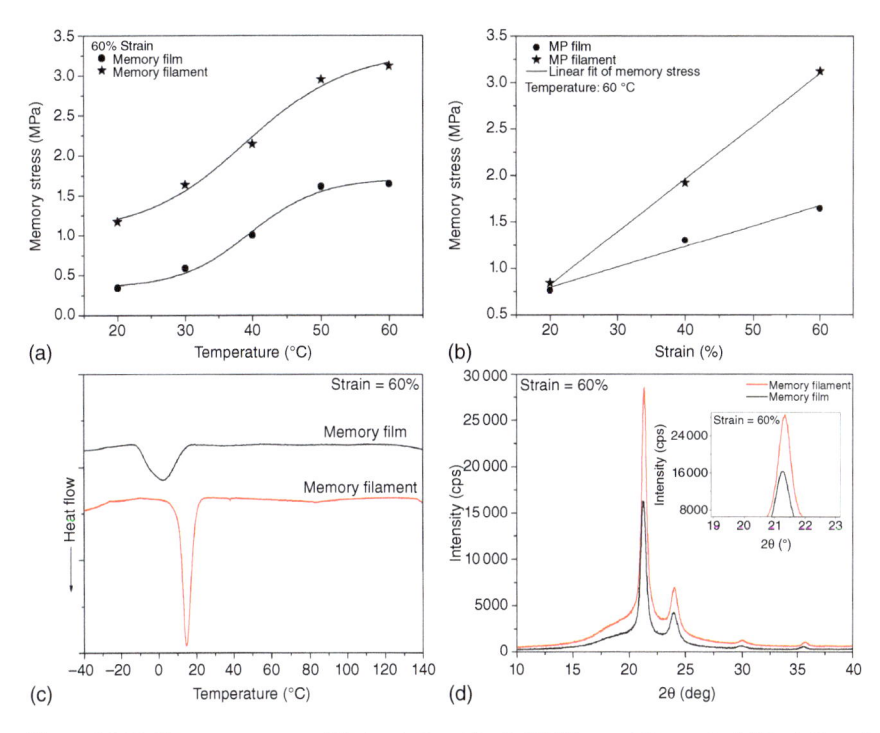

Figure 16.12 Memory stress and their relationships in MP film and filaments. (a) Evolution of memory stress with temperature. (b) Response of memory stress with strain at constant temperature. (c) DSC thermograms of stress memory programmed specimens. (d) WAXD profiles of stress memory programmed specimens.

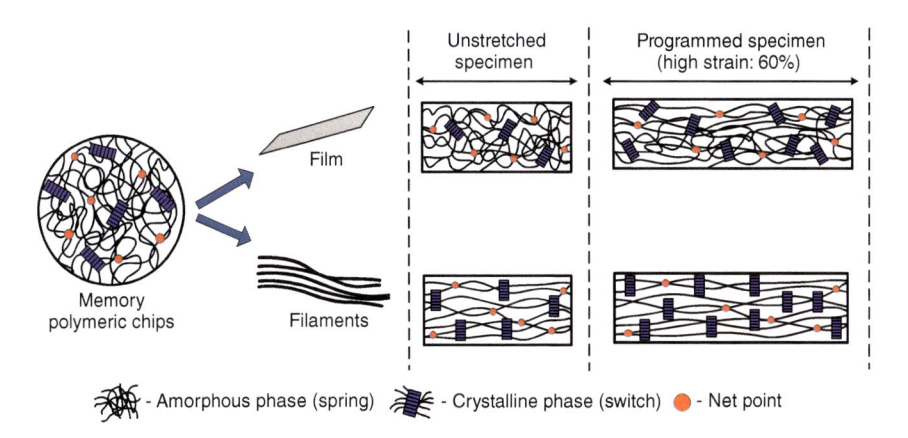

Figure 16.13 Schematic scheme showing structural changes in MP film and filaments.

16.7 Techniques of Characterization for Memory Fibers

Choosing the right method and equipment for characterization of memory polymeric fiber properties is very much important. Figure 16.14 shows the various structural property characterization methods and equipment to be chosen in one chart.

16.8 Potential Application of Stress Memory Fiber/Filaments

Novel stress memory behavior was first discovered in the memory polymeric film [9, 42], and then it is revealed in fiber or filament [45, 46]. As discussed in the previous section, memory fibers could memorize and retrieve maximum amount of stress reversibly upon an external heat stimulus. Memory stress can be controlled with strain and temperature. If the stress control is possible in the fibers, it would help to control the stress in other smart structures such as textile fabrics for numerous application. Narayana et al. [45] have utilized the stress memory filament potential to implicate them into textile knitted fabric structure for smart compression stockings. The memory polymeric composition can be tailor-made to obtain a residual stress upon cooling, and this is very much helpful in compression stockings. Compression stocking is considered as a gold standard for compression therapy to treat phlebological and lymphological deceases such as venous ulcers, varicose veins, venous stasis, and deep vein thrombosis (DVT) [47–50]. It has been a great challenge to maintain the constant interfacial pressure in the stockings due to various attributes such as material

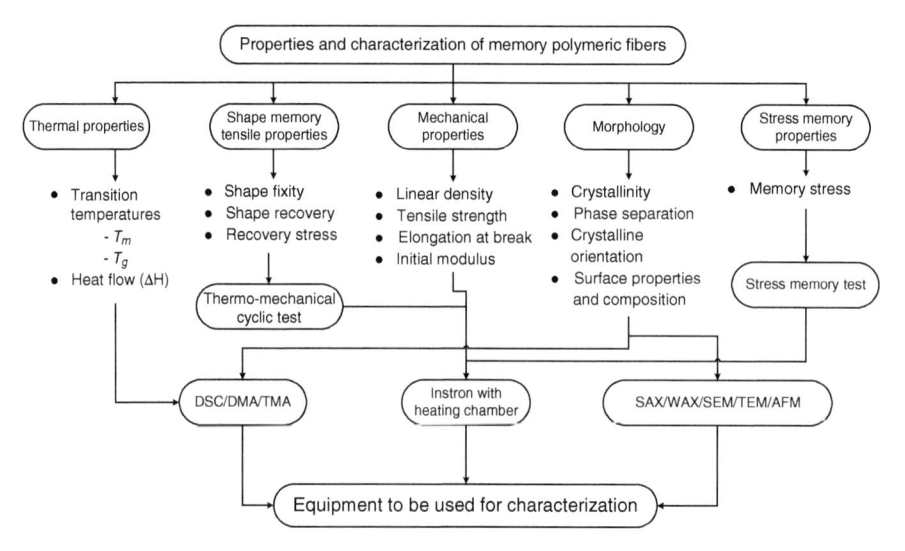

Figure 16.14 Method of characterization for memory polymeric fibers.

characteristics (stress relaxation), different size of legs, and pressure loss over time. Usage of memory polymeric filaments in the compression stockings could solve the current real problems by maintaining the stable memory stress over time. Stockings with smart stress memory filaments could provide a pressure gradient [9], massage effect, and selective pressure control just by triggering the memory stress. Compression stockings should not reach to zero pressure upon application on human leg and must retain with some base pressure around 20 mmHg; this could be achieved using the residual stress present in the stress memory filaments. The experimental investigation of massage effect in the compression stockings is shown in Figure 16.15. The potential of stress memory filaments could bring the groundbreaking revolutionary changes in the field of smart compression management.

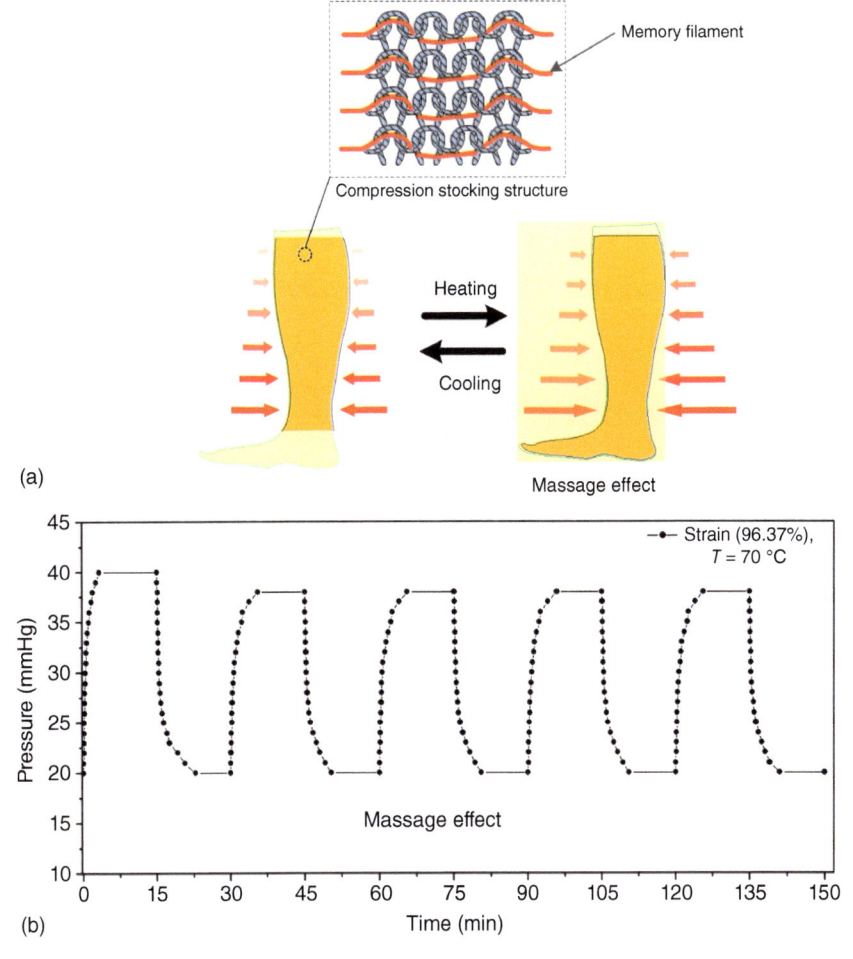

Figure 16.15 Stress memory filaments integrative smart compression stocking. (a) Schematic of smart compression stockings. (b) Massage effect in smart stockings with heat stimulus.

16.9 Recent Advances in MP Fibers

Researchers and industrial technocrats have been continuously working in the arena of MP fibers to scientifically apply them into vivid applications in multidisciplinary areas such as fiber supercapacitors, vibration damping structures, biomimetic fibrous scaffolds, fiber assembly for artificial muscles, and self-healing composites. Zhang and coworkers have fabricated a biodegradable nanofibrous scaffolds with shape memory properties and implied into bone tissue engineering offering shape fixity and recovery ratios more than 90% [51]. Electrochemical performance of MP fiber-based supercapacitors could offer an excellent stability suitable for shape programming and recovery [52, 53]. MP fibers possess higher toughness (276–289 MJ/m^3) compared with spider dragline silks (160 MJ/m^3) due to their excellent stretchability and have higher vibration damping capability [54, 55]. Two-way shape memory behavior-based MP fibers can be utilized in making hierarchically chiral structured artificial muscles [56]. The negative coefficient of thermal expansion of MP fibers is one order higher than those made of polyethylene fibers [57]. The great potential of memory fibers would further enable unveiling of untapped novel behaviors to prepare them as a futuristic smart material for broad horizon of applications to the mankind.

16.10 Future Trends

Undoubtedly, memory fibers could be a futuristic material for broad applications in multidisciplinary areas. MPs offers great advantages such as less density, cheaper price, and easily available. Memory fibers can be processed via spinning methods such as melt spinning, dry spinning, wet spinning, and electrospinning into different diameters from nano- to macrosize. They possess excellent platform to fine-tune their transition temperature range suitable for vivid specific applications. The revelation of novel stress memory behavior in smart MP fibers could inspire researchers to unveil other untapped unique behaviors to prepare for the future. Controlling the stress via stress memory programming in MP fibers could help them integrate into other novel structures for versatile applications. MP fiber integrative smart compression stockings would bring the revolutionary changes in the field of compression management for chronic disorders. At present, nowhere we could see commercially manufacturing plants of memory fibers except research institutes or universities. Commercialization of memory fiber manufacturing and adopting them into different end uses may uncover their great potentials by solving many real problems such as compression stockings.

References

1 Lendlein, A. and Langer, R. (2002). Biodegradable, elastic shape-memory polymers for potential biomedical applications. *Science* 296 (5573): 1673–1676.

2 Hu, J.L., Zhu, Y., Huang, H.H., and Lu, J. (2012). Recent advances in shape-memory polymers: structure, mechanism, functionality, modeling and applications. *Progress in Polymer Science* 37 (12): 1720–1763.

3 Leng, J.S., Lan, X., Liu, Y.J., and Du, S.Y. (2011). Shape-memory polymers and their composites: stimulus methods and applications. *Progress in Materials Science* 56 (7): 1077–1135.

4 Sun, L., Huang, W.M., Ding, Z. et al. (2012). Stimulus-responsive shape memory materials: a review. *Materials and Design* 33: 577–640.

5 Lester, B.V.B. and Vernon, H.M. (1941). Process of manufacturing articles of thermoplastic synthetic resins. Patent No. US2234994A.

6 Weigel, T., Mohr, R., and Lendlein, A. (2009). Investigation of parameters to achieve temperatures required to initiate the shape-memory effect of magnetic nanocomposites by inductive heating. *Smart Materials and Structures* 18 (2): 1–21.

7 Choi, W., Lahiri, I., Seelaboyina, R., and Kang, Y.S. (2010). Synthesis of graphene and its applications: a review. *Critical Reviews in Solid State and Materials Sciences* 35 (1): 52–71.

8 Kim, H., Abdala, A.A., and Macosko, C.W. (2010). Graphene/polymer nanocomposites. *Macromolecules* 43 (16): 6515–6530.

9 Hu, J.L., Kumar, B., and Narayana, H. (2015). Stress memory polymers. *Journal of Polymer Science Part B: Polymer Physics* 53 (13): 893–898.

10 Kratz, K., Madbouly, S.A., Wagermaier, W., and Lendlein, A. (2011). Temperature-memory polymer networks with crystallizable controlling units. *Advanced Materials* 23 (35): 4058–4062.

11 Wu, Y., Hu, J.L., Huang, H.H. et al. (2014). Memory chromic polyurethane with tetraphenylethylene. *Journal of Polymer Science Part B: Polymer Physics* 52 (2): 104–110.

12 Yuan, J.K., Zakri, C., Grillard, F. et al. (2014). Temperature and electrical memory of polymer fibers. *Times of Polymers (Top) and Composites*, Volume 1599, pp. 198–201.

13 Hu, J.L, Lu, J., Meng, Q. et al. (2007). Shape memory fibers prepared via wet, reaction, dry, melt, and electro spinning. US Patent US20090093606 A1, filed 7 October 2007 and issued 9 April 2009.

14 Ji, F.L., Hu, J.L., Li, T.C., and Wong, Y.W. (2007). Morphology and shape memory effect of segmented polyurethanes. Part I: With crystalline reversible phase. *Polymer* 48 (17): 5133–5145.

15 Lendlein, A. and Kelch, S. (2002). Shape-memory polymers. *Angewandte Chemie International Edition* 41 (12): 2034–2057.

16 Kim, B.K., Lee, S.Y., and Xu, M. (1996). Polyurethanes having shape memory effects. *Polymer* 37 (26): 5781–5793.

17 Liu, C.D., Chun, S.B., Mather, P.T. et al. (2002). Chemically cross-linked polycyclooctene: synthesis, characterization, and shape memory behavior. *Macromolecules* 35 (27): 9868–9874.

18 Voit, W., Ware, T., Dasari, R.R. et al. (2010). High-strain shape-memory polymers. *Advanced Functional Materials* 20 (1): 162–171.

19 Tobushi, H., Hara, H., Yamada, E., and Hayashi, S. (1996). Thermomechanical properties in a thin film of shape memory polymer of polyurethane series. *Smart Materials and Structures* 5 (4): 483–491.

20 Takahashi, T., Hayashi, N., and Hayashi, S. (1996). Structure and properties of shape-memory polyurethane block copolymers. *Journal of Applied Polymer Science* 60 (7): 1061–1069.

21 Hu, J.L. (2014). *Shape Memory Polymers: Fundamentals, Advances and Applications*. Shropshire, United Kingdom: Smithers Rapra.

22 Lin, J.R. and Chen, L.W. (1998). Study on shape-memory behavior of polyether-based polyurethanes. I. Influence of the hard-segment content. *Journal of Applied Polymer Science* 69 (8): 1563–1574.

23 Chun, B.C., Cho, T.K., and Chung, Y.C. (2006). Enhanced mechanical and shape memory properties of polyurethane block copolymers chain-extended by ethylene diamine. *European Polymer Journal* 42 (12): 3367–3373.

24 Ji, F.L., Zhu, Y., Hu, J.L. et al. (2006). Smart polymer fibers with shape memory effect. *Smart Materials and Structures* 15 (6): 1547–1554.

25 Zhu, Y., Hu, J.L., Yeung, L.Y. et al. (2006). Development of shape memory polyurethane fiber with complete shape recoverability. *Smart Materials and Structures* 15 (5): 1385–1394.

26 Wang, W.S., Ping, P., Chen, X.S., and Jing, X.B. (2007). Biodegradable polyurethane based on random copolymer of L-lactide and epsilon-caprolactone and its shape-memory property. *Journal of Applied Polymer Science* 104 (6): 4182–4187.

27 Min, C.C., Cui, W.J., Bei, J.Z., and Wang, S.G. (2005). Biodegradable shape-memory polymer-polylactideco-poly(glycolide-co-caprolactone) multiblock copolymer. *Polymers for Advanced Technologies* 16 (8): 608–615.

28 Li, F.K., Zhang, X., Hou, J.N. et al. (1997). Studies on thermally stimulated shape memory effect of segmented polyurethanes. *Journal of Applied Polymer Science* 64 (8): 1511–1516.

29 Zhu, Y., Hu, J., Yeung, K.W. et al. (2007). Effect of cationic group content on shape memory effect in segmented polyurethane cationomer. *Journal of Applied Polymer Science* 103 (1): 545–556.

30 Jeong, H.M., Kim, B.K., and Choi, Y.J. (2000). Synthesis and properties of thermotropic liquid crystalline polyurethane elastomers. *Polymer* 41 (5): 1849–1855.

31 Anthamatten, M., Roddecha, S., and Li, J.H. (2013). Energy storage capacity of shape-memory polymers. *Macromolecules* 46 (10): 4230–4234.

32 Hu, J.L, Han, J., Lu, J. et al. (2010). Items of clothing having shape memory. US Patent US20120000251 A1, filed 30 June 2010 and issued 5 January 2012.

33 Meng, Q.H. and Hu, J.L. (2008). Study on poly(epsilon-caprolactone)-based shape memory copolymer fiber prepared by bulk polymerization and melt spinning. *Polymers for Advanced Technologies* 19 (2): 131–136.

34 Meng, Q.H., Hu, J.L., Zhu, Y. et al. (2007). Morphology, phase separation, thermal and mechanical property differences of shape memory fibers prepared by different spinning methods. *Smart Materials and Structures* 16 (4): 1192–1197.

35 Meng, Q.H., Hu, J.L., Yeung, L.Y., and Hu, Y. (2009). The influence of heat treatment on the properties of shape memory fibers. II. Tensile properties, dimensional stability, recovery force relaxation, and thermomechanical cyclic properties. *Journal of Applied Polymer Science* 111 (3): 1156–1164.

36 Meng, Q.H., Hu, J.L., Zhu, Y. et al. (2007). Polycaprolactone-based shape memory segmented polyurethane fiber. *Journal of Applied Polymer Science* 106 (4): 2515–2523.

37 Meng, Q.H., Liu, J.L., Shen, L.M. et al. (2009). A smart hollow filament with thermal sensitive internal diameter. *Journal of Applied Polymer Science* 113 (4): 2440–2449.

38 Meng, Q.H., Hu, J.L., and Yeung, L. (2007). An electro-active shape memory fiber by incorporating multi-walled carbon nanotubes. *Smart Materials and Structures* 16 (3): 830–836.

39 Zhuo, H.T., Hu, J.L., and Chen, S.J. (2011). Study of water vapor permeability of shape memory polyurethane nanofibrous nonwovens. *Textile Research Journal* 81 (9): 883–891.

40 Zhuo, H.T., Hu, J.L., and Chen, S.J. (2008). Electrospun polyurethane nanofibers having shape memory effect. *Materials Letters* 62 (14): 2074–2076.

41 Zhuo, H.T., Hu, J.L., and Chen, S.J. (2011). Coaxial electrospun polyurethane core-shell nanofibers for shape memory and antibacterial nanomaterials. *Express Polymer Letters* 5 (2): 182–187.

42 Narayana, H., Hu, J.L., Kumar, B., and Shang, S.M. (2016). Constituent analysis of stress memory in semicrystalline polyurethane. *Journal of Polymer Science Part B: Polymer Physics* 54 (10): 941–947.

43 Wang, A.Q. and Li, G.Q. (2015). Stress memory of a thermoset shape memory polymer. *Journal of Applied Polymer Science* 132 (24): 1–11.

44 Liu, Y.P., Gall, K., Dunn, M.L. et al. (2006). Thermomechanics of shape memory polymers: uniaxial experiments and constitutive modeling. *International Journal of Plasticity* 22 (2): 279–313.

45 Narayana, H., Hu, J.L., Kumar, B. et al. (2017). Stress-memory polymeric filaments for advanced compression therapy. *Journal of Materials Chemistry B* 5 (10): 1905–1916.

46 Hu, J.L. and Narayana, H. (2016). Review of memory polymeric fibers and its potential applications. *Advance Research in Textile Engineering* 1 (2): 1–7.

47 Sue, J. (2002). Compression hosiery in the prevention and treatment of venous leg ulcers. *Journal of Tissue Viability* 12 (2): 67–74.

48 Labropoulos, N., Gasparis, A.P., Caprini, J.A., and Partsch, H. (2014). Compression stockings to prevent post-thrombotic syndrome. *Lancet* 384 (9938): 129–130.

49 Partsch, H. (2013). Compression therapy in leg ulcers. *Reviews in Vascular Medicine* 1 (1): 9–14.

50 Partsch, H. (2014). Compression for the management of venous leg ulcers: which material do we have? *Phlebology: The Journal of Venous Disease* 29: 140–145.

51 Bao, M., Lou, X.X., Zhou, Q.H. et al. (2014). Electrospun biomimetic fibrous scaffold from shape memory polymer of PDLLA-co-TMC for bone tissue engineering. *ACS Applied Materials & Interfaces* 6 (4): 2611–2621.

52 Deng, J.E., Zhang, Y., Zhao, Y. et al. (2015). A shape-memory supercapacitor fiber. *Angewandte Chemie International Edition* 54 (51): 15419–15423.

53 Zhong, J., Meng, J., Yang, Z.Y. et al. (2015). Shape memory fiber supercapacitors. *Nano Energy* 17: 330–338.

54 Yang, Q.X. and Li, G.Q. (2014). Spider-silk-like shape memory polymer fiber for vibration damping. *Smart Materials and Structures* 23 (10): 1–14.

55 Sharafi, S. and Li, G.Q. (2016). Multiscale modeling of vibration damping response of shape memory polymer fibers. *Composites Part B: Engineering* 91: 306–314.

56 Fan, J. and Li, G. (2017). High performance and tunable artificial muscle based on two-way shape memory polymer. *RSC Advances* 7 (2): 1127–1136.

57 Yang, Q.X., Fan, J.Z., and Li, G.Q. (2016). Artificial muscles made of chiral two-way shape memory polymer fibers. *Applied Physics Letters* 109 (18).

17

Textile Mechanics: Fibers and Yarns

Zubair Khaliq[1] and Adeel Zulifqar[2]

[1] National Textile University, Faculty of Engineering and Technology, Faisalabad, Pakistan
[2] The Hong Kong Polytechnic University, Institute of Textile and Clothing, Hung Hom, Kowloon, Hong Kong, China

17.1 Introduction

The yarn is a long and fine structure that consists of very fine fibers. The yarns are usually intermediate products that are further used to make ropes, braids, cords, and fabrics (knitted and woven). The fiber is the building block of these products, so the physical properties of these textile products are highly dependent on the physical properties of the fiber. Practically, the mechanical properties of textile products are important to meet certain standards. The structure of textile products contributes to these properties. However, fiber and yarn mechanical properties remain a fundamental element. Since fabrics can be used in multiple end products like carpets, towels, sheets, apparel, and woven clothing, the requirement of fiber and yarn mechanical properties can be varied with its requirement. This results in desired fiber and yarn properties instead of general requirements. Indeed, if we consider yarn, the required yarn properties are different according to the end user. The requirements of a spinner can be uniformity, low level of imperfections, optimum hairiness, and high strength. A knitter might not be sensitive about high strength but bulkiness, flexibility, and softness of yarn, so the yarn can bend easily during the knitting process. However, a weaver might require the high strength of yarn to withstand the larger stresses during the weaving process to make fabric. Better pilling resistance, lower fluff, lower hairiness, higher abrasion resistance, least contamination, and higher evenness are the general requirements of yarn. Conclusively, it becomes very difficult to quantify general yarn standards. The variable requirements of end users lead to different yarn structures to meet their specific needs. This becomes important to understand these yarn structures to study the textile product scientifically. This chapter describes the general properties of fibers along with the yarn structure to give a better understanding of mechanical properties of the final textile product.

Handbook of Fibrous Materials, First Edition. Edited by Jinlian Hu, Bipin Kumar, and Jing Lu.
© 2020 Wiley-VCH Verlag GmbH & Co. KGaA. Published 2020 by Wiley-VCH Verlag GmbH & Co. KGaA.

17.2 Fiber

Fiber is the smallest and fundamental element of yarn and a textile product. A significant factor of the mechanical properties of a textile product originated from the mechanical properties of a fiber. However, the fibers are very small. The length of a textile fiber should be at least 1000 times longer than its width. If the length of the fiber is too short, then it will be difficult to process and to make a textile product. Also, the fibers should be flexible, which is a prerequisite to process and make a textile product. Otherwise, the stiffer fibers are difficult to process and uncomfortable to wear. In the past, the natural fibers were the first to be used as textile fibers. The mechanical properties of textile fibers are largely dependent on their building polymer molecules. The nature of covalent bonding along with the physical interactions between fiber molecules contribute to the mechanical behavior of the fiber. The physical interaction includes van der Waals forces, polar interaction, and hydrogen bonding between polymer molecules. These interactions define the amorphous and crystalline structure of a fiber and eventually the mechanical behavior.

17.2.1 Fiber Consumption

Figure 17.1 shows the classification of textile fibers. Yarn spinning consumes the largest share (90%) of total fiber volume produced. Nonwovens have a share of 7%, and the remaining share of 3% is being used as fillings and cigarette filters. According to science from the 1960s, the world's population is increasing leading to the increased demand of fibers for clothing. This demand was satisfied by synthetic fibers. As the production of synthetic fibers is time saving, people moved

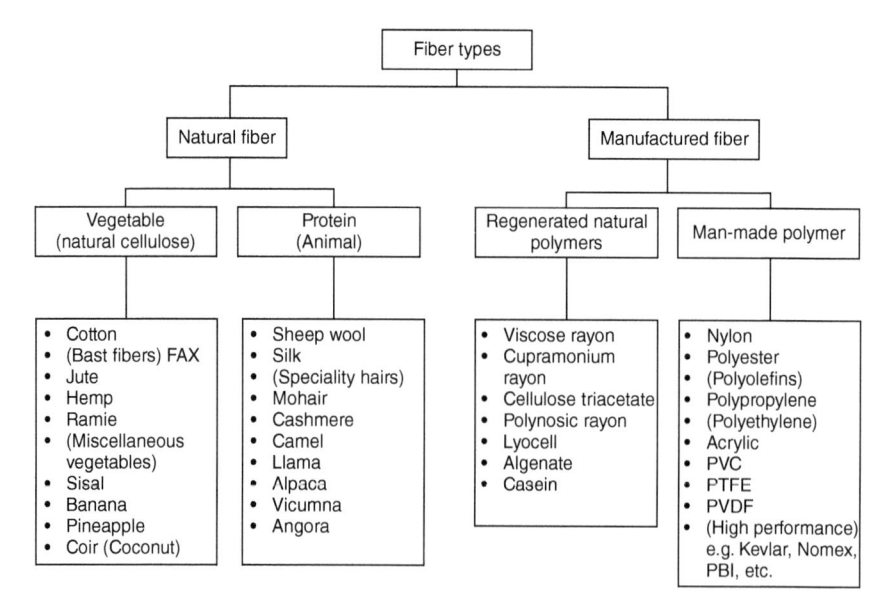

Figure 17.1 Flow chart of different textile fibers.

toward it, and its share rose up to 57%. The natural fibers have the share of only 43% of total fiber production. Among synthetic fibers, polyester has the highest share of 59.3%, then followed polyolefin fibers (polypropylene + polyethylene) having a share of 18.4% and polyamide (Nylon) fiber having a share of 13.1%. The acrylic has the least share of 8.5%. The manufactured cellulosic fibers may also be included into the class of synthetic fibers, which have a share of about 8% in the global production of synthetic fibers.

Cotton is being used in textiles from the beginning. It has been used in history for over 5000 years. Almost every human on earth was using it to cover his body. Cotton is still very important fiber till now because it is largely used to produce fabric, apparel, and composites. It occupies the second largest share of 33% after polyester, followed by jute that is most widely used in the production of sacks for vegetable or other food item storage and transportation. It is also widely used in composite manufacturing as cheap and environment-friendly reinforcement. Wool is also an important client for fashion and textile design, having a share of 2.3% of total fiber production. Other natural fibers have relatively less share in the market.

Spun yarn spinning technique may be used to produce yarn from these remaining organic natural fibers (except bast fibers). These techniques include spun on short-staple, worsted, woolen, or the exceptional spinning techniques. The bast fibers, on the other hand, cannot be spun by conventional spinning techniques but with special techniques [1].

Asbestos is also a natural fiber from the mineral origin, but it is not widely used in textiles due to its carcinogenic effects and environmental hazards. However, scientists are developing its environment-friendly formulation with concrete to produce heat-resistant building structures. Other mineral fibers like glass fiber, bast fibers, and metal fibers are largely used as a filament to produce yarn. Nonwoven fabrics may also be produced from these yarns. Metal staple fibers may also be spun into yarn to produce conductive clothing. It was used to produce protective clothing for soldiers in traditional wars.

The inorganic fibers have a very low application that's why they are not being produced in large quantity. So, they are quite significant in yarn spinning and textile clothing. But on the other hand synthetic fibers like polyester, acrylic, and regenerated cellulosic fibers and natural fibers like cotton, jute, wool, etc. are quite significant for spinning and clothing. Many of them are very useful for medical applications. Cotton is used as a wound dressing from the beginning. Many other fibers are also discovered to be inherently antimicrobial and also cure the wound more quickly. Silk and hairs are used to produce luxury clothing, which is very expensive and exotic. But theses fibers have highest potential growth in the field of textiles.

Fiber growth rate strongly depends upon population, fiber cost, and the ability of personnel to buy them. The world's population is about 6 billion peoples, which is increasing by a ratio of 1.7% every year. Sixteen percent of world's population consumes more than 10 kg fibers per capita. And the remaining 84% of world's population is using 3–10 kg of fibers per capita. It is supposed that the average annual fiber consumption will increase up 12 kg/capita, up to the middle of the twenty-first century, in which natural fibers will have their share of

40% and synthetic fibers will occupy a huge share of 60% in the global market. Sixty-two percent of total fibers produced will be converted into spun yarn, and the volume of filament yarn will 30%. The remaining 8% will be used to nonwoven fabrics, filters, and cigarettes.

17.3 Strength Contributing Fiber Parameters

The components of a textile fiber that are observable to the human eye are its diameter, length, and crimp. The diameter of a fiber is the average width of the fiber along the fiber's length. The diameter and length of a fiber determine its size, which is one of the most important quality parameters of fiber. It is described as the linear density of fiber. The linear density of fiber can be described in different ways like the weight of fiber per unit length (direct system) or length per unit weight of fiber (indirect system). A direct system is usually used to define the fineness of fiber and filament yarns, while the indirect system is used for the staple spun yarns. Depending on the fineness of the fibers, linear density can be measured in denier or tex. Denier is the weight of fiber in grams per 9000 m of fiber length, while tex is the weight in grams of 1000 m of fiber length. One denier is equal to nine times of tex. Fiber size has an important impact on fiber stiffness, which then influences the stiffness of the yarn and fabric made from that fiber. Also, it significantly influences the drape of the fabric. In addition, fiber stiffness is an important factor of how comfortable or uncomfortable the fabric feels when it is worn next to the skin. Fiber length is another most significant property of fiber. Fiber length is analytical in the processing of fibers and yarns and the transformation of fiber strength to yarn strength. Usually, a longer fiber length is favored. Textile fibers are either staple or filament length. The length of staple fibers can be from 2 to 46 cm, and that of filament fibers is the infinite length. Except silk all natural fibers are of staple length. The manufactured fibers may be staple or filament fibers. Fiber crimp indicates waves, bends, twists, or curls along the fiber length. It is measured as crimps per unit length. Synthetic fibers have a smooth surface, which increases its slippage during fiber spinning. A desired and controlled crimp is introduced in synthetic fibers to attain the desired properties of yarns and fabrics. Crimped fibers contribute to having higher elongation than linear fibers.

17.4 Mechanical Properties of Fiber

17.4.1 Stress–Strain Curve

Owing to the linear shape of the fiber, the tensile properties are the most significant and are the most studied parameter. This is because of the reason that most forces and deformations occur along the fiber length. A gradually increasing force is applied along the fiber axis until the fiber breaks. Stress–strain curve is shown in Figure 17.2, which shows many important mechanical characteristics of the fiber. Specific stress is often used in textile industry instead of the general

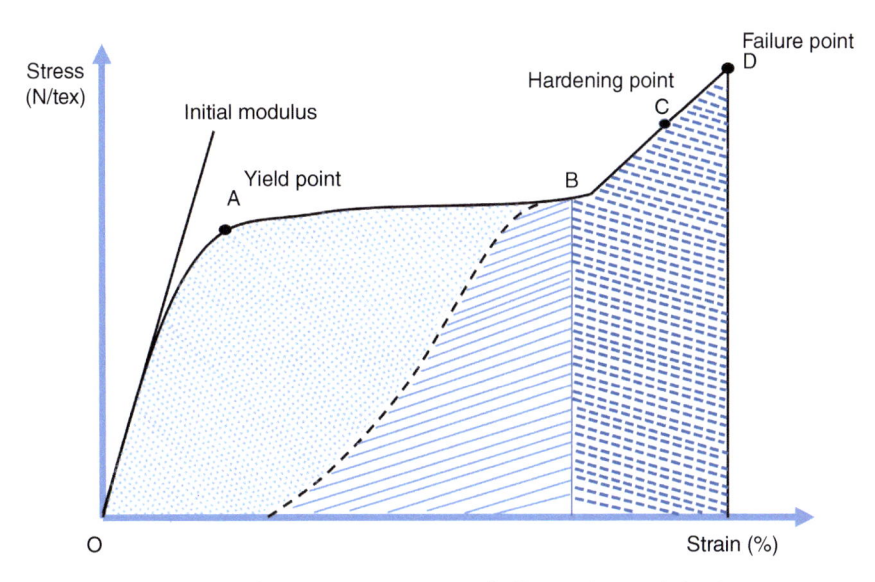

Figure 17.2 Schematic of the stress–strain curve of a fiber under tensile loading.

stress. The specific stress describes the general stress regarding linear density of the fiber.

In general engineering,

Tensile stress = Force/Area

$$s = F/A$$

Tensile strain = Change in length/Original length

$$e = \Delta l/l$$

Specific stress = Force/Linear density

$$ss = F/dl$$

Figure 17.2 shows schematic stress–strain curve. The curve begins with a straight line segment that rises as stress is increased (OA) and then suddenly flattens and rises at a slower rate (AB). Close to the failure point, the curve rises steeply (BC). The details of each of the regions are addressed as follows:

In region OA, the deformation occurs due to the bond stretching and flexing. It is reversible as long as Hooke's law is obeyed: $s = Ee$, where E is the slope of the line and called Young's modulus. In this region, it is elastic, and the deformation is recoverable. After the yield point, it is usually a plastic region, and the deformation becomes nonlinear. The deformation, results of the polymers skidding by each other, is partially recoverable. In region AB, the fiber can extend easily. The intra-fiber bonding is represented by the slope of the segment AB. When the polymers become more compact, the fiber reaches a deformation limit, harden point. Following the hardening point, the internal structure of the fiber begins to give way, and the failure point is reached. The stress and strain at this point give the tenacity and the elongation at break, respectively.

17.4.2 Elastic Recovery, Work of Rupture, and Resilience

In the initial segment of the stress–strain curve (BC), the fiber behaves like an elastic spring. When the load is applied, the polymer chains are being uncurled, perhaps becoming more oriented to the fiber direction. The fibers will retain their original length when the load is removed at any point from O to A. The elastic recovery is 100% in this case. The fiber will not recover 100% when they are stretched beyond the yield point before its hardening point. At this stage, if the load is removed, then the polymer chains will partially recover.

Work of rupture or toughness is a measure of the ability of a fiber to withstand sudden shocks of energy. The area under the stress–strain curve determined the total amount of work required to deform a fiber up to the failure point, the sum of dotted, lined, and a dashed area in Figure 17.2. The work of recovery is also known as resilience. It is the ratio of energy returned to energy absorbed when a fiber is under load or the applied load is removed. It may be extensional, flexural, compressional, or torsional. In Figure 17.2, the fiber resilience of extension is the ratio of lined area to the sum of the lined and dotted area.

17.4.3 Effects of Time, Temperature, and Moisture

The stress–strain curve behavior of a textile fiber also depends on some other factors. One of the important factors is the earlier mechanical history of the fiber. Another important factor is temperature. Also, the mechanical characteristics of the fiber depend on the aging time with temperature. It is usually called viscoelastic (combined viscous and elastic) behavior. This is because of the reason that polymers contain elastic as well as viscous nature. The mechanical history along with aging time and temperature affects this viscoelastic behavior [2–5]. Fiber may perform poorly if it suffers from these conditions. Creep and stress relaxation are the tests developed to probe their time-dependent behavior. These characteristics are measured by other mechanical factors like creep and stress relaxation. In the creep test, the sample is under constant load, and the strain increases with time. In the stress relaxation test, the stress reduces with time after the sample is given an instantaneous strain. The mechanical behavior of fibers can also be affected due to moisture. The moisture camps in the noncrystalline regions and plasticizes them results in reducing the modulus.

17.5 Yarn Classification

Figure 17.3 shows that yarn is usually expressed in the form of fiber strand made of either intertwine staple fibers or consistent parallel filaments that are proficient to interweave into a woven structure or intermeshed into a knit structure.

```
                          ┌─────────────────────┐
                          │  Yarn classification│
                          └──────────┬──────────┘
                    ┌────────────────┴────────────────┐
        ┌───────────────────────┐         ┌───────────────────┐
        │Continuous filament yarns│       │  Staple spun yarn │
        └───────────┬───────────┘         └─────────┬─────────┘
```

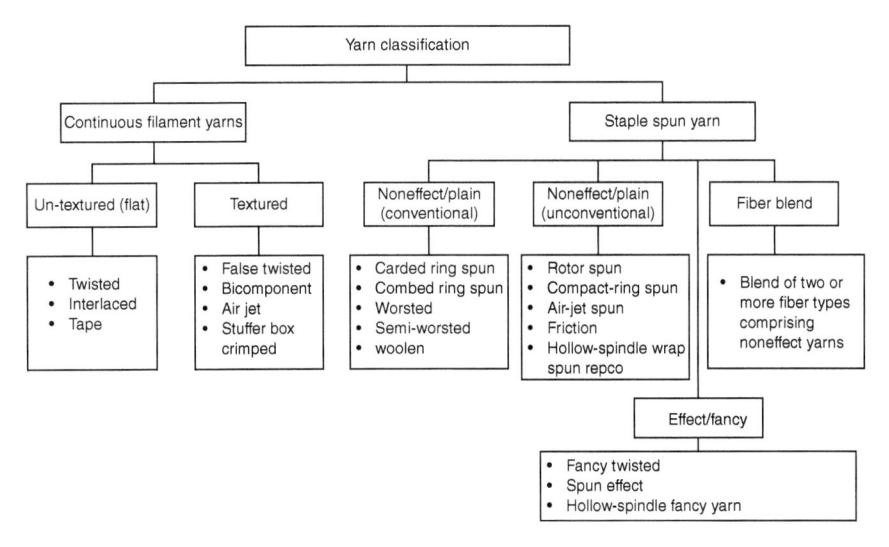

Figure 17.3 Flow chart of different yarn types.

Continuous filament yarns:
- Un-textured (flat): Twisted, Interlaced, Tape
- Textured: False twisted, Bicomponent, Air jet, Stuffer box crimped

Staple spun yarn:
- Noneffect/plain (conventional): Carded ring spun, Combed ring spun, Worsted, Semi-worsted, woolen
- Noneffect/plain (unconventional): Rotor spun, Compact-ring spun, Air-jet spun, Friction, Hollow-spindle wrap spun repco
- Fiber blend: Blend of two or more fiber types comprising noneffect yarns
- Effect/fancy: Fancy twisted, Spun effect, Hollow-spindle fancy yarn

The yarn can be intertwisted into non-fabric materials like braids and ropes. Figure 17.3 indicates that there are two important types of yarn. One is continuous filament yarns, and second is spun yarns. A continuous filament yarn is usually obtained by using extruding polymer liquid to form liquid filaments through a spinneret that are solidified into a continuous fiber strand. As shown in Figure 17.4, when a continuous filament yarn consists of single filament, then it is called a monofilament yarn, and when it consist of many filaments, then it is called multifilament yarn. The texturizing process can be used for the purpose of generating bulky yarn. The continuous filaments can also be converted into another structural form via deliberate entanglement or geometrical reconfiguration. The spun yarn can be generated from staple fibers after using many connective processes like the opening, blending, cleaning, drawing, and spinning. These fibers are stabilized into a yarn structure via twisting. These processes are summarized in Figure 17.5. These yarns can be produced with fibers, which are obtained from natural or synthetic sources. In addition, the staple spun yarn structure depends on the method that is used for fiber preparation like carded, combed, or worsted yarns. Also, it depends on spinning methods like ring-spun, rotor-spun, air-jet, and friction spun yarn.

17.6 Yarn Construction

Yarn structure had been discussed in multiple research work, and it is difficult to add further strength contributing factors. However, an understanding of the strength contributing factors should be considered. These factors include (a) fiber arrangement in the yarn, (b) the binding force of fibers (twist), and (c) fiber compactness in the yarn. Different yarn spinning methods are used to convert fibers into yarn. We will discuss the most commonly used techniques in the industry.

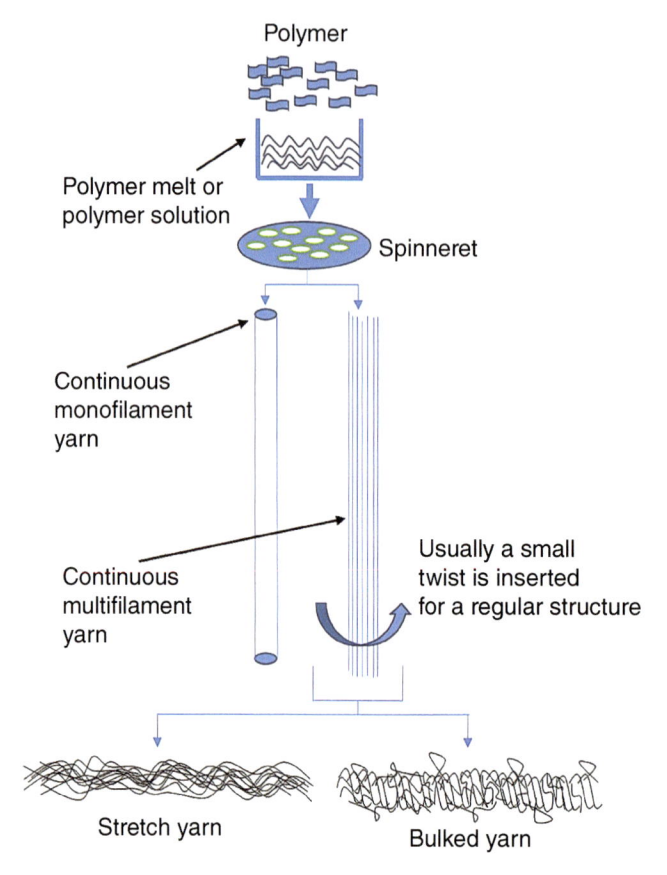

Figure 17.4 Schematic of a continuous filament yarn.

17.6.1 Ring-Spun Yarn

This yarn is produced on a ring frame, and this technology is known as "ring spinning." Ring spinning was first discovered by an American scientist John Thorp. Then another American scientist introduced the rotating traveler on the ring. Ring-spun yarn is used in all types of fabrics from knitting to woven with the most diverse yarn type. Ring spinning is relatively an expensive process due to its slower speed of production. Roving and winding are needed additionally in the process. In ring spinning, the yarn linear density ranges from fine to medium count range. Also, it has some small production range in the coarser count. In this process, the final yarn fineness is led by the process of blow room, carding, drawing and then winding the yarn on a bobbin after inserting the twist to the fibers with the rotating traveler. These stages take place continuously. In ring-spun yarn, the twist is inserted uniformly on all the yarn structure up to a certain limit. As the thickness of the yarn is uneven, the thinner section has the concentrated twisting. Comparatively a loose twisting comes at the thicker section, which results in the hairiness of yarn. Ring-spun yarn has different types depending upon the

Figure 17.5 Schematic of a spinning process for a staple spun yarn.

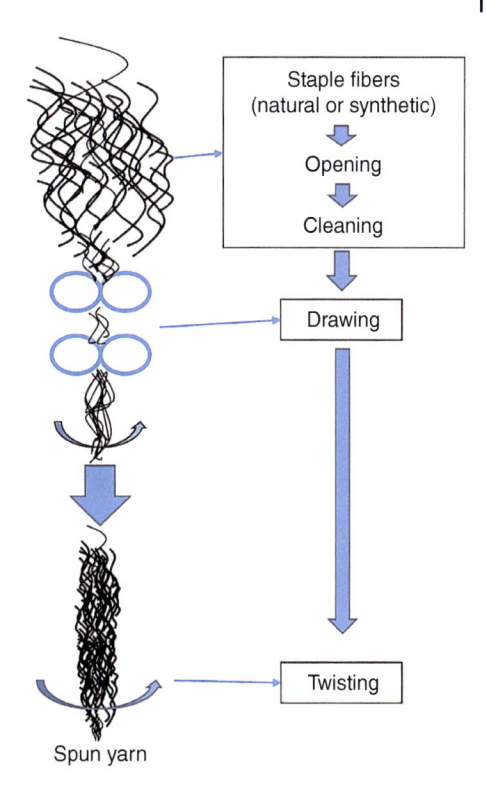

```
Staple fibers
(natural or synthetic)
        ↓
     Opening
        ↓
     Cleaning
        ↓
     Drawing
        ↓
     Twisting
        ↓
    Spun yarn
```

requirement of the final product. These types include carded yarn, combed yarn, and compact yarn.

17.6.2 Yarn Structure

17.6.2.1 Carded and Combed Yarns

Carded yarn is relatively cheaper than combed yarn due to its smooth and simplest structure. It can be manufactured by ordinary spinning tools, after opening, cleaning, carding, and drawing of fibers. Cheap tooling and easy manufacturing make carded yarn very economical. Carded yarn, especially when made up of cotton, may have a high level of trash due to poor fiber alignment and may have a high number of neps. This leads the carded yarn poor in mechanical properties. However, a wide range of courser counts may be produced and applied to make a wide range of apparel from heavy denim to shirts and blouses. Both weaving and knitting techniques may be used for fabrication. A schematic diagram of carded yarn is shown in Figure 17.6a.

Combed yarn, on the other hand, is also a single yarn. The simple tools used to make carded yarn may also be used to manufacture combed yarn, but the process of combing is to be performed additionally. Combing process usually eliminates the neps and short fibers, which results in the reduction of overall trash in yarn and improvement in physical properties of the yarn. It improves the fiber mechanical properties, and hence higher quality yarns are produced.

(a)

Hairiness

(b)

Hairiness

(c)

Hairiness

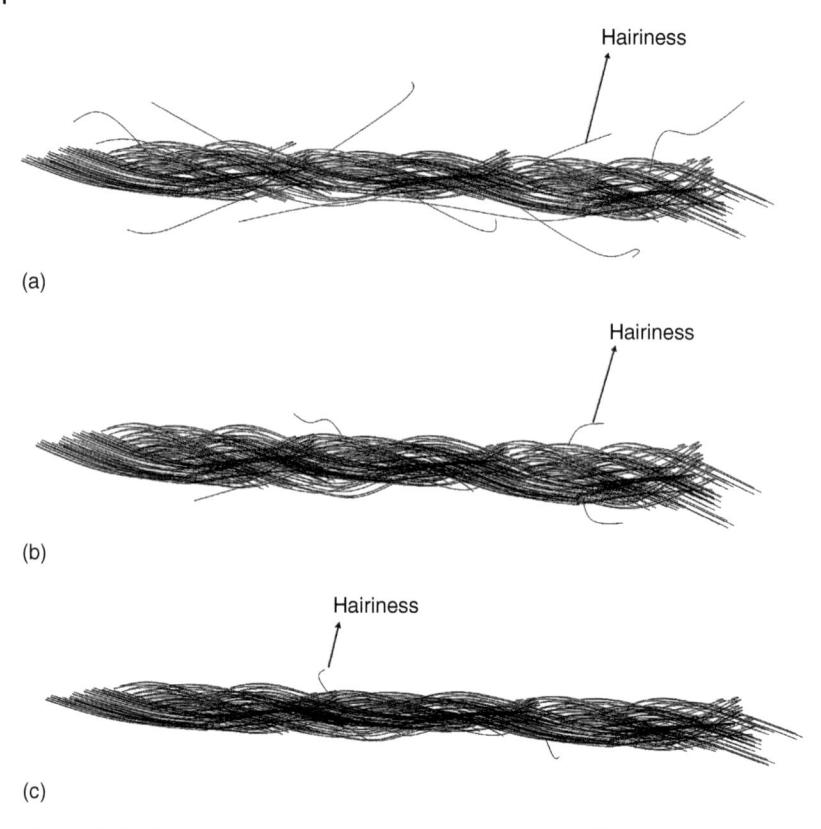

Figure 17.6 Different types of ring-spun yarn. (a) Carded yarn. (b) Combed yarn. (c) Compact yarn.

As many of raw material is wasted in the form of short fibers and neps, cost rises. That is why combed yarns are relatively expensive then carded yarns. A schematic diagram of combed yarn is shown in Figure 17.6b. After the combing process, upper medium counts, as well as very fine counts, may be produced, which have superior properties along with smooth fiber orientation and high comfort. These combed yarns have great potential to be used for highly attractive and fashionable textiles with excellent comfort and durability on behalf of exclusive performance amalgamation.

17.6.2.2 Compact Yarn

Yarn strength may also be improved by fiber condensation. A modified ring spinning technique so-called compact spinning is used to condense fibers aerodynamically. This process not only improves yarn strength but also reduces hairiness. Very fine count yarns of high strength may be produced through this technique. High-quality fabrics and apparels may be produced. A schematic diagram of compact yarn is shown in Figure 17.6c.

Prominent features:

- The ring yarn has better parallelization and orientation than other yarn types.

Figure 17.7 Modern ring spinning machines.

- It has larger hairiness, pilling, and unevenness.
- It has better strength, surface smoothness, and core fibers as compared with the other yarn types.
- It has a high breaking strength.
- The elongation at break of ring-spun yarn is lower than other yarn.
- Ring-spun yarn has a high imperfection index.
- Ring-spun yarn has lower abrasion resistance and stiffness.

Figure 17.7 indicates a modern ring spinning frame.

17.6.3 Rotor-Spun Yarn

Rotor spinning is a technique that is used for creating yarn without the use of a spindle. It is also known as open-end (OE) spinning, break spinning, or free fiber spinning. It was invented and developed in Czechoslovakia. Unlike the ring spinning, rotor spinning combines two process stages of spinning and winding in a single machine. Usually, the coarser counts are spun on the rotor. Mostly the count is below 20s, but it may as high as 40s. Rotor spinning is more successful method than many other open-end spinning methods. The fiber bundle that can be obtained from the sliver feedstock are separated into individual fibers by using the opening roller in an air stream, and the fibers that are separated are recollected using rotor groove. In rotor spun the yarn needs more twisting than ring-spun yarn. It has much application in knitwear and woven cloth, but

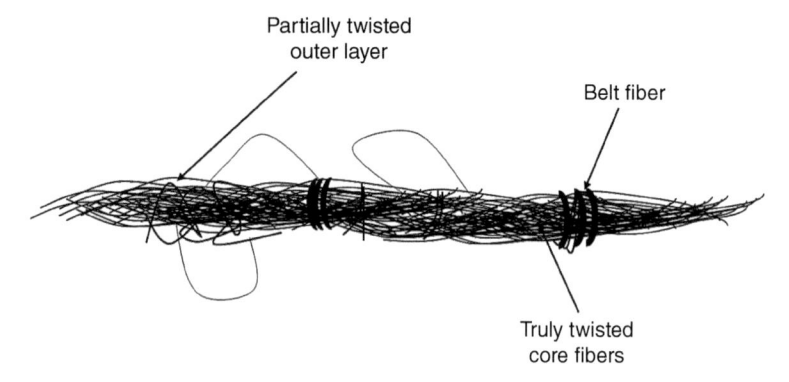

Figure 17.8 Schematic structure of a rotor yarn.

mostly its use remains in denim. It is also used in medical and industrial fabrics. The fabric is absorbent and contains a longer lifetime that is made by using open-end yarn.

17.6.3.1 Yarn Structure

The rotor yarn can adopt a carded or a combed process before it is fed to the rotor spinning system. The rotor yarn contains three-layer structures, which include truly twisted core fibers, partially twisted outer layer, and belt fibers. The schematic illustration of a rotor-spun yarn is shown in Figure 17.8. Usually coarser yarns are made on rotor spinning, which requires higher twist than ring spun. Comparatively the rotor yarn is weaker than ring-spun yarns. However, it has a lower mass variation as compared with ring spun. In addition, unlike the ring-spun yarn, the twist is inserted from center to outer side and in some cases in opposite direction. Figure 17.8 shows the schematic of a rotor yarn structure, and Figure 17.9 indicates a modern rotor machine.

Prominent features:

- The rotor yarn is more extensible as compared with ring yarn.
- Rotor spinning system takes less power than ring spinning system.
- The imperfection index of rotor yarn is lower than ring yarn.
- Rotor spinning requires more maintenance than ring machine.

17.6.4 Air Vortex Spun Yarn

This technology was developed in 1997 in Japan by Murata Company. In this process, the yarn spinning is carried out through a stream of the air jet. After its invention, a lot of improvement and modification had been introduced with the latest improvement in 2008. Murata Vortex Spinning technology is used for a wider range of fiber lengths.

17.6.4.1 Yarn Structure

In this spinning system, the drafted fibers are introduced in a spindle through an air vortex. The twist is inserted through a swirling air when the fibers are entering

Figure 17.9 Rotor machine.

and passing through the orifice. Also the lengthier fiber moves to the core, and the smaller fiber departs to the outer layer with the application of air. This procedure divides the vortex yarn into two parts in which one is twisted and other is untwisted as shown in Figure 17.10. The air vortex spinning requires less maintenance than other spinning systems due to the lesser moving parts. The yarn production can be delivered up to 400 m/min. Also, the roving formation stage is eliminated with the addition of fully automatic piecing system. Air vortex yarn has lesser hairiness than the ring-spun yarn. This is due to the air singed and air combed process, which results in reduced fabric pilling.

Prominent features:

- Vortex yarn has more abrasion resistance than other yarns.
- Vortex yarn is stiffer than other yarns, which indicate that it has a high value of tenacity.

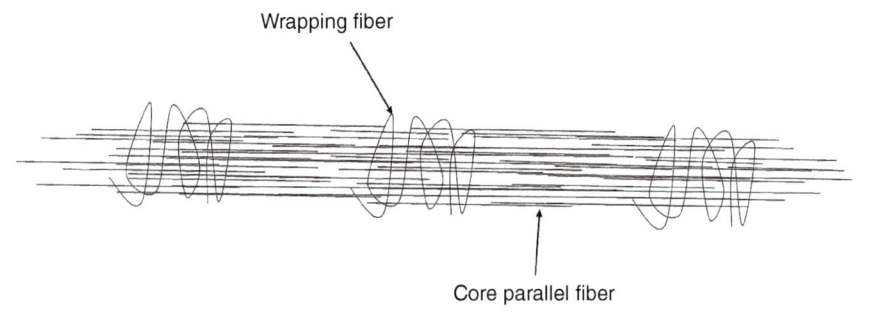

Figure 17.10 The schematic diagram of an air vortex yarn.

- Vortex yarn has minimum lint shedding and can hold its bulk as compared with ring-spun yarn, which makes a potential use as high-quality towels.
- It is used in fancy or high-fashioned fabrics due to lesser hairiness, which enhances its appearance.

The yarn structure of ring, rotor, and vortex differs significantly in terms of twist imparted. The twist insertion in the yarn structure is different in these techniques. Figure 17.11 shows this difference graphically. Ring-spun yarn has a uniform twist from core to outer layer. Rotor-spun yarn has twisted fibers in the core along with the belt fibers on the outer layer of yarn. The abrupt change in the twist level indicates the belt fibers for OE yarn in Figure 17.11b. Vortex yarn has a minimum twist in the core parallel fibers. However, this twist increases abruptly for the wrapping fibers at the outer layer of the yarn structure.

17.6.5 Frictional Spun Yarn

Frictional spun yarns are made using friction spinning. The yarn illustrates truly twisted fibers, but a great several fiber loops are present. It can be made only in very coarse yarn for the use of industrial applications. It is largely used for making industrial yarns of various structures including core/sheath yarns. Blends of raw fiber and waste fiber can also be used for its preparation. Figure 17.12 shows a schematic diagram of a frictional spun yarn.

17.6.6 Mechanical Properties of the Yarn

The mechanical properties of yarn include its behavior under the application of forces along with its recovery behavior when the forces are removed. As mentioned before, the yarn mechanical behavior depends on the fiber mechanics and yarn structure. We have discussed the yarn structure in different processing techniques. Almost all of these types include twist as a binding force of fibers. However, the level of twist in yarn varies on the final desired characteristics in the fabric.

17.6.7 Important Parameters of Yarn Tensile Strength

Mechanical properties of yarn are of core interest in the field of textile research. The scientists have discovered so many relations that relate mechanical properties of yarn with structural features. Yarn mechanics may also be addressed greatly by using computer simulating software. Many softwares are now available in the market. Studies reveal that an ideal model of twisted yarn is that which bear a resemblance to filament yarn. This means the greater the resemblance with a filament yarn, the better the twisted yarn. Ring-spun yarn provides great resemblance due to its structure. Fibers are uniformly distributed and combined in yarn under tension. This structure made the behavior of fibers easily foreseeable and understandable. Hearle et al. [6] discovered a theoretical relationship for determining the mechanical properties of continuous filament yarn. Spun yarns are more complex to understand because they have less tension in the surface layer as

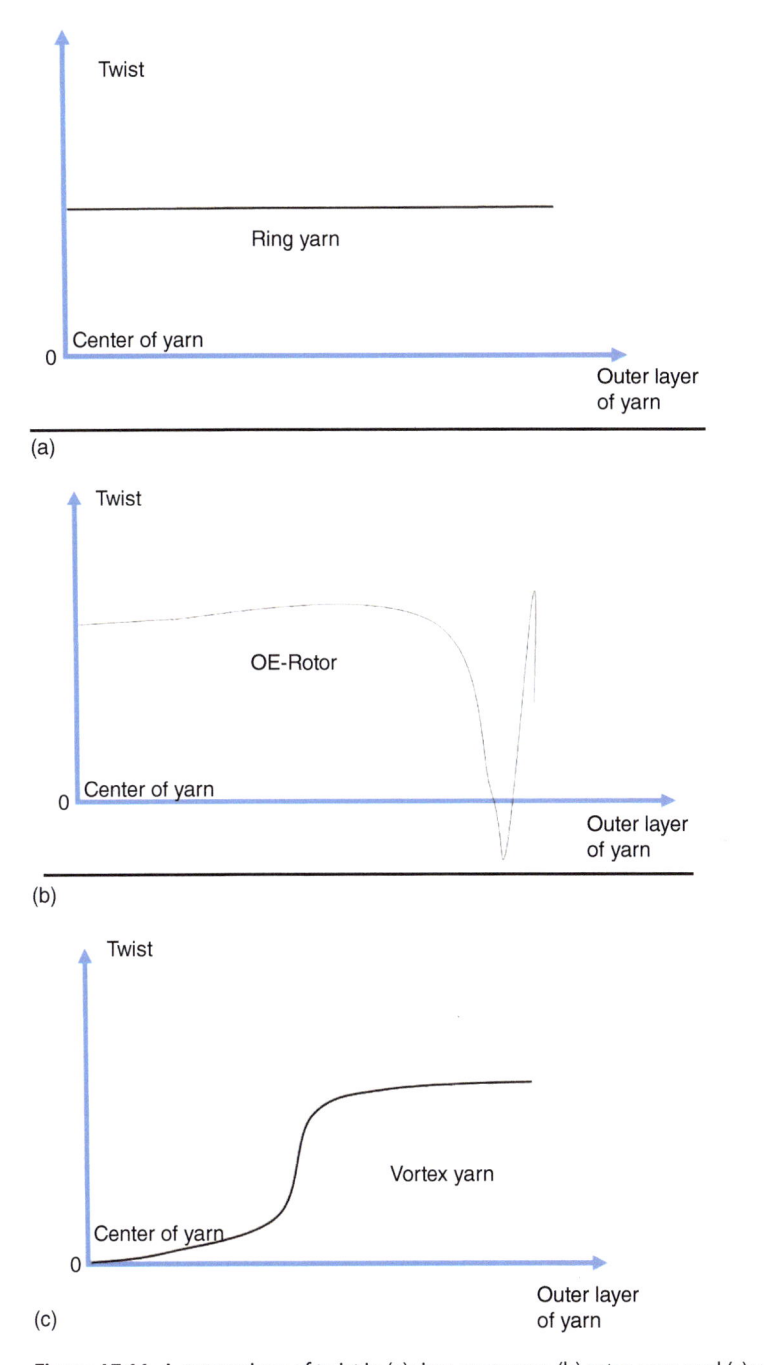

Figure 17.11 A comparison of twist in (a) ring-spun yarn, (b) rotor yarn, and (c) vortex yarn.

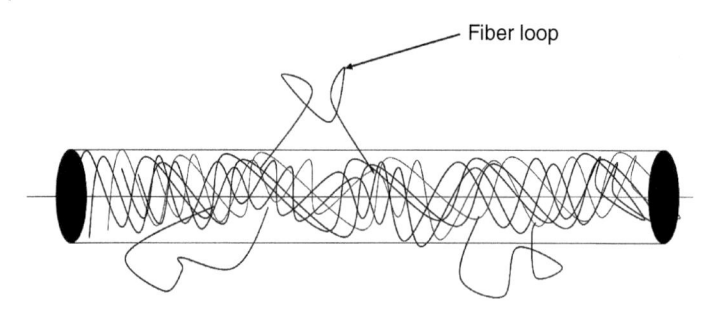

Figure 17.12 Schematic structure of friction spun yarn.

compared with inner layers due to discontinuous fiber segments or staple fibers. So, some empirical and analytical amendments were made to accommodate such complex yarns. This discovered relationship is written below:

$$\frac{\text{Yarn strength (or modulus)}}{\text{Fiber strength (or modulus)}} = \frac{E_y}{E_f} = \cos^2\alpha\,[1 - K\csc\alpha]$$

where

E_y	yarn modulus
E_f	fiber modulus
α	twist angle

and "K" may be defined by the following expression:

$$K = \frac{\sqrt{2}}{3L_f}\left(\frac{aQ}{\mu}\right)^{1/2}$$

where

L	fiber length
a	fiber radius
Q	margin period
μ	coefficient of friction

The above relationship relates to the strength of yarn on twist by two basic parameters; first is "$\cos^2\alpha$," and second is "$1 - K\csc\alpha$."

"$\cos^2\alpha$" can only be used when studying the continuous filament yarn, and it cannot be used for staple fiber spun yarn, whereas "$1 - K\csc\alpha$" is the parameter used to measure twist angle of fibers in spun or staple yarn. The previous parameter shows a trend of yarn strength with a twist angle. In the beginning, the yarn strength is inversely related to twist angle, but after it shows a linear relationship with a twist angle. The yarn strength may also have a great dependence on "K" value; the lower the K value, the higher the strength of yarn. K value increases as the length of fiber contracts, which means long fibers will impart more strength into yarn as compared with short fibers.

Fiber diameter also has an impact on yarn strength. As the diameter increases, the K value also increases, which leads to lower strength of overall yarn. The above expression also relates the yarn strength with fiber friction. Yarn strength may also be improved by using fibers having high friction. The greater the friction between fibers, the lesser the value of "K," and hence, the higher the strength.

Scientists have revealed that the above expression is based on estimation. However, the key parameters of yarn structure like fiber length, fiber diameter, twist angle, friction, and consolidation may be easily addressed through it. That's why it may be proved very beneficial while predicting the properties of newly developed yarn or developing a new yarn of specified properties.

17.6.8 Strength–Comfort–Twist Relationship

While designing the yarn strength, other parameters, which may have an impact on yarn integrity, should not be ignored. For example, yarn strength increases as a number of twists per unit length increase. However, this effect may not be true after a certain level says optimum twist. After this point, an increase in the number of twists per unit length will decrease the strength of yarn. That's why some twists must be optimized. A trend of yarn strength on twist factor is shown in Figure 17.13. The reason behind twist–strength relationship is "friction." For example, when there is no twist in yarn, fibers are oriented parallel to the axis of yarn, but due to lack of cohesive force, slippage of fibers may reduce the overall strength of yarn. A small twist may provide the necessary cohesive forces in terms of friction and consolidation. Whenever somebody is going to stretch the yarn, the transverse pressure will build up, causing the frictional forces to increase.

Further increase in twist may cause more friction between fibers, which will provide more intertwining and cross-linking points to facilitate fiber to fiber attraction by joining fibers at their ends with other fibers on the same axis. Another important factor is fiber strength. The overall strength of yarn is distributed on fiber level. The greater the fiber strength, the higher the load bearing property of yarn, and hence the higher the overall strength of yarn. Both

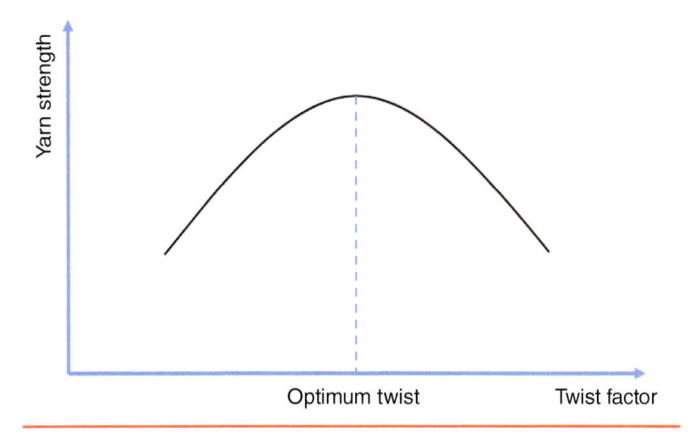

Figure 17.13 Schematic of a yarn strength with increasing twist.

the frictional forces (the number of twists per unit length) and fiber strength play their integral role in the overall strength of yarn. Most of the time, the role of fiber strength is more significant than frictional forces while determining tensile resistance. However, the role of frictional forces due to twist is also very important to bind the fibers together. The frictional force is a key parameter for spun yarn to bind short, discrete fibers together and to strengthen the yarn. As discussed before, yarn strength can only be increased by twisting up to a certain level, which is called an optimum twist. After that optimum value, any further increase in twist will cause inclination of fiber to the yarn axis. That will obviously reduce the influence of fiber strength to overall yarn strength. Therefore, after a critical value, any further increase in twist will cause in the reduction of yarn strength. From the above discussion, it is concluded that there are two leading possessions, overriding the strength–twist relationship. The first procession is the "binding force between fibers," and the second procession is "angle of inclination." Yarn strength will increase as a binding force between fibers increase, which is the result of twist before critical (optimum) value. The angle of inclination of fibers to the yarn axis also plays a vital role in terms of contribution of fiber strength to overall yarn strength.

This phenomenon happens after a critical point. As twist increases, the angle of inclination increases, causing a decrease in the contribution of fiber strength to overall yarn strength. An interpretation of strength–twist relationship is shown in Figure 17.14. The red curve shows the increasing trend for the binding force between fibers as the result of a twist. In the beginning, all fibers will slip as there is no binding force. As the number of twists per unit length increases, the fiber slippage decreases. The blue curve shows the contribution of fiber strength to overall yarn strength. The contribution decreases as a number of twists per unit length increase. The point at which two curves coincide is the critical value of twists per unit length at which maximum yarn strength is obtained. The value of twist below this point is called an optimum twist. The twist level may also affect the properties of fabric like fabric hand and skew. Yarns that have a high level of twist are dynamic, and they have the potential to untwist. In the case of spun yarn,

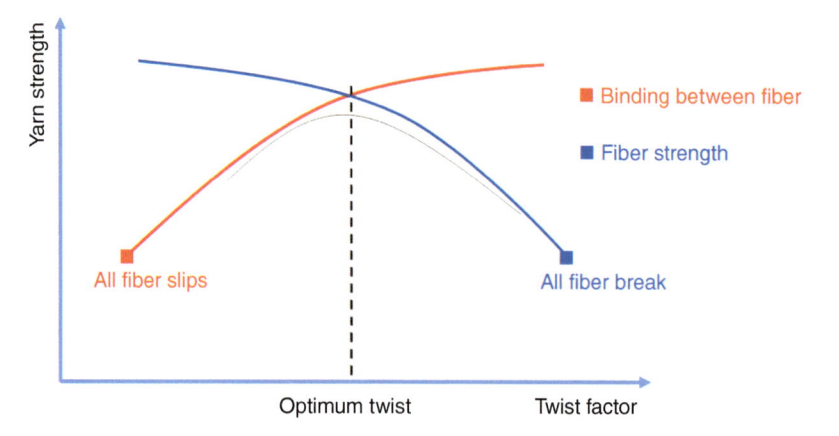

Figure 17.14 Schematic of an interpretation of strength–twist relationship.

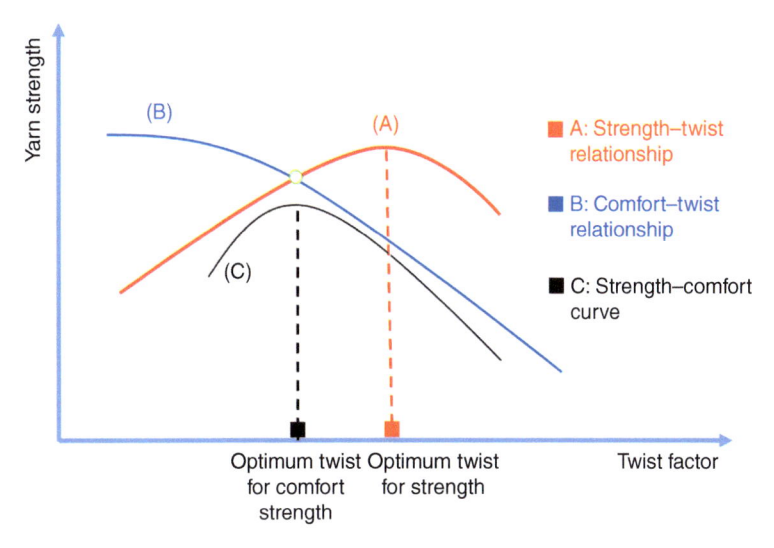

Figure 17.15 Schematic of a comfort–strength–twist relationship.

strength may be increased by the high level of twist (up to critical value), but it will reduce the softness, which is highly undesirable in fabrics [7, 8]. Fabric should not be too hard to wear. That's why not only strength but also the comfort of yarn and fabric should also be addressed properly. Y. EL Moghzy et al. examined the effect of twist on the harshness of yarn and ultimately on the discomfort of fabric [7].

They concluded that both yarn softness and yarn strength should be made under consideration to knit (or weave) wearable fabric. The twist–strength relationship was overlaid by a third parameter, which is "comfort index," which also depends on twist index. In other words, the greater the twist index, the higher the yarn strength, but adversely the lower the comfort (or softness) of yarn. As shown in Figure 17.15, the optimum twist is the point on the x-axis at which two curves (A and B) coincides. The curve A is the strength–twist relationship, which shows the increasing trend of yarn strength with twist index. The curve B is a twist–comfort relationship, which shows that comfort will decrease as twist index increases. So, an optimum level of twist was suggested. For the sake of convenience, comfort index may be denoted by numbers from zero to one. Zero indicates discomfort, and one indicate maximum possible comfort. The strength index may be described as the ratio of actual strength obtained at a certain level of twist to the maximum possible strength. The curve C is the optimization of these two active parameters. Both parameters are twist dependent. So, both strength and comfort are optimized by the twist. That's why this curve is called a strength–comfort characteristic curve. Estimation of yarn strength and prediction of yarn strength are two key factors that are addressed by yarn mechanics. Empirical methods may be employed to optimize yarn strength regarding processing parameters. But these empirical models are very limited to specific processing conditions. However, they may be modified by other processing methods. But still, they cannot be employed in generalized form to address all the processing conditions and methods.

References

1 Stout, H. and Harrison, P. (1988). *Fibre and Yarn Quality in Jute Spinning*. Textile Institute.
2 Kumar, S. and Gupta, V. (1978). A nonlinear viscoelastic model for textile fibers. *Textile Research Journal* 48 (7): 429–431.
3 Cai, Z. (1995). A nonlinear viscoelastic model for describing the deformation behavior of braided fiber seals. *Textile Research Journal* 65 (8): 461–470.
4 Jianchun, Z. (1999). The model of viscoplastoelastic behavior of tencel fibre. *Journal-Northwest Institute of Textile Science and Technology* 13 (4): 399–402.
5 Baltussen, J. and Northolt, M. (2004). The Eyring reduced time model for viscoelastic and yield deformation of polymer fibres. *Polymer* 45 (5): 1717–1728.
6 Hearle, J.W., P. Grosberg, and Backer, S. (1969). *Structural Mechanics of Fibers, Yarns, and Fabrics*. USA: Wiley-Interscience.
7 El Mogahzy, Y., Kilinc, F.S., and Hassan, M. (2005). Developments in measurement and evaluation of fabric hand. In: *Effect of Mechanical and Physical Properties on Fabric Hand* (ed. H. Behery), p. 45. Elsevier.
8 El-Mogahzy, Y. (2008). Structure and mechanics of yarns. In: *Structure and Mechanics of Textile Fibre Assemblies* (ed. Peter Schwartz), pp. 190–212. Cambridge: Woodhead Publishing Limited.

18

Textile Mechanics: Woven Fabrics

Adeel Zulifqar[1], Zubair Khaliq[2], and Hong Hu[1]

[1] The Hong Kong Polytechnic University, Institute of Textile and clothing, Hung Hom, Kowloon, Hong Kong, China
[2] National Textile University, Faculty of Engineering and Technology, Faisalabad, Pakistan

18.1 Introduction

Woven fabric is a textile material that is fabricated by interlacement of two sets of constituent orthogonal yarns called the warp (along length of fabric) and the weft (along width of fabric). The interlacement of warp and weft with each other results in an integrated and firm structure. The inter-yarn friction is responsible for holding the constituent yarns in place. The properties of the woven fabric depend greatly on the fabric structural constraints like weave, thread density per unit area, crimp, and linear densities of warp and weft yarns. The basic components of woven fabric are weave, yarns, and fibers. All these components together with post-fabrication treatments influence the ultimate fabric properties and mechanical behavior in real use. Therefore, in woven fabric (with desired mechanical behavior) manufacturing, selection of fibers with individual properties, arrangement of fibers in the yarn structure, and shaping the interlacements of warp and weft yarns in different fashions within the fabric are some vital factors to be considered. This gives the woven fabric engineer a great liberty to control and modify the woven fabric structure for desired mechanical behavior.

This section aims to describe woven fabric mechanics and factors in relation to the constituent yarns and fabric structure, which may affect the ultimate mechanical properties of woven fabrics. Factors including weave, crimp, weave factor, and structural geometry are discussed. The importance of woven fabric geometry in terms of mechanical behavior cannot be ignored; therefore, the geometries of woven fabric proposed by various fabric scientists are discussed. In order to develop a better understanding of woven fabric mechanics, different modeling approaches including Pierce's geometric model of woven fabric, numerical models, constitutive laws, energy methods, and modern modeling methodologies like computational mechanics are also included in the scope of this section.

Handbook of Fibrous Materials, First Edition. Edited by Jinlian Hu, Bipin Kumar, and Jing Lu.
© 2020 Wiley-VCH Verlag GmbH & Co. KGaA. Published 2020 by Wiley-VCH Verlag GmbH & Co. KGaA.

18.1.1 Woven Fabric Structures

The structural unit or weave of a fabric is termed as repeating unit, and it is the interlacement of warp and weft in a predefined regular or irregular configuration. The repeating unit contains the smallest number of different interlacements between warp and weft, which, when repeated in either direction, gives the fabric structure. It is usually presented by a grid of squares in which horizontal and vertical lines depict the weft and warp yarns, respectively. Each square shows the interlacement point of warp and weft. The filled square means that the warp yarn is over the weft yarn, while blank means that weft is over the warp at the corresponding place in the fabric. The weave of the fabric influences greatly the properties of fabric including its mechanical behavior in response to tensile loads [1].

18.1.1.1 Regular Woven Structures (Basic Weaves)

The warp and weft yarns intersect/interlace in same manner, and the interlacement pattern of consecutive warp and weft yarns relative to each other is displaced in equal steps in one repeat. The basic regular weaves including plain, twill, satin, and sateen are explained briefly in the following sections [2].

Plain Weave The most frequently used and simplest weave is the plain weave. It is the most rigid and stronger network of warp and weft yarns owing to the fact that it has maximum number of possible interlacements of warp and weft. It is usually written as 1/1, which means that warp yarn runs over and under the weft yarn in alternate fashion. The numerator represents number of weft yarn under warp yarn, while the denominator represents number of weft yarns over warp yarns. The repeat is the sum of two numbers $(1 + 1 = 2)$, which states that after every two warp yarns or weft yarns, the interlacement pattern is repeated as shown in Figure 18.1a.

Twill Weaves Twill weaves are described by continuous diagonal lines created by the interlacements of warp and weft yarns. Twill weaves are less rigid than plain weaves with same thread density per unit area owing to the fact that they have less number of interlacements per repeating unit. The smallest number of threads that can be used to construct twill is three. It is written as 2/1 (which means that the warp yarn runs over two weft yarns and under one weft yarn in diagonal fashion and that it is warp-faced twill) or 1/2 (which means that the warp yarn runs over one weft yarn and under two weft yarns in diagonal fashion and that it is weft

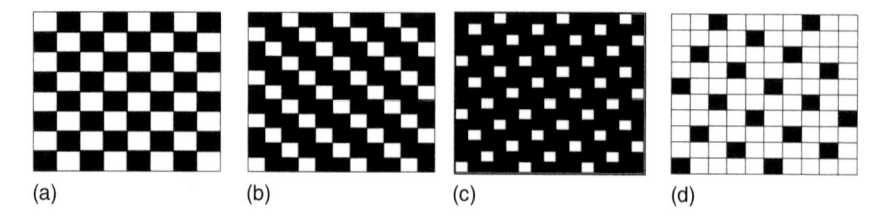

Figure 18.1 Unit cells of regular weaves. (a) Plain, (b) twill, (c) satin, and (d) sateen.

faced twill). The repeat is $1+2 = 2+1 = 3$. Therefore, after every three warp yarn or weft yarns, the interlacement patterns are repeated. The direction of diagonal can be up to the right (Z twill) or up to the left (S twill) depending upon the starting point of twill as shown in Figure 18.1b.

Warp and Weft Satins In satin or sateen weaves, the interlacement points of warp and weft yarns are distributed such as they are never adjacent. The smallest number of threads that can be used to create a satin or sateen is five. A warp satin is warp faced written as 4/1 as shown in Figure 18.1c, while a weft satin (also termed as sateen) is weft faced written as 1/4 as shown in Figure 18.1d. Warp satins are usually created by higher number of warp yarns than weft yarns. Although an infinite number of interlacements are possible, but five 4/1 and eight 7/1, end satin weaves are most common.

18.1.1.2 Irregular Woven Structures

Unlike regular or basic weaves, the warp and weft yarns interlacements in irregular weaves are not in same manner, and the interlacement pattern of consecutive warp and weft yarns relative to each other is not displaced in equal steps in one repeat as shown in Figure 18.2 [3].

18.1.2 Weave Factor

Weave factor is a number that accounts for the number of interlacements of warp and weft in a given repeat [3, 4]. It is also equal to average float and is expressed as in Eq. (18.1):

$$M = \frac{E}{I} \tag{18.1}$$

where M is the weave factor, E is number of threads per repeat, and I is number of intersections per repeat. The weave interlacing patterns of warp and weft yarns may be different. In such cases, weave factors are calculated separately with suffix 1 and 2 for warp and weft, respectively. Therefore,

$M_1 = \frac{E_1}{I_2} E_1$ and I_2 can be found by observing individual weft pick in a repeat, and

$M_2 = \frac{E_2}{I_1} E_2$ and I_1 can be found by observing individual warp end in a repeat.

As in the case of irregular weaves, the number of intersections of each thread in the weave repeat is not equal. In such cases the weave factor is obtained as in

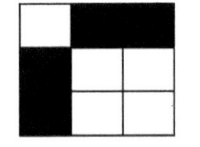

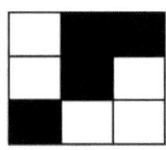

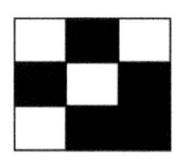

Figure 18.2 Unit cell of some irregular weaves. Source: Behera et al. 2012 [3]. https://www.intechopen.com/books/woven-fabrics/modeling-of-woven-fabrics-geometry-and-properties. Licensed under CCBY 3.0.

Eq. (18.2):

$$M = \frac{\sum E}{\sum I} \qquad (18.2)$$

18.1.3 The Myth of Crimp

Crimp is the waviness in the yarn produced due to the interlacement of warp and weft yarns. The amount of crimp is a result of various factors that include the tension in the warp yarn, linear densities of warp and weft yarns, type of fibers in warp and weft yarn, twist level in both yarns, and the weave. The crimp is calculated by Eq. (18.3):

$$C = ((L_e - L_s)/L_s) \times 100 \qquad (18.3)$$

where L_s is the length of yarn removed from the fabric sample and L_e is the extended length of the removed yarn. The crimp is expressed as percentage of sample length of yarn [4], generally, the warp faced.

18.2 Woven Fabrics Geometrical Models

The geometric structure of a woven fabric is extremely complicated and differentiates greatly from a continuum structure such as a metal sheet. The yarns in a woven fabric are crimped at the interlacement point with irregular cross-sectional shapes and having frictional contact with each other. The porosity of fabric depends upon the distance between two parallel adjacent yarns. The simplest theoretical model of yarn configuration is developed by Peirce in 1937, but in actual scenario the model contradicts with reality because in the theoretical study, the cross-sectional shape and physical properties of a yarn are always simplified and idealized. Peirce studied the shape taken up by the warp and weft yarn in a woven fabric and proposed a geometrical model [5]. The mathematical relations derived from the Peirce's geometrical model and physical characteristics of yarn can be used to understand woven fabric behavior under various conditions. The Peirce's geometrical model is shown in Figure 18.3. It exemplifies a unit cell of a plain woven fabric and satisfies to some degree loom state of the fabric. Peirce considered that the yarns as inextensible and flexible

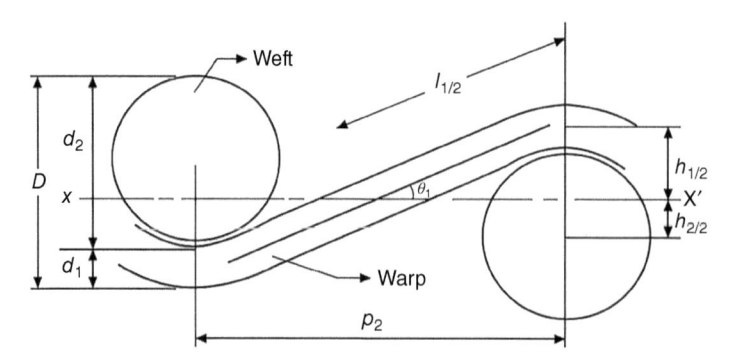

Figure 18.3 Geometrical model of plain woven fabric. Source: Peirce 1937 [5]. Reproduced with permission from Taylor & Francis.

having a circular cross section consisting of straight and curved segments. This simple geometrical model helps to study the relationship between various geometrical parameters. The parameters that can be studied by using Peirce's geometrical model of a plain woven fabric includes resistance of the fabric to mechanical deformation (initial extension, bending, and shear in terms of the resistance to deformation of individual fibers), relative resistance of the fabric to the passage of air, water or light, and maximum yarn packing density possible in the fabric. The geometrical parameters of a unit cell of plain woven fabric Peirce studied are as follows:

c = crimp
h = crimp height (maximum displacement of thread axis normal to the plane of fabric)
q = weave angle in radians (angle of thread axis to the plane of fabric)
d = diameter of thread, fabric thickness, and
p = thread spacing and derived a set of equations
l = modular length (length of thread axis between the planes through the axes of consecutive cross-threads)

$$D = d_1 + d_2$$

The suffixes 1 and 2 represent warp and weft threads, respectively. The estimation of yarn axis parallel and normal to the fabric plane provides the following equations:

$$c_1 - \frac{l_1}{p_2} - 1 \tag{18.4}$$

$$p_2 = (l_1 - Dq_1)\cos q_1 + D\sin q_1 \tag{18.5}$$

$$h_1 = (l_1 - Dq_1)\sin q_1 + D(1 - \cos q_1) \tag{18.6}$$

Three similar equations are obtained for the weft direction by interchanging suffixes 1 and 2 or vice versa as follows:

$$c_2 - \frac{l_2}{p_1} - 1 \tag{18.7}$$

$$p_1 = (l_2 - Dq_2)\cos q_2 + D\sin q_2 \tag{18.8}$$

$$h_2 = (l_2 - Dq_2)\sin q_2 + D(1 - \cos q_2) \tag{18.9}$$

$$d_1 + d_2 = h_1 + h_2 = D \tag{18.10}$$

The above equations provide relationship between eleven variables. If any four variables are known, then the equations can be solved for unknown variables [5].

Whereas the assumption of a circular thread cross section is mostly valid in the case of very open structures, due to the inter-yarn pressures that are set up during weaving process especially in the case of tightly woven fabrics, there might be yarn flattening normal to the plane of the fabric. Peirce took this into consideration and assumed the possibility of fabric geometry based on elliptical sections as shown in Figure 18.4, by replacing the circular thread diameter by the minor diameter of the appropriate elliptical section. This approximate treatment was

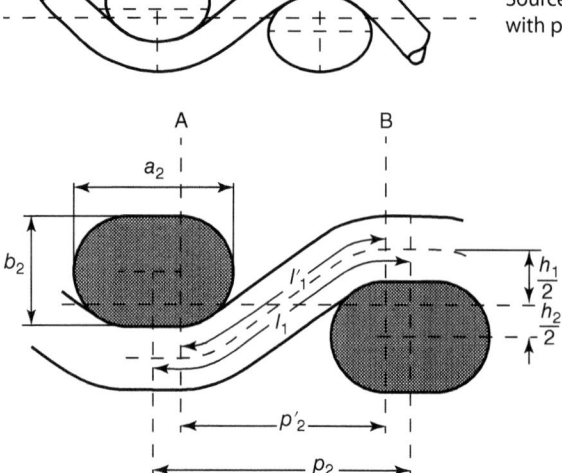

Figure 18.4 Elliptical cross section. Source: Peirce 1937 [5]. Reproduced with permission from Taylor & Francis.

Figure 18.5 Kemp's racetrack cross section. Source: Kemp 1958 [6]. Reproduced with permission from Taylor & Francis.

still invalid for tightly woven fabrics, but this model is practically valid for fabrics after dry and wet relaxation treatments. To overcome this difficulty, Kemp proposed new fabric geometry by considering yarn cross-sectional shape as racetrack section [6], which encloses a rectangle between semicircular ends as shown in Figure 18.5. This allowed the application of simple relations of circular thread geometry for the treatment of flattened thread. Generally speaking, the racetrack model is reasonably valid for pressed and calendared fabrics. Further, Hamilton used Kemp's racetrack approach of the plain woven fabrics for other weaves [7]. Olofsson and Grosberg considered yarn geometry as a function of external and reaction forces in the fabric and assumed a relation between the curvature of the yarn in the fabric and in the released state [8, 9]. Dorkin et al. developed a simpler model by assuming that the path of the yarn is made up of two equal circular arcs [10]. It has been found that the yarn shapes within the woven fabric is influenced by the weaving force, the warp and weft setts, and the finishing conditions [11–13].

According to the study of woven fabric geometrical models, it can be concluded that Peirce's flexible model, Peirce's elliptical model, and Kemp's racetrack model are true representative of the actual fabric structure.

18.3 Woven Fabric Mechanics, Theories, and Methodologies

Woven fabric mechanics gained interest of researchers in 1912 when development of airships was under focus of engineers and scientists. In 1937, Peirce first established the geometrical model of plain weave structure [5]. Other researchers contributed toward the further advancement in the area [8, 9, 14–21]. Up till today, the researchers adopted three different approaches to carry out studies in this area. Firstly, the component-oriented approach to predict the mechanical behavior of woven fabrics directed by Hearle, Grosberg, and Postle. They have studied the physical theories together with yarn properties, inter-yarn interactions, and fabric structures on the basis of Newton's third law, minimum energy principles, and mathematical analysis of construction. Secondly, the phenomenon-oriented approach that involves the behavior of woven fabrics in response to external loads including elastic, plastic, viscoelastic, and frictional aspects. Thirdly, the result-oriented approach that involves a hypothesis, a function, and a statement to describe the results and then finding a relationship of the function with fabric behavior. Theoretically, this approach involves pure mathematics and statistics; therefore, the wrong assumptions can be avoided. Woven fabrics possess complexity in terms of geometrical configuration of interlaced constituent yarns; this is the major difficulty of computational modeling of woven fabrics. The tensile deformation of a woven fabric is obviously related to that of constituent yarns, which is the combination of bending, tensile, compression, and also the intrafiber frictional sliding. In order to simplify the modeling, some assumptions must be made, for example, it is assumed that the yarn is an elastic homogeneous material and would exhibit a viscoelastic behavior. Therefore, the elastic property of homogeneous yarn could be either linear or nonlinear. To study the mechanics of woven fabric structure, it is essential to include the contacts of warp and weft yarns with each other in the modeling because the frictional contact of both at cross over point greatly influences the stress and strain relationship for a woven fabric, and during frictional contact the force would be transferred from one yarn to the contacting and adjacent yarn [22].

18.3.1 Tensile Deformation of Woven Fabrics

Unlike traditional engineering materials (TEMs), woven fabrics are inhomogeneous, discrete, and anisotropic. They can undergo large strains even at low stresses and can easily be deformed into shapes with double curvatures through buckling [14, 15]. The study of woven fabric tensile behavior reveals that small tensile forces produce fairly large membrane strain as shown in Figure 18.6a. The reason for this large deformability might be the straightening of the crimped constituent yarns within the fabric. The initial tensile modulus of a typical woven fabric is observed to be of the order of 10 MPa, compared with elastic modulus of steel, i.e. of 2×10^5 MPa. The woven fabrics under transverse loading are more prone to bending deformations in contrast to tensile deformation as sown by Figure 18.6b. Unlike stiff materials, the large deformations cannot be neglected in the engineering of woven fabrics. Woven fabrics are prone to move through these large deformations and buckle even at very small compressive stresses within fabric plane under its own weight and/or external forces. TEMs undergoes small strains at low stresses, and the stress and strain relationship is linear, but for woven fabrics this relationship is nonlinear at low stresses and becomes almost linear beyond a critical value of stress. This critical value is higher in the case of tensile loading and almost zero in other modes of loading like shear and bending. This behavior of woven fabrics could be explained by the fact that under tensile loading or tension, the crimped yarns get straight at

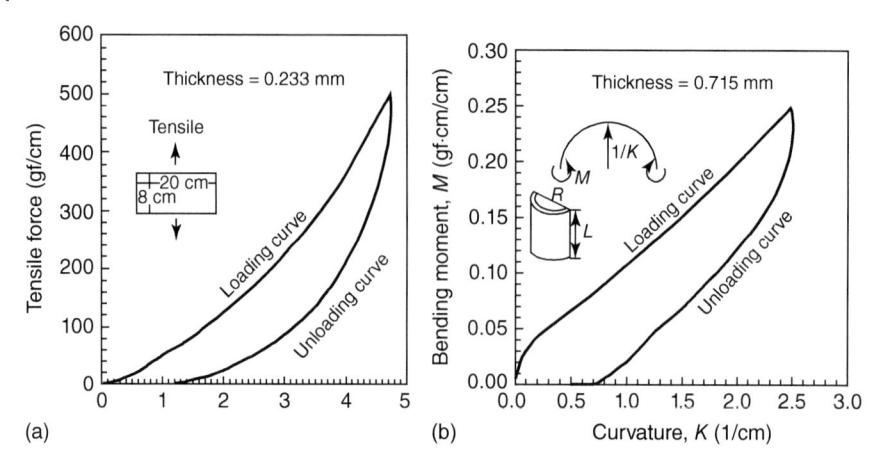

Figure 18.6 (a) Tensile stress–strain curve of a woven fabric. (b) Moment–curvature curve of a woven fabric. Source: Hu 2004 [22]. Reproduced with permission from Elsevier.

low stresses, and when they became fully straight at high stresses, the interfiber friction is increased.

At this stage, the fabric is more consolidated, and the fibers are in better orientation than before resulting in a linear relationship of stress and strain. The nonlinearity occurs at intermediary stage of fabric consolidation. An interesting observation from this explanation for tensile behavior of woven fabrics can be made in the case of TEMs when the applied stress increases the microstructure moves toward disorder. Conversely, in the case of woven fabrics, the applied stresses bring about order in the microstructure as the fiber orientation is improved [22].

18.3.1.1 Woven Fabric Behavior When Extended in Principal Directions

When a plain woven fabric is extended in principal direction (either along warp or weft direction), the crimped yarns will get straight along the direction of applied force. At this stage, the yarn amplitude and the weave angle will decrease as the contact between warp and weft yarn increases, and they tend to become more circular [19]. Furthermore, the crimp interchange will take place at the interlacement point, increasing the crimp in one set of yarns while decreasing in the other. Further extension of fabric is followed by yarn extension with decrimping and fiber extension within the yarn at the interlacement point, but the yarn extension is relatively small as compared with total extension. The movement of fiber restricts the development of high strains. During this process some energy loss will also occur due to interfiber friction. De Jong and Postle reported that the following six independent dimensionless parameters must be considered in order to study uniaxial tensile properties for a balanced plain woven fabric produced from identical warp and weft yarns [16]. The parameters include (i) the ratio of warp to weft yarn length per crossing yarn, (ii) the ratio of yarn diameter to yarn modular length, (iii) the ratio of yarn compression rigidity to bending rigidity, (iv) the yarn compression index, (v) the ratio of yarn extension rigidity to bending rigidity, and (vi) the degree of set. They also stated that the fabric extension can be explained by yarn extension when the ratio of yarn compression rigidity to bending rigidity is lower. Furthermore, the Poisson's ratios can also be explained; during extension the inter-yarn distance will increase, and yarns tend to become more circular. Conversely, a higher magnitude of tensile properties will result when fabric is extended in the bias directions than in the warp and weft directions.

18.3.1.2 Woven Fabric Behavior When Extended in Bias Direction

When a woven fabric is extended in a bias direction as shown in Figure 18.7, the maximum elongation will occur along the direction of applied force and with a deviation in the direction ($\theta_1 > \theta_2$). The magnitude of strain will be associated with the deviation of the angle to the warp direction. The maximum change in length will be associated with the diagonal direction, i.e. ($\pm 45°$) resulting in maximum elongation along one diagonal direction with maximum contraction along other diagonal direction. When extended in bias direction, there will be very little tension in the yarns, and due to frictional forces at interlacement points, some tension will also exist between two adjacent interlacements and a bending couple, which is responsible for deforming the interlaced yarns. When a fabric is extended in bias direction, it must be taken into account whether both ends of the warp or weft yarns are clamped for extension in respective directions or only one end is clamped from $\pm 15°$ to $\pm 75°$ bias direction to the warp direction or both ends of the yarn are free for extension [23].

18.3.1.3 Effect of Yarn Crimp and Yarn Friction Coefficient on Fabric Mechanical Properties

The crimp in warp and weft yarns greatly influences the fabric behavior under tensile loads. As higher crimp is usually obtained in the warp yarns, the values of tensile properties in the warp direction are comparatively higher than those in the weft direction. In the fabrics with warp-faced plain weave, the weft yarn lies near to straight, and warp is bent resulting in high warp crimp and ultimately higher shrinkage in warp direction. On the other hand, in the fabrics with weft-faced plain weave, the weft yarns have higher crimp than the warp. Therefore, it is evident that increasing crimp in one direction of the fabric reduces it in the direction. This phenomenon is known as crimp imbalance [12]. When a woven fabric is

Figure 18.7 Extension in bias direction arrows inside square indicates warp direction. (a) Initial position before extension. (b) Final position after extension. Source: Hu 2004 [22]. Reproduced with permission from Elsevier.

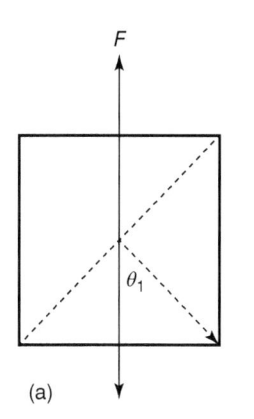

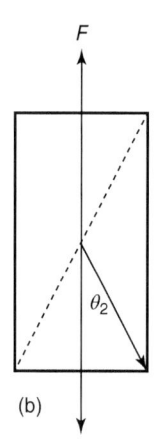

subjected to uniaxial tensile load, the crimp decreases in the direction of applied load while increases in the cross direction. This phenomenon can be explained as when the applied load in the tensile direction is increased, the yarns in the loading direction get more straight, and the diameter of the yarns decreases (consolidation), resulting in more circular cross section. Conversely, the yarns in the cross direction gets more flatten, and the amount of crimp increases to maximum. This phenomenon is responsible for the initial low-resistance portion of the stress strain curve of a woven fabric. Therefore, fabric elongation at break depends on the amount of crimp in the constituent yarns. When the load is applied biaxially, the load buildup will occur in one direction before the other. The yarns in two directions undergo consolidation, i.e. they tend to move toward more circular cross section and take the position of maximum crimp [19]. Realff, with the help of series of fabric cross-sectional images taken at various strain levels, reported [24] that yarn consolidation (diameter decrease), yarn flattening, yarn bending, and fabric shearing are potentially important factors that influence the fabric deformation and failure. It was further reported that the shearing behavior of the fabric depends on the yarn frictional properties and normal forces at the yarn interlacement points, and the yarns with higher friction coefficient will undergo less shearing.

18.3.1.4 Anisotropy of Woven Fabric Tensile Properties

Hearle et al. reported that the tensile performance of a woven fabric is an integration of a multidirectional effect, which means that the woven fabric tensile properties are influenced by loading direction [25, 26]. For example, when a woven fabric is subjected to bias extension, shear deformation will occur and influence the tensile behavior of the fabric, and the tensile behavior of the fabric will be different from the behavior of the fabric, when it is extended in the warp and weft directions. They named this phenomenon the "anisotropy" of tensile properties of woven fabrics. They have studied all the four parameters measured on Kawabata Evaluation System of Fabric (KES-F), including tensile work (WT), tensile elongation (EMT), tensile linearity (LT), and tensile resilience (RT) based on Kilby's Young's Modulus model [27].

Tensile Work (WT) In order to describe the stress strain curve of a woven fabric, an exponential function with two parameters can be written as

$$f = \frac{e^{\alpha \varepsilon} - 1}{\beta} + e_r \tag{18.11}$$

where f is stress, ε is strain, α and β are unknown parameters, and e_r is the error term. The f and ε can be obtained from tensile stress strain curves of the fabric tested on the KES system. The unknown parameters α and β can be estimated by using a nonlinear regression technique using data obtained from the curves of woven fabrics. WT is a parameter to calculate the amount of work done during tensile loading process. It is also termed as tensile energy and can be from Eq. (18.12)

$$\mathrm{WT} = \frac{E}{\alpha^2} - \left[\frac{\ln(\beta F_m + 1) + 1}{\alpha \beta} \right] \tag{18.12}$$

WT depends upon several factors that contribute to the amount of energy loss, including the ratio of the yarn counts in the warp and weft yarns, the ratio of the yarn spacing, the average yarn spacing, the type of weaves, the direction of force applied to the fabrics, and frictional force between the interlacement points of the warp and weft yarns. The experimental data of WT indicates that for plain woven fabrics, the WT is higher than twill and satin woven fabrics. As the ratio of yarn spacing and the average yarn spacing of plain woven fabrics is comparatively smaller than those of twill and satin woven fabrics, also greater energy is needed to overcome the frictional forces at the interlacement points of the warp and weft yarns. Therefore, higher energy intake occurs in the extension of plain woven fabric in bias directions. In contrast with this, lower yarn crimp is exhibited in the looser weave construction of twill and satin woven fabrics, especially in the weft direction This also explains why the tensile work in the weft direction is found to be higher than that in the warp direction [22].

Tensile Elongation (EMT) EMT is the measure of the extensibility of a fabric. It explains a fabric's ability to be stretched under tensile load. The higher EMT value means the fabric is more extensible. Kilby [27] established the model for the prediction of Young's modulus in any direction other than the warp and weft directions. EMT can be obtained by following Eq. (18.13) [22]:

$$\text{EMT}_\theta = \text{EMT}_1 \cos^4\theta + \frac{\cos^2\theta \sin^2\theta}{G'} + \text{EMT}_2 \sin^4\theta \qquad (18.13)$$

Linearity (LT) LT is the measure of the degree of nonlinearity of the tensile stress–strain curve. It depends on the ratio between WT and EMT from the stress–strain curve. The model of LT can thus be given as follows [22]:

$$\text{LT}_\theta = \frac{\text{WT}_\theta}{\text{Constant } t * \text{EMT}_\theta} \qquad (18.14)$$

Tensile Resilience (RT) RT is the ratio of work recovered to the work done in tensile deformation, and it is expressed as a percentage. Work recovered is the tensile force at the recovery process, while tensile deformation is obtained by the area under the stress–strain curve during loading process. Therefore RT of woven fabrics can be obtained by Eq. (18.15) [22]:

$$\text{RT} = \frac{\text{WT}'}{\text{WT}} \times 100\% \qquad (18.15)$$

18.3.2 Compression Deformation of Woven Fabrics

The response of a woven fabric to applied forces normal to its plane is known as fabric compressional behavior. A change in cross-sectional area of yarn occurs when a fabric is under compressive load. This change affects the void ratio, which has a profound effect on the mechanical properties of the fabric. Pierce's geometrical model assumes that yarn cross section in a woven fabric is circular or elliptical. The yarn cross-sectional areas under different loading conditions can be calculated by assuming a lenticular model. The Kemp's racetrack model or lenticular cross-sectional model is more suitable for fabrics made up of filament yarns under compressive loading.

When a woven fabric is under compression, two major changes occur: (i) Warp and weft yarns are compressed, and (ii) the less crimped yarn system increases its crimp and vice versa due to additional bending forces resulting from interaction of more crimped parts of the yarn with compression surface. This change of crimp in the yarn systems results in reduction of fabric thickness. Figure 18.8 shows a typical compression curve recorded on the KES system. It can be observed from the curve that relationship between pressure and thickness under a pressure larger than 20 gf/cm² is very close to linear.

This part of curve also has a very steep slope, which shows that woven fabrics are really incompressible. Therefore, the shape of the curve is largely governed by pressure in the range from 0 to 20 gf/cm². The values of t_m and WC provided by the KES system are not very reliable for predicting the whole curve because they do not usually match the data read off the curves. In order to examine the mechanics of woven fabric compression, De Jong et al. [35] considered the woven fabric as a three-layer structure as shown in Figure 18.9. They considered that there is an incompressible core layer in contact with more compressible surface layers that consists of protruding fibers on both sides of the fabric as shown in Figure 18.9. These two compressible surface layers follow Van Wyk's law that is actually the relationship of applied compressive load and thickness of the fabric. They developed the equation as the following:

$$P = \frac{\lambda}{(t - t')^3}$$

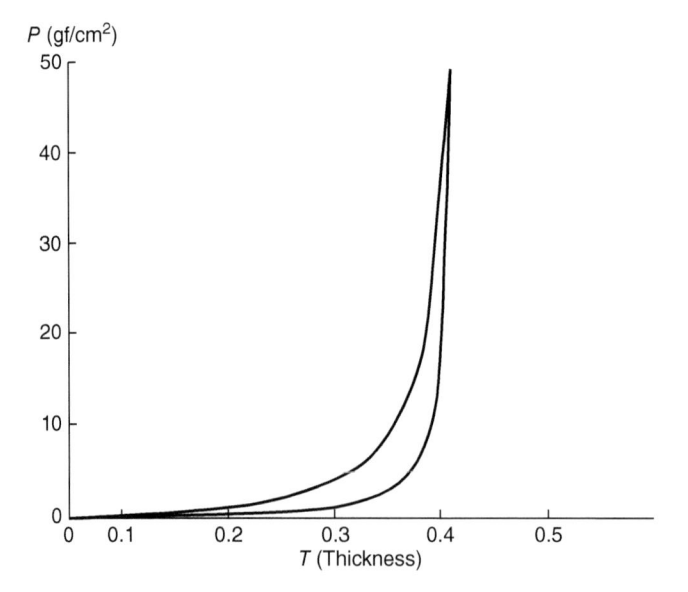

Figure 18.8 Typical compression (pressure–thickness) curve of woven fabrics.

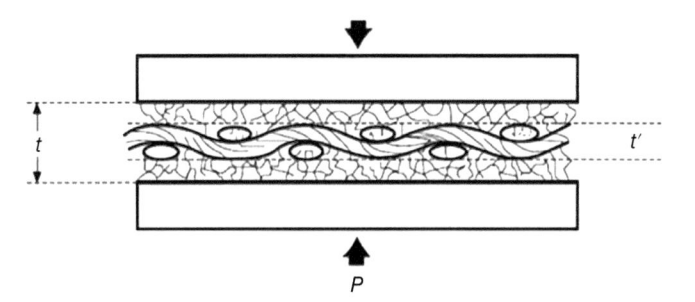

Figure 18.9 The proposed model of a woven fabric under lateral compression.

where P = compressive load, λ = a constant of proportionality, t = thickness of fabric at zero load, and t' = thickness of the incompressible core or limiting thickness. The value of t' and λ can be obtained by the following equations:

$$t' = t_m \frac{2E}{P}, \lambda = \frac{8E^3}{P^2}$$

where E = energy absorbed by the fabric with pressure between 0 and 50 gf/cm² and equals to WC that can be obtained by KES system.

They have applied this equation to wool fabrics. Wool fabric is very different from cotton fabric in structure, and cotton fabrics have very few protruding fibers on their surface than wool. The applicability of the above equation to cotton fabric was tested by Jinlian Hu [22]. Nonlinear regression method was used to improve the fitting of the compression curve of cotton fabric obtained by KES; also a five-layer structure is suggested as shown in Figure 18.10. The thickness of fabric t at zero compressive load can also be regarded as the geometrical thickness and can be calculated from measured geometrical parameters, i.e. crimp height and minor diameter of yarn. The mechanical thicknesses of fabric T_0 at 0.5 gf/cm² pressure and T_m at 50 gf/cm² can be obtained from the KES system. The limiting thickness t' can be obtained from above equation.

The geometrical thickness of all woven fabrics lies between T_0 and T_m, or T_0 and t'. The geometrical thickness is actually much smaller than T_0. This is beyond expectation because the geometrical thickness was measured principally under zero pressure, but T_0 was measured under a pressure of 0.5 gf/cm². Therefore,

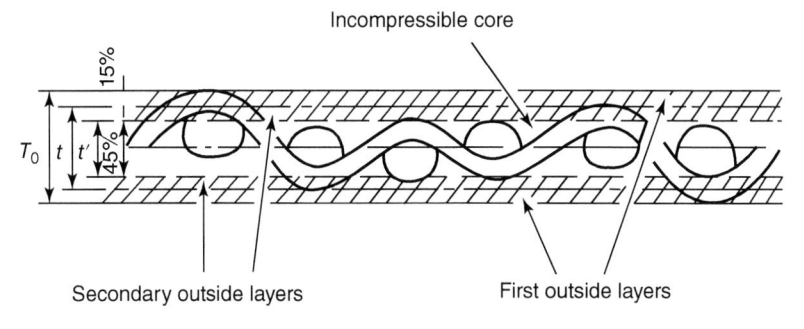

Figure 18.10 Five-layer structure of woven fabrics.

theoretically, T_0 should be smaller than the geometrical thickness t. This phenomenon can possibly be explained by the fact that during the measurement of the geometrical parameters, crimp height h, and minor diameter b, the protruding fibers of the yarn surface were ignored. Therefore, the geometrical thickness ignores hairs on the yarn surface and the crimp crowns above the average thickness. Conversely, the KES compression tester cannot ignore it and can sense anything including the hairs and the crimp crowns above the average height of woven fabrics. Therefore, the actual structure of woven fabrics can be considered as a five-layer structure. The outer layers on both sides of the fabric contain hairy fibers and crowns above the average geometrical thickness and retain about 40% of the whole fabric thickness. The secondary layers on both sides of the fabric represent another two compressible layers that form the firm structure of the fabric and retain about 20% of the whole fabric thickness. The t' represents the incompressible core of a fabric and retains about 40% of the whole fabric thickness, which confirms that fabrics are highly incompressible. The outer layers and secondary layers of this structure obey Van Wyk's law [22].

18.3.3 Shearing Deformation of Woven Fabrics

When a woven fabric is subjected to shear deformed at lower strains, the initial shear stiffness is large as shown in the OA region of the typical shear stress–strain curve in Figure 18.11. Initially, the frictional forces at the interlacement points are higher, and the incremental stiffness is decreased due to the progressive movement of the frictional elements. When the stress overcomes the smallest frictional forces at the crossover points of warp and weft, the yarns start to slip at crossover points, and incremental shear stiffness decreases as shown in the AB region of the curve. The point B in the curve represents the minimum value of incremental stiffness corresponding to a particular value of stress. Beyond this point, the incremental stiffness is linear and controlled by the deformation of the purported

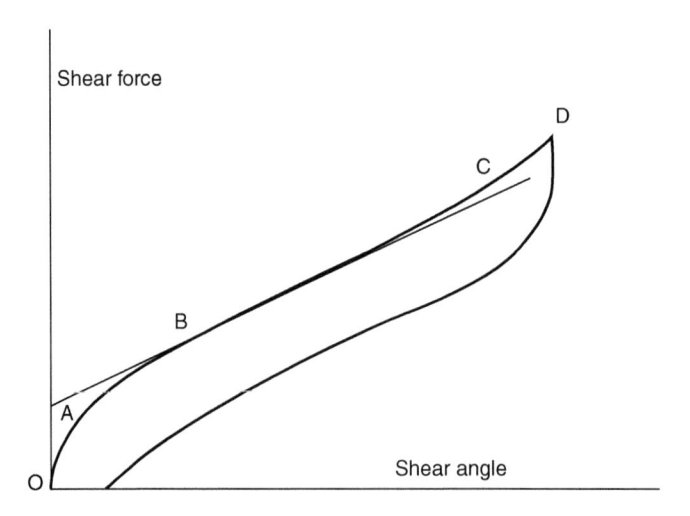

Figure 18.11 Stress–strain curve of woven fabrics during shear deformation.

elastic elements in the fabric. The increase in shear stiffness is observed above a low level of shear strains (5–10°). At stress levels higher than a certain amount, the incremental stiffness again increases, and the closed curve becomes wider as the shear angle increases. This is caused by the steric hindrance between the warp and weft yarns, which results in transverse distortion of yarns or dislocation of the crossover. Two important parameters are responsible for controlling the extent of nonlinear region of the curve. One is the point where the slope reaches its minimum value, which is attributed by the elastic elements in the fabric structure, and at this point it is assumed that the frictional contact points are in motion or dislocated, and the other is the region where stiffness decreases, which can be used to determine the hysteresis loss. The shear stiffness is also influenced by the direction of shearing force and fabric cover factor. In the case of square fabrics, those with similar yarns in warp and weft direction and similar cover factor, the shear stiffness is higher in the warp direction and also higher for the large cover factor. The hardening in the warp direction has little influence on the fabric shear properties but more on bending properties. This can be explained by the fact of jamming due to friction at crossover points in warp direction in the case of fabrics with high cover factor, but the jamming effect is less influential in the case of fabrics with low cover factor or similar cover factor in both warp and weft directions. The friction in the case of shear deformation exists mainly at the crossover of the two yarns, but interfiber friction within a yarn occurs in the case of bending deformation. The interfiber friction in the case of shear deformation is less important than in the case of bending deformation. Therefore, the hardening of warp yarns of woven fabrics, which affects the internal friction, has less effect on shear behavior [22].

18.4 Mathematical Modeling of Woven Fabric Constitutive Laws

A mathematical model is based on a large number of assumptions. However, remarkable work is done in this area, but most of the work is of theoretical nature and not sufficient to deal with problems of real fabric. Up till today three kinds of modeling techniques including predictive, descriptive, and fitting or numerical models are established by textile scientists [22]. The predictive models developed by Hearle et al. [19], Postle et al. [28] are based on consideration of the most important factors, and the effect of others is shown by some assumptions. The number of assumption for this kind of modeling is high, and it involves too much mathematics. A detailed study of the mechanisms of fabric deformation is therefore possible, yielding relationships between the structural parameters of a woven fabric and its important mechanical properties. The descriptive models developed by Paipetis [29] on the other hand are largely empirical and reflect the need for simple mathematical relations, expressing the phenomenological behavior of a fabric from the point of view of a particular property. However, such models completely ignore the physics of the material, need adjustment to reality through a number of experimental values, and operate within a specific range of the relevant parameters only. Still, they are undoubtedly useful, if no rigorous models are available. The fitting or numerical models are based on statistical considerations. Such models emphasize on numerical relationship of two variables such as stress and strain relation, ignoring the exact mechanism. This method needs fewer assumptions and provides an approach that is more relevant to real situations. There exist various methods for fitting a curve in many industrial or science fields. Constitutive laws are often estimated by using a polynomial, which contains the appropriate variables and approximates to the true function over some limited range of the variables involved. Spline, especially the cubic spline interpolation method, is also widely used for this purpose [22].

18.4.1 Constitutive Laws of Woven Fabric

The complex mechanical behavior of a woven fabric can be explored by basic frame of the infinitesimal elastic theory of a sheet. In the most general case, the stress–strain relationships, or constitutive laws, of a linearly elastic plate (an initially flat) sheet are as follows in Eqs. (18.16) and (18.17), where T_1 and T_2 are the tensile stresses, ε_1 and ε_2 are the tensile strains, T_{12} and ε_{12} are the shear stress and shear strain in the fabric plane, M_1 and M_2 are the bending stresses, K_1 and K_2 are the bending curvatures, M_{12} and K_{12} are twisting stress and strain, and the submatrices A_{ij} and D_{ij} represent the membrane and bending (and twisting) stiffness, respectively:

$$\begin{Bmatrix} T_1 \\ T_2 \\ T_{12} \\ M_1 \\ M_2 \\ M_{12} \end{Bmatrix} = \begin{bmatrix} A_{11} & A_{12} & A_{13} & B_{14} & B_{15} & B_{16} \\ & A_{22} & A_{23} & B_{24} & B_{25} & B_{26} \\ & & A_{33} & B_{34} & B_{35} & B_{36} \\ & & & D_{44} & D_{45} & D_{46} \\ & & & & D_{55} & D_{56} \\ & & & & & D_{66} \end{bmatrix} \begin{Bmatrix} \varepsilon_1 \\ \varepsilon_2 \\ \varepsilon_{12} \\ K_1 \\ K_2 \\ K_{12} \end{Bmatrix} \quad (18.16)$$

Total elements = 21

$$[\sigma] = [S][\varepsilon] \quad (18.17)$$

The B_{ij} is coupling stiffness that connects the membrane and bending modes of deformation. $[\sigma]$ is the stress matrix, $[S]$ the stiffness matrix, and $[\varepsilon]$ the strain matrix. Therefore, in order to specify the elastic behavior of an originally flat sheet, a total of 21 stiffnesses (6 for membrane deformations, 6 for bending and twisting, and 9 for coupling between the two modes) are required. For orthotropic fabric where directions 1 and 2 are assumed to coincide with the principal directions of orthotropy, i.e. the warp and weft directions, 13 stiffnesses are required. When the coupling matrix is symmetric, 12 stiffnesses are required. For symmetric fabric 8 stiffnesses, for square fabric (with the same yarns in each direction) 6 stiffnesses, for an isotropic sheet with bending behavior unrelated to planar behavior 4 stiffnesses, and for an isotropic solid sheet 2 stiffnesses plus the thickness are required. If the relationship is nonlinear, many of the interaction terms would reappear. The tangential elasticity matrix given in Eq. (18.18) can be used in continuum analysis for nonlinear material properties, provided that the form of nonlinear stress–strain laws is already known. If the initial stresses are 0 at zero displacement, then the nonlinearities can be contained in $[\sigma_0]$ and used to apply the necessary corrections. This is known as the initial stress method in such analysis as finite element methods [22]:

$$[S_T] = \frac{d[\sigma]}{d[\varepsilon]} \tag{18.18}$$

18.4.2 Computational Woven Fabric Mechanics

Computational fabric mechanics deals with the challenging numerical problems with the help of an intelligent CAD system. Hearle et al. initiated the use of computer programs to approach woven fabric mechanics problems [30]. One of the significant algorithms used in computational techniques is the finite element method to predict the behavior of a fabric under certain loading conditions.

18.4.2.1 Continuum Models

Continuum models assume woven fabric as a continuum regardless of its discrete microstructure. Previously developed mathematical methods (numerical solutions) of continuum mechanics are applied to the fabric deformation problem. Finite element method is one such example of continuum technique; in this method the fabric is divided into numerous small areas called the finite elements. The fabric is then modeled using flat or curved shell elements in order to consider both bending and stretching. So far this method has only been able to study simple draping behavior of fabric by nonlinear finite element analysis considering the fabric as two-dimensional orthotropic sheets with both bending and membrane stiffnesses. The nonlinear stress–strain relationships on fabric deformations are an area that is still unaddressed by this method due to absence of unambiguous nonlinear constitutive equations of woven fabrics. Recently, Hu and Newton established a whole set of nonlinear constitutive equations for woven fabrics under tension, bending, shear, and lateral compression [31, 32]. Use of these equations in finite element method will further improve predictability of nonlinear stress–strain relationships during fabric deformations [22].

18.4.2.2 Discontinuum Models

Unlike continuum models, the discontinuum models consider fabric as discrete assembly of their constituent yarns. The mechanical responses of fabrics are predicted by combining yarn properties, inter-yarn interactions, and fabric structures. By considering the yarns as curved or straight rod elements with frictional connections at the interlacement points between the warp and weft yarns, the finite element method can be further extended to study fabrics using discontinuum models. A discontinuum model is useful in predicting fabric mechanical properties from yarn properties, because only a small patch of cloth needs to be modeled. Therefore, a discontinuum model is useful for modeling of fabric deformations before they are actually manufactured but has limitations in predicting complex fabric deformations due to the prohibitive number of yarns present [22].

18.4.3 Energy Methods for Woven Fabric Mechanics

Hearle et al. described an energy-based approach for development of CAD in order to deal with fabric mechanics [33]. They have introduced appropriate energy terms to deal with yarn extension, bending, flattening, and frictional forces at interlacement points in a woven fabric:

$$U = -(F_1 x_1 + F_2 x_2) \tag{18.19}$$

where F_1 and F_2 are the forces in the warp and weft directions and x_1 and x_2 are the lengths of the unit cell in corresponding directions. The principle of energy minimization will lead to Eq. (18.20) [34]:

$$\frac{F_1}{F_2} = -\frac{dx_2}{dx_1} \tag{18.20}$$

This equation can be applied to any fabric model; in some suitable circumstances it may be useful to study the state of crimp interchange in the fabric. They not only have studied the elastic response of plain woven fabrics but also suggested the methods to deal with other weaves. According to their study, when fabric is under deformation, interfiber and intrafiber energy changes may occur and are associated with yarn deformations and displacements and can be expressed as

$$U = -\sum (F_x) + \sum (U_y) \tag{18.21}$$

where F and x may include other directions like shear and bending or biaxial directions and the sign convention shows the reduction in potential energy of the applied forces due to increase in strain energy. If the fabric is perfectly elastic, the state of the fabric can be given by minimum value of U. They have reported that in the case of biaxial deformation, the energies involved are yarn extension U_e, yarn bending U_b, and energy associated with change in yarn cross section U_x:

$$\sum (U_y) = \sum_{1,2} (U_e) + \sum_{1,2} (U_b) + \sum_{1,2} (U_x) \tag{18.22}$$

The energy associated to yarn extension can easily be stated for linear elasticity. It can be expressed as

$$U_e = \sum \left(\frac{1}{2}\right)(tEe^2)l \tag{18.23}$$

where t is the linear density, E is the specific modulus, e is the strain, and l is an increment of the yarn length. $\left(\frac{1}{2}\right)(Ee^2)$ can be replaced by the area under the stress–strain curve, which is conveniently expressed by a polynomial function. The other two energy terms can be expressed as

$$U_b = \sum \frac{1}{2}(Mc)\delta l = \sum \frac{1}{2}(Bc^2)\delta l \tag{18.24}$$

where M is the bending moment, c is the yarn curvature (reciprocal of radius of curvature), and B is the bending stiffness. For a nonlinear response, $(1/2)(Mc)$ could be replaced by the area under a plot of bending moment against curvature. In order to describe the energy term associated with change in yarn cross section, two effects must be taken into account: change in yarn volume and change in yarn shape; therefore it can be expressed as

$$U_x = U_v + U_s \quad (18.25)$$

where U_v and U_s are the energies related to volume and shape, respectively, and can be formulated as shown in Eqs. (18.26) and (18.27):

$$U_v = \frac{1}{2}K_2[(V-v)/V]^2 \quad (18.26)$$

$$U_s = \frac{1}{2}K_3[(D_v-d)/D_v]^2 \quad (18.27)$$

where K_2 is the bulk modulus related to the reduction in yarn volume from V to v, K_3 is the shape change modulus, and D_v is the diameter of the circle for the volume v. For weaves other than a plain weave, two additional features need to be taken into account in formulating the energy terms:

(1) The contact zones will include straight yarn sections as well as curved sections at the edge of crossovers. This will influence the bending energy terms.
(2) When the yarns are in contact, the change of shape due to flattening will be different, and this will influence the flattening energy terms.

By taking these two features into account, energy terms for other weaves like twill and satin can also be formulated. The energy methods will make it possible to develop CAD programs in order to predict the mechanical properties of fabric in terms of structure and yarn properties.

18.5 Conclusion

From the above study it can be concluded that the mechanical response of woven fabrics in the case of tensile deformation depends greatly on certain factors including the constituent yarn properties, the amount of crimp, the weave of the fabric, or the number of interlacements per unit area and density of yarns (warp and weft). The mechanical response is also influenced greatly by the direction of loading like in the case of tensile stress in principle direction or biased direction. When a woven fabric is subjected to uniaxial tensile load, the crimp imbalance will occur, i.e. crimp decreases in the direction of applied load while increases in the cross direction. When a woven fabric is subjected to bias extension, shear deformation will occur and influence the tensile behavior of the fabric, and anisotropy will increase. Additionally, in the case of biased extension, the maximum change in length will be associated with the diagonal direction, i.e. ($\pm 45°$) resulting in maximum elongation along one diagonal direction with maximum contraction along other diagonal direction. From the study of modeling techniques, it can be concluded that in the case of continuum analysis of woven fabric, if the forms of nonlinear stress–strain laws are already known, the tangential elasticity matrix given in Eq. (18.18) can be used in continuum nonlinear material properties. A discontinuum can be used for modeling of fabric deformations before they are actually manufactured by using yarn properties, but not useful in the case of complex fabric deformations. Set of nonlinear constitutive equations for woven fabrics under tension, bending, shear, and lateral compression formulated by Hu and Newton [31, 32] can be used in finite element method to improve predictability of nonlinear stress–strain relationships during fabric deformations. Energy methods can be successfully applied for the development of CAD to deal yarn extension, bending, flattening, and frictional forces at interlacement points in a plain woven fabric. By taking into account the straight yarn section and curved sections at the edges of interlacement point together with the change of shape due to flattening when yarns are in contact, energy terms can also be described for weaves other than plain like twill and satin. The energy methods are capable of developing CAD programs in order to predict the mechanical properties of fabric in terms of structure and yarn properties.

References

1 Gandhi, K. (2012). *Woven Textiles: Principles, Technologies and Applications.* Elsevier.
2 Robinson, A. T. C., & Marks, R. (1967). *Woven Cloth Construction.* New York: Plenum Press.
3 Behera, B., Kremenakova, D., Militky, J., and Mishra, R. (2012). *Modeling of Woven Fabrics Geometry and Properties.* INTECH Open Access Publisher.
4 Behera, B.K. and Hari, P. (2010). *Woven Textile Structure: Theory and Applications.* Elsevier.
5 Peirce, F.T. (1937). 5—The geometry of cloth structure. *Journal of the Textile Institute Transactions* 28 (3): T45–T96.
6 Kemp, A. (1958). An extension of Peirce's cloth geometry to the treatment of non-circular threads. *Journal of Textile Institute Transactions* 49 (1): T44–T48.
7 Hamilton, J. (1964). 7—A general system of woven-fabric geometry. *Journal of the Textile Institute Transactions* 55 (1): T66–T82.
8 Grosberg, P. and Park, B. (1966). The mechanical properties of woven fabrics Part V: the initial modulus and the frictional restraint in shearing of plain weave fabrics. *Textile Research Journal* 36 (5): 420–431.
9 Olofsson, B. (1964). 49—A general model of a fabric as a geometric-mechanical structure. *Journal of the Textile Institute Transactions* 55 (11): T541–T557.
10 Dorkin, C., Munden, D., and Whewell, C. (1973). The rôle of science in a modern clothing industry. *Journal of the Royal Society of Arts* 121 (5208): 799–812.
11 Ellis, P. and Munden, D. (1973). 52—A theoretical analysis and experimental study of the plain square weave. Part III: The geometry of the plain square weave. *Journal of the Textile Institute* 64 (10): 565–571.
12 Liao, T. and Adanur, S. (1998). A novel approach to three-dimensional modeling of interlaced fabric structures. *Textile Research Journal* 68 (11): 841–847.

13 Liao, T. and Chen, W. (1990). An investigation of wool worsted yarn relaxation shrinkage and its effect on fabric structure and performance. In: *Proceedings of the 8th International Wool Textile Research Conference* (7–14 February 1990), 128. Christchurch, New Zealand: Fibre Assemblies and Product Properties.
14 Amirbayat, J. (1991). The buckling of flexible sheets under tension part I: Theoretical analysis. *Journal of the Textile Institute* 82 (1): 61–70.
15 Amirbayat, J. and Hearle, J. (1989). The anatomy of buckling of textile fabrics: drape and conformability. *Journal of the Textile Institute* 80 (1): 51–70.
16 De Jong, S. and Postle, R. (1978). A general energy analysis of fabric mechanics using optimal control theory. *Textile Research Journal* 48 (3): 127–135.
17 Hearle, J.W. (1980). *Mechanics of Flexible Fibre Assemblies*. Kluwer Academic Publishers.
18 Hearle, J. (1980). The mechanics of dense fibre assemblies. In: *The Mechanics of Flexible Fibre Assemblies* (eds. J.W. Hearle, J.J. Thwaites, and J. Amirbayat), pp. 51–86. UK: Springer.
19 Hearle, J., Grosberg, P., and Backer, S. (1969). *Structural Mechanics of Yarns and Fabrics*, vol. 1. USA: Wiley-Interscience.
20 Hu, J. and Teng, J. (1996). Computational fabric mechanics: present status and future trends. *Finite Elements in Analysis and Design* 21 (4): 225–237.
21 Moghe, S., Hearle, J., Thwaites, J., and Amirbayat, J. (1980). From fibers to woven fabrics. In: *Mechanics of Flexible Fiber Assemblies* (eds. J.W. Hearle, J.J. Thwaites, and J. Amirbayat). Sijthoff and Noordhoff.
22 Hu, J. (2004). *Structure and Mechanics of Woven Fabrics*. Elsevier.
23 Spivak, S. and Treloar, L. (1968). The behavior of fabrics in shear part III: the relation between bias extension and simple shear. *Textile Research Journal* 38 (9): 963–971.
24 Realff, M.L. (1994). Identifying local deformation phenomena during woven fabric uniaxial tensile loading. *Textile Research Journal* 64 (3): 135–141.
25 Amirbayat, J. and Hearle, J. (1986). The complex buckling of flexible sheet materials—Part I. Theoretical approach. *International Journal of Mechanical Sciences* 28 (6): 339–358.
26 Hearle, J. and Amirbayat, J. (1986). Analysis of drape by means of dimensionless groups. *Textile Research Journal* 56 (12): 727–733.
27 Kilby, W. (1963). 2—Planar stress–strain relationships in woven fabrics. *Journal of the Textile Institute Transactions* 54 (1): T9–T27.
28 Postle, R., Carnaby, G., and De Jong, S. (1987). Woven-fabric structure and tensile properties. In: *Mechanics of Wool Structures* (eds. R. Postle, G.A. Carnaby, and S. de Jong). Wiley.
29 Paipetis, S. (1981). Mathematical modelling of composites. In: *Developments in Composite Materials- 2: Stress Analysis (A 82-28568 13-39)*, 1–37. London: Applied Science Publishers.
30 Hearle, J., Konopasek, M., and Newton, A. (1972). On some general features of a computer-based system for calculation of the mechanics of textile structures. *Textile Research Journal* 42 (10): 613–626.
31 Hu, J. (1994). Structure and low-stress mechanics of woven fabrics-a comprehensive study based on the KES system. Ph.D.

32 Hu, J. and Newton, A. (1993). Modelling of tensile stress-strain curve of woven fabrics. *Journal-China Textile University-English Edition* 10: 49–49.

33 Hearle, J., Potluri, P., and Thammandra, V. (2001). Modelling fabric mechanics. *Journal of the Textile Institute* 92 (3): 53–69.

34 Hearle, J. and Shanahan, W. (1978). 11—An energy method for calculations in fabric mechanics part I: principles of the method. *Journal of the Textile Institute* 69 (4): 81–91.

35 De Jong, S., Snaith, J.W., and Michie, N.A. (1986). A mechanical model for the lateral compression of woven fabrics. *Textile Research Journal* 56 (12): 759–767.

19

Fabric Making Technologies

Tao Hua

The Hong Kong Polytechnic University, Institute of Textiles and Clothing, ST 703, Hung Hom, Kowloon, Hong Kong, China

19.1 Introduction

Fabric can be defined as the manufactured assembly of fibers and/or yarns. A wide variety of textile materials are used to produce fabrics including advanced materials. In the conversion of materials such as fibers and yarns to fabric, the fabric making technology that involves the processing machine, procedures, and parameters for fabric formation has significant influence on the properties and appearance of the resulting fabrics. There are many ways to make fabrics. The most commonly used techniques include weaving, knitting, nonwoven manufacturing, and braiding. Each fabric making method constructs a unique fabric structure that partially determines the fabric properties and appearance. Figure 19.1 shows the schematic of fabric structures formed by these methods.

Weaving is a traditional method used for fabric production wherein the resultant woven fabrics are manufactured by interlacing at least two sets of thread systems – warp and weft threads at right angles to each other. In a simple woven structure, the warp threads, which are called the warp ends in weaving terms, always run along the length of the fabric or up and down, while the weft threads, also known as picks or filling yarn in weaving terms, run in the horizontal direction or left to right and vice versa. The warp and weft threads are perpendicular to each other. More thread systems can be inserted into weaving machines to produce woven fabrics with different structures. In the past, they were mostly in one layer. With time, double-layer, triaxial, and three-dimensional (3D) woven fabrics have been developed for the textile market. They have various properties and applications for daily use.

Knitting is the second most popular technique of fabric making in the textile industry. In the knitting process, knitted fabrics are produced by interlooping yarns. There are two ways of knitting based on the direction of the yarn movement, which are known as weft and warp knitting. In weft knitting, the loops are constructed in the horizontal direction, while in warp knitting, the yarns are knitted along the vertical or diagonal direction. Compared with woven fabrics, knitted fabrics have better extensibility and formability due to their loop

Handbook of Fibrous Materials, First Edition. Edited by Jinlian Hu, Bipin Kumar, and Jing Lu.
© 2020 Wiley-VCH Verlag GmbH & Co. KGaA. Published 2020 by Wiley-VCH Verlag GmbH & Co. KGaA.

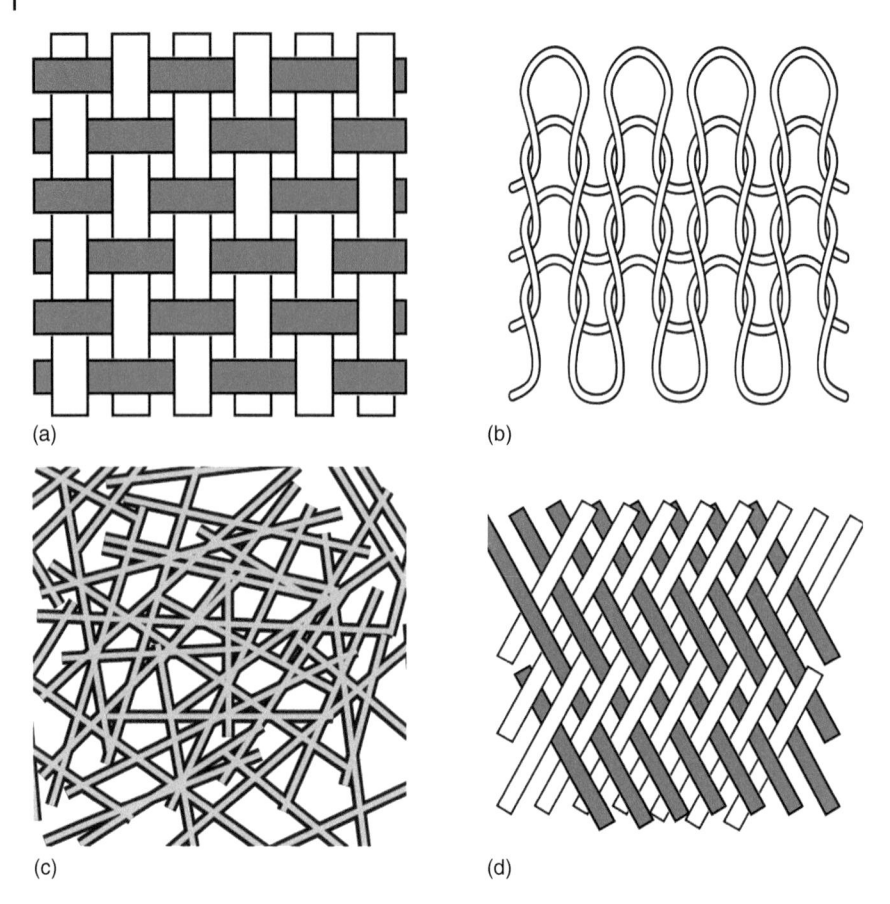

Figure 19.1 Schematic of structures of main fabric types. (a) Woven. (b) Knit. (c) Nonwoven. (d) Braided.

structure. Nowadays, knitted fabrics are widely used in apparel, home furnishings, and technical textiles.

Nonwoven fabric technology is one of the major technologies for the manufacture of fabric. Nonwoven fabrics are a manufactured sheet of web, sheet, or batt, which are produced with staple fibers or filaments through various bonding techniques that hold them together. In terms of processing by machinery, nonwovens do not need to be prepared through the yarn spinning process and can be directly made from fibers or even polymers. One of the greatest benefits of the production of nonwoven fabrics is that the production rate is much higher than that of fabrics produced by weaving or knitting technologies since there are fewer yarn procedures and the process can be more automated. A wide range of raw materials can be applied, regardless if they are natural or synthetic fibers. Nowadays, most synthetic or bicomponent fibers, or even special fibers such as nanofibers, glass fibers, and biodegradable materials, can be used to produce nonwoven fabrics since they have unique characteristics, uniformity, and consistency. These textiles can be applied toward different end uses, especially for hygiene, civil engineering, and building materials.

Braiding is a common process for constructing narrow fabrics or products, which are known as braided or plaited textiles. Braided fabrics are produced by interlacing three or more separate threads in such a way that they can cross one another in a diagonally overlapping direction. Based on the unique constructed structures through braiding, the resultant fabrics exhibit good strength in the primary loading direction and shear resistance as well as complex-shaped parts. Moreover, by varying the braided structures, different types of braided fabrics can be fabricated with different properties and shapes, including two-dimensional (2D) and 3D braids, biaxial and triaxial braids, and flat and circular braids. Traditionally, braiding was used to produce ropes and cords with a simple structure. Braids with complicated structures were normally thicker and stronger with high physical strength. Nowadays, braided fabrics are used for various industrial applications since they have good extension and are very pliable with nicely curved edges.

19.2 Weaving

19.2.1 Weaving Machines

Woven fabrics are manufactured by interlacing lengthwise warps and crosswise wefts into designed patterns on weaving machines. There are many different types of weaving machines that have been developed for the production of woven fabrics in the textile industry. Weaving machines are divided into shuttle and shuttleless based on how the weft yarns are inserted. Shuttle looms are the oldest and most simple type of weaving machines, where the weft yarn is inserted with a shuttle. Nowadays, shuttle looms are not commonly used for the production of woven fabric except for some of the specialty markets. They have been replaced by shuttleless looms, which started from the mid-twentieth century and include rapier, projectile, air-jet, and water-jet looms [1–6]. Compared with shuttle looms, shuttleless looms have a much higher production rate and are more flexible for manufacturing different types of woven fabrics.

Rapier weaving machines are a primary type of shuttleless weaving machine used for making woven fabric with a versatile insertion system and excellent control of the filling yarn in the process. In rapier looms, flexible or rigid rapiers are used to insert the weft yarn across the shed. There are two types of rapier weaving machines: single or double rapier looms. In the former, only one long and rigid rapier, which is made of metal or a composite material, is used to feed the filling yarn and then retracted, while there are two rapiers in the latter: a giver, which picks up and carries the filling yarn to the shed center, and a taker, which takes the yarn from the giver and transports the yarn to the end of the shed. The advantage of a two-rapier system is increased efficiency as opposed to the use of a single rapier loom. In addition, there are rigid and flexible double rapier machines. Nowadays, flexible double rapier looms are preferred over single or rigid rapier looms due to their advantages, such as space saving and a higher machine speed.

Projectile weaving machines involve a small bullet-like gripper that holds and then brings the weft yarn through the shed without the use of a heavy shuttle. During the insertion of the weft yarn, a picking lever beats the gripper with the yarn to slide through a channel composed of the thin prongs of a rake. When the gripper arrives the other side of the machine, it is braked in the receiving unit. Several grippers are usually used in projectile looms. After one gripper reaches the other side of the loom with the weft yarn and then returns to the starting position by the conveyor system, the other grippers are ready for the next picking. Compared with the shuttle loom, the grippers of the projectile loom can move faster and farther because of the reduced dimensions and the mass of the carrier. Therefore, projectile weaving machines are more suitable for producing wider woven fabrics and thus commonly used for the production of technical fabrics.

Jet looms take the weft yarn across the loom by using the force of air or water at high pressure. Thus, jet looms can be operated at higher speeds as opposed to projectile or rapier looms. However, they cannot accommodate heavy or bulky yarns, nor can they produce fabric with wider dimensions as they have less yarn-carrying power. In air-jet weaving machines, the filling yarn is propelled by compressed air that come from several nozzles to weave fabric. The main and tandem nozzles provide weft yarns with an initial propulsion force, while the relay nozzles found in the entire shed offer additional high-velocity air to assist with conveying the yarn through the warp shed. Another main feature of air-jet looms is that a specially designed profiled reed is used for the weft yarn insertion wherein the profiled reed builds a channel to guide the air, which prevents the abrasion of the inserted wefts with the warps. They are one of the most popular types of weaving machines since the weft insertion rate is very high with low production requirements and thus offers high productivity for mass production, relative to other shuttleless machines. Many standard types of fabrics or fabrics with a variety of styles can be produced through air-jet weaving, such as denim, glass, or terry fabric.

The operation principle of water-jet looms is similar to that of air-jet looms. During the weaving process, the nozzle provides highly pressurized water to insert the weft yarn. A small amount of water is required for the fabric production, and the weaving process is carried out at extremely high speed. Therefore, water-jet looms are very efficient. However, efficient drying units need to be equipped on the water-jet looms to dry the wet fabric. The main disadvantage of the water-jet weaving machine is that it is more suitable for weaving hydrophobic filament yarns such as polyester and nylon [1, 4–9].

In addition to these different weaving machines with weft yarn insertion systems, there are three types of shedding systems, namely, the cam, dobby, and jacquard machines. Their function is to separate the warp sheet into at least two layers for weft yarn insertion. For producing common types of fabrics, weaving machines are normally equipped with a dobby shedding system, which include negative, positive, and rotary dobbies. The latest manufactured weaving machines use rotary dobbies as they are suitable for the high speed of these machines and offer flexibility for changing patterns. Weaving machines equipped with a dobby shedding system control the movement of the warp yarn by using the harnesses, which means warps on the same harness have the same movement. In order to vary the movement of the warp yarn for weaving a pattern, more harnesses are used, but the number of harnesses that can be accommodated in a dobby system is limited. Therefore, when the patterning capacity of dobbies fails to meet the requirements of the weaving design, a jacquard shedding system must be used. The main feature of jacquard shedding is that this mechanism can control the movement of the warp yarn individually. In recent years, jacquard shedding machines are becoming more and more ubiquitous for manufacturing woven fabric since they can produce extraordinary and intricate patterns or figures, such as brocades, tapestry, and damask. There are mechanical and electronic jacquard looms with single or double lift mechanisms [10, 11].

There are also special types of weaving machines that have been developed for high productivity or creating special woven fabrics, including multiphase, circular, and triaxial weaving machines. The multiphase weaving machine can insert several weft yarns simultaneously so that more than one shed can be formed at a time. Thus, this machine offers high productivity. In circular weaving machines, the shuttles run continuously and circularly around the periphery in a ripple or wave shed. The produced fabrics are in a tubular form and mainly used for sacks and tubes. In triaxial weaving machines, two sets of warp threads and one set of weft threads are interlaced to form a multitude of equilateral triangles. The constructed fabric structures are unique and endow the resultant fabrics with good tear, burst, and abrasion resistance. Triaxial fabrics are largely used for industrial purposes, such as conveyor belts, reinforcements for plastics, and aerospace accessories.

19.2.2 Woven Structures

A woven fabric is formed by the interlacement of the warp and weft threads. This interlacing pattern is called the weave of a woven fabric. The final fabric properties and appearance depend on the weave and other structural parameters such as the fabric density and yarn count. Weaves are planned on point paper wherein each vertical column and horizontal row means a warp end and a weft pick, respectively. In a weave diagram, each square represents an intersection point between a warp and a weft yarn, and if the warp yarn crosses over the weft yarn at this point, the square is filled or shaded with a mark. On the other hand, if the weft yarn is above the warp yarn in the point of intersection, the square will be left blank. An unlimited variety of weaves can be designed for woven fabrics. The most commonly used weaves in the production of woven fabric include fundamental weaves and their derivatives, combined weaves as well as compound weaves. Each has their own characteristics.

Plain, twill, and satin/sateen weaves are three types of fundamental weaves, as shown in Figure 19.2. They have a simple structure but form the basis of even the most complex weaves where every warp and weft thread within a repeat is interlaced by only one thread of the opposite system. Plain weaves are the simplest and most commonly used weave of all woven fabrics. In this structure, weft filling yarn is alternately passed through the warp yarn, which follows the order of one up and one down. If the plain weave is extended in the warpwise or weftwise or

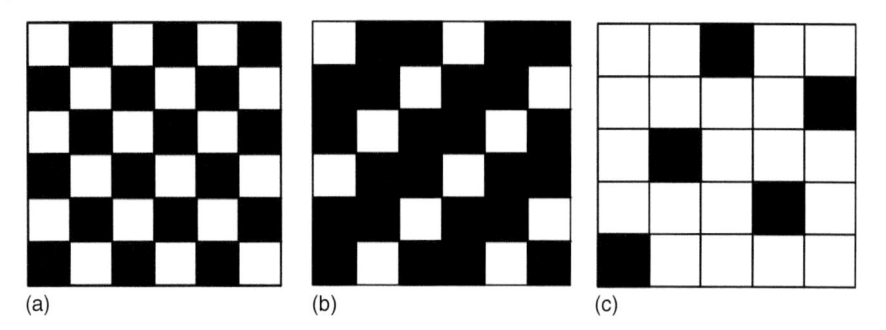

Figure 19.2 Fundamental weaves. (a) Plain. (b) Twill. (c) Sateen.

both directions, new structures with rib or basket effects are further created, such as warp rib, weft rib, and basket weaves in a regular or an irregular form.

Twill weaves also have a simple structure but can generate pronounced diagonal lines along the fabric width. Twill lines are created by the stepwise progression of yarn interlacing patterns that appear on both sides of the fabric, and the fabric constructed with a twill weave is usually strong and durable. In terms of the direction of the twill lines, there are two types: right-hand and left-hand twill. In the former, the twill lines run from lower left to upper right, while the twill lines run from lower right to upper left in the latter. There are several factors that affect the prominence of twill weaves, including the nature of the yarn, yarn floats, fabric density, and relative direction of the twill and yarn twist. By extending the floats and changing the move number and direction of the diagonal lines on the twill weaves, various twill weave derivatives can be created in woven fabrics. Commonly used twill derivatives are pointed, herringbone, curved, broken, and elongated twills as well as diamond and diaper weaves.

Sateen/satin weaves have a higher luster on one side of the fabric since there are many long floats on the fabric surface that reflect light. Sateen is a weft-faced weave wherein the long weft float is on the top surface of the fabric while the satin weave consists of long warp float on the fabric surface, thus a warp-faced weave. During the weave design, both repeat and move numbers are important parameters. The satin/sateen weaves may be modified by subtracting or adding marks or changing move number or both [12–14]. Consequently, a series of satin/sateen weave derivatives can be produced such as reinforced and rearranged sateen/satin.

Combined weaves are constructed on the basis of two or more fundamental weaves and their derivatives, which results in irregular or uneven fabric surface or small woven figures on the fabric. Commonly used combined weaves include crepe, honeycomb, and mock leno weaves, Bedford cords, simple spot figure designs, and stripe and check weave combinations. Crepe weaves give fabric the appearance of being covered by minute spots or seeds through several construction methods. Honeycomb weaves are a very special type of weave that can create a 3D cell-like structure for the fabric with excellent moisture and water absorbent properties. Mock leno weaves endow fabrics with an open structure and small holes or gaps similar to leno weave fabrics. By floating the ordinary weft or warp threads on the surface of the fabric in an order, spots or small figure designs can be created for decorative purposes. By combining two, three, or more weaves or weave variations in the warpwise or both the warp and weft directions, stripe and check effects form on the fabric surface.

Compound weaves are often referred to as a group of weaves that feature more than one layered structure to make multilayered, 3D, and other specialized types of woven fabrics. These fabrics are constructed with more than one system of warp and weft threads that are often arranged in different planes by forming two or more layers. Common types of woven fabrics fabricated with compound weaves are warp and weft backed, double, treble, gauze and leno, warp pile and weft pile fabrics, carpets, and other types of 3D fabrics. Figure 19.3 illustrates several compound weaves and structures for woven fabrics. Backing fabric is based on backed weaves, in which a second series of either weft or warp threads back single layer fabric in order to increase fabric weight and enhance warmth while maintaining a smooth surface. When double weaves are constructed, two layers of threads in which one is woven above the other and stitched together form double fabrics. There are three common types of double fabrics based on the stitching methods: self-stitched, center-stitched, and interchanging double fabrics. By using conventional or specialized weaving machines, different types of 3D woven fabrics can be manufactured. In general, 3D fabrics are produced in several layers. Solid, hollow, and shell structures have been developed in 3D woven fabrics. 3D solid woven fabrics are classified as having multilayer, orthogonal, and angle interlock architectures. They feature compound construction with different fabric thicknesses so as to provide better mechanical properties, such as fabric strength. A 3D hollow structure can be created with multiple layers of woven fabrics with even and uneven surfaces. The hollows can be triangular, rectangular, or cell shaped, which can improve the air permeability and thermal conductivity of fabrics. 3D shell woven fabrics are manufactured with a curved shell structure, which can maintain fiber continuity. Moreover, they can also feature spherical or cubic shells [15, 16].

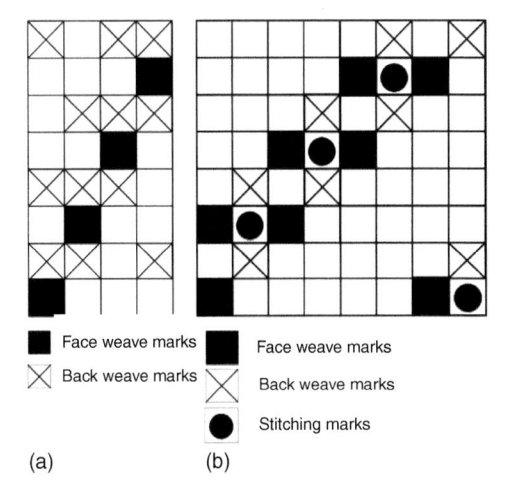

Figure 19.3 Compound weaves and structures. (a) Weft-backed weave. (b) Self-stitched double weave.

In addition to the fabric weave, there are other important structural parameters that have great impacts on the properties and appearance of woven fabrics. The main parameters are yarn linear density (count), fabric sett (density), cover factor, yarn crimp, and fabric weight (area density). The term "sett" is normally expressed as the number of threads per unit length (inches or cm) including the warp and weft densities, which are used to indicate the spacing of threads in a fabric. The cover factor is the fraction of the fabric area that is covered by the component warp or weft yarns. It shows the relative closeness of the yarns within the fabric by considering both the yarn count and fabric density. In a woven fabric, yarns are not straight, but in a wavy shape due to the interlacing of the warps with the wefts. This waviness of the yarn is called yarn crimp. The yarn crimp is an important parameter because it affects the property and appearance of fabric as well as the yarn length required for an intended fabric. Fabric weight is expressed in mass per square area, e.g. g/m^2 or oz/yd^2, which can be calculated based on the yarn linear density, fabric density, and yarn crimp.

19.2.3 Properties

The properties and appearance of woven fabrics depend on the fiber and yarn and their structure, which are commonly evaluated in terms of tensile strength and elongation, tearing strength, abrasion resistance, air permeability, fabric bow and skew, and fabric dimensional changes in accordance with ASTM, AATCC, or other testing method standards. Fabric tensile strength and tearing strength are crucial for the durability of garments, especially for finished woven fabrics such as denim fabric. Fabrics with a plain weave and its derivatives normally have maximum binding points with short yarn floats, which result in their high tensile strength but low tearing strength. Fabrics with a twill weave and its derivatives can be fabricated with more warps and wefts per unit area than those with a plain weave, so they have a higher thickness and density. There are many prominent diagonal wale lines on twill fabrics. Twill fabrics are usually durable with adequate tensile and tearing strength. Fabrics fabricated with a satin or sateen weave structure have fewer binding points and longer floating lengths, which can provide a smooth and luster surface. However, such a structure causes the yarns to more easily ravel and fray. In addition, satin/sateen fabrics have good ability to resist tearing but have low tensile strength [1, 7, 12, 16, 17].

19.2.4 Applications

As a primary type of fabric, wovens are widely used in daily life. Their applications include the three main areas of apparel, home furnishing, and technical products. Compared with knitted fabrics, woven fabrics have high stability and a tighter structure due to yarn interlacing that makes woven fabrics suitable for more formal garments. Plain weave fabrics can be made thick or thin and are widely used for shirts, blankets, canvases, and organza and chiffon fabric. Twill weave fabrics are mainly for the production of suits and jeans. Sateen or satin fabrics are usually found in evening and wedding dresses and for decorations. Besides, in order to meet the growing demand of customers for comfort and function in terms of fabric for home furnishing, more and more woven fabrics are applied in this area. Woven products include bedspreads, floor coverings, cushion covers, curtains, towels, sheets, pillowcases, etc. More complex structures are commonly employed in these types of textiles for the required thickness, color, and pattern.

Woven technical textiles are becoming universal today, which include those for medical use, sportswear, filtration, industrial use, protection, civil engineering applications, transport, and packaging and in geotextiles [1, 2, 12, 17]. Woven fabrics can be used as medical and biological textiles for hygienic and clinical purposes. The construction of such fabrics needs to be extremely stable and durable. For hygiene purposes, woven fabrics are commonly used in nappies, incontinence pads, or tampons. Besides, woven fabrics can be applied for protective and healthcare textiles, such as operation dresses and staff uniforms for doctors and nurses. Moreover, woven fabrics can be used as implantable materials, which are utilized as vascular grafts, sutures, and artificial limbs. In the latest developments, they are applied in extracorporeal devices that include artificial livers, kidneys, and lungs. Woven fabrics are the primary textile product used for automobiles since they have a highly stable structure and high mechanical resistance. They are usually applied in seat covers, headrests, door panels, and reinforcements. Woven fabrics are also commonly found in airbags with or without a coating. The purpose of airbags is to reduce the forward motion of passengers when an accident takes place in a split second. Therefore, the fabric used for airbags should be able to resist the force of the movement and have substantial strength. Woven technical textiles are usually used in sports applications. Waterproof and breathable fabrics are some of the popular fabrics in sportswear, which are mainly applied in outdoor apparel, especially for winter coats because not only can this prevent liquid penetration but also allow water vapor transfer to the outside environment and provide a comfortable temperature for the human body. This type of fabric is manufactured with a high fabric density or application of coating and laminating. Another important application of woven fabrics is as 3D fabrics, which are widely used as reinforcements and preforms for advanced textile composites due to their properties and shapes.

19.3 Knitting

19.3.1 Knitting Machines

Weft and warp knitting machines are two types of knitting machines used in the textile industry. The former is the most commonly used knitting machine due to the advantages of versatility, ability to quickly change patterns, requirement of little floor space, and low investment. The key components of weft knitting machines include knitting needles, needle bed, cam box, sinker, and yarn feeder. The needles are the main knitting element that is responsible for loop formation and subsequent interlooping with the help of the sinker. The sinker that is positioned in between the needles performs functions such as loop formation, holding down, and knocking over. The needle bed is a metal plate with a slot for needle insertion and movement. Cams provide the needles and other elements with a suitable reciprocating action for the knitting action. Weft knitting machines can be generally grouped into flat bed or circular bed machines in accordance with the needle bed arrangement. There are typically two types of flat needle beds that are organized in an inverted V formation. Latch needles are commonly used and placed in the grooves on the machine, which are also known as needle tricks. The machine gauge is normally 3–18 needles per inch. The yarn can change direction with each course on a flat bed machine, while the yarn runs continuously and horizontally in the same direction on a circular machine. In the circular machine, two sets of needles are arranged in a circular form at right angles to each other wherein one set is in the vertical and the other is in the horizontal. Latch, bearded, and occasionally compound needles can be used. By using the multiple yarn feeds, hundreds of yarns can be knitted in one rotation, so weft knitted fabric can be quickly produced. Single jersey, rib, interlock, and purl fabrics can be manufactured by using circular machines [18–23].

In terms of warp knitting, the yarns from the warp beam in the form of a parallel sheet are used to form fabric through loop formation. Each yarn is fed to an individual needle. During the same knitting cycle, every needle in the needle bar conducts the action of yarn feeding and loop forming simultaneously. A typical warp knitting machine consists of several functional elements, including needles and needle bar, presser bar, sinkers and sinker bar, guides and guide bars and warp beams. With regard to the production of warp knitted fabrics, there are two popular classes of warp knitting machines with one or two needle bars employed, which are called tricot and Raschel machines. In the tricot machine, bearded or compound needles and sinkers are normally used. Warp beams are positioned at the back of machine, and also, the guide bars are moved from the back to the front of machine in fabric knitting. Moreover, the machine gauge used in a tricot machine is high, which ranges from 28 to 40 needles per inch, and the machine speed can be up to 3500 courses per minute or even higher. Continuous filament yarns are commonly used in fabric production. Latch needles are generally used in Raschel machines, and sometimes, compound needles are now common for the production of warp knitted fabric. The warp beams are placed on the top of the machine, and the guide bars run from the front side toward the back side to knit warp knitted fabrics in overlapping and underlapping arrangements. Furthermore, the machine gauge is coarser from 12 to 32 needles per inch. The machine speed is lower compared with that of the tricot. More yarn guide bars need to be mounted on the machine. Besides, a wide range of yarns can be used, and various products can be manufactured [18–22].

19.3.2 Knitted Fabric Structures

By controlling the motion of the needle on a knitting machine, different types of knitted fabrics can be produced with different fabric structures. There are three fundamental knitting stitches for the production of weft knitted fabric: knit, tuck, and miss stitches. The majority of knitted fabrics are produced with the knit stitch. Knit stitches are a normal loop formed when a needle knocks over the old loop and receives a new yarn. Then, the loop formation is carried out by the old loop. The function of the needles is to rise to the clearing position, carry the yarn down, and then form the loop. In the formation of a tuck stitch, a needle receives the new loop but still grips the old loop, so there are two stitches on the needle. To form a tuck loop, a needle is only raised to the tuck position where the old loop is still on the latch of the needle. A miss stitch is also known as a float stitch, which is created when a needle that is holding the old loop fails to receive the new yarn and then feeds the yarn into the two nearest needle loops. The purpose of miss stitches is to reduce space widthwise and increase fabric stability [18].

Depending on whether a single knit stitch or a combination with other stitches is used, weft knitted structures can be created with different effects. Weft knitted structures can be mainly categorized as single or double jersey structures. Single jerseys are produced by using one set of needles of the machine, while double jerseys are made by using two sets of needles. Single plain jersey knitting is the simplest knitting construction in that all the loops can be knitted by using V-bed, circular, or fully fashion machines. Each loop has the same shape, which is formed on the face side of the fabric because only one set of needles is applied. The "technical face" of the fabric has a V-shape appearance while the Ω-shape appearance forms on the "technical back" of the fabric.

In the case where two set of needles are used for double jersey production, each set can produce their own loops on one side of fabric. Therefore, complicated structures with special effects can be easily constructed on double jersey machines. There are three basic structures and their derivatives, including rib, interlock, and purl. Rib structures are one of the common knitted structures in which loops are constructed in the opposite directions by using two sets of needles. The 1×1 rib is the simplest rib fabric wherein one wale of face loops is alternated with two wales of back loops. In the same way, more rib structures such as 2×2, 3×3, and 3×2 can be constructed. No curling problem may occur on the edge of fabric and fabric with a rib structure can acquire good elastic recovery. Furthermore, there are a variety of rib derivatives, such as tubular, cable, half Milano, and full Milano. Interlocking is fabricated by using both long and short needles. The row of interlocks comprises joining the two half gauge 1×1 ribs by interloping sinker loops. It has a perfectly balanced structure and the same appearance on both sides of the fabric, like the face of a single jersey. Moreover, it can be elongated to 30–40% in the crosswise direction. The interlock of 2×2, 3×3, and 4×4 can also be fabricated according to its basis.

In weft knitted jacquard fabrics, a diversity of colors and designs can be found on the fabric surface by selecting needles and colored yarns to knit for desired colors and figures. Jacquard fabric can be classified into two types: single jersey jacquard and double jacquard. In single jersey jacquard, a tie-in technique called "accordion" is used wherein specific needles are cleared to tuck the height. Consequently, the length of the floats can be minimized at the fabric back. By using one more set of needles to knit the back loop, the double jacquard can overcome the floats in a single jacquard. Moreover, it is commonly used to produce different types of backings, such as birdseye and twill backings [18, 22, 25].

Basic warp knitted structures are constructed by using warp knitting machines with only one yarn guide bar and one set of needle bar, including pillar stitches, half-tricot and variation, atlas, and double loop stitches. Pillar stitches are produced by feeding the same yarn on the same needle. They are usually used with other types of structures to produce fabric with particular effects. Half-tricot is the structure formed by alternatively feeding the same yarn on two adjacent needles. The basic half-tricot is in 1×1 lapping movement, and the loops are easily inclined because of the production with the single guide bar. The structures in the 2×1 and 3×1 lapping movement are common structures in half-tricot variation, which have a longer underlap space. The atlas structure is constructed where the yarn guide bar laps increasingly in the same direction for a minimum of two consecutive courses. In addition, double loop stitches are constructed by overlapping two needles for the mixing of the weft and warp loops [18, 25, 26].

Nowadays, warp knitting machines with two yarns guide bars are universal. Through the use of two guide bars, advanced fabric structures can be developed by combining two or more basic structures. The simplest structure of the advanced fabrics is the full tricot, which can be constructed with two half-tricots in symmetrical lapping movements. Besides, locknit and reverse locknit are the most commonly used structures for warp knitted fabrics, which are developed based on the half-tricot in 1×1 and 2×1 lapping movements, but the movement of the guide bars is reversed. Satin is the reverse structure with sharkskin that can be constructed by half-tricot and its variation in a 3×1 or 4×1 lapping movement. For queenscord, it is also composed of pillar stitches and half-tricot variation in a 3×1 or 4×1 lapping movement.

19.3.3 Properties

Modern knitted fabrics can be produced in a wide range of structures that endow the resultant fabrics with different properties. In general, knitted fabrics have high resilience and wrinkle resistance and are light in weight, better form fitting, soft, and comfortable. However, they may shrink easily after washing and pill after rubbing. Compared with weft knitted fabric, warp knitted fabric has less stretchability but better dimensional stability.

Usually, knitted fabrics have the distinct property of high stretchability and recovery. The 1×1 rib and 2×2 rib structures can provide substantial stretch. However, the stretch can be reduced by introducing horizontal linear yarns into the fabric structure. The recovery property means that when the fabric is stretched to a considerable length and then released, it will gradually return to its original form, which allows the fabric to extend in different directions – diagonal, vertical, and horizontal at the same time – and then the fabric recovers. Rib fabrics normally exhibit a much better recovery property than plain fabrics. In order to further enhance the fabric stretchability and recover property, elastomeric yarn elements under tension can be incorporated into the structure for elastic knitted fabrics.

Similar to the property of elastic recovery, wrinkle resistance is one of the characteristics of knitted fabrics. Generally, knitted fabrics are thicker, the yarns are more mobile in the fabric structure, and the yarn twist is lower compared to woven fabrics, which results in better wrinkle resistance of the knitted fabrics. It is also beneficial for garment production since it leads to ease-of-care garments.

In general, knitted fabrics offer good wear comfort that result from their high stretch properties for better conformability, good insulation property, and sweat transportation property.

The dimensional stability of fabric is its ability to resist permanent changes in its dimensions. Common knitted fabrics do not have a good performance in dimensional stability if they do not have any aftertreatment. Mostly, the dimensional changes of fabrics occur after the first laundering and drying cycle wherein the fabrics shrink or grow at different percentages lengthwise or widthwise. After several washing cycles, the knitted fabrics finally become steady. The dimensional stability of knitted fabric depends on the materials as well as the knitting structures. Interlock, double knit, and other fabrics with weft or warp inserted yarns are unusually more stable and exhibit less shrinkage. The dimensional stability of knitted fabric can be evaluated in accordance with ASTM and AATCC standards.

19.3.4 Applications

Knitted fabrics are widely applied for apparel and domestic products, such as T-shirts, sweaters, socks, stockings, panty hose, lingerie, sportswear, swimwear, sheets, and towels. In garment manufacturing, knitted fabrics are the second largest fabric type used for garments. In addition to traditional knitted fabrics, more specialized knitted fabrics are also developed for garments, including fleecy, raised, highly elastic, and plated fabrics. Nowadays, there is the rapid development of knitted fabrics applied in the technical textiles and composites. Based on the certain properties of knitted fabrics such as extensibility, moldability, openwork, and lightness, not only warp knitted fabrics but also weft knitted fabrics find more applications in these areas. With these two fabrics in different structures, many technical products can be manufactured by advanced knitting machines and techniques [18–21, 24–27].

Spacer fabrics have been developed for technical applications by using both warp and weft knitting technologies. Concerning the structure of space fabrics, two surface layers are interconnected by relatively thick monofilaments, which makes the fabric 3D, elastic, and compressible in the thickness direction. The main features of space knitted fabric include excellent compression elasticity and cushioning, high air permeability and thermal insulation, and good flexibility in surface design. The properties of the space fabric can be adjusted by varying three components: yarn material, fabric construction, and finishing. The space fabric can replace foam in shoes, bras, beds, and seats. The spacer fabrics also can be used for the manufacturing of mattresses for beds and tables.

Knitted fabric structures have the role of reinforcement in polymer composites. That is, warp and weft knitted fabrics with inlay yarns enhance the stiffness and strength of these composites. When inserted and warp stitch yarns are used to reinforce in-plane and through-thickness, this multiaxial warp knit increases the ability of fabric to conform to complex shapes and increases its tolerance to damage. Applications of these kinds of knitted fabric composites include bumper bars and door members for automobiles, rudder-tip fairings for passenger aircrafts, and aircraft radomes, helmets, and body armors.

With regard to medical applications, knitting technology is very flexible and versatile so that medical textile products and tubular fabrics for medical devices can be easily developed and produced through warp knitting. Also, the apparel worn by doctors and nurses in hospitals and clinics are normally produced by using knitting technology, such as their undershirts and socks. Most medical appliances such as artificial blood vessels, surgical meshes, and coverings of artificial heart valves are also made with weft and warp knitted fabrics in recent years. Moreover, the various types of bandages, surgical stockings, and certain parts of orthopedic equipment are produced with knitted fabrics as they have high extensibility. In addition, knitted fabrics can serve as important components of functional clothes. For example, spacer fabric is constructed as two layers – a top and a bottom layer that are kept apart by a pile layer, which has pile threads that allow elasticity. Thus the flexible structure of spacer fabrics means that they can be used as electric switches or pressure sensors when electrically conductive yarns are used to form the layers. These sensors can also be applied onto undershirts, trousers, and socks to fabricate functional clothing.

19.4 Nonwovens

19.4.1 Manufacture of Nonwovens

The basic steps for making a nonwoven fabric are first to form fibers into a web, which are then bonded and finally undergo the processes of drying, curing, and finishing, as shown in Figure 19.4. There are a variety of techniques that can be applied in the formation and bonding of a fiber web, which thus result in different fabric structures and properties. Web formation is the creation of a loosely joined sheet structure by laying down fibers through four basic techniques: drylaying, wetlaying, polymer laying, and other specialized techniques. Drylaying involves carding or airlaying of staple fibers. Carding blends and combs fibers into a web or batt, which can be parallel laid, that is, formed by feeding carded fiber layers parallel into a conveyor belt through carding machines, or random/cross laid, in which each fiber layer after carding is mounted at right angles to the main conveyor. The former results in a web with high strength and orientation in the machine direction but not in the cross direction. The latter results in a web with high strength in the cross direction but not oriented to the machine.

Airlaying allows isotropic web properties and is the aerodynamic formation of a fiber web, which involves feeding fibers into an airstream and then transported to a conveyor belt where the fibers are randomly deposited in the form of a web; thus the web is randomly oriented.

Wetlaying involves the use of a slurry of water and fibers or an aqueous suspension of fibers that are deposited into a perforated drum or a screen belt and then

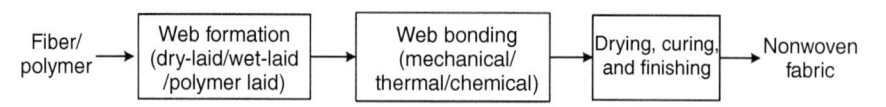

Fiber/polymer → Web formation (dry-laid/wet-laid/polymer laid) → Web bonding (mechanical/thermal/chemical) → Drying, curing, and finishing → Nonwoven fabric

Figure 19.4 Basic steps for making a nonwoven fabric.

dewatered, dried, and cured with heat, much like the papermaking process. The fibers in the web are also randomly oriented. Wetlaying is appropriate for mass production at high speeds [28–34].

Direct or polymer laying means that nonwoven webs are produced from thermoplastic polymers in which they are spunlaid, meltblown, and electrospun. These are also known as direct laid webs, which are produced directly from filaments or short fibers during an extrusion spinning process. This method is one of the most cost-efficient ways of producing fabrics. In spunbonding, molten polymer is extruded through a spinneret, and the filament extruded is drawn into a specially designed aerodynamic device. The filaments formed are deposited onto a moving conveyor or screen drum, and then the web structure is formed. Following that is the web bonding and winding for making the fabrics. Polymers with high and broad molecular weight distribution are commonly applied for producing uniform spunbond webs, such as polypropylene, polyester, and polyamide. During the meltblowing process, molten polymer is extruded through a linear die, and then the extruded polymer streams are rapidly attenuated into extremely fine diameter fibers by using a high-velocity hot airstream. The attenuated fibers are subsequently transported toward a collector conveyor by high-velocity air blowing, thus creating a self-bonded meltblown web that consists of fine fibers. In electrospinning, the molten polymer or polymer solution is drawn from the tip of a capillary to a collector by using an electric field wherein the jet emerges from the charged surface and then electrical forces increases the polymer liquid to stretch the jet. This method is mainly used to produce nanofibers with diameters in the range of 40–2000 nm and their webs with appropriate polymer and solvent systems [30–32].

After the fiber web is formed, the bonding between fibers must be strengthened and stabilized by applying different bonding techniques [27]. These include chemical, mechanical, and thermal bonding, which are commonly used web bonding techniques with adhesives or application of heat and pressure. Mechanical bonding interlocks or entangles fibers in the web through needle punching or water jets, that is, through needle punching or hydroentangling, which are the most commonly used methods. During the needle punching process, the fiber web is fed into a space between a bottom bed plate and a top stripper plate. The barbed needles on a needle board punch the fed fiber web through the plates. Consequently, the fibers in the web are mechanically interlocked or bonded. The geometry and density of the needles are important because they greatly affect the properties of the final nonwoven fabric. In hydroentangling, the needles in needle punching are replaced with high-velocity water jets that entangle the fibers. In the production process, the fibrous web is fed onto a moving conveyor in the form of a flat bed or cylindrical surface. Then high-velocity water jets entangle the fiber web with turbulent water and fully bond the fibers. The de-energized water is then drawn through the permeable conveyor sleeve into the vacuum box for recycling and reuse [30, 33–35].

In chemical bonding, also known as adhesive bonding, adhesive binders are applied onto the fiber webs to hold and bond the fibers together through various techniques such as saturating, spraying, printing, and foaming. Following that, the water in the web is evaporated through drying, and the binders thus bond the fibers. There are a variety of binders available that not only bond the fibers but also affect the properties of the resultant nonwoven fabrics. For example, spray bonding adhesives can provide some strength but high bulk, while saturation bonding provides high rigidity and stiffness. Latex polymers have now become the most commonly used binder for the chemical bonding of nonwoven fabrics due to the widely available different varieties, their good versatility, ease of application, and cost effectiveness.

The thermal bonding process fuses thermoplastic components in the form of homofil fibers, powder, film, web, etc. by using a heat supply on fiber to fiber crossover points with calender rollers and oven and ultrasonic bonding machines. The function of heat is to cause the thermoplastic component to become viscous or to melt. Consequently, the polymer flows to the fiber-to-fiber crossover points to form bonding regions followed by cooling. Hot calender bonding applies heat energy to the fiber web by passing the web through the nip of a pair of heated rollers. During this process, area bonding or point bonding can be formed by varying the surface pattern of the rollers. The ultrasonic bonding process generates thermal energy and transfers this to well-defined, restricted areas in the web through a mechanical hammering action onto the web surface by using an ultrasonic horn.

19.4.2 Properties

Nonwoven fabrics can have a wide diversity of properties, which depend on the composition and fabric structure. The latter is determined by the methods used to produce the nonwoven fabrics, including web formation, bonding, and finishing. Moreover, these properties can instill and realize specific functions, such as absorbency, liquid repellence, resilience, stretchability, softness, or flame retardancy. These properties are usually combined for specific applications.

Needle punching that uses barbed needles to entangle fibers can produce very thick textiles for various applications. Thin fabrics that are lint-free and have an excellent soft hand feel and high in quality can be produced for medical textiles with hydroentanglement. Products made by thermal bonding can be relatively soft and textile-like, while chemical-bonded nonwoven fabrics can be used for low-cost carpets, scraps, and highly stiff products [30, 31, 35].

Spunbond fabrics are very popular nonwoven fabrics because they can provide particular properties that act somewhere like between paper and woven fabric. The diameter ranges between 15 and 35 μm, whereas the weight ranges typically between 10 and 200 g/m^2, which means such fabrics range from very light with a flexible structure to heavy with a stiff structure. They have planar isotropic properties due to a random web structure. These fabrics offer excellent chemical and physical stability, such as good tearing strength, shear resistance, fray and crease resistance, and high liquid retention capacity. Furthermore, they have a high strength-to-weight and strength-to-cost ratios. Meltblown fabrics have a random fibrous structure and are constructed by using microfibers with a fiber diameter that ranges between 2 and 7 μm. Their structure and constituted fibers endow meltblown fabrics with a high surface area and smooth surface texture that result in good insulation and filtering characteristics [30, 33].

19.4.3 Applications

Nonwoven fabrics can be endowed with a variety of chemical and physical properties and thus can be used to produce a spectrum of products that fit a wide range of applications. They can be used alone or as components of apparel, home furnishings, and healthcare and engineering products as disposable or durable products, depending on the application. Disposable nonwovens can only be used once or reused a few times, whereas durable nonwovens are made to be used for a longer period of time. General applications of disposable nonwovens include personal hygiene products, such as diapers and feminine hygiene products, or medical products, such as surgical gowns and drapes. The main applications of durable nonwovens cover a wide range of areas including hygiene products, wipes, medical and surgical products, protective clothing, interlinings and garments, upholstery, furniture and bedding, floor coverings, buildings and roofings, civil engineering products and geosynthetics, filters (gas and liquids), and automotive items. According to a report from the European Disposables and Nonwovens Association (EDANA), nonwovens are mostly used for hygiene products, followed by civil engineering products and building materials and then wipes [30, 35–37].

Around 33% of nonwoven textiles are used for hygiene products including diapers and feminine hygiene and adult incontinence products. There are two major characteristics of nonwovens that make them exceptional as absorbent hygiene material, which are their high bulk for absorbing and retaining a large amount of fluid at a low product cost and their lightness in weight prior to use. Spunbonded and thermal-bonded nonwovens are also widely used for hygiene products due to their favorable structural characteristics and comfort. They are also low in cost as well as environmentally friendly. Chemically bonded nonwovens are still used for sanitary napkins but have the disadvantages of causing discomfort and environmental problems.

In the nonwoven industry, nearly 15% of nonwovens are found in the wipes market due to their absorbency, versatility, uniformity, and durability that traditional wiping materials cannot compete. Some of the nonwoven wipes are developed for consumers, including those for personal care, babies, and household cleaning. They are convenient to use and dispose, provide exceptional performance, and are low in price, which are the key demands of customers. Then there are wipes for industry applications, such as food services and medical applications. Their disposability and low cost are key for their popularity. They are therefore found in different industry segments, including food services, factories and shops, automotive industry, and hospitals.

19.5 Braiding

19.5.1 Braiding Processes and Machines

Different kinds of braiding machines have been developed to produce different kinds of braids including 2D and 3D structurally braided fabrics. Maypole braiding machines are popular 2D braiding machines that consist of a carrier motion system, track plate, take-off, and other additional components. In carrier motion systems, the most important part is the carriers with bobbins that move following the track driven by horn gears. Bobbins are installed in the carrier, and the yarn length is adjusted by the carrier during the production process. The track that determines the path of the carriers can define the type and structure of the braid. A classical braiding machine has a base and a track plate wherein one, two, or more tracks can be used for producing flat, tubular, or other types of braids. The take-off mechanism is also important in a braiding machine because this determines the braided angle of the product, one of the most important parameters of braided structures, and thus significantly influence the properties of braids. In the Maypole braiding machines, there are two sets of yarn carriers that interlace and rotate yarn on a circular track at different preset angles. One set travels in the clockwise direction and the other set rotates in the anticlockwise direction. Consequently, two sets of yarns on the carriers are interlaced together at a biased angle to the machine axis to form braided fabrics. Then the produced braid is continuously moved forward by a take-up mechanism [38–40].

In the fabrication of 3D braids, the production principle is similar to that of 2D braiding, but the standing end yarns, also known as the core end, are an addition to the movable braiding threads. With the use of static yarns, they can be interlaced to the braids. There are four types of machines for producing 3D braids: circular, four-step, two-step, and 3D rotary braiding machines.

In the braiding process on circular braiding machine, the carriers with yarn bobbins move in two concentric orbits in the opposite directions. The braided yarns are intersected at a certain angle, and then the carriers change at the crossing point after one cycle of interlacing. Concerned on the machine in two-step process, the threads can move on the base plate only in two movement directions that move through the cross section formed by stationary standing ends. The productivity is high and efficient. Besides, 3D rotary braiding machines are universal and user friendly. There are mobile carriers on the base plate with horn gears in a square form. In the operation, each horn gear and carrier with bobbins can be shifted in different directions or stopped individually. Therefore, carriers with bobbins are free to move to any braiding point in the braiding area and also easy to move in a diagonal path. Thus, variable cross sections and scale over the length of the braid can be achieved [38–42].

19.5.2 Braided Structures and Properties

Braided structures can be classified as 2D or 3D. The former include two types of yarn intertwining configurations: biaxial and triaxial constructions. Biaxial constructions are the most commonly used braided structure wherein only two sets of yarns are interlaced together in a diagonal formation. In order to enhance the tensile and compression strength as well as modulus of the braid for its applications in fiber-reinforced composites, a third set of longitudinal yarns can be introduced to form the braid in a triaxial configuration, together with the biaxial interlaced yarns. The latter have multilayered interconnected structures and have been mainly developed through composite structures. Three dimensional braided structures can be produced by using the newly developed braiding machines that are computer controlled. Since they have thicker and 3D architectures, they can be fabricated with different cross sections. Commonly, 3D braids can be categorized as tubular or rectangular-shaped structures. In 3D tubular braids, the yarns not only move in helical clockwise and anticlockwise directions but also pass under and over each other in the tube wall thickness direction. Braided fabrics with rectangular-shaped structures are produced by using a rectangular loom [39, 40, 42, 43]. Compared with 2D braids, 3D braiding can produce complex near-net-shape preforms, and the resultant composites have better impact damage tolerance and delamination resistance.

The properties of braided fabrics depend on the interlacing patterns of the braided yarns. Commonly used patterns include diamond, regular, and Hercules stitches, which are similar to the 1/1 plain, 2/2 twill, and 3/3 twill weaves of woven fabrics, respectively. Depending on how the carriers and horn gear are arranged, various patterns can be created. Besides, the geometry of the braided structure is also important for the resultant braid properties, including the line, stitch, and braiding angle. When interlacing the yarns in the diagonal direction, there is a braiding angle, which is usually in the range of 30–80°. The braiding angle is the most critical geometrical parameter for the production of braided architectures.

The characteristics of braided fabrics rely on the type of braiding yarns and architecture of the braid. Normally, braids have good dimensional flexibility since they can be controlled shapewise. In addition, braids have good porosity, ability to resist shear, and strength. Braided composites exhibit good impact resistance, efficient reinforcement for parts where torsional loads are applied, improved interlaminar shear properties, and reduced manufacturing cost of composite structures.

19.5.3 Applications

Braided fabrics have mainly been applied in the fields of engineering, medicine, and automotive in the previous century. Braided fabrics feature greater burst strength and flex fatigue and appropriate for use as hydraulic and fuel hoses in cars, trucks, earth-moving equipment, marine engines, and aeroplanes. In recent years, they are widely employed for technical purposes, such as fiber-reinforced composites, medical implants, and sports applications [42, 43].

Compared with other composites, braided structure composites have a more stable structure with isotropic properties, outstanding strength, and crack resistance. Therefore, braids can be applied to reinforce composites. In aerospace applications, braided composites have been used as rocket nozzles, rotor blades for helicopters, helicopter drive shafts, missile casing, stator vanes, and composite ducting, Braided structures are also applied in the transportation and civil engineering field, including as ropes, cables, drive shafts, belts, hoses, and concrete and wall reinforcements. In recreation, braided composites are used in hockey sticks, baseball bats, tennis racquets, and sports bicycle frames.

Medical applications of braids include sutures, stents, scaffolds, braided pillar implants, artificial limbs, and cartilage. Since braided textile has excellent agility, pliability, and surgical ease of use, high tensile strength, and a silky smooth feel, it is very suitable for polyethylene sutures. The braiding technique can also be used for the fabrication of stents because this technique provides a versatile design for different applications. Moreover, braided fibrous scaffolds can serve as a base for tissue regeneration, so it is appropriately used as ligament scaffolds. In addition, the fracture resistance and longevity of braided structures mean that they can be used in dental posts for tooth restoration and repair [44].

In terms of sports, braided products can be used for lines in fishing since their structure can provide adequate toughness and abrasion resistance. Besides, polyester or polyethylene braided cords can be applied as archery strings for reinforcement. Furthermore, braided ropes can be manufactured for sail masts, which can maintain its stiffness and retain a constant diameter.

19.6 Future Trends

In the conversion of different materials to fabrics, various fabric making technologies including weaving, knitting, nonwoven manufacturing, and braiding are widely used in the production of fabrics for applications in apparel, furnishing, and technical textiles. With the advancement of materials, high-performance and functional fibers are now being realized and used to make fabrics that meet the demands for comfort, health, and safety as well as create aesthetics in fashion and domestic textiles and endow particular properties and functionalities in technical products. Fabric making technologies are continuously progressing, targeting to enhance fabric quality, advance production flexibility, increase productivity and automation, and enable environment-friendly processing as well as improving responses to consumer demands through innovations in machinery and optimization of the manufacturing process. It is clear that complex and functional fabrics will need to be further explored for technical applications since technical textiles are the fastest-growing sector in the textile and clothing industries. In order to accommodate this rapidly growing demand, advanced fabric making technologies and their produced fabrics will be further developed with advanced materials, machinery, and processes, such as manufacturing technologies for 3D woven, knitted, nonwoven, and braided fabrics, technologies for integrating different functions into fabrics, multiaxial warp knitting machines, and large braiding machines.

References

1 Adanur, S. (2001). *Handbook of Weaving*. Lancaster, PA: Technomic Publishing Co. Inc.
2 Gries, T., Veit, D., Wulfhorst, B., and Lenz, C. (2006). Principles and machinery for production of woven fabrics. In: *Textile Technology* (ed. Wulfhorst, B.), 124–151. Munich: Hanser Publishers.
3 Fannin, A. (1998). *Handloom Weaving Technology*. New York: Design Books.
4 Fox, T.W. (2010). *The Mechanism of Weaving*. Charleston, SC: Nabu Press.

5 Hua, T. (2016). Weaving machines. *Textile Asia* 47 (1): 17–19.
6 Hua, T. (2011). Weaving machines. *Textile Asia* 42 (11): 4–7.
7 Seyam, A.M. (2016). Weaving. *Textile World* 166 (3): 38–42.
8 Seyam, A.M. (2004). Weaving technology. *Textile World*: 34–39.
9 Choogin, V., Bandara, P., and Chepelyuk, E. (2013). *Mechanisms of Flat Weaving Technology*. Cambridge: Woodhead Publishing Ltd.
10 Seyam, A.M. (2011). *Developments in Jacquard Woven Fabrics*. Cambridge: Woodhead Publishing Ltd.
11 Jiang, S.Q. and Hua, T. (eds.) (2013). *Digital Jacquard: Mythologies*. The Hong Kong Polytechnic University.
12 Sondhelm, W.S. (2016). Technical fabric structures - 1. Woven fabrics. In: *Handbook of Technical Textiles* (ed. Horrocks, A.R. and Anand, S.C.), 63–106. Cambridge: Woodhead Publishing Ltd.
13 Naik, N.K. (1994). *Woven Fabric Composites*. Lancaster, PA: Technomic Publishing.
14 Selby, M. (2011). *Contemporary Weaving Patterns*. London: A&C Black.
15 Chen, X., Taylor, L.W., and Tsai, L.J. (2016). Three dimensional fabric structures. Part 1 an overview on fabrication of three dimensional woven textile preforms for composites. In: *Handbook of Technical Textiles* (ed. Horrocks, A.R. and Anand, S.C.), 107–162. Cambridge: Woodhead Publishing Ltd.
16 Hu, J.L. (2004). *Structure and Mechanics of Woven Fabrics*. Cambridge: Woodhead Publishing Ltd.
17 Behera, B.K. and Hari, P.K. (2010). *Woven Textile Structure Theory and Applications*. Cambridge: Woodhead Publishing Ltd.
18 Anand, S.C. (2016). Technical fabric structures – 2. knitted fabrics. In: *Handbook of Technical Textiles* (ed. Horrocks, A.R. and Anand, S.C.), 107–162. Cambridge: Woodhead Publishing Ltd.
19 Gligorijevic, V. (2016). *Technology of Knitting: Theoretical and Experimental Analysis*. Saarbrucken: Lambert Academic Publishing.
20 Au, K.F. (ed.) (2011). *Advances in Knitting Technology*. Cambridge: Woodhead Publishing Ltd.
21 Kadolph, S. (2010). *Textiles*. Upper Saddle River, NJ: Prentice Hall.
22 Raz, S. (1993). *Flat Knitting Technology*. Westhausen: Universal Maschinenfabrik.
23 Gries, T., Veit, D., Wulfhorst, B., Schrank, V., Hehl, A., and Weber, K.P. (2006). Processes and machines for knitwear production. In: *Textile Technology* (ed. Wulfhorst, B.), 152–166. Munich: Hanser Publishers.
24 Brackenbury, T. (1992). *Knitted Clothing Technology*. Oxford: Blackwell Scientific Publications.
25 Spencer, D.J. (2001). *Knitting Technology*, 3e. Cambridge: Woodhead Publishing Limited.
26 Yu, W. (ed.) (2016). *Advances in Women's Intimate Apparel Technology*. Duxford: Woodhead Publishing Ltd.
27 Hale, D. (2001). New developments in the manufacture of circular knitting machines for the production of medical textiles. In: *Medical Textiles* (ed. Anand, S.), 3–11. Cambridge: Woodhead Publishing Ltd.

28 Batra, S.K. (2012). *Introduction to Nonwovens Technology*. Lancaster, PA: Destech Publications.

29 Gries, T., Veit, D., Wulfhorst, B., and Gräber, A. (2006). Processes and machines for nonwoven production. In: *Textile Technology* (ed. Wulfhorst, B.), 167–187. Munich: Hanser Publishers.

30 Russell, S.J. (ed.) (2007). *Handbook of Nonwovens*. Cambridge: Textile Institute.

31 Banerjee, P.K. (2015). *Principles of Fabric Formation*. Boca Raton, FL: CRC Press.

32 Thompson, R. (2014). *Manufacturing Processes for Textile and Fashion Design Professionals*. New York: Thames & Hudson.

33 Das, D. and Pourdeyhimi, B. (2014). *Composite Nonwoven Materials: Structure, Properties and Applications*. Cambridge: Woodhead Publishing Ltd.

34 Turbak, A.F. (ed.) (1993). *Nonwovens—Theory, Process, Performance and Testing*. Atlanta, GA: TAPPI Press.

35 Anand, S.C. (2016). Technical fabric structures – 2. Knitted fabrics. In: *Handbook of Technical Textiles* (ed. Horrocks, A.R. and Anand, S.C.), pp. 163–188. Cambridge: Woodhead Publishing Ltd.

36 Walker, I.V. (2001). Nonwovens - the choice for the medical industry into the next millennium. In: *Medical Textiles* (ed. Anand, S.), 12–19. Cambridge: Woodhead Publishing.

37 Chapman, R. (ed.) (2010). *Applications of Nonwovens in Technical Textiles*. Cambridge: Woodhead Publishing Ltd.

38 Gries, T., Veit, D., and Wulfhorst, B. (2006). Braiding processes and machines. In: *Textile Technology* (ed. Wulfhorst, B.), 188–204. Munich: Hanser Publishers.

39 Kyosev, Y. (2015). *Braiding Technology for Textiles*. Cambridge: Woodhead Publishing Ltd.

40 Kyosev, Y. (2016). *Advances in Braiding Technology*. Cambridge: Woodhead Publishing Ltd.

41 Kyosev, Y. (2012). Simulation of wound packages, woven, braided and knitted structures. In: *Simulation in Textile Technology* (ed. Veit, D.), pp. 266–309. Cambridge: Woodhead Publishing Ltd.

42 Carey, J.P. (2016). *Handbook of Advances in Braided Composite Materials: Theory, Production, Testing and Applications*. Duxford: Woodhead Publishing.

43 Rana, S. (2016). *Composite Materials: Braided Structures and Composites: Reduction, Properties, Mechanics, and Technical Applications*, vol. 3. Boca Raton, FL: CRC Press.

44 Potluri, P. and Nawaz, S. (2011). Developments in braided fabrics. In: *Specialist Yarn and Fabric Structures: Developments and Applications* (ed. Gong, R.H.). Cambridge: Woodhead Publishing Ltd.